Methods in Enzymology

Volume 282

VITAMINS AND COENZYMES

Part L

METHODS IN ENZYMOLOGY

EDITORS-IN-CHIEF

John N. Abelson   Melvin I. Simon

DIVISION OF BIOLOGY
CALIFORNIA INSTITUTE OF TECHNOLOGY
PASADENA, CALIFORNIA

FOUNDING EDITORS

Sidney P. Colowick and Nathan O. Kaplan

Methods in Enzymology

Volume 282

Vitamins and Coenzymes

Part L

EDITED BY

Donald B. McCormick

DEPARTMENT OF BIOCHEMISTRY
EMORY UNIVERSITY
ATLANTA, GEORGIA

John W. Suttie

DEPARTMENTS OF BIOCHEMISTRY AND NUTRITIONAL SCIENCES
UNIVERSITY OF WISCONSIN–MADISON
MADISON, WISCONSIN

Conrad Wagner

DEPARTMENT OF VETERANS AFFAIRS MEDICAL CENTER
AND DEPARTMENT OF BIOCHEMISTRY
VANDERBILT UNIVERSITY SCHOOL OF MEDICINE
NASHVILLE, TENNESSEE

ACADEMIC PRESS

San Diego London Boston New York Sydney Tokyo Toronto

This book is printed on acid-free paper. ∞

Academic Press
15 East 26th Street, 15th Floor, New York, New York 10010, USA
http://www.apnet.com

Academic Press Limited
24-28 Oval Road, London NW1 7DX, UK
http://www.hbuk.co.uk/ap/

International Standard Book Number: 0-12-182183-8

PRINTED IN THE UNITED STATES OF AMERICA
97 98 99 00 01 02 MM 9 8 7 6 5 4 3 2 1

Table of Contents

Section I. Vitamin A

Section II. Vitamin D

Section III. Vitamin E

Section IV. Vitamin K

Contributors to Volume 282

Article numbers are in parentheses following the names of contributors.
Affiliations listed are current.

MEGUMI AKIYOSHI-SHIBATA (19), *Biotechnology Laboratory, Sumitomo Chemical Co., Ltd., Hyogo 665, Japan*

ELIZABETH A. ALLEGRETTO (3), *Ligand Pharmaceuticals, Department of Retinoid Research, San Diego, California 92121*

BRUCE A. ANDRIEN (13), *Analytica of Branford, Inc., Branford, Connecticut 06405*

MATTHEW J. BECKMAN (15, 18), *Department of Biochemistry, University of Wisconsin–Madison, Madison, Wisconsin 53706*

ALISON BEHARKA (22), *Nutritional Immunology Laboratory, Jean Mayer USDA Human Nutrition Research Center on Aging at Tufts University, Boston, Massachusetts 02111*

MARGARET E. BENTON (32), *Department of Human Oncology, Clinical Sciences Center, University of Wisconsin–Madison, Madison, Wisconsin 53792*

KATHLEEN L. BERKNER (27), *Department of Molecular Cardiology, Cleveland Clinic Research Institute, Cleveland, Ohio 44195*

SARAH L. BOOTH (38), *Vitamin K Laboratory, Jean Mayer USDA Human Nutrition Center on Aging at Tufts University, Boston, Massachusetts 02111*

J. THOMAS BRENNA (12), *Division of Nutritional Sciences, Cornell University, Ithaca, New York 14853*

PAUL M. BRICKELL (4), *Leukaemia Research Fund Centre for Childhood Leukaemia, Molecular Haemotology Unit, Institute of Child Health, University College London Medical School, London WC1N 1EH, United Kingdom*

REGINA BRIGELIUS-FLOHÉ (26), *German Institute of Human Nutrition, D-14558 Potsdam-Rehbrücke, Germany*

FRANCIS J. CASTELLINO (31), *Department of Chemistry and Biochemistry, University of Notre Dame, Notre Dame, Indiana 46556*

MARGARET CLAGETT-DAME (2), *Pharmaceutical Science Division, School of Pharmacy and Departments of Biochemistry and Agricultural and Life Sciences, University of Wisconsin–Madison, Madison, Wisconsin 53706*

JOHN M. CONLY (39), *Department of Medicine, University of Toronto, Toronto, Ontario M5G 2C4, Canada*

KENNETH W. DAVIDSON (34), *Vitamin K Laboratory, Jean Mayer USDA Human Nutrition Center on Aging at Tufts University, Boston, Massachusetts 02111*

HECTOR F. DELUCA (10, 15, 18), *Department of Biochemistry, University of Wisconsin–Madison, Madison, Wisconsin 53706*

GREGORY G. DOLNIKOWSKI (13), *United States Department of Agriculture, Jean Mayer Human Nutrition Research Center on Aging at Tufts University, Boston, Massachusetts 02111*

ASIM K. DUTTA-ROY (25), *Rowett Research Institute, Aberdeen AB21 9SB, Scotland, United Kingdom*

ANGELIKA ELSNER (26), *German Institute of Human Nutrition, D-14558 Potsdam-Rehbrücke, Germany*

MARY C. FARACH-CARSON (21), *Department of Basic Science, University of Texas–Houston, Dental Branch, Houston, Texas 77030*

NICOLETTA FERRARI (5), *Laboratory of Molecular Biology, National Cancer Institute Genoa, c/o Advanced Biotechnology Center, 16132 Genoa, Italy*

B. FURIE (28), *Department of Medicine and Department of Biochemistry, Tufts University School of Medicine and Division of Hematology/Oncology, New England Medical Center, Boston, Massachusetts 02111*

B. C. FURIE (28), *Department of Medicine and Department of Biochemistry, Tufts University School of Medicine and Division of Hematology/Oncology, New England Medical Center, Boston, Massachusetts 02111*

JIE-PING GENG (31), *Department of Chemistry and Biochemistry, University of Notre Dame, Notre Dame, Indiana 46556*

KEITH J. GOODMAN (12), *Metabolic Solutions, Inc., Merrimack, New Hampshire 03054*

THOMAS M. GUENTHNER (33), *Department of Pharmacology, University of Illinois College of Medicine, Chicago, Illinois 60612-3796*

D. J. HARRINGTON (35), *Haemophilia Centre, St. Thomas' Hospital, London SE1 7EH, United Kingdom*

SHIN-ICHI HAYASHI (17), *Department of Biochemistry, Saitama Cancer Center Research Institute, Saitama 362, Japan*

RICHARD A. HEYMAN (3), *Ligand Pharmaceuticals, Department of Retinoid Research, San Diego, California 92121*

STEPHEN J. HODGES (36), *Department of Human Metabolism and Clinical Biochemistry, University of Sheffield, Northern General Hospital, Sheffield S5 7AU, United Kingdom*

MICHAEL F. HOLICK (14), *Vitamin D, Skin, and Bone Research Laboratory, Section of Endocrinology, Department of Medicine, Boston University School of Medicine, Boston, Massachusetts 02118*

BRUCE W. HOLLIS (16), *Departments of Pediatrics, Biochemistry and Molecular Biology, Medical University of South Carolina, Charleston, South Carolina 28425*

ROGER J. T. J. HOUBEN (30), *Department of Biochemistry, University of Limburg, 6200 MD Maastricht, The Netherlands*

EIJIRO JIMI (20), *Department of Biochemistry, School of Dentistry, Showa University, Shinagawa-ku, Tokyo 142, Japan*

AKIRA KAKIZUKA (8), *Osaka Bioscience Institute, Osaka 565, Japan*

NORMAN I. KRINSKY (11), *Department of Biochemistry, Tufts University School of Medicine, Boston, Massachusetts 02111*

A. KULIOPULOS (28), *Department of Medicine and Department of Biochemistry, Tufts University School of Medicine and Division of Hematology/Oncology, New England Medical Center, Boston, Massachusetts 02111*

CATHERINE Y. LAU (6), *R. W. Johnson Pharmaceutical Research Institute, Don Mills, Ontario M3C 1L9, Canada*

MARCEL LEIST (26), *German Institute of Human Nutrition, D-14558 Potsdam-Rehbrücke, Germany*

LYNETTE LEKA (22), *Nutritional Immunology Laboratory, Jean Mayer USDA Human Nutrition Research Center on Aging at Tufts University, Boston, Massachusetts 02111*

ELLEN LI (1), *Departments of Medicine and Biochemistry and Molecular Biophysics, Washington University School of Medicine, St. Louis, Missouri 63110*

BONNIE MARMOR (12), *Tufts University School of Medicine, Boston, Massachusetts 02111*

P. T. MCCARTHY (35), *Haemophilia Centre, St. Thomas' Hospital, London SE1 7EH, United Kingdom*

BETH A. MCNALLY (27), *Department of Molecular Cardiology, Cleveland Clinic Research Institute, Cleveland, Ohio 44195*

J. GARY MESZAROS (21), *Department of Basic Science, University of Texas–Houston, Dental Branch, Houston, Texas 77030*

SIMIN NIKBIN MEYDANI (22), *Nutritional Immunology Laboratory, Jean Mayer USDA Human Nutrition Research Center on Aging at Tufts University, Boston, Massachusetts 02111*

VASANTHA P. MUTUCUMARANA (29), *Department of Biology, University of North Carolina–Chapel Hill, Chapel Hill, North Carolina 27599-3280*

ICHIRO NAKAMURA (20), *Department of Biochemistry, School of Dentistry, Showa University, Shinagawa-ku, Tokyo 142, Japan*

ETSUO NIKI (24), *Research Center for Advanced Science and Technology, University of Tokyo, Meguro, Tokyo 153, Japan*

NORIKO NOGUCHI (24), *Research Center for Advanced Science and Technology, University of Tokyo, Meguro, Tokyo 153, Japan*

ANDREW W. NORRIS (1), *Department of Biochemistry and Molecular Biophysics, Washington University School of Medicine, St. Louis, Missouri 63110*

MITSUHIDE NOSHIRO (17, 19), *Department of Biochemistry, University School of Dentistry, Hiroshima University, Hiroshima 734, Japan*

YOSHIHIKO OHYAMA (17, 19), *Graduate Department of Gene Science, Faculty of Science, Hiroshima University, Higashi-Hiroshima 724, Japan*

KYU-ICHIRO OKUDA (17, 19), *Department of Surgery, Miyazaki Medical College, Kiyotake, Miyazaki 889-16, Japan*

GAIL OTULAKOWSKI (6), *Respiratory Research Division, Hospital for Sick Children, Toronto, Ontario M5G 1X8, Canada*

ROBERT S. PARKER (12), *Division of Nutritional Sciences, Cornell University, Ithaca, New York 14853*

ULRICH PFEFFER (5), *Laboratory of Molecular Biology, National Cancer Institute Genoa, c/o Advanced Biotechnology Center, 16132 Genoa, Italy*

RAHUL RAY (14), *Vitamin D, Skin, and Bone Research Laboratory, Section of Endocrinology, Department of Medicine, Boston University School of Medicine, Boston, Massachusetts 02118*

SUSAN REDICAN (22), *Nutritional Immunology Laboratory, Jean Mayer USDA Human Nutrition Research Center on Aging at Tufts University, Boston, Massachusetts 02111*

JOYCE J. REPA (2), *Pharmaceutical Science Division, School of Pharmacy and Departments of Biochemistry and Agricultural and Life Sciences, University of Wisconsin–Madison, Madison, Wisconsin 53706*

D. A. ROTH (28), *Department of Medicine, Division of Hematology/Oncology, New England Medical Center, Boston, Massachusetts 02111*

ANNIE ROWE (4), *Department of Academic Therapeutics, Chelsea and Westminster Medical School, London SW10 9NH, United Kingdom*

ROBERT M. RUSSELL (13), *United States Department of Agriculture, Jean Mayer Human Nutrition Research Center on Aging at Tufts University, Boston, Massachusetts 02111*

JAMES A. SADOWSKI (34, 38), *Vitamin K Laboratory, Jean Mayer USDA Human Nutrition Center on Aging at Tufts University, Boston, Massachusetts 02111*

MITINORI SAITOU (8), *Department of Cell Biology, Kyoto University Faculty of Medicine, Kyoto 606-01, Japan*

MANFRED SCHULTZ (26), *German Institute of Human Nutrition, D-14558 Potsdam-Rehbrücke, Germany*

M. J. SHEARER (35), *Haemophilia Centre, St. Thomas' Hospital, London SE1 7EH, United Kingdom*

BERRY A. M. SOUTE (30), *Department of Biochemistry, University of Limburg, 6200 MD Maastricht, The Netherlands*

DARREL W. STAFFORD (29), *Department of Biology, University of North Carolina–Chapel Hill, Chapel Hill, North Carolina 27599-3280*

MANFRED STEINER (23), *Department of Medicine, Division of Hematology/Oncology, East Carolina University, School of Medicine, Greenville, North Carolina 27858-4354*

TATSUO SUDA (20), *Department of Biochemistry, School of Dentistry, Showa University, Shinagawa-ku, Tokyo 142, Japan*

I. SUGIURA (28), *Nagoya University School of Medicine, Showa-ku, Nayoga 467, Japan*

J. W. SUTTIE (32), *Department of Biochemistry, University of Wisconsin–Madison, Madison, Wisconsin 53706-1569*

JOY E. SWANSON (12), *Division of Nutritional Sciences, Cornell University, Ithaca, New York 14853*

PRAVEEN K. TADIKONDA (10), *Department of Biochemistry, University of Wisconsin–Madison, Madison, Wisconsin 53706*

NAOYUKI TAKAHASHI (20), *Department of Biochemistry, School of Dentistry, Showa University, Shinagawa-ku, Tokyo 142, Japan*

TOSHIHIRO TANAKA (8), *Department of Dermatology, Kyoto University Faculty of Medicine, Kyoto 606-01, Japan*

GUANGWEN TANG (13), *United States Department of Agriculture, Jean Mayer Human Nutrition Research Center on Aging at Tufts University, Boston, Massachusetts 02111*

EMIKO USUI (17, 19), *Department of Biochemistry, University School of Dentistry, Hiroshima University, Hiroshima 734, Japan*

YUJI USUI (37), *Department of Surgery, Tazuke Kofukai Medical Research Institute, Osaka-city, Osaka 530, Japan*

CEES VEERMER (30), *Department of Biochemistry, University of Limburg, 6200 MD Maastricht, The Netherlands*

GIORGIO VIDALI* (5), *Laboratory of Molecular Biology, National Cancer Institute Genoa, c/o Advanced Biotechnology Center, 16132 Genoa, Italy*

MICHAEL A. WAGNER (9), *State University of New York, Health Science Center at Brooklyn, Brooklyn, New York 11203*

REIDAR WALLIN (33), *Rheumatology, Department of Internal Medicine, Bowman Gray School of Medicine, Wake Forest University, Winston-Salem, North Carolina 27157-1058*

C. T. WALSH (28), *Department of Biological Chemistry and Molecular Pharmacology, Harvard Medical School, Boston, Massachusetts 02115-5718*

XIANG-DONG WANG (11), *United States Department of Agriculture, Jean Mayer Human Nutrition Research Center on Aging at Tufts University, Boston, Massachusetts 02111*

LESZEK WOJNOWSKI (7), *Section on Genetics, National Institute of Mental Health, National Institutes of Health, Bethesda, Maryland 20892*

SHEUE-MEI WU (29), *Department of Biology, University of North Carolina–Chapel Hill, Chapel Hill, North Carolina 27599-3280*

YOSHIYASU YABUSAKI (19), *Biotechnology Laboratory, Sumitomo Chemical Co., Ltd., Hyogo 665, Japan*

LUBING ZHOU (6), *R. W. Johnson Pharmaceutical Research Institute, Raritan, New Jersey 08869*

ANDREAS ZIMMER (7), *Section on Genetics, National Institute of Mental Health, National Institutes of Health, Bethesda, Maryland 20892*

* Deceased.

Preface

From 1970 to 1986, eight "Vitamins and Coenzymes" volumes were published in the *Methods in Enzymology* series. Volumes XVIII A, B, and C appeared in 1970–1971 and Volumes 62 (D), 66 (E), and 67 (F) in 1979–1980. These volumes were edited by D. B. McCormick and L. D. Wright. Volumes 122 (G) and 123 (H), published in 1986, were edited by F. Chytil and D. B. McCormick. In the decade that has elapsed since the last volume was published, considerable progress has been made, so it was reasonable to update the subject of "Vitamins and Coenzymes."

In this current set of volumes (279, 280, 281, and 282) we attempted to collect and collate many of the newer techniques and methodologies attendant to assays, isolations, and characterizations of vitamins, coenzymes, and those systems responsible for their biosynthesis, transport, and metabolism. There are examples of procedures that are modifications of earlier ones as well as of those that have newly evolved. As before, there has been an attempt to allow such overlap as would offer flexibility in the choice of methods, rather than presume any one is best for all laboratories. Where there is no inclusion of a particular subject covered in earlier volumes, we believe the subject was adequately treated and the reader should refer to those volumes.

The information provided reflects the efforts of our numerous contributors to whom we express our gratitude. We are also grateful to our secretaries at our academic home bases and to Shirley Light and the staff of Academic Press. Finally, one of us (D. B. M.) recalls fondly the encouragement proffered years ago by Drs. Nathan O. Kaplan and Sidney P. Colowick who saw the need for "Vitamins and Coenzymes" within the *Methods in Enzymology* series, which they initiated.

Donald B. McCormick
John W. Suttie
Conrad Wagner

METHODS IN ENZYMOLOGY

VOLUME I. Preparation and Assay of Enzymes
Edited by SIDNEY P. COLOWICK AND NATHAN O. KAPLAN

VOLUME II. Preparation and Assay of Enzymes
Edited by SIDNEY P. COLOWICK AND NATHAN O. KAPLAN

VOLUME III. Preparation and Assay of Substrates
Edited by SIDNEY P. COLOWICK AND NATHAN O. KAPLAN

VOLUME IV. Special Techniques for the Enzymologist
Edited by SIDNEY P. COLOWICK AND NATHAN O. KAPLAN

VOLUME V. Preparation and Assay of Enzymes
Edited by SIDNEY P. COLOWICK AND NATHAN O. KAPLAN

VOLUME VI. Preparation and Assay of Enzymes (*Continued*)
Preparation and Assay of Substrates
Special Techniques
Edited by SIDNEY P. COLOWICK AND NATHAN O. KAPLAN

VOLUME VII. Cumulative Subject Index
Edited by SIDNEY P. COLOWICK AND NATHAN O. KAPLAN

VOLUME VIII. Complex Carbohydrates
Edited by ELIZABETH F. NEUFELD AND VICTOR GINSBURG

VOLUME IX. Carbohydrate Metabolism
Edited by WILLIS A. WOOD

VOLUME X. Oxidation and Phosphorylation
Edited by RONALD W. ESTABROOK AND MAYNARD E. PULLMAN

VOLUME XI. Enzyme Structure
Edited by C. H. W. HIRS

VOLUME XII. Nucleic Acids (Parts A and B)
Edited by LAWRENCE GROSSMAN AND KIVIE MOLDAVE

VOLUME XIII. Citric Acid Cycle
Edited by J. M. LOWENSTEIN

VOLUME XIV. Lipids
Edited by J. M. LOWENSTEIN

VOLUME XV. Steroids and Terpenoids
Edited by RAYMOND B. CLAYTON

VOLUME XVI. Fast Reactions
Edited by KENNETH KUSTIN

Volume XXXVI. Hormone Action (Part A: Steroid Hormones)
Edited by Bert W. O'Malley and Joel G. Hardman

Volume XXXVII. Hormone Action (Part B: Peptide Hormones)
Edited by Bert W. O'Malley and Joel G. Hardman

Volume XXXVIII. Hormone Action (Part C: Cyclic Nucleotides)
Edited by Joel G. Hardman and Bert W. O'Malley

Volume XXXIX. Hormone Action (Part D: Isolated Cells, Tissues, and Organ Systems)
Edited by Joel G. Hardman and Bert W. O'Malley

Volume XL. Hormone Action (Part E: Nuclear Structure and Function)
Edited by Bert W. O'Malley and Joel G. Hardman

Volume XLI. Carbohydrate Metabolism (Part B)
Edited by W. A. Wood

Volume XLII. Carbohydrate Metabolism (Part C)
Edited by W. A. Wood

Volume XLIII. Antibiotics
Edited by John H. Hash

Volume XLIV. Immobilized Enzymes
Edited by Klaus Mosbach

Volume XLV. Proteolytic Enzymes (Part B)
Edited by Laszlo Lorand

Volume XLVI. Affinity Labeling
Edited by William B. Jakoby and Meir Wilchek

Volume XLVII. Enzyme Structure (Part E)
Edited by C. H. W. Hirs and Serge N. Timasheff

Volume XLVIII. Enzyme Structure (Part F)
Edited by C. H. W. Hirs and Serge N. Timasheff

Volume XLIX. Enzyme Structure (Part G)
Edited by C. H. W. Hirs and Serge N. Timasheff

Volume L. Complex Carbohydrates (Part C)
Edited by Victor Ginsburg

Volume LI. Purine and Pyrimidine Nucleotide Metabolism
Edited by Patricia A. Hoffee and Mary Ellen Jones

Volume LII. Biomembranes (Part C: Biological Oxidations)
Edited by Sidney Fleischer and Lester Packer

Volume LIII. Biomembranes (Part D: Biological Oxidations)
Edited by Sidney Fleischer and Lester Packer

Volume LIV. Biomembranes (Part E: Biological Oxidations)
Edited by Sidney Fleischer and Lester Packer

VOLUME 73. Immunochemical Techniques (Part B)
Edited by JOHN J. LANGONE AND HELEN VAN VUNAKIS

VOLUME 74. Immunochemical Techniques (Part C)
Edited by JOHN J. LANGONE AND HELEN VAN VUNAKIS

VOLUME 75. Cumulative Subject Index Volumes XXXI, XXXII, XXXIV–LX
Edited by EDWARD A. DENNIS AND MARTHA G. DENNIS

VOLUME 76. Hemoglobins
Edited by ERALDO ANTONINI, LUIGI ROSSI-BERNARDI, AND EMILIA CHIANCONE

VOLUME 77. Detoxication and Drug Metabolism
Edited by WILLIAM B. JAKOBY

VOLUME 78. Interferons (Part A)
Edited by SIDNEY PESTKA

VOLUME 79. Interferons (Part B)
Edited by SIDNEY PESTKA

VOLUME 80. Proteolytic Enzymes (Part C)
Edited by LASZLO LORAND

VOLUME 81. Biomembranes (Part H: Visual Pigments and Purple Membranes, I)
Edited by LESTER PACKER

VOLUME 82. Structural and Contractile Proteins (Part A: Extracellular Matrix)
Edited by LEON W. CUNNINGHAM AND DIXIE W. FREDERIKSEN

VOLUME 83. Complex Carbohydrates (Part D)
Edited by VICTOR GINSBURG

VOLUME 84. Immunochemical Techniques (Part D: Selected Immunoassays)
Edited by JOHN J. LANGONE AND HELEN VAN VUNAKIS

VOLUME 85. Structural and Contractile Proteins (Part B: The Contractile Apparatus and the Cytoskeleton)
Edited by DIXIE W. FREDERIKSEN AND LEON W. CUNNINGHAM

VOLUME 86. Prostaglandins and Arachidonate Metabolites
Edited by WILLIAM E. M. LANDS AND WILLIAM L. SMITH

VOLUME 87. Enzyme Kinetics and Mechanism (Part C: Intermediates, Stereochemistry, and Rate Studies)
Edited by DANIEL L. PURICH

VOLUME 88. Biomembranes (Part I: Visual Pigments and Purple Membranes, II)
Edited by LESTER PACKER

VOLUME 89. Carbohydrate Metabolism (Part D)
Edited by WILLIS A. WOOD

VOLUME 90. Carbohydrate Metabolism (Part E)
Edited by WILLIS A. WOOD

Volume 126. Biomembranes (Part N: Transport in Bacteria, Mitochondria, and Chloroplasts: Protonmotive Force)
Edited by Sidney Fleischer and Becca Fleischer

Volume 127. Biomembranes (Part O: Protons and Water: Structure and Translocation)
Edited by Lester Packer

Volume 128. Plasma Lipoproteins (Part A: Preparation, Structure, and Molecular Biology)
Edited by Jere P. Segrest and John J. Albers

Volume 129. Plasma Lipoproteins (Part B: Characterization, Cell Biology, and Metabolism)
Edited by John J. Albers and Jere P. Segrest

Volume 130. Enzyme Structure (Part K)
Edited by C. H. W. Hirs and Serge N. Timasheff

Volume 131. Enzyme Structure (Part L)
Edited by C. H. W. Hirs and Serge N. Timasheff

Volume 132. Immunochemical Techniques (Part J: Phagocytosis and Cell-Mediated Cytotoxicity)
Edited by Giovanni Di Sabato and Johannes Everse

Volume 133. Bioluminescence and Chemiluminescence (Part B)
Edited by Marlene DeLuca and William D. McElroy

Volume 134. Structural and Contractile Proteins (Part C: The Contractile Apparatus and the Cytoskeleton)
Edited by Richard B. Vallee

Volume 135. Immobilized Enzymes and Cells (Part B)
Edited by Klaus Mosbach

Volume 136. Immobilized Enzymes and Cells (Part C)
Edited by Klaus Mosbach

Volume 137. Immobilized Enzymes and Cells (Part D)
Edited by Klaus Mosbach

Volume 138. Complex Carbohydrates (Part E)
Edited by Victor Ginsburg

Volume 139. Cellular Regulators (Part A: Calcium- and Calmodulin-Binding Proteins)
Edited by Anthony R. Means and P. Michael Conn

Volume 140. Cumulative Subject Index Volumes 102–119, 121–134

Volume 141. Cellular Regulators (Part B: Calcium and Lipids)
Edited by P. Michael Conn and Anthony R. Means

Volume 142. Metabolism of Aromatic Amino Acids and Amines
Edited by Seymour Kaufman

VOLUME 143. Sulfur and Sulfur Amino Acids
Edited by WILLIAM B. JAKOBY AND OWEN GRIFFITH

VOLUME 144. Structural and Contractile Proteins (Part D: Extracellular Matrix)
Edited by LEON W. CUNNINGHAM

VOLUME 145. Structural and Contractile Proteins (Part E: Extracellular Matrix)
Edited by LEON W. CUNNINGHAM

VOLUME 146. Peptide Growth Factors (Part A)
Edited by DAVID BARNES AND DAVID A. SIRBASKU

VOLUME 147. Peptide Growth Factors (Part B)
Edited by DAVID BARNES AND DAVID A. SIRBASKU

VOLUME 148. Plant Cell Membranes
Edited by LESTER PACKER AND ROLAND DOUCE

VOLUME 149. Drug and Enzyme Targeting (Part B)
Edited by RALPH GREEN AND KENNETH J. WIDDER

VOLUME 150. Immunochemical Techniques (Part K: *In Vitro* Models of B and T Cell Functions and Lymphoid Cell Receptors)
Edited by GIOVANNI DI SABATO

VOLUME 151. Molecular Genetics of Mammalian Cells
Edited by MICHAEL M. GOTTESMAN

VOLUME 152. Guide to Molecular Cloning Techniques
Edited by SHELBY L. BERGER AND ALAN R. KIMMEL

VOLUME 153. Recombinant DNA (Part D)
Edited by RAY WU AND LAWRENCE GROSSMAN

VOLUME 154. Recombinant DNA (Part E)
Edited by RAY WU AND LAWRENCE GROSSMAN

VOLUME 155. Recombinant DNA (Part F)
Edited by RAY WU

VOLUME 156. Biomembranes (Part P: ATP-Driven Pumps and Related Transport: The Na,K-Pump)
Edited by SIDNEY FLEISCHER AND BECCA FLEISCHER

VOLUME 157. Biomembranes (Part Q: ATP-Driven Pumps and Related Transport: Calcium, Proton, and Potassium Pumps)
Edited by SIDNEY FLEISCHER AND BECCA FLEISCHER

VOLUME 158. Metalloproteins (Part A)
Edited by JAMES F. RIORDAN AND BERT L. VALLEE

VOLUME 159. Initiation and Termination of Cyclic Nucleotide Action
Edited by JACKIE D. CORBIN AND ROGER A. JOHNSON

VOLUME 160. Biomass (Part A: Cellulose and Hemicellulose)
Edited by WILLIS A. WOOD AND SCOTT T. KELLOGG

VOLUME 161. Biomass (Part B: Lignin, Pectin, and Chitin)
Edited by WILLIS A. WOOD AND SCOTT T. KELLOGG

VOLUME 162. Immunochemical Techniques (Part L: Chemotaxis and Inflammation)
Edited by GIOVANNI DI SABATO

VOLUME 163. Immunochemical Techniques (Part M: Chemotaxis and Inflammation)
Edited by GIOVANNI DI SABATO

VOLUME 164. Ribosomes
Edited by HARRY F. NOLLER, JR., AND KIVIE MOLDAVE

VOLUME 165. Microbial Toxins: Tools for Enzymology
Edited by SIDNEY HARSHMAN

VOLUME 166. Branched-Chain Amino Acids
Edited by ROBERT HARRIS AND JOHN R. SOKATCH

VOLUME 167. Cyanobacteria
Edited by LESTER PACKER AND ALEXANDER N. GLAZER

VOLUME 168. Hormone Action (Part K: Neuroendocrine Peptides)
Edited by P. MICHAEL CONN

VOLUME 169. Platelets: Receptors, Adhesion, Secretion (Part A)
Edited by JACEK HAWIGER

VOLUME 170. Nucleosomes
Edited by PAUL M. WASSARMAN AND ROGER D. KORNBERG

VOLUME 171. Biomembranes (Part R: Transport Theory: Cells and Model Membranes)
Edited by SIDNEY FLEISCHER AND BECCA FLEISCHER

VOLUME 172. Biomembranes (Part S: Transport: Membrane Isolation and Characterization)
Edited by SIDNEY FLEISCHER AND BECCA FLEISCHER

VOLUME 173. Biomembranes [Part T: Cellular and Subcellular Transport: Eukaryotic (Nonepithelial) Cells]
Edited by SIDNEY FLEISCHER AND BECCA FLEISCHER

VOLUME 174. Biomembranes [Part U: Cellular and Subcellular Transport: Eukaryotic (Nonepithelial) Cells]
Edited by SIDNEY FLEISCHER AND BECCA FLEISCHER

VOLUME 175. Cumulative Subject Index Volumes 135–139, 141–167

VOLUME 176. Nuclear Magnetic Resonance (Part A: Spectral Techniques and Dynamics)
Edited by NORMAN J. OPPENHEIMER AND THOMAS L. JAMES

VOLUME 177. Nuclear Magnetic Resonance (Part B: Structure and Mechanism)
Edited by NORMAN J. OPPENHEIMER AND THOMAS L. JAMES

VOLUME 178. Antibodies, Antigens, and Molecular Mimicry
Edited by JOHN J. LANGONE

VOLUME 179. Complex Carbohydrates (Part F)
Edited by VICTOR GINSBURG

VOLUME 180. RNA Processing (Part A: General Methods)
Edited by JAMES E. DAHLBERG AND JOHN N. ABELSON

VOLUME 181. RNA Processing (Part B: Specific Methods)
Edited by JAMES E. DAHLBERG AND JOHN N. ABELSON

VOLUME 182. Guide to Protein Purification
Edited by MURRAY P. DEUTSCHER

VOLUME 183. Molecular Evolution: Computer Analysis of Protein and Nucleic Acid Sequences
Edited by RUSSELL F. DOOLITTLE

VOLUME 184. Avidin–Biotin Technology
Edited by MEIR WILCHEK AND EDWARD A. BAYER

VOLUME 185. Gene Expression Technology
Edited by DAVID V. GOEDDEL

VOLUME 186. Oxygen Radicals in Biological Systems (Part B: Oxygen Radicals and Antioxidants)
Edited by LESTER PACKER AND ALEXANDER N. GLAZER

VOLUME 187. Arachidonate Related Lipid Mediators
Edited by ROBERT C. MURPHY AND FRANK A. FITZPATRICK

VOLUME 188. Hydrocarbons and Methylotrophy
Edited by MARY E. LIDSTROM

VOLUME 189. Retinoids (Part A: Molecular and Metabolic Aspects)
Edited by LESTER PACKER

VOLUME 190. Retinoids (Part B: Cell Differentiation and Clinical Applications)
Edited by LESTER PACKER

VOLUME 191. Biomembranes (Part V: Cellular and Subcellular Transport: Epithelial Cells)
Edited by SIDNEY FLEISCHER AND BECCA FLEISCHER

VOLUME 192. Biomembranes (Part W: Cellular and Subcellular Transport: Epithelial Cells)
Edited by SIDNEY FLEISCHER AND BECCA FLEISCHER

VOLUME 193. Mass Spectrometry
Edited by JAMES A. MCCLOSKEY

VOLUME 194. Guide to Yeast Genetics and Molecular Biology
Edited by CHRISTINE GUTHRIE AND GERALD R. FINK

VOLUME 195. Adenylyl Cyclase, G Proteins, and Guanylyl Cyclase
Edited by ROGER A. JOHNSON AND JACKIE D. CORBIN

Volume 230. Guide to Techniques in Glycobiology
Edited by William J. Lennarz and Gerald W. Hart

Volume 231. Hemoglobins (Part B: Biochemical and Analytical Methods)
Edited by Johannes Everse, Kim D. Vandegriff, and Robert M. Winslow

Volume 232. Hemoglobins (Part C: Biophysical Methods)
Edited by Johannes Everse, Kim D. Vandegriff, and Robert M. Winslow

Volume 233. Oxygen Radicals in Biological Systems (Part C)
Edited by Lester Packer

Volume 234. Oxygen Radicals in Biological Systems (Part D)
Edited by Lester Packer

Volume 235. Bacterial Pathogenesis (Part A: Identification and Regulation of Virulence Factors)
Edited by Virginia L. Clark and Patrik M. Bavoil

Volume 236. Bacterial Pathogenesis (Part B: Integration of Pathogenic Bacteria with Host Cells)
Edited by Virginia L. Clark and Patrik M. Bavoil

Volume 237. Heterotrimeric G Proteins
Edited by Ravi Iyengar

Volume 238. Heterotrimeric G-Protein Effectors
Edited by Ravi Iyengar

Volume 239. Nuclear Magnetic Resonance (Part C)
Edited by Thomas L. James and Norman J. Oppenheimer

Volume 240. Numerical Computer Methods (Part B)
Edited by Michael L. Johnson and Ludwig Brand

Volume 241. Retroviral Proteases
Edited by Lawrence C. Kuo and Jules A. Shafer

Volume 242. Neoglycoconjugates (Part A)
Edited by Y. C. Lee and Reiko T. Lee

Volume 243. Inorganic Microbial Sulfur Metabolism
Edited by Harry D. Peck, Jr., and Jean LeGall

Volume 244. Proteolytic Enzymes: Serine and Cysteine Peptidases
Edited by Alan J. Barrett

Volume 245. Extracellular Matrix Components
Edited by E. Ruoslahti and E. Engvall

Volume 246. Biochemical Spectroscopy
Edited by Kenneth Sauer

Volume 247. Neoglycoconjugates (Part B: Biomedical Applications)
Edited by Y. C. Lee and Reiko T. Lee

VOLUME 248. Proteolytic Enzymes: Aspartic and Metallo Peptidases
Edited by ALAN J. BARRETT

VOLUME 249. Enzyme Kinetics and Mechanism (Part D: Developments in Enzyme Dynamics)
Edited by DANIEL L. PURICH

VOLUME 250. Lipid Modifications of Proteins
Edited by PATRICK J. CASEY AND JANICE E. BUSS

VOLUME 251. Biothiols (Part A: Monothiols and Dithiols, Protein Thiols, and Thiyl Radicals)
Edited by LESTER PACKER

VOLUME 252. Biothiols (Part B: Glutathione and Thioredoxin; Thiols in Signal Transduction and Gene Regulation)
Edited by LESTER PACKER

VOLUME 253. Adhesion of Microbial Pathogens
Edited by RON J. DOYLE AND ITZHAK OFEK

VOLUME 254. Oncogene Techniques
Edited by PETER K. VOGT AND INDER M. VERMA

VOLUME 255. Small GTPases and Their Regulators (Part A: Ras Family)
Edited by W. E. BALCH, CHANNING J. DER, AND ALAN HALL

VOLUME 256. Small GTPases and Their Regulators (Part B: Rho Family)
Edited by W. E. BALCH, CHANNING J. DER, AND ALAN HALL

VOLUME 257. Small GTPases and Their Regulators (Part C: Proteins Involved in Transport)
Edited by W. E. BALCH, CHANNING J. DER, AND ALAN HALL

VOLUME 258. Redox-Active Amino Acids in Biology
Edited by JUDITH P. KLINMAN

VOLUME 259. Energetics of Biological Macromolecules
Edited by MICHAEL L. JOHNSON AND GARY K. ACKERS

VOLUME 260. Mitochondrial Biogenesis and Genetics (Part A)
Edited by GIUSEPPE M. ATTARDI AND ANNE CHOMYN

VOLUME 261. Nuclear Magnetic Resonance and Nucleic Acids
Edited by THOMAS L. JAMES

VOLUME 262. DNA Replication
Edited by JUDITH L. CAMPBELL

VOLUME 263. Plasma Lipoproteins (Part C: Quantitation)
Edited by WILLIAM A. BRADLEY, SANDRA H. GIANTURCO, AND JERE P. SEGREST

VOLUME 264. Mitochondrial Biogenesis and Genetics (Part B)
Edited by GIUSEPPE M. ATTARDI AND ANNE CHOMYN

VOLUME 265. Cumulative Subject Index Volumes 228, 230–262

VOLUME 266. Computer Methods for Macromolecular Sequence Analysis
Edited by RUSSELL F. DOOLITTLE

VOLUME 267. Combinatorial Chemistry
Edited by JOHN N. ABELSON

VOLUME 268. Nitric Oxide (Part A: Sources and Detection of NO; NO Synthase)
Edited by LESTER PACKER

VOLUME 269. Nitric Oxide (Part B: Physiological and Pathological Processes)
Edited by LESTER PACKER

VOLUME 270. High Resolution Separation and Analysis of Biological Macromolecules (Part A: Fundamentals)
Edited by BARRY L. KARGER AND WILLIAM S. HANCOCK

VOLUME 271. High Resolution Separation and Analysis of Biological Macromolecules (Part B: Applications)
Edited by BARRY L. KARGER AND WILLIAM S. HANCOCK

VOLUME 272. Cytochrome P450 (Part B)
Edited by ERIC F. JOHNSON AND MICHAEL R. WATERMAN

VOLUME 273. RNA Polymerase and Associated Factors (Part A)
Edited by SANKAR ADHYA

VOLUME 274. RNA Polymerase and Associated Factors (Part B)
Edited by SANKAR ADHYA

VOLUME 275. Viral Polymerases and Related Proteins
Edited by LAWRENCE C. KUO, DAVID B. OLSEN, AND STEVEN S. CARROLL

VOLUME 276. Macromolecular Crystallography (Part A)
Edited by CHARLES W. CARTER, JR., AND ROBERT M. SWEET

VOLUME 277. Macromolecular Crystallography (Part B)
Edited by CHARLES W. CARTER, JR., AND ROBERT M. SWEET

VOLUME 278. Fluorescence Spectroscopy
Edited by LUDWIG BRAND AND MICHAEL L. JOHNSON

VOLUME 279. Vitamins and Coenzymes, Part I
Edited by DONALD B. MCCORMICK, JOHN W. SUTTIE, AND CONRAD WAGNER

VOLUME 280. Vitamins and Coenzymes, Part J
Edited by DONALD B. MCCORMICK, JOHN W. SUTTIE, AND CONRAD WAGNER

VOLUME 281. Vitamins and Coenzymes, Part K
Edited by DONALD B. MCCORMICK, JOHN W. SUTTIE, AND CONRAD WAGNER

VOLUME 282. Vitamins and Coenzymes, Part L
Edited by DONALD B. MCCORMICK, JOHN W. SUTTIE, AND CONRAD WAGNER

VOLUME 283. Cell Cycle Control
Edited by WILLIAM G. DUNPHY

VOLUME 284. Lipases (Part A: Biotechnology)
Edited by BYRON RUBIN AND EDWARD A. DENNIS

VOLUME 285. Cumulative Subject Index Volumes 263, 264, 266–289 (in preparation)

VOLUME 286. Lipases (Part B: Enzyme Characterization and Utilization)
Edited by BYRON RUBIN AND EDWARD A. DENNIS

VOLUME 287. Chemokines
Edited by RICHARD HORUK

VOLUME 288. Chemokine Receptors
Edited by RICHARD HORUK

VOLUME 289. Solid Phase Peptide Synthesis
Edited by GREGG B. FIELDS

VOLUME 290. Molecular Chaperones (in preparation)
Edited by GEORGE H. LORIMER AND THOMAS O. BALDWIN

Section I

Vitamin A

[1] Generation and Characterization of Cellular Retinoic Acid-Binding Proteins from *Escherichia coli* Expression Systems

By ANDREW W. NORRIS and ELLEN LI

Introduction

The cellular retinoic acid-binding proteins (CRABPs) are small (~15 kDa), cytosplasmic proteins that bind all-*trans*-retinoic acid with very high affinity and are found in a variety of vertebrate tissues. Two highly homologous isoforms, CRABP-I and CRABP-II, have been identified. Full-length CRABP cDNAs have been isolated from a number of organisms including humans,[1,2] mice,[3,4] and *Xenopus*.[5,6]

Study of the CRABPs has been facilitated by recombinant generation and purification of these proteins from *Escherichia coli* as first reported by Fiorella and Napoli.[7] This allows the rapid isolation of milligram quantities of pure, functional CRABP. Purification of CRABP from *E. coli* avoids some of the difficulties associated with isolation of native CRABP from animal tissues. These difficulties include low abundance of CRABP in tissues,[8] the presence of similar proteins such as cellular retinol-binding protein (CRBP),[8] and the copurification of endogenous retinoic acid, which must be removed if apo-CRABP is to be studied.[9] These difficulties are avoided by recombinant expression in *E. coli,* which does not contain endogenous retinoids or retinoid-binding proteins.[10] The properties of purified recombinant CRABP are nearly identical to those of native CRABP.[11]

[1] A. Åström, A. Tavakkol, U. Pettersson, M. Cromie, J. T. Elder, and J. J. Voorhees, *J. Biol. Chem.* **266,** 17662 (1991).
[2] M. S. Eller, M. F. Oleksiak, T. J. McQuaid, S. G. McAfee, and B. A. Gilchrest, *Exp. Cell Res.* **198,** 328 (1992).
[3] C. M. Stoner and L. J. Gudas, *Cancer Res.* **49,** 1497 (1989).
[4] T. M. MacGregor, N. G. Copeland, N. A. Jenkins, and V. Giguère, *J. Biol. Chem.* **267,** 7777 (1992).
[5] E.-J. Dekker, M.-J. Vaessen, C. van der Berg, A. Timmerman, S. Godsave, T. Holling, P. Nieuwkoop, A. Geurts van Kessel, and A. Durston, *Development* **120,** 973 (1994).
[6] L. Ho, M. Mercola, and L. J. Gudas, *Mech. Dev.* **47,** 53 (1994).
[7] P. D. Fiorella and J. L. Napoli, *J. Biol. Chem.* **266,** 16572 (1991).
[8] D. Ong and F. Chytil, *Methods Enzymol.* **67,** 288 (1980).
[9] D. E. Ong and F. Chytil, *J. Biol. Chem.* **253,** 4551 (1978).
[10] M. S. Levin, E. Li, and J. I. Gordon, *Methods Enzymol.* **189,** 506 (1990).
[11] P. D. Fiorella, V. Giguère, and J. L. Napoli, *J. Biol. Chem.* **268,** 21545 (1993).

METHODS IN ENZYMOLOGY, VOL. 282
0076-6879/97 $25.00

Many groups have now expressed and purified CRABP from *E. coli*, including laboratories that have purified both isoforms.[11–13] The methods described here have been adapted from the original report of recombinant CRABP purification[7] as well as from techniques used to express and purify the CRBPs from *E. coli*.[10]

Recombinant Expression

Expression Constructs

Expression vectors for murine CRABP-I and CRABP-II have been created by using the plasmid pMON2670. This plasmid uses the *recA* promoter to direct expression from a unique *Nde*I site located at an initiator ATG.[14] The cDNAs for mouse CRABP-I and CRABP-II[3,4] were amplified by PCR (polymerase chain reaction) creating two new restriction sites: an *Nde*I site at the initiator codon and a second restriction site, *Sma*I for CRABP-I and *Sac*I for CRABP-II, immediately downstream of the stop codon. These restriction sites were used for directional cloning of the CRABP cDNAs into the expression site of pMON2670.

Other expression vectors have been used for expression of the CRABPs in *E. coli* including pET3a[7,11,12] and pT7.[15,16] Human and bovine CRABPs have been expressed in *E. coli*,[7,12] as have several engineered CRABP fusion proteins.[17–19]

Bacterial Strains

CRABP-II is expressed in the protease-deficient *E. coli* strain BL21(DE3) (*lon*$^-$ *ompT*$^-$). This prevents proteolysis of the CRABP-II, which has been observed in other *E. coli* strains such as JM101.[13] CRABP-I has been successfully expressed in *E. coli* strains BL21(DE3) and JM101.[7,13]

Expression Media

Significant expression can be obtained in a variety of media. We observe enhanced solubilization of CRABP expressed in a supplemented, M9-based

[12] K. Fogh, J. J. Voorhees, and A. Åström, *Arch. Biochem. Biophys.* **300,** 751 (1993).

[13] A. W. Norris, L. Cheng, V. Giguère, M. Rosenberger, and E. Li, *Biochim. Biophys. Acta* **1209,** 10 (1994).

[14] E. Li, B. Locke, N. C. Yang, D. E. Ong, and J. I. Gordon, *J. Biol. Chem.* **262,** 13773 (1987).

[15] Z.-P. Liu, J. Rizo, and L. M. Gierasch, *Biochemistry* **33,** 134 (1994).

[16] R. S. Jamison, M. E. Newcomer, and D. E. Ong, *Biochemistry* **33,** 2873 (1994).

[17] S. Sanquer and B. A. Gilchrest, *Arch. Biochem. Biophys.* **311,** 86 (1994).

[18] C. P. F. Redfern and K. E. Wilson, *FEBS Lett.* **321,** 163 (1993).

[19] L. X. Chen, Z.-P. Zhang, A. Scafonas, R. C. Cavalli, J. L. Gabriel, K. J. Soprano, and D. R. Soprano, *J. Biol. Chem.* **270,** 4518 (1995).

TABLE I
SUPPLEMENTED M9 MEDIUM

Supplemented M9 medium (1 liter)	25× M9 salts (1 liter)	10,000× Trace minerals (100 ml)
900 ml Autoclaved H_2O	283 g $Na_2HPO_4 \cdot 7H_2O$	5.4 g $FeCl_3$
40 ml 25× M9 salts	75 g KH_2PO_4	0.4 g $ZnSO_4$
1.2 ml 1 *M* $MgSO_4$	12.5 g NaCl	0.7 g $CoCl_2$
250 μl 0.1% Thiamin	25 g NH_4Cl	0.7 g Na_2MoO_4
6.4 ml 40% Glucose	pH to 7.4	0.8 g $CuSO_4$
250 μl 0.1 *M* $CaCl_2$		0.2 g H_2BO_3
50 ml 20% (w/v) Casamino acids		0.5 g $MnSO_4$
100 μl 10,000× Trace minerals		

medium (Table I). The water and M9 salts are sterilized by autoclaving. The remainder of the reagents are sterilized by filtration, with the exception of the 10,000× trace minerals solution, which does not require a sterilization procedure. The 0.1% (w/v) thiamin solution should be stored in the dark at 4°. The 20% (w/v) casamino acid solution should be prepared fresh.

Expression Procedure

A fresh overnight culture, grown in Luria broth plus ampicillin (100 μg/ml), of *E. coli* harboring the appropriate expression plasmid is diluted 1 : 15 into freshly prepared supplemented M9 medium. The expression cultures are incubated at 37° with shaking. Expression should be induced while the culture is in log phase growth, for example, when its OD_{600} is 1.5–3 cm^{-1}, by adding nalidixic acid (1 : 200 dilution of a 10 mg/ml stock prepared in 0.1 *N* NaOH). The culture is then incubated for two more hours at 37° with shaking. The culture should be maintained in log phase growth. If growth ceases, addition of glucose to 0.1% (w/v) may restore growth. If this fails, the addition of other nutrients, such as NH_4Cl, may restore growth. Following the 2-hr induction period, the bacteria are pelleted by centrifugation at ~3750 *g* for 15 minutes at 4°. The cell paste is frozen at −80°.

Assay Method

The expression and purification of CRABP can be evaluated by denaturing SDS–polyacrylamide gel electrophoresis (SDS–PAGE) analysis[20] on a 15% (w/v) polyacrylamide gel followed by Coomassie staining. Bacterial samples for SDS–PAGE analysis should be processed quickly (i.e., pelleted

[20] U. K. Laemmli, *Nature* **227,** 680 (1970).

and resuspended in SDS–PAGE sample buffer) because insults to the bacteria may artifactually induce the *recA* promoter. Following induction, CRABP (~15 kDa) is the darkest band detected by Coomassie-stained SDS–PAGE.

Purification Procedures

Step 1. Lysis

The frozen bacterial pellet is thawed and resuspended in 1–2 volumes of freshly prepared lysis buffer (50 m*M* Bis–Tris-HCl, pH 7.0, 10% (w/v) sucrose, 1 m*M* EDTA, 0.05% (w/v) sodium azide, 2 m*M* 2-mercaptoethanol, 10 m*M* $MnCl_2$, 2.5 m*M* phenylmethylsulfonyl fluoride, 2.5 m*M* benzamidine, 6 μg/ml DNase I). Greater solubilization of CRABP-II occurs at pH 8.0 compared to pH 7.0. To this end, substitute 50 m*M* Tris-HCl, pH 8.0, for the Bis–Tris when lysing CRABP-II. DNase I can be stored as a stock solution [50% (v/v) glycerol, 20 m*M* Tris–HCl, pH 7.5, 1 m*M* $MgCl_2$, 3 mg/ml DNase I] at −20°.

One freeze/thaw cycle is usually adequate to release ~30–40% of the total expressed CRABP into the soluble fraction. Further lysis will release up to ~50–60% of the CRABP, as well as other unwanted proteins. This can be accomplished by either sonication or French press of the thawed, resuspended pellet. Sonication, six 30-sec bursts each followed by a 30-sec delay, should be performed on ice. French press, at 18,000 psi, should be performed at 4°. Following lysis, the lysate should be incubated for 30 min at room temperature.

Step 2. Fractionation

The insoluble fraction of the lysate is removed by centrifugation at ~27,000*g* for 30 min. The supernatant and pellet should be separated and checked for CRABP content by SDS–PAGE. Additional soluble CRABP may be extracted from the pellet by resuspending it in lysis buffer, followed by an additional centrifugation step. The supernatants obtained from these steps may be turbid. As long as the pellet is not disturbed during removal of the supernatant, the turbidity will not interfere with the next purification step.

Step 3. Gel Filtration on Sephadex G-50

The collected supernatants from the fractionation step should be concentrated by ultrafiltration to a volume appropriate for loading onto a Sephadex G-50 column. Ultrafiltration may be performed using an Amicon concentrator, Type YM3 filter (Amicon, Beverly, MA). For preparations created from 1 to 6 liters of bacterial culture, the supernatant is concentrated to 25–50 ml, and loaded onto a 5- × 80-cm column of Sephadex G-50. Collected fractions are checked for ultraviolet absorbance at 260 and 280 nm, and are assayed for CRABP content by SDS–PAGE. Any turbidity should elute at the void volume. CRABP should elute soon thereafter. Later fractions are typically high in A_{260} indicating the presence of nucleotides. Buffer A (for CRABP-I: 20 m*M* Bis–Tris-HCl, pH 7.0, 2 m*M* 2-mercaptoethanol, 0.05% sodium azide; for CRABP-II: 20 m*M* Tris-HCl, pH 8.0, 2 m*M* 2-mercaptoethanol, 0.05% sodium azide) is used as the column buffer.

Step 4. Anion-Exchange Chromatography

The gel filtration fractions containing CRABP are then loaded onto a quaternary amine anion-exchange column. FPLC (fast protein liquid chromatography)-type Mono Q (Pharmacia, Piscataway, NJ) or Q (Waters Millipore, Milford, MA) columns have both proved successful for this step. Buffer A (see earlier description) serves as the column buffer. Under these conditions, both CRABPs should be retained on these columns. Owing to its higher p*I*, CRABP-II will not be retained on these columns at pH 7.0, and thus pH 8.0 should be used. The CRABPs are then eluted using a gradient from 100% buffer A to 100% buffer B (buffer A + 250 m*M* NaCl) during a 30-min period. The CRABP should elute as a single peak as monitored by A_{280}. CRABP-I elutes at approximately 125–150 m*M* NaCl; CRABP-II, at 50–75 m*M* NaCl. CRABP-I may elute as two peaks at this step, due to heterogeneity in the N-terminal amino acid sequence[7] (see later discussion).

Typically there will be more sample than can be loaded onto the column in a single run. The column should be washed with buffer A containing 1 *M* NaCl between runs. Additional purification, if necessary, may be obtained by dialyzing the CRABP against buffer A, and repassaging it over the anion-exchange column. The purified CRABP should show high levels of homogeneity when analyzed by SDS–PAGE (Fig. 1). CRABP at this stage may be stored at 4°.

Step 5. Delipidation

CRABP purified by these methods is partially complexed with an endogenous bacterial lipid. For this reason, purified CRABP should be delipidated prior to study. This can be accomplished by passage over a column of hydroxyalkylpropyldextran (type IV, Sigma, St. Louis, MO) or Lipidex-

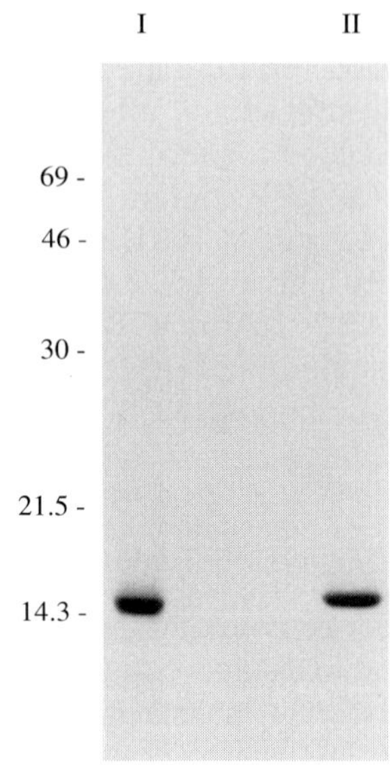

FIG. 1. SDS–PAGE analysis of purified recombinant mouse CRABP-I (**I**) and CRABP-II (**II**).

1000 (Packard, Downers Grove, IL) at 37°.[21,22] CRABP will not be retained on this column and will elute in the void volume. A column containing 20 ml of packed Lipidex gel is adequate to delipidate 10 mg of CRABP. The shelf life of CRABP is reduced by delipidation.

Properties

UV-VIS Absorption Spectra

Both isoforms of CRABP exhibit an absorption peak at 280 nm typical of tryptophan-containing proteins. The presence of high absorbance at 260 nm may indicate the presence of contaminating nucleotides. A second peak is present for holo-CRABP, at 350 nm, due to the bound retinoic acid. For pure holo-CRABP-I the A_{350} to A_{280} ratio has been reported to be 1.8, for both native and recombinant protein.[7,9] This value has been reported to be 1.3 for native CRABP-II,[23] and 1.8 for recombinant CRABP-II.[11]

[21] J. B. Lowe, J. C. Sacchettini, M. Laposata, J. J. McQuillan, and J. I. Gordon, *J. Biol. Chem.* **262,** 5931 (1987).

[22] J. F. C. Glatz and J. H. Veerkamp, *J. Biochem. Biophys. Meth.* **8,** 57 (1983).

[23] J. S. Bailey and C.-H. Siu, *J. Biol. Chem.* **263,** 9326 (1988).

Quantitation

The CRABPs may be accurately quantitated by measurement of their absorbance at 280 nm. The extinction coefficients of recombinant murine CRABP-I and CRABP-II in 20 m*M* KPO_4, pH 7.4, 100 m*M* KCl have been determined to be 21,270 cm^{-1} M^{-1} and 19,990 cm^{-1} M^{-1}, respectively.[13] The extinction coefficient for CRABP in other buffers or from other species may be simply determined by comparing its absorbance in the native, A_n, and denatured, A_d, states.[24,25] The absorbance in the denatured state can be obtained in the presence of buffer plus 6 *M* guanidine chloride. The extinction coefficient in the native state, ε_n, is related to the extinction coefficient in the denatured state, ε_d, as

$$\varepsilon_n = A_n \varepsilon_d / A_d \tag{1}$$

The extinction coefficient in the denatured state can be accurately calculated from the tryptophan, tyrosine, and cysteine content of the protein.[24]

Amino Acid Content

Analysis of purified CRABP by sequential Edman degradation can be a measure of protein purity, and should reproduce the expected sequence. Heterogeneity in purified CRABP-I due to the presence or absence of a blocked amino-terminal methionine has been reported.[7] The two forms of CRABP-I are distinguished by different isoelectric points and different retention times on anion-exchange chromatography.[7] Preparations of recombinant CRABP created in this laboratory have not contained the initiator methionine, and elute as a single peak from anion-exchange chromatography.

Fluorescence Spectra of Apo-CRABP

The three tryptophans found in the CRABPs are fluorescent. When selectively excited at 290 nm they exhibit an emission maximum of 327 and 334 nm for recombinant mouse apo-CRABP-I and -II, respectively.[13]

Fluorescence Spectra of Holo-CRABP

The fluorescence spectra of holo-CRABP is more complex than that of apo-CRABP because retinoic acid becomes fluorescent when bound to the CRABPs. The excitation spectra of holo-CRABP, when monitored at

[24] S. C. Gill and P. H. von Hippel, *Anal. Biochem.* **182,** 319 (1989).
[25] T. M. Lohman and D. P. Mascotti, *Methods Enzymol.* **212,** 424 (1992).

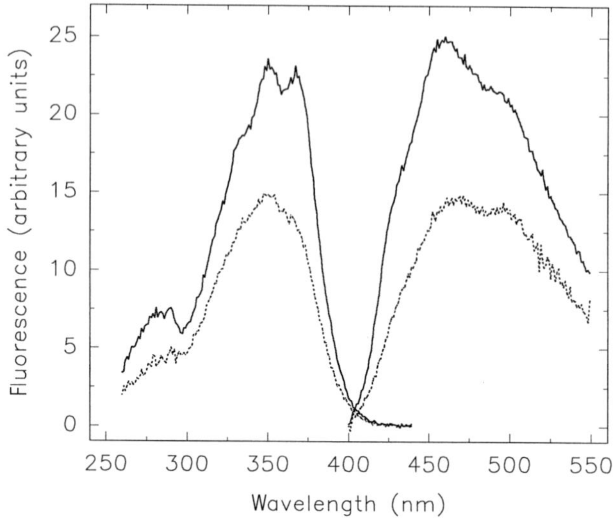

FIG. 2. Comparison of the fluorescence spectral properties of *E. coli*-derived mouse CRABP-I (solid lines) and CRABP-II (broken lines). Excitation spectra are the leftmost curves, and were recorded using an excitation wavelength of 475 nm. Emission spectra are the rightmost curves, and were recorded using an emission wavelength of 475 nm. All measurements were made on 10 μM holo-CRABP samples at 25° in 20 mM KH_2PO_4, pH 7.4, 100 mM KCl. Reference channel normalization was used to correct for variations in excitation energies. The spectra have been corrected for fluorescence intrinsic to apo-CRABP/buffer. Excitation and emission slit widths were set to 2 nm.

475 nm, has several bands including one at 280–290 nm (see Fig. 2). This band, which is believed to represent energy transfer from tryptophan residues to retinoic acid, is found for both native and recombinant CRABP-I and CRABP-II.[7,9,11,23] A second, much more intense excitation band is found at 350 and 349 nm in native and recombinant CRABP-II, respectively.[11,23] This correlates well with the absorbance maximum of retinoic acid at ~350 nm. This band at ~350 nm is also present in CRABP-I, but has an additional strong, prominent shoulder located at ~369 nm for both native[9,26] and recombinant (Fig. 2) CRABP-I.

The emission spectra, exciting at 350 nm, of native holo-CRABP-I and CRABP-II show a peak at 475 nm.[9,23] This band is well conserved in recombinant CRABP-I and CRABP-II at 460 and 469 nm (Fig. 2), respectively.

[26] M. Okuno, M. Kato, H. Moriwaki, M. Kanai, and Y. Muto, *Biochim. Biophys. Acta* **923,** 116 (1987).

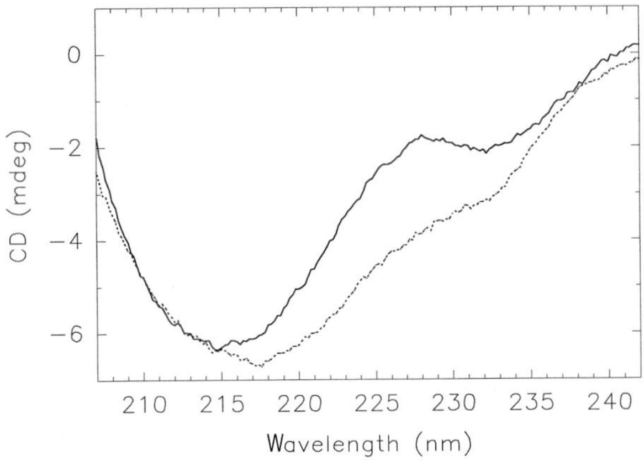

FIG. 3. Circular dichroism spectra of *E. coli*-derived mouse CRABP-I (solid line) and CRABP-II (broken lines). All spectra were recorded on 7 μM holo-CRABP samples at 25° in 20 mM KH_2PO_4, pH 7.4, 100 mM KCl.

Circular Dichroism Spectra

The circular dichroism (CD) spectrum of both CRABPs is typical of β-sheet proteins. Apo-CRABP-I has a minimum CD band at 216 nm.[27] This minimum has a shoulder at 230 nm, which is presumably due to an aromatic residue.[27] Apo-CRABP-II exhibits a very similar CD spectra, except that the shoulder at 230 nm is less pronounced (Fig. 3). There is a minimum CD band centered at 350 nm in the CD spectrum of holo-CRABP-I, due to bound retinoic acid.[27] The presence of this feature has not been investigated for holo-CRABP-II.

Isoelectric Point

The isoelectric point of recombinant bovine CRABP-I (which has the same amino acid sequence as murine CRABP-I) is lower than that of murine CRABP-II: 5.0 versus 5.46.[11] This is consistent with the lower p*I* of CRABP-I compared to CRABP-II found for native CRABPs purified from a number of species.[28,29]

[27] J. Zhang, Z.-P. Liu, T. A. Jones, L. M. Gierasch, and J. F. Sambrook, *Proteins* **13,** 87 (1992).

[28] W. J. Scott, Jr., R. Walter, G. Tzimas, J. O. Sass, H. Nau, and M. D. Collins, *Dev. Biol.* **165,** 397 (1994).

[29] G. Siegenthaler, I. Tomatis, D. Chatellard-Gruaz, S. Jaconi, U. Eriksson, and J.-H. Saurat, *Biochem. J.* **287,** 383 (1992).

TABLE II
ALL-*trans*-RETINOIC ACID AFFINITIES MEASURED FOR CRABP FROM VARIOUS SOURCES

Species	Source	K_d (nM) CRABP-I	K_d (nM) CRABP-II	Method[a]
Rat	Native	—	65	Fluorescence[d]
Mouse	Recombinant	—	106	Fluorescence[e]
Bovine/mouse	Recombinant	7	14	Fluorescence[f]
Human	Recombinant	6.8	39	Separation[g]
Mouse	Recombinant	<0.4	2	Fluorescence[h]
Rat	Native	4.2	—	Fluorescence[i]
Human	Recombinant[b]	—	1.2	Fluorescence[j]
Human	Native	1.5, 69[c]	4.7, 101[c]	Separation[k]
Chick	Native	13.8	48.2	Separation[l]
Mouse	Native	11	43.5	Separation[l]
Human	Native	16.6	50	Separation[m]
Mouse	Recombinant	39	—	Fluorescence[n]

[a] Techniques for measuring retinoic acid-binding affinities are broadly characterized into methods based on fluorescence and methods that rely on physical separation of bound from unbound ligand.

[b] CRABP-II was expressed as a glutathione *S*-transferase fusion protein.

[c] Two classes of sites were demonstrated by Scatchard analysis.

[d] J. S. Bailey and C.-H. Siu, *J. Biol. Chem.* **263,** 9326 (1988).

[e] L. X. Chen, Z.-P. Zhang, A. Scafonas, R. C. Cavalli, J. L. Gabriel, K. J. Soprano, and D. R. Soprano, *J. Biol. Chem.* **270,** 4518 (1995).

[f] P. D. Fiorella, V. Giguère, and J. L. Napoli, *J. Biol. Chem.* **268,** 21545 (1993).

[g] K. Fogh, J. J. Voorhees, and A. Åström, *Arch. Biochem. Biophys.* **300,** 751 (1993).

[h] A. W. Norris, L. Cheng, V. Giguère, M. Rosenberger, and E. Li, *Biochim. Biophys. Acta* **1209,** 10 (1994).

[i] D. E. Ong and F. Chytil, *J. Biol. Chem.* **253,** 4551 (1978).

[j] C. P. F. Redfern and K. E. Wilson, *FEBS Lett.* **321,** 163 (1993).

[k] S. Sanquer and B. A. Gilchrest, *Arch. Biochem. Biophys.* **311,** 86 (1994).

[l] W. J. Scott, Jr., R. Walter, G. Tzimas, J. O. Sass, H. Nau, and M. D. Collins, *Dev. Biol.* **165,** 397 (1994).

[m] G. Siegenthaler, I. Tomatis, D. Chatellard-Gruaz, S. Jaconi, U. Eriksson, and J.-H. Saurat, *Biochem. J.* **287,** 383 (1992).

[n] J. Zhang, Z.-P. Liu, T. A. Jones, L. M. Gierasch, and J. F. Sambrook, *Proteins* **13,** 87 (1992).

Mobility on Nondenaturing Polyacrylamide Gel Electrophoresis

Both CRABPs migrate as a single peak when analyzed by nondenaturing PAGE.[30] The apo and holo forms are not resolved by this process. As expected by their different isoelectric points, CRABP-I and CRABP-II have different electrophoretic mobilities, with CRABP-I migrating faster.

[30] G. Siegenthaler, *Methods Enzymol.* **189,** 299 (1990).

The R values on 7.5% PAGE of native human CRABP-I and CRABP-II are 0.65 and 0.44, respectively.[29] We have observed similar results for recombinant mouse CRABP-I and CRABP-II, with R_f on 8% PAGE of 0.7 and 0.4, respectively.

Binding of all-trans-Retinoic Acid

Both isoforms of native and recombinant CRABP bind all-*trans*-retinoic acid with one-to-one stoichiometry.[7,11,13,23,31] Both native and recombinant CRABPs bind all-*trans*-retinoic acid with extremely high affinity. However, the actual binding affinities measured are highly dependent on the technique and protein concentration used. This is illustrated by Table II, which shows the range of dissociation constants obtained for native and recombinant CRABP. As can be seen, it is generally agreed that CRABP-I exhibits a higher affinity for all-*trans*-retinoic acid than does CRABP-II. Note that the techniques used for measurement of binding affinities can be divided into two categories: assays that separate bound from unbound retinoic acid[32] and assays based on fluorescence.[7,13,33] We have found that both CRABPs, once purified, are subject to a gradual loss of binding affinity over the span of months.

Acknowledgments

Ellen Li is a Burroughs Wellcome Scholar in toxicology. This work was supported by National Institutes of Health grant RO1 DK49684-01.

[31] J. C. Saari, S. Futterman, and L. Bredberg, *J. Biol. Chem.* **253,** 6432 (1978).
[32] A. K. Daly and C. P. F. Redfern, *Biochim. Biophys. Acta* **965,** 118 (1988).
[33] A. W. Norris and E. Li, *in* "Methods in Molecular Biology" (C. P. F. Redfern, ed.), in press. Humana Press, Clifton, New Jersey (1997).

[2] Generating and Characterizing Retinoid Receptors from *Escherichia coli* and Insect Cell Expression Systems

By MARGARET CLAGETT-DAME and JOYCE J. REPA

Introduction

The vitamin A metabolite, retinoic acid (RA), supports a number of biologic functions, including animal growth and the differentiation of a variety of cell types. RA exerts its effects by binding to nuclear retinoid

0076-6879/97 $25.00

receptors (RAR and RXR), which, in turn, modulate gene expression. cDNAs encoding three subtypes of RAR and RXR proteins (α, β, and γ) have been isolated. The RAR and RXR families belong to the steroid/thyroid hormone receptor superfamily. These proteins contain a well-conserved DNA-binding domain as well as a carboxy-terminal ligand-binding domain. The investigation of binding of natural and synthetic compounds to individual types of RAR and RXR proteins is of interest because of the importance of ligand binding in receptor function. This article focuses on the production and characterization of recombinant retinoid receptors in bacteria and insect cells, analysis of receptors using equilibrium saturation and competition binding analyses, and the use of sucrose density gradients (SDGs) to examine how DNA influences ligand binding to RAR and RXR proteins.

Advantages of Recombinant Receptors

Large quantities of individual RAR and RXR proteins can be generated using recombinant technology, whereas receptors occur at very low concentrations in animal cells or tissue and multiple subtypes of receptor are present. Animal cells also express a number of other retinoid-binding proteins, including the cellular retinoic acid-binding proteins (CRABP-I and -II) and the cellular retinol-binding proteins (CRBP-I and -II). Bacteria, yeast, and insect cells are devoid of CRABP and CRBP activity, which makes these advantageous host systems in which to generate receptors for ligand-binding studies.

Choice of Expression System

Expression of receptor proteins in bacteria is fast and inexpensive and hence large quantities of protein can be produced. However, bacteria do not perform many of the posttranslational protein modifications of eukaryotic cells, and the desired protein may be produced as an insoluble aggregate requiring renaturation before use in functional studies. For experiments in which a high percentage of functional receptor is desired, or in which receptor posttranslational modification is an issue, expression in insect cells or yeast may be preferred. Table I illustrates the wide range of uses for RAR and RXR proteins produced by bacterial and/or insect cell systems. This table presents the initial description of a given vector system and/or a novel application of the RAR or RXR protein expressed in bacteria or insect cells. Receptor expression in yeast is covered elsewhere in this volume.[1]

[1] E. A. Allegretto and R. A. Heyman, *Methods Enzymol.* **282** [3], 1997, (this volume).

Production of Recombinant RAR and RXR Proteins

Expression of RAR Fusion Proteins in Bacteria

pATH Vector System. This system has been described in detail by Koerner *et al.*[1a] Briefly, the receptor sequence of interest is cloned in-frame to the *trpE* gene of *Escherichia coli* under the control of the *trp* operon. The resulting hybrid protein contains 323 amino acid residues of anthranilate synthase (37 kDa) at the amino-terminal end of the receptor. Full-length and deletion mutants of RAR proteins have been produced in our laboratory using the pATH plasmid expression system (Fig. 1A).

Screening for Receptor Fusion Proteins. Following the expression of receptor fusion protein in bacteria, the total cell lysate and/or the soluble and insoluble pellet fractions are analyzed. The hybrid receptor protein is detected after resolution on sodium dodecyl sulfate (SDS)–polyacrylamide gels and staining with Coomassie Blue dye. Bacterial extracts from cells transformed with the parent vector (e.g., pATH20, Fig. 1B, lane 2); and/or extracts prepared from noninduced RAR/pATH-transformed cells should be examined as negative controls. Receptor fusion proteins are generally found in the insoluble fraction; however, we have observed small receptor fusion products (<45 kDa) in the soluble fraction. An example of the electrophoretic analysis of RAR/pATH fusion proteins is shown in Fig. 1B.

Hybrid protein can also be detected by immunoblotting. Antibodies to some RAR and RXR proteins are commercially available (Affinity Bioreagents, Golden, CO; Biomol Research Labs, Plymouth Meeting, PA; Santa Cruz Biotechnology, Santa Cruz, CA). Antibodies to anthranilate synthase can be purchased from Oncogene Research Products (Cambridge, MA).

RAR/pATH fusion proteins containing the intact ligand-binding domain can also be detected by specific radioligand binding. However, this requires denaturation and refolding of the expressed protein prior to assay. This is accomplished by combining the receptor-containing fraction (generally the resuspended pellet, 70 μl) with 70 μl of denaturing buffer [6 *M* guanidine hydrochloride, 50 m*M* Tris-HCl, pH 7.4 (at 25°), 0.3 *M* KCl, 1.5 m*M* EDTA, 5 m*M* dithiothreitol (DTT), 20% (v/v) glycerol, 5 μg/ml soybean trypsin inhibitor (STI), and 0.5 m*M* phenylmethylsulfonyl fluoride (PMSF)] and incubating the mixture at 4° for 30 min with shaking. The mixture is then subjected to centrifugation for 30 min at 16,000*g* in a refrigerated microfuge. The supernatant (70 μl) is combined with 3.5 ml

[1a] T. J. Koerner, J. E. Hill, A. M. Myers, and A. Tzagoloff, *Methods Enzymol.* **194,** 477 (1991).

TABLE I
RECOMBINANT RETINOID RECEPTOR EXPRESSION AND USE[a]

Protein	Vector	Application	Ref.[b]
E. coli			
his-h**RAR**αDEF and βDEF	pDS56/RSB	Ligand binding	a
AS-h**RAR**α,β,γ	pATH	Ligand binding	b
h**RAR**αEF and βDEF and γ	pET3d	Ligand binding	c
MBP-h**RAR**β	pMAL-cR1	Ligand binding	d
GST-h**RAR**γ	pGEX-2T	Ligand binding	e
his-h**RAR**αDEF mutants	pQE-9	Ligand binding	f
his-h**RAR**α/β chimeras	pET15b	Ligand binding	g
h**RAR**α and αC	pET-8c	Ligand/DNA binding	h
h**RAR**α	pT7-7	Ligand/DNA binding	i
h**RAR**βC	pET3	NMR spectroscopy	j
AS-n**RAR**δ fragments	pATH	Ab production	k
GST-h**RAR**α	pGEX-2T	Phosphorylation	l
GST-h**RAR**α	pGEX-2T	Associated proteins	m
his-h**RAR**γDE	pET15b	CD/fluorescence	n
MBP-h**RAR**αE	pMAL-c2	Photoaffinity label	o
his-h**RAR**γDE	pET15b	X-ray crystallography	p
his-m**RXR**αCDE	pET15b	DNA binding	q
GST-h**RXR**αDEF and EF	pGEX-KG	Fluorescence	r
his-m**RXR**αCDE	pET14b	Limited proteolysis	s
his-h**RXR**αDE	pET15b	X-ray crystallography	t
his-h**RXR**αDE, **RAR**αDEF	pET15b	Fluorescence (dimers)	u
Insect cell			
h**RAR**α,β,γ	pVL1392	Receptor purification	v
h**RAR**γ	pVL1392	Receptor purification	w
m**RAR**γ	pVL1393	Ligand/DNA binding	x
h**RAR**α	pVL1392	Ligand/DNA binding	y
c**RAR**β, m**RAR**β	pVL1392&3	Ligand/DNA binding	z
h**RAR**γ	pVL1393	Receptor purification/Ab	aa
h**RAR**α,β,γ, m**RXR**α,β,γ	na	Ligand binding/Ab	bb
m**RXR**β (H2RIIBP)	pAc436	DNA binding	cc
his-m**RXR**αCDE	pVL1392	DNA binding	q

[a] A–F denote receptor domains; his, polyhistidine tag; AS, anthranilate synthase; MBP, maltose-binding protein; GST, glutathione *S*-transferase; na, not available; Ab, antibody; CD, circular dichroism; c, chick; h, human; m, mouse; n, newt.

[b] *Key to references:* (a) M. Crettaz, A. Baron, G. Siegenthaler, and W. Hunziker, *Biochem. J.* **272,** 391 (1990); (b) J. J. Repa, K. K. Hanson, and M. Clagett-Dame, *Proc. Natl. Acad. Sci. U.S.A.* **90,** 7293 (1993); (c) H. Fukasawa, T. Iijima, H. Kagechika, Y. Hashimoto, and K. Shudo, *Biol. Pharm. Bull.* **16,** 343 (1993); (d) A. Lombardo, E. Costa, W.-R. Chao, L. Toll, P. D. Hobbs, L. Jong, M.-O. Lee, M. Pfahl, K. R. Ely, and M. I. Dawson, J. Biol. Chem. **269,** 7297 (1994); (e) M. I. Dawson, W.-R. Chao, P. Pine, L. Jong, P. D. Hobbs, C. K. Rudd, T. C. Quick, R. M. Niles, X.-K. Zhang, A. Lombardo, K. R. Ely, B. Shroot, and J. A. Fontana, *Cancer Res.* **55,** 4446 (1995); (f) B. Lefebvre, C. Rachez, P. Formstecher, and P. Lefebvre, *Biochemistry* **34,** 5477 (1995); (g) J. Ostrowski,

(continued)

TABLE I (*continued*)

L. Hammer, T. Roalsvig, K. Pokornowski, and P. R. Reczek, *Proc. Natl. Acad. Sci. U.S.A.* **92,** 1812 (1995); (h) N. Yang, R. Schüle, D. J. Mangelsdorf, and R. M. Evans, *Proc. Natl. Acad. Sci. U.S.A.* **88,** 3559 (1991); (i) S. Keidel, E. Rupp, and M. Szardenings, *Eur. J. Biochem.* **204,** 1141 (1992); (j) R. M. A. Knegtel, M. Katahira, J. G. Schilthuis, A. M. J. J. Bonvin, R. Boelens, D. Eib, P. T. van der Saag, and R. Kaptein, *J. Biomol. NMR* **3,** 1 (1993); (k) D. S. Hill, C. W. Ragsdale, Jr., and J. P. Brockes, *Development* **117,** 937 (1993); (l) J. I. Huggenvik, M. W. Collard, Y.-W. Kim, and R. P. Sharma, *Mol. Endocrinol.* **7,** 543 (1993); (m) A. J. Hörlein, A. M. Näär, T. Heinzel, J. Torchia, B. Gloss, R. Kurokawa, A. Ryan, Y. Kamei, M. Söderström, C. K. Glass, and M. G. Rosenfeld, *Nature* **377,** 397 (1995); (n) J. A. Lupisella, J. E. Driscoll, W. J. Metzler, and P. R. Reczek, *J. Biol. Chem.* **270,** 24884 (1995); (o) T. Sasaki, R. Shimazawa, T. Sawada, T. Iijima, H. Fukasawa, K. Shudo, Y. Hashimoto, and S. Iwasaki, *Biochem. Biophys. Res. Commun.* **207,** 444 (1995); (p) J.-P. Renaud, N. Rochel, M. Ruff, V. Vivat, P. Chambon, H. Gronemeyer, and D. Moras, *Nature* **378,** 681 (1995); (q) Z.-P. Chen, L. Shemshedini, B. Durand, N. Noy, P. Chambon, and H. Gronemeyer, *J. Biol. Chem.* **269,** 25770 (1994); (r) L. Cheng, A. W. Norris, B. F. Tate, M. Rosenberger, J. F. Grippo, and E. Li, *J. Biol. Chem.* **269,** 18662 (1994); (s) M. Leid, *J. Biol. Chem.* **269,** 14175 (1994); (t) W. Bourguet, M. Ruff, P. Chambon, H. Gronemeyer, and D. Moras, *Nature* **375,** 377 (1995); (u) S. Kersten, M. I. Dawson, B. A. Lewis, and N. Noy, *Biochemistry* **35,** 3816 (1996); (v) W. Bourguet, B. Sablonnière, P. Formstecher, J.-Y. Chen, J.-L. Bernier, and J.-P. Hénichart, *Biochem. Biophys. Res. Commun.* **187,** 711 (1992); (w) A. P. Reddy, J.-Y. Chen, T. Zacharewski, H. Gronemeyer, J. J. Voorhees, and G. J. Fisher, *Biochem. J.* **287,** 833 (1992); (x) T. K. Ross, J. M. Prahl, I. M. Herzberg, and H. F. DeLuca, *Proc. Natl. Acad. Sci. U.S.A.* **89,** 10282 (1992); (y) T. C. Quick, A. M. Traish, and R. M. Niles, *Receptor* **4,** 65 (1994); (z) J. J. Repa, L. A. Plum, P. K. Tadikonda, and M. Clagett-Dame, *FASEB J.* **10,** 1078 (1996); (aa) J. J. Repa, J. A. Berg, M. E. Kaiser, K. K. Hanson, S. A. Strugnell, and M. Clagett-Dame, *Prot. Expr. Purif.* **9,** 319 (1997); (bb) M. W. Titcomb, M. M. Gottardis, J. W. Pike, and E. A. Allegretto, *Mol. Endocrinol.* **8,** 870 (1994); (cc) M. S. Marks, B.-Z. Levi, J. H. Segars, P. H. Driggers, S. Hirschfeld, T. Nagata, E. Appella, and K. Ozato, *Mol. Endocrinol.* **6,** 219 (1992).

of cold renaturation/binding buffer [denaturing buffer lacking guanidine hydrochloride] and the protein is renatured for 12 hr on ice. This procedure generates sufficient material for a one-point ligand-binding assay (described later), and can be easily scaled up for larger assays.

Expression of RAR and RXR Protein Using Baculovirus Insect Cell System

The expression system for the generation of recombinant retinoid receptors described below is based on the lytic virus, *Autographa californica* nuclear polyhedrosis virus (AcNPV) developed by Summers and Smith.[2]

[2] M. D. Summers and G. E. Smith, "A Manual of Methods for Baculovirus Vectors and Insect Cell Culture Procedures." Texas Agricultural Experimental Station Bulletin No. 1555 (1987).

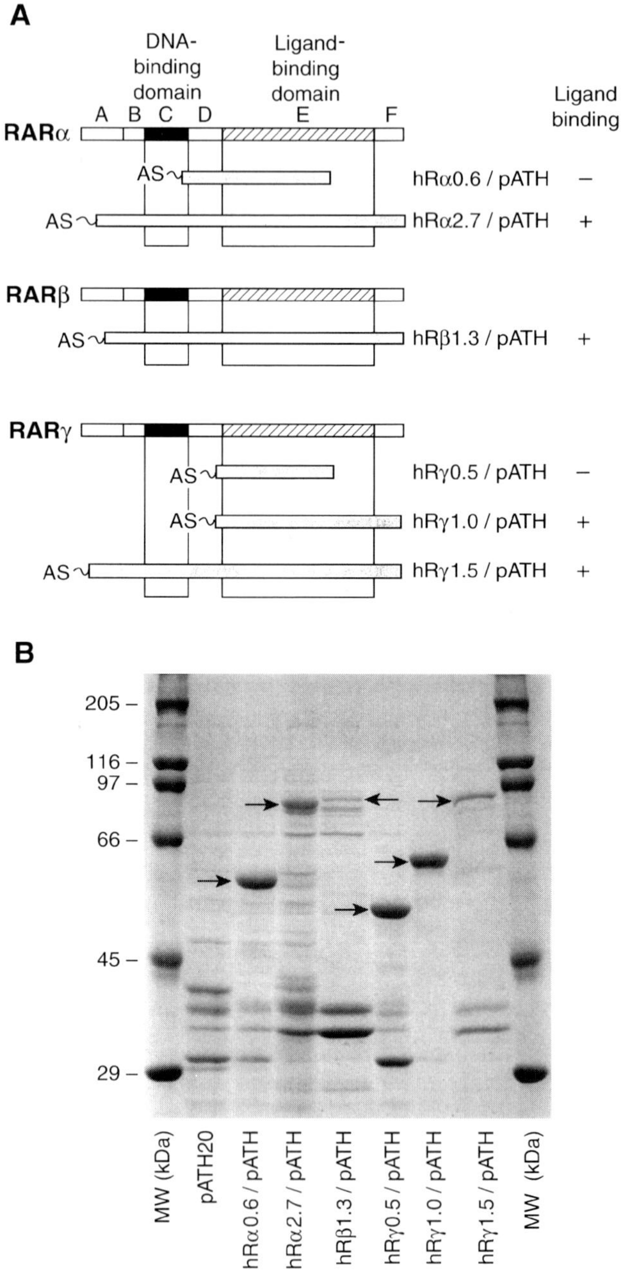
A
DNA-binding domain
Ligand-binding domain
Ligand binding
A B C D E F
RARα
AS
hRα0.6 / pATH −
AS
hRα2.7 / pATH +
RARβ
AS
hRβ1.3 / pATH +
RARγ
AS
hRγ0.5 / pATH −
AS
hRγ1.0 / pATH +
AS
hRγ1.5 / pATH +
B
205 –
116 –
97 –
66 –
45 –
29 –
MW (kDa)
pATH20
hRα0.6 / pATH
hRα2.7 / pATH
hRβ1.3 / pATH
hRγ0.5 / pATH
hRγ1.0 / pATH
hRγ1.5 / pATH
MW (kDa)

The DNA sequence encoding for the full-length receptor is cloned into the multiple cloning region of the transfer vector (pVL1392/1393). The recombinant vector is then transfected along with wild-type viral DNA into *Sf*21 insect cells (reviewed in Ref. 3). Generation of recombinant virus places the receptor coding sequence downstream of the strong polyhedrin protein promoter. In this system, recombinant virus is first identified based on a change in the refractile property of cells lacking polyhedrin-containing occlusion bodies. Once potential positives are identified, the presence of recombinant virus can be verified by Southern blotting[4] or, alternatively, by assaying directly for the presence of retinoid receptor protein in cell isolates using immunoblotting and/or specific binding to [^{3}H]RA. Once a pure viral stock is obtained, large numbers of cells can be infected to generate receptor for biochemical studies. Our laboratory has used this system to generate full-length RAR and RXR proteins. An example of detection of baculovirus expressed human RARγ by immunoblotting is shown in Fig. 2A. The majority of the human RARγ is present in the nuclear fraction, and migrates at the expected molecular size (~51 kDa). The two immunoreactive bands can be accounted for on the basis of differences in the extent of receptor phosphorylation (Repa and Clagett-Dame, unpublished manuscript).

Analysis of Radioligand Binding to RAR and RXR Proteins

Types of Binding Assays

Saturation equilibrium binding and Scatchard analysis is used to determine the affinity of a receptor for radiolabeled ligand (K_d) and the number of receptor-binding sites in a given sample (B_{max}). A variation on this method is to label the receptor at a single concentration of ligand (one-point binding assay). This is useful to screen samples rapidly for ligand-

[3] D. R. O'Reilly, L. K. Miller, and V. A. Luckow, "Baculovirus Expression Vectors: A Laboratory Manual." W. H. Freeman and Company, New York, 1992.

[4] B. B. Goswami and R. I. Glaser, *BioTechniques* **10,** 626 (1991).

FIG. 1. Bacterial expression of RAR/pATH fusion proteins. (A) Schematic representation of RARα, RARβ, and RARγ proteins fused at the N terminus to a portion of anthranilate synthase (AS). Only fusion proteins containing the entire ligand-binding domain exhibit specific all-*trans*-[^{3}H]RA binding. (B) SDS–PAGE analysis of RAR/pATH fusion proteins. The pellet fraction of each expression was solubilized and proteins were resolved on a 10% reducing SDS–polyacrylamide gel. Hybrid receptor proteins (indicated by arrows) were detected by Coomassie Blue staining.

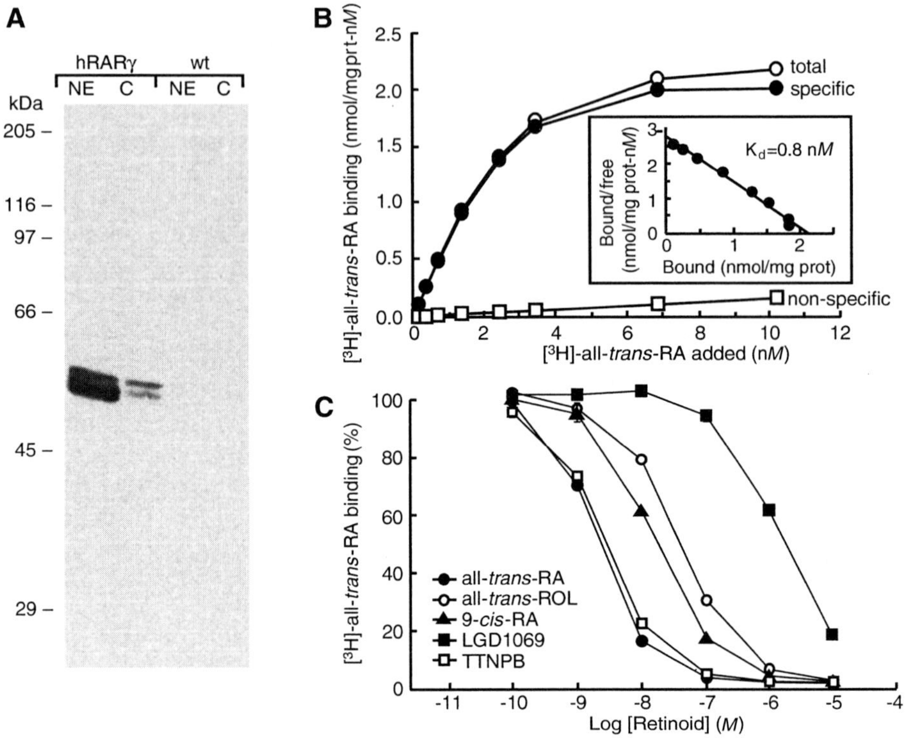

FIG. 2. Characterization of human RARγ produced by the baculovirus expression vector system. (A) Immunoblot of nuclear (NE) and cytosolic (C) extracts from *Sf*21 cells infected with hRARγ recombinant virus (hRARγ) or wild-type AcNPV virus (wt). Extracts (25 μg protein/lane) were resolved on a 10% SDS-reducing polyacrylamide gel and proteins were transferred to a nitrocellulose membrane. Human RARγ was detected using a hRARγ-specific monoclonal antibody (A10) developed in our laboratory using the hRγ1.0/pATH protein as antigen. (B) Equilibrium saturation binding of all-*trans*-[11, 12-^{3}H]RA (50 Ci/mmol, DuPont-NEN, Boston, MA) to hRARγ nuclear extract. Data were transformed by Scatchard analysis using the LIGAND program (shown in the inset). (C) Competition of retinoids for all-*trans*-[20-methyl-^{3}H]RA (67 Ci/mmol, DuPont-NEN) binding to baculovirus-expressed hRARγ. Nuclear extract was incubated with 1.2 n*M* all-*trans*-[^{3}H]RA in the absence (100% binding) or presence of unlabeled retinoids followed by HAP assay.

binding activity, or to check a novel compound for its ability to compete with radiolabeled ligand for binding to receptor. Competition binding studies are used to determine relative affinities of unlabeled ligands for binding to a given receptor (IC_{50}). K_i values can be determined if the K_d for radiolabeled drug is known. Reviews covering the theoretical aspects of these methods

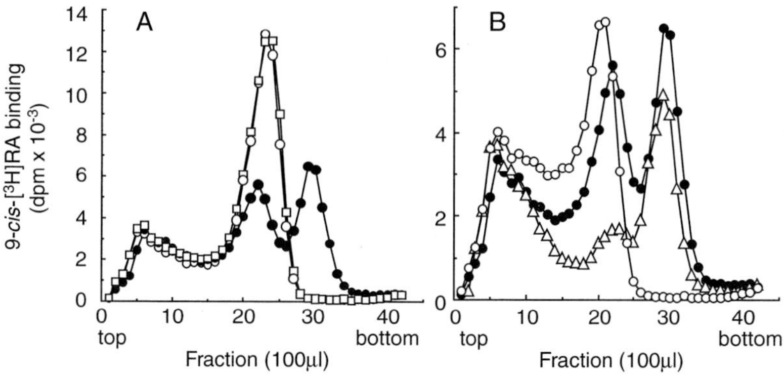

FIG. 3. Sedimentation analyses of human RARγ and murine RXRγ with and without added DNA. (A) Differential sedimentation of mRXRγ + hRARγ versus mRXRγ + hRARγ bound to βRARE. Baculovirally expressed hRARγ and mRXRγ were incubated with the pan-agonist, 9-*cis*-[20-*methyl*-^{3}H]RA (5 n*M*, 80 Ci/mmol, DuPont-NEN) for 3 hr on ice. The βRARE (●), mutant RE (□), or vehicle (o, no DNA) was added for the last hour of the incubation, and the mixtures were analyzed on SDG (4–18%). hRARγ + mRXRγ generate a single radiolabeled peak (~fraction 22, ~4S). The addition of specific (βRARE) but not mutant DNA leads to the formation of a larger radiolabeled complex (fraction 30). (B) The effect of DNA on specific 9-*cis*-[^{3}H]RA binding to hRARγ and mRXRγ proteins. hRARγ + excess mRXRγ was incubated with 9-*cis*-[^{3}H]RA (5 n*M*) in the absence (●) or presence of the unlabeled RXR-selective ligand, LGD1069 (Δ, 500 n*M*) or RAR-specific ligand, TTNPB (o, 500 n*M*). βRARE was added to all tubes for the last hour of the incubation. The peak sedimenting at ~fraction 22 corresponds to mRXRγ protein, as evidenced by competition by LGD1069 but not TTNPB. Under conditions in which radioligand is limiting, only TTNPB competes for the radioactivity specifically bound to the receptor-DNA complex sedimenting at fraction 30.

can be found in Refs. 5 and 6. In this chapter, we also describe the use of sedimentation analysis to study the formation of receptor DNA complexes, and we describe a way in which this technique can be used to delineate how DNA influences the ligand-binding properties of RAR and RXR proteins (Fig. 3).

General Considerations

Retinoids. Retinoids should be manipulated under amber lighting to prevent isomerization and oxygen must be excluded during storage to prevent oxidation. The concentration of nonradiolabeled compounds in

[5] E. C. Hulme and N. J. M. Birdsall, *in* "Receptor–Ligand Interactions. A Practical Approach" (E. C. Hulme, ed.), p. 63. IRL Press, Oxford, 1992.

[6] D. B. Bylund and H. I. Yamamura, *in* "Methods in Neurotransmitter Receptor Analysis" (H. I. Yamamura, S. J. Enna, and M. J. Kuhar, eds.), p. 1. Raven Press, New York, 1990.

solution (ethanol) is determined spectrophotometrically and the purity of all compounds is verified by high-performance liquid chromatography. A useful table of extinction coefficients can be found in Ref. 7. Retinoid solutions are flushed with argon or other inert gas before storage at −70°.

Receptor Proteins. Recombinant receptor-containing extracts are frozen in liquid N_2 and stored at −70° until use. These preparations generally require dilution prior to use in binding studies. Ideally, less than 10% of added radioligand should be bound to receptor (specific binding). There should be a linear decrease in binding with dilution.

Separation of Free Ligand from That Bound to Receptor. Various methods have been described in the literature including the use of hydroxylapatite (HAP),[8,9] filtration,[10] dextran-coated charcoal,[11] and size-exclusion chromatography.[12] The use of hydroxylapatite is fast, yields consistent results and gives a low percentage of nonspecific binding.

Detergent. The inclusion of CHAPS (0.5%, w/v) in all buffers is recommended when studying binding of 9-*cis*-[^{3}H]RA to RAR and RXR proteins.[13] We observe little or no specific binding of 9-*cis*-[^{3}H]RA to RXR unless CHAPS is present in the assay. We have determined that the K_d for all-*trans*-[^{3}H]RA binding to RAR is not altered in the presence or absence of CHAPS. However, we have not used the CHAPS method in studies of renatured RAR/pATH fusion proteins described here, nor is CHAPS included in buffers for sedimentation studies.

Binding Assay

Materials

Binding buffer for insect cell-expressed receptors: 50 m*M* Tris-HCl, pH 7.4, 1.5 m*M* EDTA, 5 m*M* DTT, 120 m*M* KCl (150 m*M* for SDG), 8% (v/v) glycerol (omit for SDG), 0.5% (w/v) CHAPS (omit for SDG) and the protease inhibitors: 1 m*M* PMSF, 10 μg/ml STI, 2

[7] H. C. Furr, A. B. Barua, and J. A. Olson, *in* "The Retinoids: Biology, Chemistry, and Medicine" (M. B. Sporn, A. B. Roberts, and D. S. Goodman, eds.), p. 179. Raven Press, New York, 1994.

[8] D. Williams and J. Gorski, *Biochemistry* **13,** 5537 (1974).

[9] M. C. Dame, E. A. Pierce, and H. F. DeLuca, *Proc. Natl. Acad. Sci. U.S.A.* **82,** 7825 (1985).

[10] Y. Hashimoto, M. Petkovich, M. P. Gaub, H. Kagechika, K. Shudo, and P. Chambon, *Mol. Endocrinol.* **3,** 1046 (1989).

[11] B. Sablonnière, N. Dallery, I. Grillier, P. Formstecher, and M. Dautrevaux, *Anal. Biochem.* **217,** 110 (1994).

[12] A. A. Levin, L. J. Sturzenbecker, S. Kazmer, T. Bosakowski, C. Huselton, G. Allenby, J. Speck, C. Kratzeisen, M. Rosenberger, A. Lovey, and J. F. Grippo, *Nature* **355,** 359 (1992).

[13] E. A. Allegretto, M. R. McClurg, S. B. Lazarchik, D. L. Clemm, S. A. Kerner, M. G. Elgort, M. F. Boehm, S. K. White, J. W. Pike, and R. A. Heyman, *J. Biol. Chem.* **268,** 26625 (1993).

μ*M* E-64 [*trans*-epoxysuccinyl-L-leucylamido(4-guanidino)butane], 1.6 m*M* benzamidine, 1 μ*M* pepstatin A, and 5 μg/ml leupeptin

HAP slurry (50%): 10 g hydroxylapatite (Bio-Gel HTP, Bio-Rad, Richmond, CA) in 60 ml 50 m*M* Tris-HCl, pH 7.4, 1.5 m*M* EDTA

HAP wash buffer: 50 m*M* Tris-HCl, pH 7.4, 1.5 m*M* EDTA, 120 m*M* KCl, 0.5% (w/v) CHAPS [for RAR/pATH proteins substitute 0.5% (v/v) Triton X-100 for CHAPS]

Methods. All procedures are performed at 4°. Prepare 25× labeling solutions containing [^{3}H]RA + ethanol or [^{3}H]RA + unlabeled competitor. For one-point binding assays a saturating concentration of [^{3}H]RA is used (>5-fold the K_d; for RAR we use a final concentration of 5 n*M* all-*trans*-[^{3}H]RA, for RXR 50 n*M* 9-cis-[^{3}H]RA) and a 100- to 200-fold excess of unlabeled ligand. To determine receptor K_d, the amount of radioligand added is varied (~10-fold above and below the predicted K_d value) in the presence and absence of unlabeled competitor (100-fold the highest concentration of radiolabel). For competition studies, a constant amount of [^{3}H]RA is used with increasing concentrations of competitor (no added competitor = 100% or total [^{3}H]RA binding).

Dilute receptor protein in binding buffer to provide ~50 fmol of receptor in 500 μl (note, RAR/pATH proteins have already been diluted in renaturation/binding buffer). Maintain total protein at ~0.1 mg/ml by adding non- or wild-type-infected nontransformed cell extracts; this prevents loss of receptor protein and ligand to the walls of the tube. Pipette 500 μl of diluted receptor into polypropylene microfuge tubes (in triplicate). Add 20 μl of each labeling solution to 500 μl receptor extract and vortex. [^{3}H]RA and competitor *must* be added simultaneously. Incubate samples on ice for 3 hr (experiments show that equilibrium binding is reached at ~2 hr under these conditions).

The HAP assay is used to separate [^{3}H]RA bound to receptor from free ligand for saturation equilibrium binding studies and competition analyses. In sedimentation studies, unbound ligand remains at the top of the gradient (some investigators use dextran-coated charcoal to remove unbound ligand before application of receptor-ligand complexes to gradients).

HAP Assay. Prepare the HAP slurry the day before the assay and store at 4°. Remove fines by centrifugation at 900*g* and resuspend the HAP (50%, v/v) in the same buffer also containing 5 m*M* DTT. Add 200 μl HAP slurry to each 500 μl binding reaction (cut ~1–2 mm off the end of a pipette tip because the HAP crystals are quite large). Incubate on ice for 15 min; mix using a vortex at 5-min intervals. Centrifuge tubes for 5 min at 1500*g*. Aspirate the supernatant and resuspend the pellet in 500 μl HAP wash buffer. Incubate 5 min, collect the pellet by centrifugation, aspirate the supernatant, and repeat the wash procedure. Transfer the HAP pellet using

$2\times$ 200 μl ethanol washes to a scintillation vial, add scintillation cocktail and measure the radioactivity.

Data Analysis. Total binding ([^{3}H]RA in the absence of competitor) minus nonspecific binding ([^{3}H]RA in the presence of excess competitor) yields specific binding ([^{3}H]RA bound to receptor). For simple, single-site binding, the Scatchard plot of specific binding (bound)/free radioligand (y axis) versus bound (x axis) yields an estimate of the equilibrium dissociation constant, ($K_d = -1/\text{slope}$) and number of receptor binding sites (B_{max} = X intercept) (see Fig. 2B, inset). In this calculation, free ligand equals the [^{3}H]RA added to the tube minus that specifically bound to receptor. Computer programs to determine binding constants are available (reviewed in Refs. 14–16). For competition binding studies, the IC_{50} is defined as the concentration of added nonradioactive competitor that competes for 50% of total binding. The K_i can be calculated[17]:

$$K_i = IC_{50} \times [1 + ([^3H]RA \text{ added}/K_d \text{ of RA})]^{-1}$$

Sedimentation Analysis. A SDG (4 ml) is prepared using 2 ml each of solutions of 4 and 18% sucrose in 50 mM Tris-HCl, pH 7.4, 1.5 mM EDTA, 150 mM KCl, 5 mM DTT, and protease inhibitors: 1 mM PMSF, 10 μg/ml STI, 2 μM E-64, 1.6 mM benzamidine, 1 μM pepstatin A, and 5 μg/ml leupeptin. Chill gradients at 4° for 1 hr before use. Apply 200 μl radiolabeled receptor mixture to each sucrose gradient. We find it convenient to include a ^{14}C-labeled protein [~2500 dpm (disintegrations per minute) in 5 μl/gradient] to each gradient as an internal standard. Gradients are centrifuged at 50,000 rpm in an SW60 Ti rotor (Beckman, Palo Alto, CA) for 20 hr at 4°. Fractionate gradients (100 μl) into scintillation tubes, add scintillation cocktail, and count radioactivity on a dual channel setting to detect both ^{3}H (ligand) and ^{14}C (molecular weight standards).

Acknowledgments

We thank Dr. R. W. Curley for synthesizing LGD1069 and Hoffmann-La Roche, Nutley, NJ, for providing TTNPB for these studies. We also thank J. Berg, H. Fukuzawa, K. Hanson, and T. Verhalen for technical assistance. This work was supported by research grant DK14881 from the National Institutes of Health.

[14] E. C. Hulme, *in* "Receptor-Ligand Interactions. A Practical Approach" (E. C. Hulme, ed.), p. 447. IRL Press, Oxford, 1992.

[15] J. R. Unnerstall, *in* "Methods in Neurotransmitter Receptor Analysis" (H. I. Yamamura, S. J. Enna, and M. J. Kuhar, eds.), p. 37. Raven Press, New York, 1990.

[16] P. J. Munson and D. Rodbard, *Anal. Biochem.* **107,** 220 (1980).

[17] Y.-C. Cheng and W. H. Prusoff, *Biochem. Pharmacol.* **22,** 3099 (1973).

[3] Expression and Characterization of Retinoid Receptors in Yeast

By ELIZABETH A. ALLEGRETTO and RICHARD A. HEYMAN

Introduction

Since the discovery and cloning of the retinoic acid receptors (RARs)[1] and the retinoid X receptors (RXRs),[1] efforts have been made to express the receptors as recombinant proteins in heterologous systems. Overexpression of the retinoid receptors has afforded an opportunity to characterize receptor function employing a broad series of biochemical and molecular approaches that has provided new understanding of retinoid receptor signaling pathways. For example, expressed receptor is useful in a variety of techniques to study receptor functional properties such as interactions with DNA (electromobility shift assays, DNase footprints), interactions with accessory proteins and/or receptors (two-hybrid system, coimmunoprecipitation), ligand-binding properties, and transcriptional activity (transactivational assays, *in vitro* transcription). In addition, receptor overexpression is a prerequisite for structural studies using techniques such as X-ray crystallography and nuclear magnetic resonance (NMR), because milligram quantities of pure protein are necessary to complete the structural analyses.

A number of host expression systems have been employed and include *Escherichia coli,* yeast, insect cells, and mammalian cells. The yeast system has certain advantages over insect cells and mammalian cells including the economical feasibility of growing yeast and the ability to produce easily large quantities of protein. In addition, protease-deficient strains of yeast have been developed that generally lead to production of nonproteolyzed, full-length, functional receptors. Proteolysis can be a significant problem with insect and mammalian cell expression systems. Another significant advantage of yeast is that it provides a unique opportunity to reconstitute the transcriptional properties of the retinoid receptor signaling pathway in a receptor-null system.[2,3] This is a clear advantage over mammalian and insect systems in that all mammalian cells studied to date seem to express at least one subtype each of the RAR (α, β, γ) and the RXR (α, β, γ) subfamilies, and insect cells contain *ultraspiracle,* the insect homolog of

[1] D. J. Mangelsdorf, K. Umesono, and R. M. Evans, *in* "The Retinoids," pp. 319–349. Raven Press, New York, 1994.

[2] D. Metzger, J. H. White, and P. Chambon, *Nature* **334,** 31 (1988).

[3] M. Schena and K. R. Yamamoto, *Science* **241,** 965 (1988).

0076-6879/97 $25.00

RXR. Also, transactivation assays in yeast are facilitated by the fact that yeast can be transformed with the appropriate receptor and reporter plasmids and propagated as permanent lines, whereas mammalian cotransactivation assays are usually performed using transient transfection incorporated into each assay.

Although the *E. coli* system is also economically feasible and provides a receptor-null background, it is not believed to perform many of the postranslational modifications of proteins that occur in mammalian cells. Yeast have been reported to support posttranslational modifications of proteins, for example, progesterone receptor expressed in yeast is phosphorylated on amino acid residues identical to those that were observed in mammalian cells.[4] Although each of the expression systems discussed is useful, yeast cells represent a versatile system with many advantages for the study of receptor structure and function.

Yeast as a Protein Expression System for Retinoid Receptors

A previously described yeast expression system that has been quite successful for other intracellular receptors[5,6] has been used. The yeast expression vector contains the copper-inducible yeast metallothionein promoter linked to an ubiquitin-receptor fusion protein, which is cleaved by the host yeast at an ubiquitin cleavage site.[5,6] This feature results in the yeast cell "recognizing" the cleaved intact receptor as an endogenous protein and therefore does not further degrade it at the amino terminus. To minimize carboxy-terminal proteolysis, a protease-deficient strain of yeast, BJ2168, which is defective in expression of three of the major protease genes is used. The six retinoid receptors (RARα, β, γ and RXRα, β, γ) are individually expressed in this system and total soluble protein extracts are prepared as described previously[7] using the glass bead method and a BeadBeater (Bartlesville, OK) to lyse the yeast cells. Care is taken to keep the chamber cold and to minimize frothing of the cell extract during the lysis procedure to help prevent protein degradation. Receptor expression is verified by immunoblot methodology using antipeptide subtype-selective antibodies

[4] A. Poletti, O. M. Conneely, D. P. McDonnell, W. T. Schrader, B. W. O'Malley, and N. L. Weigel, *Biochemistry* **32,** 9563 (1993).

[5] D. P. McDonnell, J. W. Pike, D. J. Drutz, T. R. Butt, and B. W. O'Malley, *Mol. Cell. Biol.* **9,** 3517 (1989).

[6] D. P. McDonnell, Z. Nawaz, C. Densmore, N. L. Weigel, T. A. Pham, J. H. Clark, and B. W. O'Malley, *J. Ster. Biochem. Mol. Biol.* **39,** 291 (1991).

[7] E. A. Allegretto, M. R. McClurg, S. B. Lazarchik, D. L. Clemm, S. A. Kerner, M. G. Elgort, M. F. Boehm, S. K. White, J. W. Pike, and R. A. Heyman, *J. Biol. Chem.* **268,** 26625 (1993).

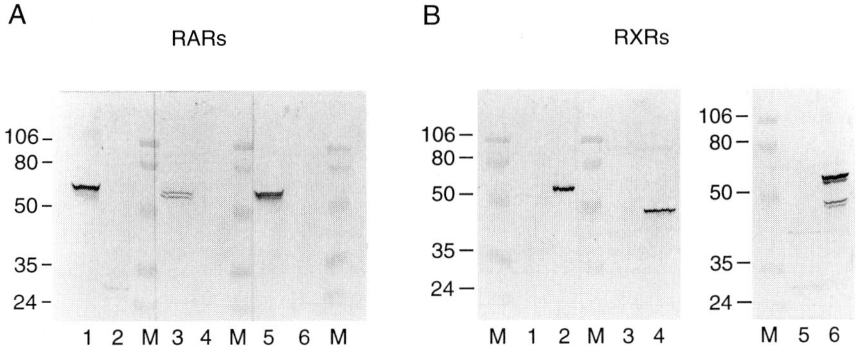

FIG. 1. Immunoblot analysis of RARs and RXRs expressed in yeast strain BJ2168. Extracts (total soluble protein) from nontransformed yeast (50 μg; A, lanes 2, 4, 6; B, lanes 1, 3, 5) or from yeast transformed with hRARα (50 μg; A, lane 1), hRARβ (50 μg; A, lane 3) or hRARγ (50 μg; A, lane 5), hRXRα (10 μg; B, lane 2), mRXRβ (50 μg; B, lane 4), mRXRγ (50 μg; B, lane 6) were subjected to immunoblot analysis as described previously.[7,8] Prestained molecular weight markers (M; Bio-Rad) were as follows: phosphorylase β, 106,000; bovine serum albumin, 80,000; ovalbumin, 49,500; carbonic anhydrase, 35,000; soybean trypsin inhibitor, 24,000.

that were generated against each of the retinoid receptors.[7,8] Figure 1 shows that each receptor has been expressed in its intact, full-length state and has not been appreciably proteolyzed. These preparations of receptor represent ~0.1–1.0% of the total soluble protein, depending on the receptor subtype. Histidine-tagged retinoid receptor expression vectors have also been constructed for ease in subsequent protein purification via metal affinity chromatography. These tagged receptors have been characterized by Western blotting and for ligand-binding activity (see following section) and are indistinguishable from the nonhistidine-tagged receptors (data not shown). These proteins are useful for applications wherein purified receptors are required including crystallization trials, *in vitro* transcription assays, antibody production, and certain analyses of protein–protein interactions, for example, via Biacore (Pharmacia, Piscataway, NJ) technology.[9]

Receptor–Ligand-Binding Analysis

One measure of retinoid receptor function is the ability of the receptors to bind to their known ligands, all-*trans*-retinoic acid (tRA) and 9-*cis*-

[8] M. W. Titcomb, M. M. Gottardis, J. W. Pike, and E. A. Allegretto, *Mol. Endocrinol.* **8,** 870 (1994).

[9] B. Cheskis and L. Freedman, *Biochemistry* **35,** 3309 (1996).

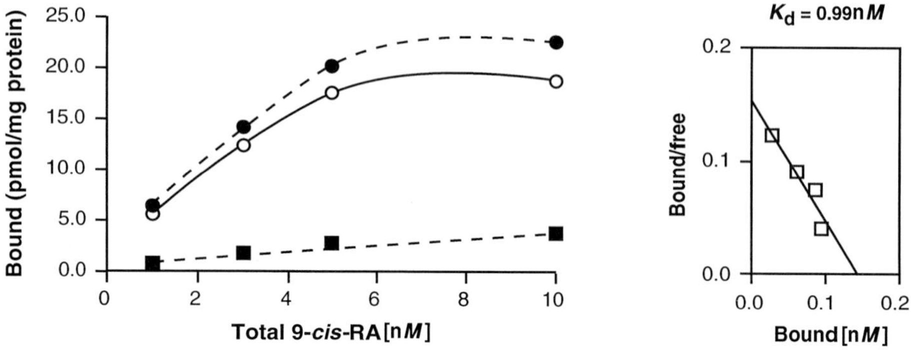

FIG. 2. Saturation binding and Scatchard analysis. Extract from yeast transformed with RXRα (5 μg) is incubated with the indicated concentrations of [^{3}H]9cRA in the absence or presence of a 200-fold molar excess of unlabeled 9cRA in duplicate at each point. Nonspecific binding activity (■) is subtracted from total binding activity (●) to generate specific binding activity (○). Scatchard analysis (□) is performed on specific binding data to yield the indicated K_d value.

retinoic acid (9cRA), with high affinity and specificity.[10,11] Therefore, each of the retinoid receptors expressed in yeast is tested for ligand-binding capability. Total soluble protein extracts (2–50 μg, depending on expression level of the individual receptor subtype) are incubated with [^{3}H]tRA or [^{3}H]9cRA in the presence and absence of a 200-fold molar excess of the unlabeled ligands, as described previously.[7] Nonspecific binding of retinoid to extract components is substantially reduced by the inclusion of CHAPS detergent (0.5%) in the binding buffer. Saturation-binding curves are generated and Scatchard analysis is used to determine equilibrium dissociation constants (K_d values) of the receptors for the ligands. Figure 2 shows a typical saturation-binding curve and Scatchard plot for yeast-expressed RXRα and 9cRA. Affinities are determined for each of the RXRs and RARs for 9cRA and of the RARs for tRA.[7] These values are in the subnanomolar range (0.2–0.8 n*M*) for the RARs, which are in good accordance with previously published values for these receptors expressed in other systems[12] and from endogenous sources.[13] The K_d values obtained

[10] R. A. Heyman, D. J. Mangelsdorf, J. A. Dyck, R. B. Stein, G. Eichele, R. M. Evans, and C. Thaller, *Cell* **68,** 397 (1992).

[11] A. A. Levin, L. J. Sturzenbecker, S. Kazmer, T. Bosakowski, C. Huselton, G. Allenby, J. Speck, C. I. Kratzeisen, M. Rosenberger, A. Lovey, and J. F. Grippo, *Nature* **355,** 359 (1992).

[12] G. Allenby, M.-T. Bocquel, M. Saunders, S. Kazmer, T. Speck, M. Rosenberger, A. Lovey, P. Kastner, J. F. Grippo, P. Chambon, and A. A. Levin, *Proc. Natl. Acad. Sci. U.S.A.* **90,** 30 (1993).

[13] C. Nervi, J. F. Grippo, M. I. Sherman, M. D. George, and A. M. Jetten, *Proc. Natl. Acad. Sci. U.S.A.* **86,** 5854 (1989).

for the RXRs expressed in yeast are 1–2 n*M*, which are in agreement with values we obtained for the RXRs expressed in insect cells[7] and are 8- to 10-fold lower than other reported values.[12]

Specific Methodology for Ligand-Binding Assay

For saturation-binding analyses, tritiated retinoid ([^{3}H]tRA, ~50 Ci/mmol (NEN); [^{3}H]9cRA ~29 Ci/mmol (Ligand Pharmaceuticals,[7] San Diego, CA) is added to final concentrations of 0.3–10 n*M* in 12 × 75 mm borosilicate glass tubes under dim light. Solvents other than ethanol are evaporated by the application of a gentle stream of nitrogen gas and the ligand is resuspended in ethanol prior to distribution to tubes at a final concentration of 5–10% ethanol. Each ligand concentration point is generally done in duplicate or triplicate. High-salt protein extracts (5–50 μg total soluble protein) from yeast cells expressing individual recombinant retinoid receptors are added to the tubes in binding buffer [0.15 *M* KCl, 10 m*M* Tris-HCl, pH 7.5, 1–5 m*M* dithiothreitol (DTT), 0.5% CHAPS detergent (Boehringer Mannheim, Germany)]. Nonspecific binding is measured at each point in the presence of a 200-fold molar excess of unlabeled ligand. The retinoid–protein mixture is incubated for 16 hr at 4°, shielded from light. Receptor–ligand binding is determined by the addition of 50 μl of 1 : 1 slurry of hydroxylapatite (HAP) resin (Bio-Rad, Richmond, CA, fined, 1 : 1 slurry in binding buffer) for 30 min at 4°, mixing by gentle vortexing every 10 min. The HAP pellets are washed three times in binding/wash buffer and centrifuged at ~1500*g* for 2 min. The HAP pellets are then transferred to scintillation vials in two 0.5-ml aliquots of H_2O, 10 ml liquid scintillation fluid is added and tritium disintegrations per minute (dpm) are determined using a Beckman liquid scintillation counter. Tritium dpm representing nonspecific binding is subtracted from total binding to yield specific binding. From the specific saturation-binding curve data, Scatchard analysis is employed to determine K_d values of the receptor–ligand interactions.

Ligand-Dependent Retinoid Receptor-Driven Transcription in Yeast

Yeast has been shown to provide the necessary transcriptional machinery to allow intracellular receptors, such as estrogen receptor[2] and glucocorticoid receptor,[3] to function in transactivation assays in a ligand-dependent manner. Yeast are transformed with the receptor expression vector of interest and a reporter plasmid containing the appropriate receptor-responsive element linked to a minimal *cyc* promoter driving β-galactosidase (β-Gal) expression.[5,6] Retinoid-induced, receptor-dependent transcriptional activity has been studied in yeast.[7,14,15] The nonprotease-deficient yeast strain BJ5409 is used for the retinoid receptor cotransactivation assays, because the higher amounts of receptor produced by the protease-deficient strain, BJ2168, are not necessary (and may even be detrimental) for these assays. Each of the retinoid receptors is transformed into yeast along with appropriate reporter vectors to analyze the activity of each receptor subtype. For example, each of the RXRs is separately transformed into yeast with a reporter vector containing the RXRE[1] from the CRBP-II (cellular retinol-binding protein II) promoter driving β-Gal expression.[7] Each of the RARs is also separately transformed with a reporter vector driven by the RARE[1] from the RARβ promoter.[7]

In addition, the combination of RAR and RXR with each of these reporters was studied.[7] Figure 3A illustrates the profile of β-Gal activity that is generated by 9cRA treatment of yeast cells expressing each of the RXRs and the CRBP-II RXRE-driven reporter. Each receptor subtype enabled 9cRA to stimulate the RXRE-driven yeast reporter, and no significant differences were observed between the subtypes. In the absence of receptor, there was no observed reporter activity. RXR activity via the CRBP-II element was specific in that RXR was not able to stimulate β-Gal activity through a VDRE (DR3)-driven reporter (Fig. 3D). Figure 3B shows that tRA does not stimulate CRBP-II RXRE-driven reporter activity in the presence of RXR. This result is in contrast to mammalian cotransactivation assays in which tRA is shown to stimulate a CRBP-II-driven reporter via RXR, which is believed to occur by conversion of tRA to 9cRA in CV-1 cells, which is not apparent in yeast.[7]

Although RXR is believed to act as a homodimer to stimulate transcription from the CRBP-II element, RXR functions as a heterodimeric partner with RAR in the activation of certain other genes, such as RARβ.[1] Therefore, we also studied RAR and RXR activity via the RARβ promoter RARE (βRE) linked to the yeast reporter. While RAR alone or RXR alone weakly stimulated β-Gal activity in a dose-dependent manner with 9cRA, the combination of RAR and RXR yielded an increased basal activity that was further stimulated with 9cRA in a dose-dependent manner (Fig. 3C). all-*trans*-Retinoic acid elicited lower β-Gal activity than did 9cRA, with either receptor alone or in combination (data not shown), leading to the conclusion that greater activation occurs through this element if both receptors are occupied with ligand.[7]

[14] B. L. Hall, Z. Smit-McBride, and M. L. Privalsky, *Proc. Natl. Acad. Sci. U.S.A.* **90,** 6929 (1993).

[15] D. M. Heery, B. Pierrat, H. Gronemeyer, P. Chambon, and R. Losson, *Nucleic Acids Res.* **22,** 726 (1994).

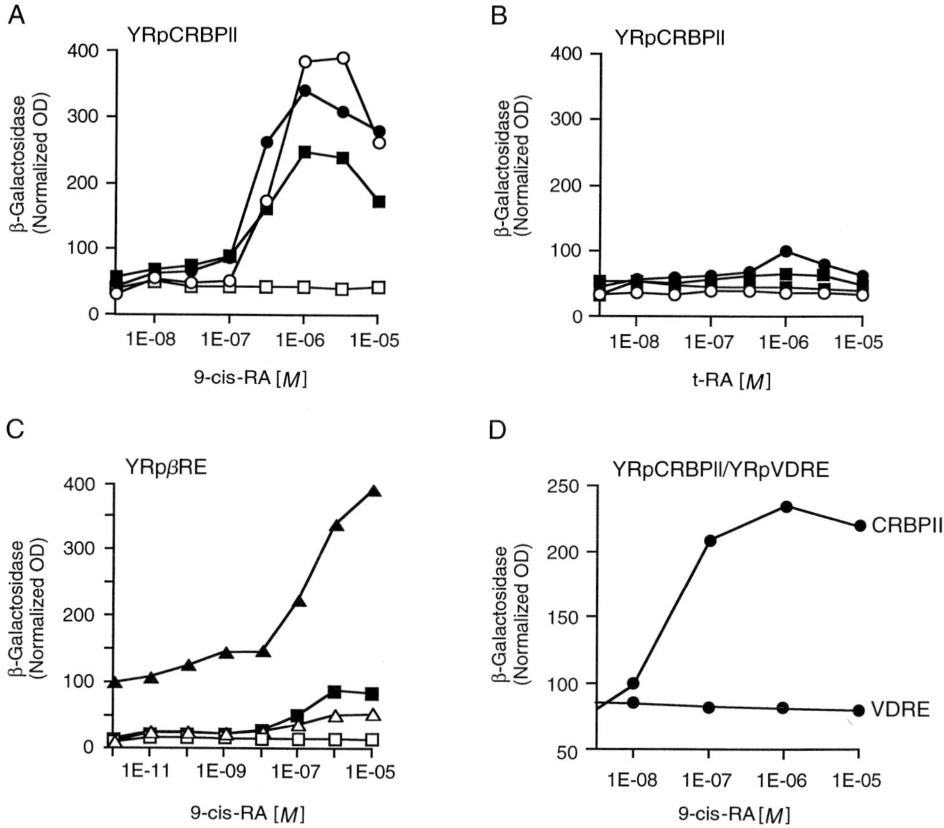

FIG. 3. Yeast RXR and RAR cotransactivation assays. (A, B, D) CRBP-II RXRE (A, B, D) or VDRE-(D) β-Gal reporter plasmids were transformed without (□; A, B) or with RXRα (●; A, B, D), RXRβ (○; A, B), RXRγ (■; A, B) expression vectors into yeast strain BJ5409. (C) βRE-β-Gal reporter plasmid was transformed without (□) or with RXRγ (■), RARγ (△), or RARγ and RXRγ (▲) expression vectors into yeast strain BJ5409. Cultures were grown and induced with 0–10 μ*M* $CuSO_4$ and with varying concentrations of 9cRA (A, C, D) or tRA (B). β-Gal readings were determined and corrected for cell density and for time of development [$(A_{415}/A_{600}) \times 1000$/min]. [Reproduced from Allegretto *et al., J. Biol. Chem.* **268,** 26625 (1993), with permission.]

Specific Methodology for Yeast Cotransactivation Assays

The retinoid receptor cDNAs are constructed into yeast expression vectors under the control of the copper-inducible yeast metallothionein gene promoter as described in detail elsewhere.[7] The RXRE from the CRBP-II gene promoter and the RARE from the RARβ gene promoter (βRE) are constructed into a yeast β-Gal reporter vector as previously described.[7] The transformed yeast are plated in 96-well tissue culture dishes (100 μl/well) at an OD_{600} of 0.05–0.2 and ligands are dispensed in water in concentrations between 10^{-12} and 10^{-5} *M* along with 0–50 μM $CuSO_4$ for 4–16 hr at 30°. After incubation, OD_{600} is determined as an indication of cell number. Yeast are then lysed with 100 μl lysis/development buffer [60 m*M* Na_2HPO_4, 40 m*M* NaH_2PO_4, 10 m*M* KCl, 1 m*M* $MgSO_4$, β-Gal substrate (ONPG) 2 mg/ml, 0.2% sodium dodecyl sulfate (SDS), 34 m*M* 2-mercaptoethanol, 2000 units/ml oxalyticase (Enzogenetics)] per well for 10–30 min and then 50 μl stop solution (2 *M* Na_2CO_3, 0.05% antifoam) is added prior to OD_{415} determination. Normalized β-Gal values are determined as follows: $(OD_{415}/OD_{600}) \times 1000$/min development time.

Conclusion

Yeast is a system that can be exploited to study many facets of receptor structure and function. Discussion within this chapter is a yeast expression system for producing receptor proteins in essentially nonproteolyzed form. These yeast-expressed retinoid receptors are also functionally intact in that they bind their cognate ligands with high affinity and specificity. In addition, the retinoid receptors function transcriptionally in yeast to activate promoter elements that they have been shown to stimulate in mammalian cell cotransfection/cotransactivation assays.[1] Yeast has also been utilized to identify proteins that interact with receptors by the use of GAL4–receptor fusion protein transcription assay. Chimeric receptors are generated by linking the GAL4 DNA-binding domain with a receptor ligand-binding domain to yield a fusion protein that responds to receptor ligands and activates transcription from GAL4-responsive elements that drive reporter activity. RXR has been used as bait in this system by a number of groups and, recently, several receptor-interacting proteins have been identified as cofactors[16–18] of receptor-activated transcription. The yeast two-hybrid screen offers a relatively simple way to identify and clone these new proteins for further study. Yeast will undoubtedly continue to serve as a valuable tool in the study of receptor structure, function, and mechanism of action in the years ahead.

Acknowledgments

We thank Mike McClurg, Sandra Kerner, Marc Elgort, Bryan Macy, Ken Henry, Bill Clevenger, and Donald McDonnell for contributions to this work.

[16] J. W. Lee, F. Ryan, J. C. Swaffield, S. A. Johnston, and D. D. Moore, *Nature* **374,** 91 (1995).
[17] J. D. Chen and R. M. Evans, *Nature* **377,** 454 (1995).
[18] A. J. Horlein, A. M. Naar, T. Heinzel, J. Torchia, B. Gloss, R. Kurokawa, A. Ryan, Y. Kamei, M. Soderstrom, C. K. Glass, and M. G. Rosenfeld, *Nature* **377,** 397 (1995).

[4] Use of *in Situ* Hybridization Techniques to Study Embryonic Expression of Retinoid Receptors and Binding Proteins

By ANNIE ROWE and PAUL M. BRICKELL

Introduction

Retinoids are believed to play important roles as signaling molecules in vertebrate embryonic development, and both retinoid excess and retinoid deficiency can cause developmental abnormalities.[1] Much effort has therefore been expended in determining the expression patterns of retinoid receptors and retinoid-binding proteins during embryogenesis.[1] This has largely focused on chick and mouse as experimental systems, although some information is available for rat, *Xenopus,* and zebra fish embryos. These studies have made a generally useful, and sometimes crucial, contribution to our understanding of retinoid signaling.

Most studies of retinoid receptor expression have relied on *in situ* hybridization to RNA because it has proved difficult to raise specific antibodies to retinoid receptor proteins that work well for immunohistochemistry. In contrast, suitable antibodies against retinoid-binding proteins are available, and so many analyses of retinoid-binding protein expression have been performed by immunocytochemistry,[2,3] as well as by *in situ* hybridization.[4–7]

We have used a variety of *in situ* hybridization methods to study retinoid receptor gene expression in developing chick embryos, and similar procedures have been used with mouse embryos.[6–8] These include (1) use of ^{35}S-labeled RNA probes on paraffin sections,[9–12] (2) use of ^{35}S-labeled probes

[1] R. Blomhoff, ed., "Vitamin A in Health and Disease," Marcel Dekker, New York, 1994.
[2] M. Maden, D. E. Ong, and F. Chytil, *Development* **109,** 75 (1990).
[3] M. Maden, P. Hunt, U. Eriksson, A. Kuroiwa, R. Krumlauf, and D. Summerbell, *Development* **111,** 35 (1991).
[4] A. V. Perez-Castro, L. E. Toth-Rogler, L.-N. Wei, and M. C. Nguyen-Huu, *Proc. Natl. Acad. Sci. U.S.A.* **86,** 8813 (1989).
[5] M. Maden, C. Horton, A. Graham, L. Leonard, J. Pizzey, G. Siegenthaler, A. Lumsden, and U. Eriksson, *Mech. Dev.* **37,** 13 (1992).
[6] P. Dollé, E. Ruberte, P. Kastner, M. Petkovich, C. Stoner, L. Dugas, and P. Chambon, *Nature* **342,** 702 (1989).
[7] P. Dollé, E. Ruberte, P. Leroy, G. Morriss-Kay, and P. Chambon, *Development* **110,** 1133 (1990).
[8] P. Dollé, V. Fraulob, P. Kastner, and P. Chambon, *Mech. Dev.* **45,** 91 (1994).
[9] A. Rowe, N. S. C. Eager, and P. M. Brickell, *Development* **111,** 771 (1991).

0076-6879/97 $25.00

on cultured cells grown on slides,[13] (3) use of nonradioactive digoxigenin-labeled (DIG-labeled) probes on frozen or paraffin sections, and (4) use of DIG-labeled probes on whole-mount preparations of embryos.[14,15]

Each of these methods is described in this chapter. Detection with radiolabeled probes is more sensitive than with DIG-labeled probes, but we have found that retinoid receptor mRNAs are generally sufficiently abundant to be detected by DIG-labeled probes. DIG-labeled probes give much better cellular resolution than radiolabeled probes when hybridized to sections, and results can be obtained more quickly because lengthy autographic exposure times are not required. However, use of radiolabeled probes permits semiquantitation of RNA levels following autoradiography because silver grains in the exposed emulsion can be counted. Although earlier studies of retinoid receptor expression, in our laboratory and others, were performed with radiolabeled probes, we now recommend DIG-labeled probes for most purposes.

All *in situ* hybridization techniques involve the following steps: synthesis of probes, harvesting and preparation of tissue/cells, pretreatment and permeabilization of tissue/cells, hybridization, and washing and detection of hybridized probe. The methods we describe are based on several protocols and have evolved in our laboratory for use with chick probes and chick embryo tissues. Detailed descriptions of other *in situ* hybridization methods and discussions of theory can be found elsewhere.[16]

Probe Design

Retinoid Receptors: Gene and Isoform Specificity

A number of features of nuclear retinoid receptors are germane to the design of suitable probes. (1) Vertebrates have two classes of nuclear retinoid receptor genes, each with three members: retinoic acid receptor (RAR) α, β, γ, and retinoic X receptor (RXR) α, β, γ.[1] (2) RARs and RXRs are members of the steroid/thyroid hormone receptor superfamily,[1] sharing the general structure shown in Fig. 1. (3) Most of the RAR and RXR genes encode a number of protein isoforms as a result of alternative splicing/promoter use, and these isoforms generally differ from each other in their

---

[10] A. Rowe, J. M. Richman, and P. M. Brickell, *Development* **111,** 1007 (1991).
[11] A. Rowe, J. M. Richman, and P. M. Brickell, *Development* **114,** 805 (1992).
[12] J. N. Schofield, A. Rowe, and P. M. Brickell, *Devel. Biol.* **152,** 344 (1992).
[13] A. Rowe, S. Sarkar, P. M. Brickell, and P. Thorogood, *Roux's Arch. Devel. Biol.* **203,** 445 (1994).
[14] E. A. P. Seleiro, A. Rowe, and P. M. Brickell, *Roux's Arch. Devel. Biol.* **204,** 244 (1995).
[15] A. Rowe and P. M. Brickell, *Anat. Embryol.* **192,** 1 (1995).
[16] D. G. Wilkinson, ed., "*In Situ* Hybridisation: A Practical Approach." IRL Press, Oxford, 1992.

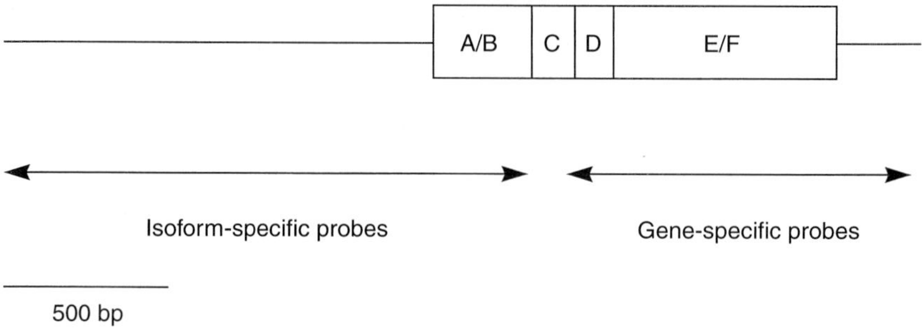

FIG. 1. Sketch of a generic RAR/RXR mRNA, showing regions used for probe preparation.

A/B domains.[1,17] The sequence of the region encoding the DNA-binding C domain (Fig. 1) is highly conserved both within and between the RAR and RXR classes, and is best avoided as probe for specific gene transcripts. The sequence of the region encoding the D–F domains is less conserved within the RAR and RXR classes, bears little homology between classes, and gives excellent probes for specific gene transcripts. However, such probes will hybridize to mRNAs for all isoforms encoded by a particular gene. For isoform-specific probes, the region encoding the A/B domains must be used (Fig. 1). It is necessary to hybridize embryos with probes from the same animal species because dropping stringency to allow the probe to hybridize across animal species also allows cross-hybridization to other RAR/RXR gene transcripts.

Retinoid-Binding Proteins

Vertebrates have two cellular retinol-binding proteins (CRBP-I and -II) and two cellular retinoic acid-binding proteins (CRABP-I and -II), encoded by four distinct genes.[1] These genes differ sufficiently throughout their length to permit gene-specific probes to be constructed without difficulty.

Preparation of Probes

Preparation of cDNA Templates

Methods described next have been optimized for use with antisense RNA probes, synthesized from template cDNAs cloned into pBluescript (Stratagene, La Jolla, CA) or pGem (Promega, Madison, WI) vectors.[18] Hybridization of parallel sections/whole mounts with matching sense strand RNA control probes is essential to control for nonspecific hybridization.

Plasmid DNA is prepared by standard methods, and linearized templates prepared by digesting 20–50 μg of plasmid DNA overnight. Linear-

[17] E. A. P. Seleiro, D. Darling, and P. M. Brickell, *Biochem. J.* **301,** 283 (1994).

[18] M. A. Alting-Mees, J. A. Sorge, and J. M. Short, *Methods Enzymol.* **216,** 483 (1992).

ized DNA is cleaned by phenol and chloroform extraction and ethanol precipitation, using buffers prepared under ribonuclease-free conditions, and is resuspended in ribonuclease-free, deionized water at 1 μg/μl.

Preparation of ^{35}S-Labeled RNA Probes

All reagents should be ribonuclease-free and of molecular biology grade. Water should be deionized and ribonuclease-free.

1. *Transcription.* Bring the following reagents to room temperature and mix in the order listed: 5 μl 5× reaction buffer (supplied by the manufacturer of T3, T7, or SP6 RNA polymerase, as appropriate), 0.5 μl 1 *M* dithiothreitol (DTT), 1.2 μl 10 m*M* GTP, 1.2 μl 10 m*M* ATP, 1.2 μl 10 m*M* UTP, 1 μl 50 μ*M* CTP, 2 μl linearized template DNA, 4.9 μl water, and 7 μl [α-^{35}S]CTP (37–55.5 TBq/mmol; DuPont-NEN, Boston, MA, NEG-064H). When using ^{35}S-labeled probes, it is essential that 10 m*M* DTT is present at all times during probe preparation and use. Failure to include DTT results in high background signals, as oxidation products bind to components of tissue sections. Add 0.5 μl RNasin and 0.5 μl of the appropriate RNA polymerase and incubate the mixture at 37°. After 45 min add 0.5 μl of fresh RNA polymerase and incubate for a further 60 min.
2. *Template removal.* Add 0.5 μl fresh RNasin, 1 μl 10 mg/ml *Escherichia coli* tRNA, 0.5 μl 1 *M* DTT and 0.5 μl ribonuclease-free DNase I to the reaction mixture and incubate at 37° for 10 min. Add 95 μl of water and 10 μl 5 *M* LiCl, mix and precipitate with ethanol. Spin down the RNA pellet, wash with 70% (v/v) ethanol containing 10 m*M* DTT and dry under vacuum. Resuspend in 50 μl water containing 50 m*M* DTT. A successful reaction should contain 1–2.5 × 10^6 counts/min/μl.
3. *Reduction of probe length by hydrolysis.* We have synthesized probes using cDNA templates of up to 1500 bp. However, for probes longer than approximately 300 bases, alkaline hydrolysis[19] should be used to reduce probe length and so enhance penetration of probe into tissue. Add 50 μl hydrolysis buffer (80 m*M* $NaHCO_3$, 120 m*M* Na_2CO_3 pH 10.2, 10 m*M* DTT) and incubate at 60° for *X* min, where *X* is calculated using the equation:

$$X = (L_o - L_f)/0.11(L_o L_f)$$

where L_o is the original transcript length and L_f is the required transcript length (both in kilobases).

Stop the reaction by adding 50 μl neutralizing buffer [0.2 *M* sodium acetate, 1% (v/v) glacial acetic acid, 10 m*M* DTT] and 15 μl 3 *M* sodium acetate, pH 5.2, and precipitate with ethanol. Spin down the pellet, wash with 70% (v/v) ethanol containing 10 m*M* DTT, and dry under vacuum. Resuspend in 50 μl water containing 50 m*M* DTT and measure incorporated

[19] K. H. Cox, L. M. Angerer, and R. C. Angerer, *Devl. Biol.* **101,** 485 (1984).

counts. Approximately 90% of unhydrolyzed probe should be recoverable. Probes can be used immediately or stored at −70° for up to 4 weeks.

Preparation of Digoxigenin-Labeled RNA Probes

1. *Transcription.* Bring the following reagents to room temperature and mix in the order listed: 5 μl 5× reaction buffer (supplied by the manufacturer of T3, T7, or SP6 RNA polymerase, as appropriate), 2 μl 10 m*M* GTP, 2 μl 10 m*M* ATP, 2 μl 10 m*M* CTP, 1.3 μl 10 m*M* UTP, 0.7 μl 10 m*M* DIG–UTP (Boehringer Mannheim), 2 μl linearized template DNA, and 9 μl water. It is not necessary to include DTT when preparing or using DIG-labeled probes. Add 0.5 μl RNasin and 0.5 μl of the appropriate RNA polymerase and incubate at 37°. After 45 min add 0.5 μl of fresh RNA polymerase and incubate for 60 min. Reaction products can be analyzed by running a 0.5-μl aliquot on an agarose gel. A complete reaction should yield approximately 10 μg of RNA probe.

2. *Template removal.* This is performed as for ^{35}S-labeled probes, except that DTT is omitted from solutions. After precipitation, care must be taken to resuspend probes fully (in 100 μl water) because DIG-labeled RNA is slightly more hydrophobic than unlabeled RNA. Heating to 70° for 10–15 min is usually necessary. Products can be checked by running a sample on an agarose gel.

3. *Reduction of probe length by hydrolysis.* For hybridization to whole mounts, we have used probes of 300–900 bases without hydrolysis. Longer probes should be hydrolyzed as described earlier.

Preparation of Paraffin Sections

Fixation

To prepare 4% (w/v) paraformaldehyde for fixation, heat 200 ml water to 65°, add a few drops of 5 *M* NaCl, swirl, add 10 g solid paraformaldehyde and swirl to dissolve. Add 25 ml 10× phosphate-buffered saline (PBS), add water to give a final volume of 250 ml, and filter through Whatman (Clifton, NJ) No. 1 paper. Use within 12 hr or store in aliquots at −20°.

Fix specimens for 4–48 hr in 4% (w/v) paraformaldehyde in PBS at 4°. This is suitable for specimens of less than 5 mm in two out of three dimensions. Larger specimens should be dissected further before fixation. Cavities, such as the cranial cavity, should be punctured carefully with tungsten needles before fixation. As an aid to orientation, some structures such as limb buds and facial primordia can be pinned out flat on a silicone rubber base (Dow Corning Silastic) cast in a glass petri dish, using stainless steel pins, and fixed *in situ*.

Embedding

Carry out all of the following steps, up to paraffin wax, in 7-ml glass vials with snap-on lids. Process pinned specimens *in situ* until they are in 70% (v/v) ethanol and then transfer to 7-ml glass vials.

Incubate specimens as follows, using ethanol diluted in water except where otherwise stated: twice in PBS at 4° for 30 min, once in 50% (v/v) ethanol in PBS at 4° for 15 min, twice in 70% (v/v) ethanol at room temperature for 15 min, and for 30 min each at room temperature in 85% (v/v) ethanol, 95% (v/v) ethanol, absolute ethanol (two changes), and toluene (two changes). Incubate specimens three times, for 20 min each, at 60° in fresh paraffin wax (e.g., Polywax, Difco, West Molesey, Surrey, or Fibrowax, BDH/Merck, Poole, Dorset), which has been filtered through Whatman No. 1 paper. Finally, transfer specimens to a fourth change of hot wax in plastic molds, and allow the blocks to cool to room temperature. The orientation of specimens during embedding can be controlled with preheated tungsten needles or watchmaker's forceps. Blocks can be stored indefinitely at 4° in sealed boxes containing desiccant.

Preparation of Sections

Prepare 3-aminopropyltriethoxysilane (TESPA)-coated microscope slides as follows. Wash slides (Premium, BDH/Merck, or Gold Seal, Chance) in washing-up liquid and rinse thoroughly, using deionized water for the final rinse. Place slides in 10% (v/v) HCl for 20 min and rinse thoroughly, finishing with a rinse in deionized water. Bake at 180° for at least 2 hr. Make a fresh solution of 2% (v/v) TESPA in acetone, and dip slides for 10 sec. Rinse slides twice in acetone and once in deionized water (10 sec for each rinse). Drain excess water from slides and incubate at 42° for a minimum of 1 hr, until dry. TESPA-coated slides are usable for several months. If in doubt, unused slides can be recoated without cleaning.

Cut 6-μm tissue sections on a standard rotary microtome and float them out on drops of sterile water on TESPA-coated slides on a slide warmer at 45–50°. As the sections unwrinkle, carefully blot off water with a tissue, using a paintbrush to control the position of the sections. Allow sections to dry down at 45° for 2–16 hr and then bake them on to slides at 60° for 6–16 hr. Store at 4° in sealed boxes containing desiccant.

Preparation of Frozen Sections

Fix embryos overnight at 4° in 4% (w/v) paraformaldehyde, 4% (w/v) sucrose, 0.12 m*M* $CaCl_2$, 0.1 *M* sodium phosphate pH 7.4, prepared in a manner similar to that described earlier. Rinse three times in the same buffer, but without paraformaldehyde. Embed embryos in 1.5% (w/v) agar, 5% (w/v) sucrose in PBS. After trimming the blocks, transfer to 30% (w/v) sucrose, 0.1% (w/v) sodium azide in PBS at 4° overnight.

Cut 15-mm cryostat sections from the blocks at −24° and collect sections on TESPA-coated slides as described earlier. Air-dry sections for at least 2 hr and store at −20° in sealed boxes with desiccant.

Processing of Cells Cultured on Slides

Cells grown on slides coated with appropriate substrata, or directly on glass, can be used. All reagents are prepared as for paraffin sections. Fix cells on slides for 20 min with 4% (w/v) paraformaldehyde in PBS at room temperature. Rinse twice, for 5 min each, in PBS. Dehydrate, for 2 min each, in 30% (v/v), 60% (v/v), 80% (v/v), 95% (v/v), and absolute ethanol. Air-dry for at least 1 hr and store at −20° in sealed boxes containing desiccant.

Preparation of Whole-Mount Specimens

Fix specimens for whole-mount *in situ* hybridization in the same way as specimens for paraffin embedding. They can be pinned out or dissected before fixation if required. Probe and detection reagents easily become trapped inside cavities such as head, heart, gut, and neural tube in whole-mount specimens, so careful dissection of specimens and puncturing of cavities is essential.

After fixation, specimens are washed twice for 5 min each in PBS containing 0.1% (w/v) Tween 20 (PBT) and dehydrated through 30% (v/v), 60% (v/v), and 80% (v/v) methanol in PBT for 5 min each, followed by two 5-min washes in 100% methanol. All these steps are carried out in 30-ml universal tubes containing 10–15 ml of solution, on a rocking platform. Specimens are finally transferred to fresh methanol and can be stored at −20° for up to 3 months.

In Situ Hybridization with ^{35}S-Labeled Probes

This protocol was developed to overcome problems we encountered when using RAR and RXR probes on embryonic chick tissue. Typical results are shown in Fig. 2, for hybridization to sections, and Fig. 3, for hybridization to cultured cells. All steps are carried out in baked 400-ml glass troughs, with up to 40 slides per rack. Ethanol and solvents can be reused four or five times, but separate stocks should be kept for dewaxing, dehydration, and posthybridization washes. Buffers and water should be autoclaved wherever possible.

Paraffin Sections

Warm sealed slide boxes to room temperature. Perform steps at room temperature unless otherwise stated. Dewax sections in fresh xylene or toluene for 15 min and dehydrate for 2 min each in absolute ethanol and

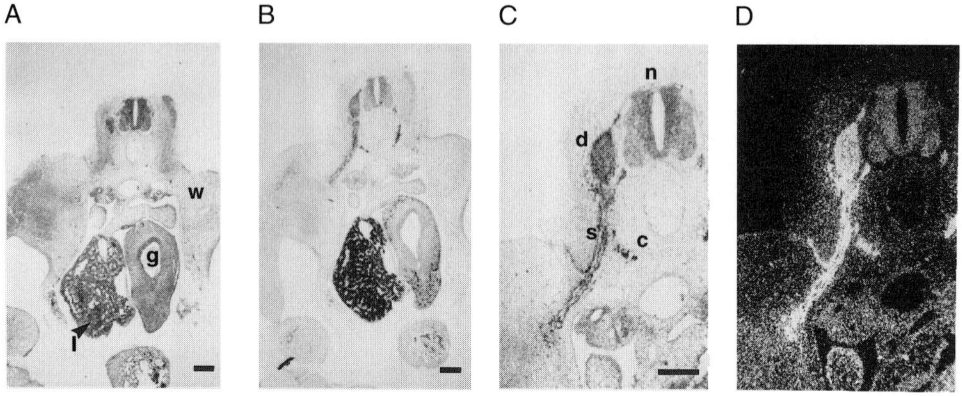

FIG. 2. *In situ* hybridization to adjacent transverse sections of a stage 27 chick embryo, using (A) a ^{35}S-labeled negative control sense strand RNA probe or (B) a ^{35}S-labeled antisense strand RNA probe, corresponding to the regions encoding the D–F domains of chicken RXR-γ. See Ref. 9 for details. Photographic emulsion was exposed for 18 days. (C and D) Higher power views of (B), photographed under (C) bright-field or (D) dark-field illumination. Hybridization is visible as black silver grains in (B) and (C), and as white silver grains in (D). c, Sympathetic chain; d, dorsal root ganglion; g, gut; l, liver; n, neural tube; s, spinal nerve; w, wing bud. Bars: 200 μm.

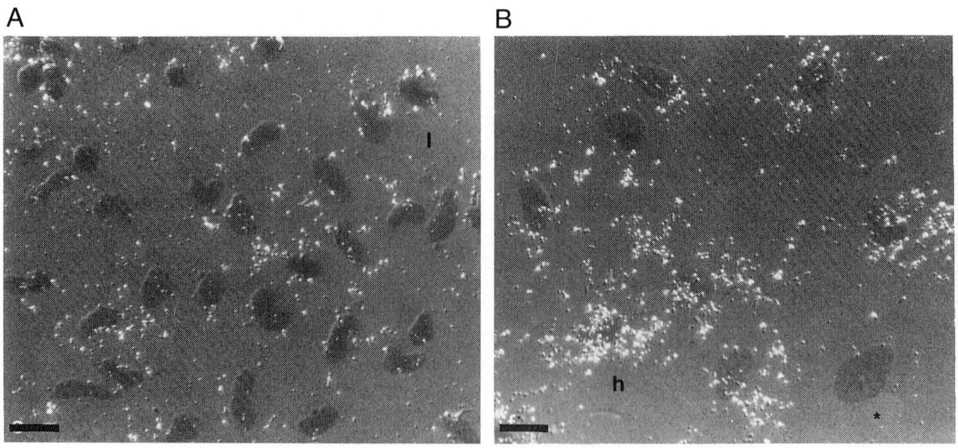

FIG. 3. *In situ* hybridization to (A) RARβ or (B) RXRγ transcripts in cultured quail neural crest explants, using ^{35}S-labeled antisense strand RNA probe corresponding to the regions encoding the D–F domains of chicken RARβ or RXRγ, respectively. See Ref. 13 for details. Photographs are taken using Nomarski DIC optics. Cell nuclei are clearly visible, but cytoplasm is hard to discern. Hybridization signal appears as white silver grains. Some cells show high levels of hybridization (h), some show low levels (l) and some show no detectable hybridization (star). Bars: 20 μm.

95% (v/v), 90% (v/v), 80% (v/v), 60% (v/v), and 30% (v/v) ethanol in water. Rinse for 2 min in water, and incubate in 2% (v/v) HCl in water for 20 min, or less if sections are small.

Neutralize for 5 min in 2× SSC (20× SSC is 3 *M* NaCl, 0.3 *M* sodium citrate, pH 7.4) and incubate for 10 min at 37° with 5 μg/ml proteinase K in 100 m*M* Tris-HCl, pH 7.5, and 50 m*M* EDTA. Incubate for 2 min with a fresh solution of 2 mg/ml glycine in PBS, and rinse twice for 1 min each in PBS. Refix for 15–20 min with 4% (w/v) paraformaldehyde in PBS, prepared as described earlier, and rinse for 2 min in PBS. Block reactive sites by acetylation as follows. Fill a baked glass trough with 0.1 *M* triethanolamine and stir continuously in a fume hood. Add 1/400 volume of acetic anhydride and immerse the rack with slides in this within 30 sec. After 10 min, rinse the slides in PBS for 2 min and in water for 2 min, and dehydrate through 30% (v/v), 60% (v/v), 80% (v/v), 95% (v/v), and absolute ethanol, for 2 min each. Air-dry under a dust cover for 1–24 hr.

Fixed Cells on Slides

Bring the sealed slide box to room temperature, incubate the slides for 20 min at 70° in 2× SSC, and proceed as for paraffin sections from the "neutralization" step onward.

Hybridization

For optimum signal-to-noise ratio, hybridization solution [1× Denhardt's solution, 20 m*M* Tris-HCl pH 8.0, 5 m*M* EDTA, pH 8.0, 10 m*M* sodium phosphate, pH 6.8, 50% (v/v) deionized formamide, 10% (w/v) dextran sulfate, 50 m*M* DTT, 500 μg/ml yeast total RNA, 50 μg/ml poly(A) RNA] should contain probe diluted to 5×10^4 to 1×10^5 counts/min/μl.

Denature RNA probes, to remove secondary structure, by heating hybridization solution to 80° for 2–3 min. Place the solution on ice for 2 min and pipette it directly on to dry sections, using 20 μl for a 22 × 22 mm coverslip. Spread the solution gently with Parafilm if sections are large. Place coverslips over sections. High-quality coverslips can be used directly without cleaning or siliconizing. Hybridize for 8–24 hr at 55° in a sealed humid box, prepared by lining the lid with dry filter paper to prevent drops of condensation forming, and lining the bottom with filter paper soaked in 2× SSC, 50% (v/v) formamide. Place slides on glass rods in the box and seal it with waterproof plastic tape.

Posthybridization Washes

Perform all steps in 400-ml baked glass troughs. Rapidly transfer slides to a rack and remove coverslips by agitating in 2× SSC, 50% (v/v) formamide, 10 m*M* DTT, at 55° for 15 min. Transfer slides to a second rack in fresh solution at 55°, ensuring that all coverslips have floated off. Wash for 20 min and, during this time, allow the waterbath and solution to warm to 65°. Wash in fresh solution for 20 min at 65°.

Rinse twice, for 15 min each, at 37° in 500 m*M* NaCl, 10 m*M* Tris-HCl, pH 8.0, and 1 m*M* EDTA pH 8.0. This removes DTT and formamide, which may inhibit ribonuclease. Incubate for 30 min at 37° in 40 μg/ml pancreatic RNase A in this buffer, and rinse for 10 min at 37° in fresh buffer. Wash twice, for 20 min each, at 65° in 2× SSC, 50% (v/v) formamide, 10 m*M* DTT. Wash once for 20 min at 65° in 0.1× SSC, 10 m*M* DTT. Rinse for 5 min at room temperature in 0.1× SSC. Dehydrate through 300 m*M* ammonium acetate in 70% (v/v) ethanol, 95% (v/v) ethanol, and absolute ethanol, for 2 min each, and air-dry for 1–48 hr.

Autoradiography

Autoradiography requires a clean darkroom, preferably with a safelight suitable for Ilford K5 emulsion (Ilford Ltd, Ilford, UK). Preheat 2% (v/v) glycerol in water in a plastic slide mailer to 42°, using 4/7 of the final volume of emulsion required. Add shreds of solid emulsion to the glycerol until the required volume is reached. As the emulsion melts, invert the tube gently to mix. When the emulsion has melted and mixed thoroughly, check the solution for air bubbles by dipping a blank slide into it. As there will be more bubbles around the edges of the tube than in the middle, dip slides with sections facing inward.

Drain off excess emulsion by standing slides vertically on paper towel for a few seconds and then stand them vertically on a paper towel for 2–16 hr to dry. Place them in a light-tight slide box with dry silica gel, wrap the box in a lightproof, black plastic bag, and store at 4° until slides are ready for developing.

Developing

Develop one or two test slides after 4–7 days. All steps can be carried out in Coplin jars. Warm slides to room temperature for 30–60 min and develop for 5 min using Kodak D19 developer (Kodak, Rochester, NY). Rinse briefly with water and fix for 5 min with Kodak Unifix (Kodak, Rochester, NY). Wash under running water for 10–15 min, rinse in water, and counterstain. A number of counterstains are suitable, but we generally use 1% (w/v) malachite green for sections and Giemsa stain for cells. After staining and rinsing in water, slides are air-dried for several hours, without alcohol dehydration, and mounted in a xylene-based mounting medium.

Whole-Mount *in Situ* Hybridization

DIG-labeled RNA probes are used in this protocol, which is based on that of Wilkinson.[16] A typical result is shown in Fig. 4.

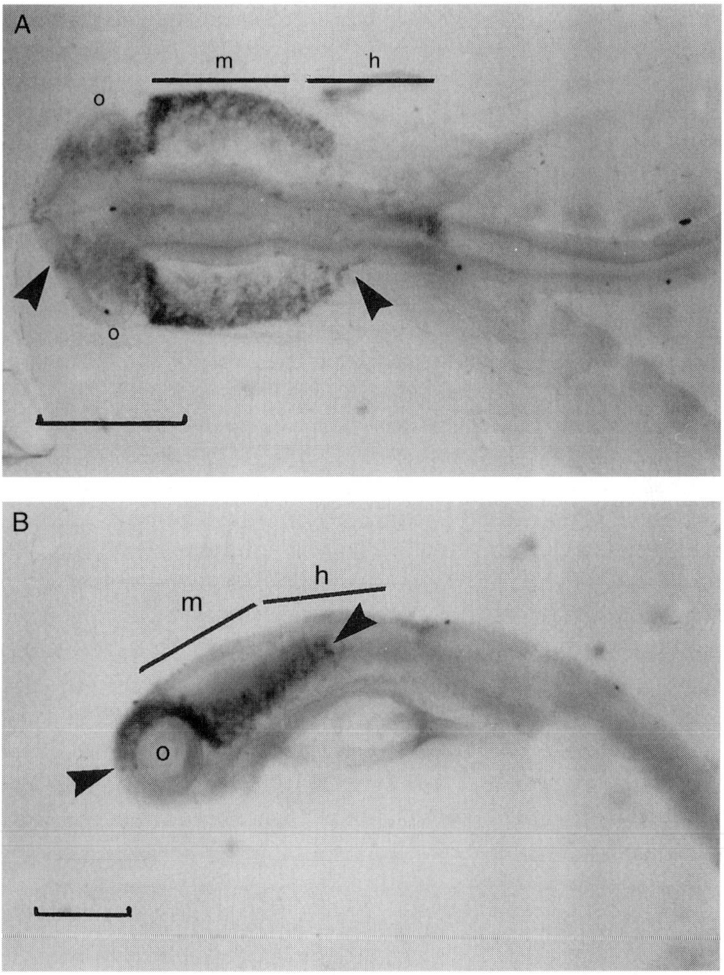

FIG. 4. *In situ* hybridization to RXRγ transcripts in whole-mount preparations of stage 10–11 chick embryos. (A) Dorsal view and (B) lateral view. Hybridization is visible as a purple stain, shown here in monochrome. The extent of the hybridizing region is indicated with arrow heads. m, Midbrain; o, optic lobe; h, hindbrain. Bars: 0.5 mm.

Pretreatment

All steps are carried out in 30-ml universal tubes containing 10–15 ml of solution, on a rocking platform. Several embryos can be processed in the same tube for hybridization together, but care must be taken not to lose embryos when changing solutions.

Bring embryos in methanol to room temperature, rehydrate through 80% (v/v), 60% (v/v), and 30% (v/v) methanol in PBT and wash twice in

PBT, taking 5 min for each step. Bleach with 6% (v/v) hydrogen peroxide in PBT for 1 hr. This reduces background and bleaches hemoglobin, but is unnecessary for small embryos, younger than about stage 14. Rinse embryos once in PBT and wash twice, for 5 min each, in PBT. Incubate with 10 μg/ml proteinase K in PBT for 5–15 min. The duration of proteinase K treatment is critical and may require optimization for embryos at different stages; smaller embryos require shorter times. Embryos become fragile and sticky after proteinase K treatment and may disintegrate during subsequent steps unless handled carefully. Wash embryos once with 2 mg/ml glycine in PBT and twice in PBT (5 min for each wash). Refix embryos for 20 min with 0.2% (w/v) glutaraldehyde and 4% (w/v) paraformaldehyde in PBT, prepared by adding glutaraldehyde solution and Tween 20 to 4% (w/v) paraformaldehyde in PBS. Wash twice, for 5 min each, in PBT.

Hybridization

Transfer embryos in PBT to a 2-ml cryotube, remove as much PBT as possible, and add 1 ml prehybridization solution [50% (v/v) deionized formamide, 5× SSC, pH 5.0, 1% (w/v) SDS, 50 μg/ml heparin, 50 μg/ml yeast RNA], preheated to 70°. Mix by inverting the tube, pipette off as much liquid as possible, and replace with another 1 ml prehybridization solution at 70°. Incubate at 70° for 1–2 hr. Agitation is not necessary. Pipette off as much solution as possible and replace with 1 ml prehybridization solution containing 1 μg/ml DIG-labeled probe. Wrap tubes in Parafilm and incubate overnight at 70°.

Posthybridization Washes

Washing steps are carried out with embryos in the hybridization tubes. Apart from gentle inversion after adding solutions, agitation is not necessary. The first wash solution can be prepared the day before and left at 70° overnight. Carefully pipette off most of the hybridization solution, but do not allow embryos to dry out. Keep the hybridization solution, as it can be used once more. Wash embryos twice, for 30 min each, in 50% (v/v) deionized formamide, 5× SSC, pH 5.0, 1% (w/v) SDS at 70°. Wash embryos for 10 min in one part of this solution plus one part RNase buffer [0.5 *M* NaCl, 10 m*M* Tris-HCl, pH 8.0, 0.1% (v/v) Tween 20] at 70°. Wash three times for 5 min in RNase buffer at 37°, and incubate in 100 μg/ml pancreatic RNase A in RNase buffer, twice for 30 min each at 37°. Wash for 5 min in RNase buffer, for 5 min at 37° in 50% (v/v) deionized formamide, 2× SSC, pH 5.0, and then twice, for 30 min each, at 65° in 50% (v/v) deionized formamide, 2× SSC, pH 5.0. Wash three times, for 5 min each, at room temperature in TBST [5× TBST in 0.7 *M* NaCl, 14 m*M* KCl, 250 m*M* Tris-HCl, pH 8.0, and 0.5% (v/v) Tween 20].

Detection of Digoxigenin

Hybridized probe is detected with alkaline phosphatase-conjugated goat anti-DIG Fab antibody (Boehringer Mannheim). Antibody should be preabsorbed with chick embryo powder to reduce nonspecific binding to embryos. Embryo powder is prepared by homogenizing 10-day-old chick embryos in 3 ml ice-cold PBS, adding five volumes ice-cold acetone, pelleting the homogenate in a low-speed centrifuge, resuspending the pellet in ice-cold acetone, and centrifuging again. After repeating the acetone wash twice, dry the pellet on a few sheets of filter paper and grind to a fine powder with a pestle and mortar. Incubate 10 mg powder with 10 μl goat serum at 70° for 30 min in 1 ml TBST. After cooling the powder mixture on ice, add 2 μl anti-DIG antibody and incubate at 4° on a roller for 1–2 hr. Microcentrifuge for 10 min at 4° and add 1% (v/v) goat serum in TBST to 4 ml.

While the antibody is preabsorbing, block embryos for 1.5 hr in 10% (v/v) goat serum in TBST. Pipette off the solution, add preabsorbed antibody, using 2 ml per cryotube, and incubate overnight at 4° on a roller.

Wash embryos in TBST containing 2 m*M* levamisole (stored at −20° as a 1 *M* solution and added to buffers on the day of use) as follows: three times for 5 min each, once for 30 min, and four times for 45 min each. Wash embryos three times, for 10 min each, in NTMT [100 m*M* NaCl, 100 m*M* Tris-HCl, pH 9.5, 50 m*M* $MgCl_2$, 0.1% (v/v) Tween 20, prepared within 12 hr of use] containing 2 m*M* levamisole. Prepare NTMT/2 m*M* levamisole containing 3.5 μl/ml nitroblue tetrazolium salt (NBT, 75 mg/ml in 70% dimethylformamide, stored at −20° and 4.5 μl/ml 5-bromo-4-chloro-3-indolyl phosphate (BCIP, 50 mg/ml in dimethylformamide, stored at −20°). Wrap tubes in foil, since the substrate is light sensitive, and incubate at room temperature with gentle agitation. Faint color should begin to develop after 20–30 min. When color has developed sufficiently, and the signal-to-noise ratio is still acceptable (usually within 2–4 hr), stop the reaction by rinsing embryos in two changes of TBST, for 5 min each. Clear specimens by gently shaking, for 20 min each, with 20% (v/v), 40% (v/v), and 80% (v/v) glycerol in PBST. Store embryos in the dark in 80% (v/v) glycerol in PBST.

In Situ Hybridization to Frozen Sections with DIG-Labeled Probes

A typical result is shown in Fig. 5. We have also successfully hybridized paraffin sections with DIG-labeled probes. The protocol is similar to those described earlier, with some variations such as the use of buffers containing maleic acid and of commercial blocking reagent in place of homemade embryo powder. We have obtained good results with all variations.

Hybridization

Bring sections to room temperature in the sealed boxes. Dilute DIG-labeled probe to 0.1–1 mg/ml in hybridization solution, as described for

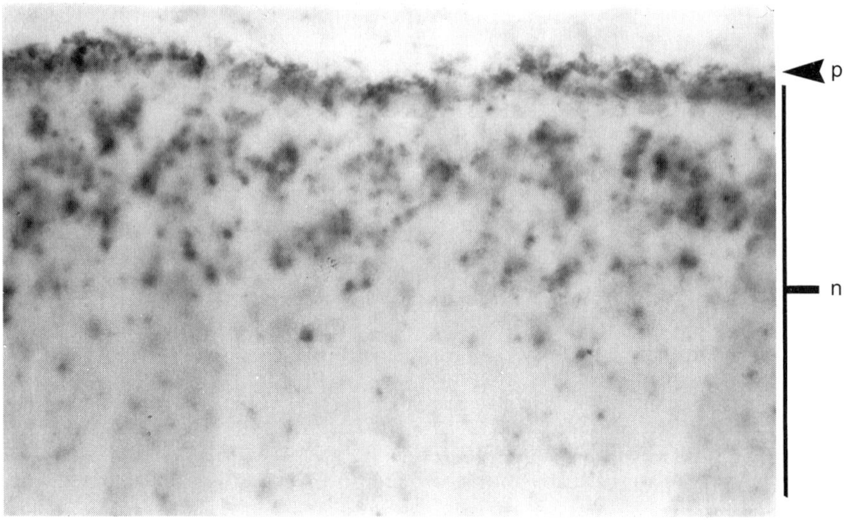

FIG. 5. Distribution of RXRγ transcripts in the neural retina of a day 7 chick embryo. The tissue section was hybridized with a DIG-labeled antisense strand RNA probe, corresponding to the regions encoding the D–F domains of chicken RXRγ. Both purple hybridizing cells within the neural retina (n) and brown cells in the pigmented cell layer (p) appear black in this monochrome.

^{35}S-labeled probes. Mix thoroughly, and incubate at 70° for 5–10 min to remove RNA secondary structure. Add 75 μl of solution to each slide. The solution might not completely cover the section at this stage, but will spread as the agar melts during hybridization. Cover sections with 22 × 50 mm coverslips and hybridize overnight at 65° in a humid box, as previously described.

Posthybridization Washes

Transfer slides to a rack in 300 ml washing solution [1× SSC, 50% (v/v) formamide, 0.1% (v/v) Tween 20] at 65°, and wash for 15 min to allow coverslips to float off. Wash twice more at 65° for 30 min each, and twice for 30 min each at room temperature with gentle rocking in MABT [100 m*M* maleic acid, 150 m*M* NaCl, pH 7.5, 0.1% (v/v) Tween 20].

Detection of Digoxigenin

Block sections for at least 1 hr at room temperature in MABT containing 2% (v/v) blocking reagent (Boehringer Mannheim) and 20% (v/v) heat-inactivated sheep serum, using approximately 0.8 ml solution per slide. Dilute anti-DIG antibody to 1/8000 in MABT containing 2% (v/v) blocking reagent and 20% (v/v) heat-inactivated sheep serum and add 75 μl to each section. Place coverslips on slides and incubate overnight at room temperature in a humid box containing PBS. Wash sections four times, for 20 min each, at room temperature in MABT with gentle rocking. Rinse sections twice, for 10 min each, in NTMT containing 5 m*M* levamisole, with gentle rocking. Transfer slides to a glass Coplin jar in NTMT containing 5 m*M* levamisole and NBT and BCIP, as described. Wrap the jar in foil and keep in a dark place. Up to 10 slides can be fitted into one jar with 30 ml of solution. Slides can be removed and checked for staining. The reaction should be checked after a few hours but is generally left overnight. However, if left too long, a precipitate may form. When color has developed sufficiently and the signal-to-noise ratio is still acceptable, stop the reaction by rinsing slides in two changes of MABT and one of PBS (for 5 min each). Mount under coverslips in glycerol, as described earlier for whole mounts.

Interpretation and Presentation of Results

Careful interpretation of results is essential. Particular attention should be paid to matched specimens/sections hybridized with control sense strand probes. One problem with whole-mount *in situ* hybridization is that products of the color reaction, generated in small regions of the embryo, may leach out of tissue and become trapped in cavities. This can give a false impression of the level and extent of specific hybridization, and is not controlled for by use of sense strand probes, which do not generate any color reaction products. The only solution is to puncture cavities before hybridization, and to wash thoroughly after color detection. Doubt as to whether a signal is genuine, or represents trapping, can be resolved by sectioning stained whole mounts.

Photographing Autoradiographs

Specimens should be photographed under both bright field and dark-field illumination. For the latter, use a dark-field condenser or pseudodark-field conditions obtained with phase contrast condenser and objectives. Other microscopy techniques such as Nomarski DIC can also be used effectively (Fig. 3). Autoradiographic data can usually be presented adequately as monochrome prints. A good quality low-speed film such as Pan-F or Technical Pan (Kodak, Rochester, NY) should be used, and test exposures are often necessary. Color photography can also be used effectively, especially when combined with fluorescent counterstaining.[20]

Photography of Whole-Mount Specimens

Whole-mount preparations are best photographed in color, although Fig. 4 has to be monochrome. We use Ektachrome 64T or 160T transparency

[20] S. M. Smith and G. Eichele, *Development* **111,** 245 (1991).

film (Kodak) and make prints from transparencies. This is more expensive than using color negative film and prints, but gives better image quality and color balance. Whole mounts can be photographed in monochrome where contrast is good. Very small embryos, younger than stage 16, are photographed in drops of 80% (v/v) glycerol on depression slides, and dark-field illumination gives best results. Larger specimens are photographed in 30-mm petri dishes using a dissecting microscope with substage illumination. It is worth experimenting with different conditions to get the best possible pictures.

Acknowledgments

We thank Amanda Barlow for Fig. 5. This work was supported by The Wellcome Trust and The British Heart Foundation.

[5] Use of Quantitative Polymerase Chain Reaction to Study Retinoid Receptor Expression

By Nicoletta Ferrari, Giorgio Vidali,* and Ulrich Pfeffer

Introduction

The retinoid receptors belong to the large superfamily of ligand-activated transcription factors. The retinoid receptor subfamily consists of two groups of receptors, the retinoic acid receptors (RARs), which bind all-*trans*- and 9-*cis*-retinoic acid, and the retinoid X receptors (RXRs), which only bind 9-*cis*-retinoic acid. These groups are composed of three members each, the receptor subtypes α, β, and γ, which are encoded by different genes. Differential usage of 5′ exons and the presence of multiple promoters in the receptor genes lead to the production of up to seven different receptor isoforms from a single gene. These isoforms differ in their amino-terminal portion, which is believed to interact with other nuclear proteins in the transactivation of genes.[1–3]

The expression analysis of these receptors is complicated by the complexity of this subfamily, by the high degree of conservation among the

* Deceased.

[1] D. J. Mangelsdorf, C. Thummel, M. Beato, P. Herrlich, G. Schütz, K. Umesono, B. Blumberg, P. Kastner, M. Mark, P. Chambon, and R. M. Evans, *Cell* **83,** 835 (1995).

[2] P. Kastner, M. Mark, and P. Chambon, *Cell* **83,** 859 (1995).

[3] D. J. Mangelsdorf and R. M. Evans, *Cell* **83,** 841 (1995).

0076-6879/97 $25.00

different members, by the fact that most cells express more than one receptor subtype, and by the relatively low level of expression. For these reasons reverse transcription–polymerase chain reaction (RT-PCR) protocols have been developed. Following the development of the gene amplification method,[4–6] the same procedure has been applied to the amplification of reverse transcribed RNA (RT-PCR).[7,8] This method consists of the combination of reverse transcription of mRNA with the amplification of the cDNA obtained using sequence-specific primers. The reverse transcription reaction can be performed following standard procedures using viral reverse transcriptases or exploiting the reverse transcriptase activity of thermostable DNA polymerases.[9] These methods allow the detection of the few specific mRNA molecules present in preparations of total RNA from very reduced biological samples such as tumor biopsies. In addition to the high sensitivity, high specificity is obtained through the use of specific oligonucleotides as primers for the amplification reaction, which, in the case of retinoid receptors, allow the unequivocal distinction between receptor subtypes and isoforms. Semiquantitative analyses can be done with RT-PCR if an appropriate control for the amount and the integrity of the starting material is included. Competitive PCR using known amounts of *in vitro* transcripts from synthetic genes can be used to determine the absolute number of receptor mRNA molecules present in the sample examined.[10] Quantitative analyses are particularly important since variations of the amount of receptor present in the cell is expected to be of physiological significance and, at least in some instances, the receptor mRNA level is altered in response to the ligand.[11]

Although the RT-PCR protocols are straightforward and easy to perform they present some drawbacks: the high sensitivity of the method also determines a high risk of contamination leading to false-positives and the results sometimes show a reduced reproducibility and a high sample-to-sample variation. These problems can however be overcome by rigorous procedures avoiding cross-contamination and by the design of "master

[4] R. Saiki, S. Scharf, E. Faloona, K. B. Mullis, G. T. Horn, H. A. Erlich, and N. Arnheim, *Science* **230,** 1350 (1985).

[5] K. B. Mullis and F. A. Faloona, *Methods Enzymol.* **155,** 335 (1987).

[6] R. K. Saiki, D. H. Gelfand, S. Stoffel, S. J. Scharf, R. Higuchi, G. T. Horn, K. B. Mullis, and H. A. Erlich, *Science* **239,** 487 (1988).

[7] E. S. Kawasaki, S. S. Clark, M. Y. Coyne, S. D. Smith, R. Champlin, O. N. Witte, and F. P. McCormick, *Proc. Natl. Acad. Sci. U.S.A.* **85,** 5698 (1987).

[8] D. A. Rappolee, D. Mark, M. J. Banda, and Z. Werb, *Science* **241,** 708 (1988).

[9] T. W. Myers and D. H. Gelfand, *Biochemistry* **30,** 7661 (1991).

[10] P. D. Siebert and J. W. Larrick, *Nature* **359,** 557 (1992).

[11] B. Hoffmann, J. M. Lehmann, X.-K. Zhang, T. Hermann, M. Husmann, G. Graupner, and M. Pfahl, *Mol. Endocrinol.* **4,** 1727 (1990).

mixes" that guarantee the homogeneity of reaction conditions throughout a series of experiments.

In this chapter we describe different methods for the expression analysis of various retinoid receptors, which we have developed in our laboratory. The protocol and the primer sequences given refer to the human retinoid receptors, the design of similar strategies for other organisms, however, should not prove particularly difficult.

RT-PCR Systems

Materials

Fast protein liquid chromatography (FPLC)-purified oligonucleotide primers were obtained from Tib-Molbiol (Genoa), RNase inhibitor and AMV (avian myeloblastosis virus) reverse transcriptase from Boehringer Mannheim (Germany), premade solutions of dNTPs from Pharmacia (Brussels), *Taq* polymerase, and RNA-extraction medium ("RNA-clean") from AGS (Heidelberg). Enzymes provided by other suppliers were occasionally tested and performed equally well. The PCR buffer at final concentration (1×) contains: 10 m*M* Tris-HCl (pH 8.3 at 25°), 50 m*M* KCl, 2 m*M* $MgCl_2$, and 0.01% (v/v) Tween 20. Amplifications were performed in a Perkin-Elmer (Norwalk, CT) 480 thermal cycler.

Methods of Sample Preparation

Total RNA can be prepared following any one of the standard protocols provided that the resulting preparation is sufficiently devoid of contaminating proteins, DNA, salts, and detergents and that the RNA remains intact. Generally RT-PCR is less sensitive to the quality of RNA preparation than other RNA analysis methods. Whenever possible, the primers for the amplification step should be selected on different exons like those described here, to allow amplification only from RNA and not from contaminating DNA. In our hands, the guanidium chloride/isothiocyanate procedure described by Chomczynski and Sacchi[12] or the commercial formulations of it work best.

Clinical samples are to be processed within a reasonable time after surgery. RNA in large specimens is relatively stable as compared to minced or homogenized tissue because RNases become active on disrupture of cells. Small fresh biopsies can be treated directly with the lysis buffer and homogenized in a blender. Larger tissue samples and frozen tissues must be disintegrated in a mortar under liquid nitrogen cooling. The powder obtained is transferred to a tube containing the lysis buffer and homoge-

[12] P. Chomczynski and N. Sacchi, *Annals Biochem.* **162,** 156 (1987).

nized in a blender. Caution must be taken that frozen samples do not thaw during the operation. The pestle is most efficient with less nitrogen present in the mortar so that the cooling liquid should be added continuously in small amounts. Powdered tissue samples can be stored frozen at −80° for extended periods. Extracted RNA samples can be stored in water at −80° for more than 1 month; for prolonged storage and for shipment RNA should be kept in 70% (v/v) ethanol containing 2 *M* LiCl. RNA can also be isolated from frozen tissue sections, which are directly transferred to the lysis solution. Amplification starting from RNA obtained from paraffin-embedded sections is possible when the primers are close to each other,[13] but the protocols described here have not been tested for this application.

One-Tube RT-PCR

Different one-tube RT-PCR reactions that assemble the reverse transcription reaction and the cyclic amplification reaction in a single tube have been described. One of the protocols exploits the reverse transcriptase activity of thermostable polymerases. In the presence of Mn^{2+} ions the enzyme isolated from *Thermus thermophilus* (*Tth* polymerase) uses RNA as a template for DNA polymerization. In the second step of the one-tube reaction, Mn^{2+} ions are sequestered by EGTA, a chelator with a high affinity for Mn^{2+} as compared to Mg^{2+}; the latter ion and PCR primers are added to the diluted reaction mix and amplification of the cDNA produced can be performed in the same tube. This technology has, to our knowledge, not been used for RT-PCR reactions of the retinoid receptors. Limitations of this method are the low processing ability of the reverse transcriptase activity of the thermostable enzyme, and, if the amplification product is to be analyzed by sequencing, the elevated misincorporation rate.[9]

The alternative one-tube RT-PCR reaction described here uses AMV transcriptase for cDNA synthesis and *Taq* polymerase for amplification. This system exploits the ability of the viral enzyme to work under the same conditions as the thermostable polymerase so that no change of buffer occurs. The protocol uses specific primers for reverse transcription instead of the more common oligo(dT) or random hexamer primers. Most of the nuclear receptor transcripts comprise a long 3′ untranslated region so that the efficiency of the cDNA synthesis is relatively low for the distant coding region. The sequence-specific primers allow cDNA synthesis to start in the immediate vicinity of the region to be amplified and introduce a certain specificity with respect to oligo(dT) and random hexamer primers. The sequence-specific RT primer is different from the PCR antisense primer, thus increasing the specificity over reactions using the same antisense primer for reverse transcription and PCR. Due to the low T_m the short sequence-

[13] C. Mies, *J. Histochem. Cytochem.* **42,** 811 (1994).

specific RT primers (12 nucleotides) do not interfere with the subsequent PCR.[14,15]

Procedure. Sample RNA is dissolved in water and denatured at 65° for 10 min and then chilled on ice and processed immediately. A blank containing water and a positive control are processed in parallel. A mix containing 1 m*M* dNTP, 25 pmol of each of the standard and the sample reverse transcription primer, 20 units RNasin, and 20 units of AMV reverse transcriptase in PCR buffer is added to the denatured RNA (final volume: 40 μl). Reverse transcription is carried out by incubation at 25° for 10 min and at 42° for 45 min. For amplification PCR mixes (90 μl per sample, each) in PCR buffer containing 0.1 m*M* dNTP, 25 pmol of either sample or standard primer pairs and 2.5 units *Taq* polymerase are prepared and added to 10 μl of cDNA without prior denaturation of the latter.

The thermal cycler is set to the following combined program: (1) denaturation at 94° for 2 min, (2) amplification in 35 cycles of 94°, 65°, and 72° for 30 sec each, and (3) final extension at 72° for 7 min followed by storage at 4°. Ten microliters of the products are loaded onto a 1.4–2.0% agarose gel containing ethidium bromide.[16] Quantitation is carried out by densitometric scanning normalizing different samples for the standard band. Different receptors (and the internal standard) may be analyzed in parallel starting from common cDNAs for either all three RARs or RXRs. In the first case, a single RT primer is used, whereas in the latter, three different RT primers are mixed together. Amplification is carried out using a common antisense primer and receptor subtype or isoform-specific sense primers. Primers to be used for the different receptors are listed in Table I, and the expected amplification products are indicated in Table II. Figure 1(a) shows the analysis of two cell lines expressing either all three RAR subtypes (SKN-BE, RARβ is induced after treating the cells with 2×10^{-8} *M* all-*trans*-retinoic acid for 24 hours) or all three RXR subtypes (Colo320DM). Figure 1(b) shows a complete retinoid receptor analysis by the one-tube protocol for two frequently used cell lines. HeLa cells express RARα and γ and RXRα and β, whereas MCF-7 cells express RARα and γ and RXRα.

Nested RT-PCR

The major advantage of RT-PCR over other methods of gene expression analysis is the potentially unlimited sensitivity that allows analyses of gene

[14] U. Pfeffer, E. Fecarotta, and G. Vidali, *BioTechniques* **18,** 204 (1995).

[15] U. Pfeffer, *in* "Methods in Molecular Biology" (R. Rapley and D. L. Manning, eds.); Vol. 86. RNA Isolation and Characterization Protocols, Humana Press, Inc., Totowa, NJ, in press.

[16] M. Ausubel, R. Brent, R. E. Kingston, D. D. Moore, J. G. Seidman, J. A. Smith, and K. Struhl (eds.), "Current Protocols in Molecular Biology." Wiley, New York, 1990.

expression in few cells or in a complex biological specimen where the cells expressing the gene analyzed constitute only a small portion of the total number of cells. The practical limitations of the method are given by the potential cross-reactivity of the primers with unrelated mRNA or cDNA sequences. At high dilution of the target sequence, many amplification cycles must be performed to obtain a signal and the risk of amplification of cDNAs containing limited sequence homology with the primers used is increased proportionally. This is generally overcome by the design of appropriate PCR primers, a task that remains prevalently empirical since most of the expressed sequences are not known. In most cases, nested reactions using different sets of primers for the amplification of a single target cDNA constitute a simple way to increase specificity, thus avoiding undesired cross-reactions at higher amplifications. The protocol presented here describes a system where the cDNA is obtained through reverse transcription of cellular RNA using a specific RT primer or a 25-mer primer identical to the antisense primer used in the following PCR reactions.

PCR is carried out in two rounds using the same antisense primer. In the first round, RAR subtype-specific sense primers are used and a small portion of the reaction is used as a template for the second round of amplification where all subtypes are amplified, again using the same common antisense primer and a sense primer common to the three subtypes of the receptor located at 3′ with respect to the sense primer of the first reaction. The subtype-specific sense primers for the first round can be chosen either in a region close to the DNA binding domain of the receptors (*B*-domain, identical isoforms) or in the region coding for the amino-terminal isoform-specific domain. So far we have developed nested RT-PCR protocols for RAR and RXR subtypes of retinoid receptors, but isoform-specific primers have been designed only for RARβ2 and RARγ1 (see Table I and Ref. 17).

Depending on the expression level of the receptors in the sample analyzed and on the amount of RNA used for the reaction, the cycle number of the first round of amplification can be adjusted and different amounts of the first round of amplification can be used as a template for the second round to calibrate the sensitivity of the system. Under the conditions indicated in the procedure described next, 0.25 μg total RNA extracted from a cell line that expresses the corresponding receptor yields a band visible by ethidium bromide staining.

Procedure. The nested RT-PCR reaction is carried out as described for the one-tube procedure with the exception that the PCR antisense primer can also be used as the RT primer. The first round of amplification

[17] N. Ferrari, U. Pfeffer, F. Tosetti, C. Brigati, and G. Vidali, *Exp. Cell Res.* **211,** 121 (1994).

TABLE I
PRIMER LIST[a]

Primer number	Receptor type	Position	Use	Sequence	Application	Mismatches
	RARs					
1	RARα_{common}	460–484	sense	5′-acc ccc tct acc ccg cat cta caa g	O/N	
2	RARβ_{common}	547–571	sense	5′-ccc atc tcc act tcc tcc ccc tcg a	O/N	
3	RARβ_2	394–418	sense	5′-ttt cta cac tgc gag tcc gtc ttc c	N	
4	RARγ_{common}	441–465	sense	5′-ctc cgc ctc ctc ggg tct aca agc c	O/N	
5	RARγ_1	314–338	sense	5′-ttc gag atg ctg agc cct agc ttc c	O/N	
6	RAR$_{common}$	504–527(α) 600–623(β) 483–506(γ)	sense	5′-aca agt cct cag ggt acc act atg	N	α:1 β:2 γ:2
7	RAR$_{common}$	1092–1070(α) 1188–1166(β) 1071–1049(γ)	antisense	5′-ccg tc(tcg) gag aa(ga) gtc atg gtg tc	O	
8	RAR$_{common}$	685–662(α) 781–758(β) 664–641(γ)	antisense	5′-cat gcc cac ttc aaa gca ctt ctg	N	α:0 β:1 γ:1
9	RAR$_{common}$	1129–1118(α) 1225–1214(β) 1108–1097(γ)	rt 12-mer	5′-gcc agc att gtg	O/N	α:1 β:1 γ:1
	RXRs				O/N	
10	RXRα	427–451	sense	5′-aag gtc ccc gcc cac ccc tca gga a	O/N	
11	RXRβ	680–704	sense	5′-gcc tgc act gtc cac ccc ctc cag g	O/N	

12	RXRγ	399–423	sense	5′-gaa cta ccc atc cac cag ccc cgg a	N	
13	RXRα/β	528–522(α)	sense	5′-ctg cga ggg ttg caa ggg ctt ctt c	N	α:1
		780–804(β)				β:1
14	RXRγ	492–516	sense	5′-ttg tga agg ctg caa agg gtt ctt c	N	
		944–921(α)				
		944–921(α)			O/N	
15	RXR$_{\text{common}}$	1199–1175(β)	antisense	5′-ga(g/a) aag tg(t/g) gg(g/a) at(c/a) c(g/t)c tt(c/g) gcc c		
		899–875(γ)				
16	RXRα	997–984	rt 13-mer	5′-gct cat tcc agc c	O/N	
17	RXRβ	1252–1240	rt 13-mer	5′-gtt cat tcc agc c	O/N	
18	RXRγ	952–941	rt 13-mer	5′-att cat tcc acc c	O/N	
	Standard					
19	β-Actin	269–290	sense	5′-ctc acc ctg aag tac ccc atc g		
20	β-Actin	1153–1132	antisense	5′-ctt gct gat cca cat ctg ctg g		
21	β-Actin	1166–1155	rt 12-mer	5′-act cgt cat act		
	Transgene					
22	RARγ	1675–1664	endo-rt	5′-tga aca cag agg	O/N	
23	SV40–poly(A)	556–545	exo-rt	5′-aca aac cac aac	O/N	

[a] Antisense primers are indicated in the transformed form (the actual primer sequence). The position indicated refers to the sequences as deposited in the EMBL gene databank under the following accession numbers: RARα: X06538; RARβ: X07282; RARγ: M57707; RXRα: X52773; RXRβ: M84820; RXRγ: U38480; β-actin: X63432. Primer 22 is obtained from the sequence M38258 (RARγ), the position of the SV40-poly(A) primer (23) refers to the sequence L36849 (pZeoSV, expression vector, Invitrogen, San Diego, CA). For the common RAR sense primers, oligonucleotides with some mismatches as indicated in the relative column have been designed. For the RXR sense primer, due to the considerable sequence divergence, two oligonucleotides have been produced. These two primers may be assembled in a single PCR mix to be used for all three receptor subtypes. The "common" RAR antisense primer and the "common" RXR antisense primer are degenerated to match the three different sequences. Applications: O, one-tube reaction; N, nested reaction. All primers are available as premade primer kit from Tib-Molbiol (Genoa, Italy).

TABLE II
PRIMER PAIRS TO BE USED AND EXPECTED LENGTHS OF AMPLIFICATION PRODUCTS[a]

	One-tube		Nested	
Source[b]	Primers[b]	Product length (bp)	Primers	Product length (bp)
RARα	1/7	633	1/8, 6/8	(226), 182
RARβ	2/7	642	2/8, 6/8	(235), 182
RARβ_1	3/7	795	3/8, 6/8	(388), 182
RARγ	4/7	629	4/8, 6/8	(224), 182
RARγ_1	5/7	757	5/8, 6/8	(351), 182
RXRα	10/15	518	10/15, 13/15	(518), 417
RXRβ	11/15	520	11/15, 13/15	(520), 420
RXRγ	12/15	501	12/15, 14/15	(501), 408
β-Actin	19/20	885	NA	NA

[a] NA, Not applicable (no nested reaction has been designed for the internal standard). The lengths of the products obtained in the first round of amplification are given in parentheses because they are not visible on ethidium bromide-stained gels.
[b] RT, primers; RAR, primer 9 for one-tube, primer 8 or 9 for nested; RXR, a mixture of primers 16–18; and β-actin, primer 21.

is performed at the cycle parameters indicated earlier for only 12 cycles; 10 μl of the first reaction is diluted with water to 100 μl, and 10 μl of the dilution is used as a template for the second round of amplification at the same conditions for 25 or 32 cycles for RAR and RXR, respectively. The inner sense primer is added only to the mix for the second round and the elongation step is reduced to 20 sec. If the internal standard (β-actin) is to be analyzed in parallel, reverse transcription may be carried out in the presence of the standard RT primer. The amplification step of the internal standard must be performed separately as indicated earlier. Figure 1(a) shows the amplification products of a nested reaction in parallel with those obtained by the one-tube reaction.

Competitive RT-PCR

Absolute quantitation of the number of specific mRNA molecules present in a sample is not an easy task. Competitive PCR quantitations are based on the observation that two templates present in a PCR reaction compete for the components of the system, including the primers if both templates contain the same primer sequences. If the amplification efficiency for the two templates is identical, the two products of amplification will be present at an equimolar ratio exactly when the two templates are at equimo-

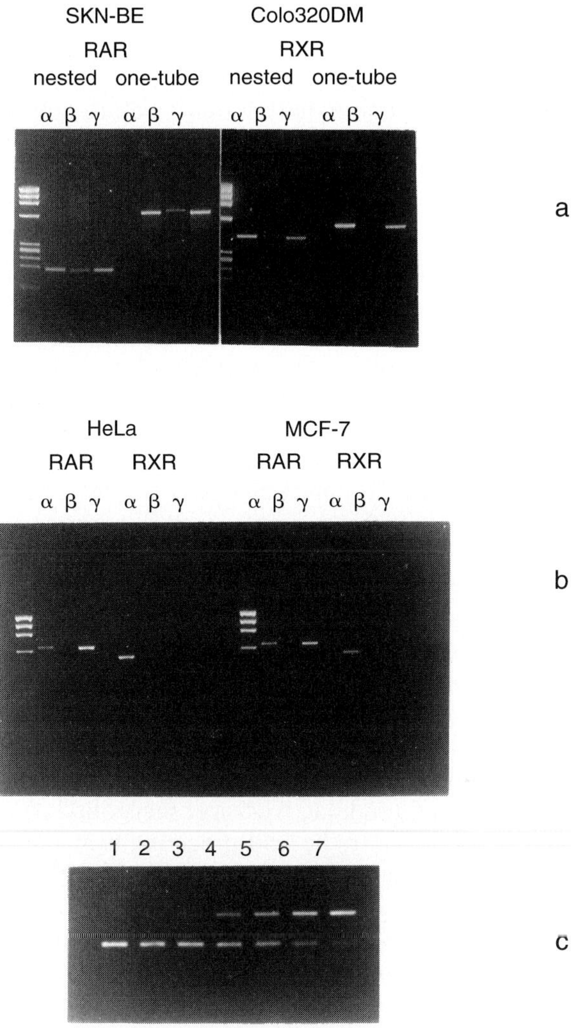

FIG. 1. (a) RAR and RXR expression analysis applying either the nested or the one-tube reaction protocols. The neuroblastoma cell line SKN-BE expresses all three RAR subtypes after induction with 2×10^{-8} *M* all-*trans*-retinoic acid for 24 hours and the colon carcinoma cell line Colo320DM constitutively expresses all three RXR subtypes. (b) Retinoid receptor expression analysis of two sample cell lines applying the one-tube procedure. The cervic carcinoma cell line HeLa expresses RARα and γ and RXRα and β, whereas the mammary carcinoma cell line MCF-7 cells expresses RARα and γ and RXRα. (c) Competitive nested RT-PCR using constant amounts (10^6 molecules) of RNA transcribed from the synthetic gene containing all necessary primer sequences and varying amounts of RARβ_1 RNA *in vitro* transcribed from a plasmid containing the complete cDNA (lane 1, 10^4; lane 2; 5×10^4; lane 3, 10^5; lane 4, 5×10^5; lane 5, 10^6; 5×10^6; lane 7, 10^7 molecules). Note that equal amounts of standard and receptor RNA yield equal amounts of amplification products.

lar ratio in the starting material. If different known quantities of standard template are used in parallel reactions also containing the sample template, the reaction that contains equimolar amounts of the two templates after amplification also contains equimolar amounts before amplification.[10]

To distinguish the sample template from the standard, the two templates must be different. The difference can occur in a restriction site, in a deletion, or in an additional sequence artificially introduced into the cloned cDNA. Alternatively, a completely synthetic gene can be synthesized. The latter has the advantage that a single template can contain many different primer sequences and thus serve as a standard for different RT-PCR reactions. In any case, the modifications of the sequence with respect to the sample sequence may determine a different reverse transcription or amplification efficiency. The efficiency of sample versus standard template amplification can be measured when equal amounts of both are amplified in separate reactions. The ratio of the products obtained constitutes a factor for which the results of competitive PCRs must be corrected. In our experience, it is easier to determine the "efficiency coefficient" rather than to design standard templates that perfectly match the amplification efficiency of the sample template. In the specific case of the RARs and RXRs, this approach avoids the construction of standard templates for each one of the many subtypes and isoforms.

Standard templates are obtained from a synthetic gene cloned into a vector containing promoter sequences for phage RNA polymerases such as T3 or T7 (an example is given in Ref. 17; submitted to the EMBL databank, accession number X97686). Transcription of the template is carried out in the presence of a radioactive tracer nucleotide of known specific activity. The activity obtained in the acid-precipitable material after *in vitro* transcription allows the calculation of the number of RNA molecules if counting efficiency, the specific activity of the tracer (considering also the "cold" nucleotide used), and the percentage of the tracer nucleotide in the transcribed sequence are taken into account. Obviously the standard RNA must be of full length (for a detailed *in vitro* transcription protocol, see Ref. 16).

According to this calculation, different amounts of standard RNA molecules are used in combination with a fixed amount of template RNA in RT-PCR reactions. The amplification products are separated on agarose gels and the ethidium bromide-stained bands are quantified by densitometry scanning. The reaction where equal amounts of the two templates are measured gives the crude amount of sample template molecules that must be corrected for the potentially different amplification efficiency and the different staining efficiency (which can be calculated from the length). Figure 1(c) shows the results of a competitive RT-PCR reaction using constant amounts of RNA transcribed *in vitro* from the synthetic gene and varying amounts of RNA transcribed *in vitro* form a plasmid carrying the RARβ_1 cDNA. At equimolar starting concentrations (10^6 molecules) equimolar end products are obtained. This indicates that the amplification efficiencies of the two templates are not drastically different or are compensated by the different staining intensity of the two products. Similar analyses using RNA samples containing unknown amounts of target messenger allow the absolute quantitation of receptor expression.

Internal Standards

Semiquantitative analyses, in which the relative amounts of specific mRNA molecules present in different samples are measured, are more common. The information about whether a specific tissue transcribes more or less of the messenger than another tissue or whether a given treatment results in an alteration of the receptor mRNA expression can be obtained without measuring the absolute numbers of the molecules just by comparing amplification reactions performed under identical conditions on different samples. To do this, however, care must be taken to standardize the reaction conditions. In our hands, the most critical features for the quantitation outcome of the RT-PCR reactions are the quantity of RNA used, the quality of the template (RNA degradation!), and the quantities of enzymes pipetted into the samples. The latter is easily overcome by preparing "master mixes" containing everything, including the enzyme, for the number of samples to be analyzed, so that only the error test can occur when pipetting the mix onto the different samples persists.

The quantity and quality of the sample RNA must be controlled by the coanalysis of an unrelated messenger that is not expected to vary between the samples. Obviously there is no one messenger that does not change its relative concentration in response to any treatment and this problem is even more severe when comparing different tissues. In practice, the analysis of different internal standards may be necessary. The most commonly used standard messengers code for β-actin (used here) or for glyceraldehyde-3-phosphate dehydrogenase (E.C. 1.2.1.12).

Internal standards are not normally suitable for coamplification in the same tube since multiple primers present in the same reaction potentiate the risk of cross-reaction with unrelated sequences and the internal standard messengers are normally present in much higher quantities than the sample messengers and the two templates would therefore compete with each other. In the extreme case, low-abundance sample messengers would thus escape detection. One possibility to overcome this limitation is to carry out a common reverse transcription and to separate the cDNA for the sample and standard PCR. In this approach the critical reverse transcription is carried out on the same RNA sample and the error is limited to the pipetting of the larger volumes of cDNA and PCR master mix.

Semiquantitative analyses must further be controlled for the linear relation between the amounts of the initial target and the final product: using the double amount of RNA the double quantity of PCR product must be obtained for the sample as well as for the internal standard. Therefore, these reactions need to be adapted to every new situation, since a higher expression level in an unknown cellular system could lead to saturation. The analytical reaction must be carried out at conditions in the linear range of the plot of starting material over product yield.

Comparison of the Systems

The two methods of retinoid receptor expression analysis by means of RT-PCR presented here correspond to different analytical situations: if enough material can be obtained, as is normally the case when cell lines are analyzed, the one-tube reaction protocol yields reliable results in a straightforward reaction easy to perform and thus applicable for the analysis of large sets of samples. Reverse transcription and PCR mixes can be prepared in large stocks, thus limiting the handling to a few pipetting steps (addition of the enzymes and application to the samples). On the contrary, the nested protocol is better suited for reduced sample quantities that often occur in the analysis of biopsies. Both systems are suitable for semiquantitative analysis, whereas the determination of absolute messenger amounts, although possible in both cases, has been developed only for the nested RT-PCR system.

Distinction between Endogenous and Transgene Expression

Many studies have used transfection assays for the functional characterization of retinoid receptor cDNAs or for the analysis of the effects of overexpression of the native receptor cDNA, of mutated receptor cDNAs, or of ectopic expression of the receptors. Much of this work has been carried out in cellular systems, such as CV-1 or Schneider cells, that do not express endogenous retinoid receptors. To analyze the effect of transfected receptor cDNAs on cells that express endogenous receptors or to evaluate the efficiency of transfection, we have designed RT primers that distinguish between transgenic and endogenous receptor mRNAs. Reverse transcription is carried out in the presence of a dodecamer primer complementary either to a sequence in the 3′-untranslated region of the endogenous receptor transcript (primer 22) or to a sequence of the SV′-40 (simian virus 40) polyadenylation signal often used in transgene constructs (primer 23). The dodecamer primer is sufficient to discriminate between the two transcripts and the following PCR can be performed following either of the two protocols described.

Pitfalls

The most frequent problem encountered when performing RT-PCR analyses is the appearance of "false-positives", that is, the occurrence of amplification products in blanks (control reactions without RNA) or in samples derived from receptor negative cells. The main source of these contaminations is the PCR products and, therefore, the risk of contamination increases with increasing numbers of analyses performed. In other words, at a certain point after a period of successful RT-PCR amplification all the samples analyzed appear to express the messenger analyzed equally and so does the blank. Once this type of contamination occurs it is difficult to "clean up" the reaction environment and equipment. Therefore, some basic protection measures should be applied before contaminations occur.

The basic recommendation consists of physically separating "pre-PCR" from "post-PCR," and this holds for equipment (pipettes, racks, etc.), as well as for solutions, reagents, primers, etc., and personal protection such as gloves, laboratory coats, and the like. If possible, RT-PCR should be performed in a one-way sequence in which the different steps are separated from each other. This is not always possible in a normal research laboratory and unless extreme numbers of samples are handled, the use of two different rooms will be sufficient. Another useful protection is the use of filter tips since a major source of contamination is the aerosol formed during pipetting. The preparation of large master mixes before a new series of amplifications is started also reduces the contamination risk by limiting the number of pipetting steps to be performed. Another source of ghost bands often encountered in research laboratories is plasmids containing related sequences, which should not be handled in the same room as RT-PCR components or by the operator who also performs RT-PCR. Many specific protocols designed to avoid contamination have been published and they are principally compatible with the RT-PCR systems described here. However, in our hands, physical separation and the use of filter tips is sufficient. If bulk quantities of primers, reagents, and enzymes are ordered, they should be aliquoted. Once a contamination occurs, all equipment, as far as possible, should be cleaned with detergent and autoclaved. Potentially contaminated solutions, including primer dilutions and enzymes should be discarded and all surfaces should be cleaned with diluted acids and detergents.

The opposite problem, the complete absence of amplification products from positive control samples, is encountered no less frequently. Apart from degraded enymes or other reaction components, RNase contaminations are most frequent, especially when RNases are used in the same laboratory.

Nonspecific amplification products are the result of mispriming. RT-PCR primers like those described here are tested on a limited number of cell lines and it is possible that new lines express different messengers with limited homology to the retinoid receptors thus allowing amplification, although less efficient, with the same primers. In this case, new primers must be designed. However, the application of specific primers for reverse transcription and eventually the performance of nested reactions should make this possibility rather unlikely. Otherwise, nonspecific bands can occur as a result of (1) incorrect temperature settings or buffer component concentrations, (2) enzyme excess, or (3) too many amplification cycles.

Comparison to Other Methods of Gene Expression Analysis

A comparison of all of the different methods of RNA analysis reveals that there is no one method superior to all the others. Specific applications require suitable methods that may be less useful for other applications. For simple expression analysis, Northern blotting, dot or slot blots, protection or primer extension assays, and RT-PCR are equally applicable, and in the laboratory, the choice is governed by which method is better established. The situation is very different if mRNA expression is to be analyzed in the context of morphological information as in the case of complex tissue samples composed of various cell types with each one a specific expression pattern. This can be obtained only by *in situ* hybridization and is particularly important for the retinoid receptors because most cells express at least one of them. The analysis of samples containing different cell types by methods other than *in situ* hybridization may be misleading since a varying composition of cells in two different samples could mimic a variation in receptor expression of one cell type. The only alternative is RT-PCR of samples obtained from frozen tissue section when adjacent sections are morphologically evaluated.

Other specific applications, for example, the detection of intron/exon borders or the mapping of 5′ ends of transcripts, although theoretically feasible by RT-PCR, will be easier to perform by well-established S1 or RNase protection assays.

However, once reliable primers for RT-PCR of a specific gene have been selected and the protocol has been adapted to the specific situation, the ease of performance, sensitivity, and speed of RT-PCRs are unmatched. Moreover, the decrease in enzyme prices makes RT-PCRs no more expen-

TABLE III
RAR AND RXR EXPRESSION IN FREQUENTLY USED CELL LINES[a]

Cell line	Cell type	RARα	RARβ	RARγ	RXRα	RXRβ	RXRγ
MCF-7	Mammary carcinoma	+	−	+	+	−	−
HeLa	Cervix carcinoma	+	−	+	+	+	−
Colo230DM	Colon carcinoma	+	+	+	+	+	+
Caki	Renal carcinoma	+	−	+	+	+	−
SKN-BE	Neuroblastoma	+	i	+	+	+	−
HepG2	Hepatoma	+	−	+	+	−	−
A549	Lung carcinoma	+	i	+	+	−	−
HL-60	Promyelocytic leukemia	+	−	+	+	+	−

[a] +, Amplification product obtained from 1 μg total RNA under the conditions described; −, no amplification product obtained from 1 μg total RNA under the conditions described; and i, amplification product obtained from 1 μg total RNA of all-*trans*-retinoic acid-induced cells under the conditions described.

sive than other methods. Dot blots are the only alternative in terms of analysis of large sample numbers. However, in this type of assay no information on the molecular entity actually hybridizing is available, whereas RT-PCR products are controlled by gel electrophoresis allowing the analysis of product size. If downstream operations such as cloning, sequencing, or mutation analysis have to be performed, RT-PCR is the method of choice, because after detection, most of the reaction product is available for further processing.

Applications

We have applied the RT-PCR systems described to the analysis of RAR and RXR expression in frequently used cell lines. The results are summarized in Table III and an example is given in Fig. 1(b). In another study we addressed the issue of correlation between RAR expression and tumor grading of neuroblastoma tumors. Although the tumors analyzed showed very heterogeneous patterns of RAR expression, there was no correlation to the established tumor stages.[18]

RT-PCR analyses that are different from those described here have been used to detect expression of PML-RARα in acute promyelocytic leukemia[19,20] where the RARα gene on chromosome 17 is fused to the

[18] N. Ferrari, G. P. Tonini, A. Briata, F. Bottini, and G. Vidali, *Intl. J. Oncol.* **5,** 1019 (1994).

[19] W. H. Miller, Jr., A. Kakizuka, S. R. Frankel, R. P. Warrell, Jr., A. DeBlasio, K. Levine, R. M. Evans, and E. Dimitrovsky, *Proc. Natl. Sci. U.S.A.* **89,** 2694 (1992).

[20] A. G. Turhan, F. M. Lemoine, C. Debert, M. L. Bonnet, C. Baillou, F. Picard, E. A. Macintyre, and B. Varet, *Blood* **85,** 2154 (1995).

PML gene on chromosome 15 by reciprocal translocation. This rearrangement and the expression of the resulting fusion proteins cause the disease, and the presence of residual cells expressing the rearranged gene is highly predictive for relapse.

Acknowledgment

This work is dedicated to the memory of Giorgio Vidali. Research was funded by grants from the Consiglio Nazionale delle Ricerche, Associazione Italiana per la Ricerca sul Cancro, Ministero della Sanità.

[6] Use of Quantitative Polymerase Chain Reaction to Study Cellular Retinoic Acid-Binding Protein-II mRNA Expression in Human Skin

By LUBING ZHOU, GAIL OTULAKOWSKI, and CATHERINE Y. LAU

Introduction

Cellular retinoic acid-binding protein I (CRABP-I) and II (CRABP-II) are homologous proteins that bind all-*trans* retinoic acid (RA) with high affinity.[1,2] The active metabolite of vitamin A, RA, has pleiotropic effects that include modulation of cellular proliferation and differentiation.[3–6] Retinoic acid is a potent teratogen that in excess can cause a characteristic pattern of congenital malformation.[7] The predominant mechanism through which RA regulates biological functions is through the activation of nuclear receptors, with resultant modulation of target gene transcription. The nuclear receptors (reviewed by Chambon[8]) are divided into two families, each encoded by three genes. Initially, the RA receptors (RARα, β, γ) were found to activate transcription following binding of all-

[1] D. E. Ong and F. Chytil, *J. Biol. Chem.* **253,** 4551 (1978).

[2] J. S. Bailey and C.-H. Siu, *J. Biol. Chem.* **263,** 9326 (1988).

[3] A. B. Roberts and M. B. Sporn, *in* "The Retinoids," Vol. 2 (D. S. Goodman, ed.), p. 209. Academic Press, Orlando, 1984.

[4] A. E. Griep and H. Westphal, *Proc. Natl. Acad. Sci. U.S.A.* **85,** 6806 (1988).

[5] D. B. Holland, G. Gowland, and W. J. Cunliffe, *Br. J. Dermatol.* **110,** 343 (1984).

[6] C. E. Brinckerhoff, H. Nagase, J. E. Nagel, and E. D. Harris, *J. Acad. Dermatol.* **6,** 591 (1982).

[7] E. J. Lammer, D. T. Chen, R. M. Hoar, N. D. Agnish, P. J. Benke, J. T. Braun, C. J. Curry, P. M. Fernhoff, A. W. Grix, Jr., I. T. Lott, J. M. Richard, and S. C. Sun, *N. Engl. J. Med.* **313,** 837 (1985).

[8] P. Chambon, *Recent Prog. Horm. Res.* **50,** 317 (1995).

0076-6879/97 $25.00

trans-RA. More recently, the homologous retinoid X receptors (RXRα, β, γ) were discovered, which form both homodimers and heterodimers with the RARs. The 9-*cis* stereoisomer of RA binds with high affinity and activates transcription through both RXRs and RARs. The interactions of various RAR and RXR dimers with the wide array of naturally occurring RA responsive elements and with various RA stereoisoforms appears to be the basis of RA's pleiotropic effects. The two binding proteins CRABP-I and CRABP-II are believed to control the availability of active retinoids within the cell.[9]

Transgenic mice expressing a dominant negative RAR specifically in the epidermis have demonstrated a requirement for RAR activity in normal epidermal maturation.[10] In adult human skin, RA alters epithelial morphology,[11] and influences the expression of high molecular weight keratins[12,13] and other differentiation-associated proteins such as filaggrin and involucrin.[14–16] All-*trans*-RA has been approved for the treatment of photodamaged skin. Among the target genes for retinoid action in skin, CRABP-II has been shown to increase markedly in a dose-related fashion in response to topical RA treatment.[17] Measurements of the steady-state levels of CRABP-II mRNA have been proposed as a selective marker for cutaneous retinoid activity *in vivo*.[18] The technology described later provides a sufficiently sensitive method for CRABP-II mRNA quantitation to allow the use of punch biopsy specimens of human skin in clinical studies.

Principle of Method for Quantitative RT-PCR

The use of linked reverse transcription–polymerase chain reactions (RT-PCR) to quantitate specific mRNAs is rapidly gaining favor over Northern blot or RNase protection analyses due to its convenience, speed, and sensitivity. However, accurate quantitative results require careful de-

[9] A. L. Gustafson, L. Dencker, and U. Eriksson, *Development* **117,** 451 (1993).

[10] M. Saitou, S. Sugai, T. Tanaka, K. Shimouchi, E. Fuchs, S. Narumiya, and A. Kakizuka, *Nature* **374,** 159 (1995).

[11] R. Lotan, *Biochem. Biophys. Acta.* **605,** 33 (1980).

[12] E. Fuchs and H. Green, *Cell* **25,** 617 (1981).

[13] C. L. Marcelo and K. C. Madison, *Arch. Dermatol. Res.* **276,** 381 (1984).

[14] P. R. Cline and R. H. Rice, *Cancer Res.* **43,** 3203 (1983).

[15] D. S. Rosenthal, C. E. Griffiths, S. H. Yuspa, D. R. Roop, and J. J. Voorhees, *J. Invest. Dermatol.* **98,** 343 (1992).

[16] T. Magnaldo, F. Bernerd, D. Asselineau, and M. Darmon, *Differentiation* **49,** 39 (1992).

[17] A. Astrom, A. Tavakkol, U. Pettersson, M. Cromie, J. T. Elder, and J. J. Voorhees, *J. Biol. Chem.* **266,** 17662 (1991).

[18] J. T. Elder, M. A. Cromie, C. E. Griffiths, P. Chambon, and J. J. Voorhees, *J. Invest. Dermatol.* **100,** 356 (1993).

sign of proper controls. The system described in this chapter uses a synthetic RNA generated by *in vitro* transcription from the cDNA of human CRABP-II, with a small deletion in the coding region. Using identical primers, the test sample and internal control amplify competitively in each reaction tube. This strategy avoids problems due to tube-to-tube variations, unequal amplification efficiencies, and the plateau effect that arise with external controls or unrelated internal controls.

A titration system is used to quantitate CRABP-II mRNA accurately. A serial dilution of the sample RNA, together with a fixed amount of internal control RNA, is co-reverse transcribed by a CRABP-II-specific primer. An aliquot of each RT reaction is subsequently amplified by PCR. Because the target and control cDNAs amplify with identical efficiencies, the ratio of the two products is cycle number independent, and the products can be allowed to accumulate to high levels. Following polyacrylamide gel electrophoresis (PAGE), the two products are visualized by ethidium bromide staining. At the equivalence point, where the intensities of the control and target bands are equal, the amount of CRABP-II mRNA in the sample RNA must equal the amount of internal control RNA.

Maximum accuracy can be achieved by the use of a computerized image analyzer system. A ratio of the integrated optical density (IOD) of the DNA bands is plotted against the amount of internal control RNA, yielding a linear relationship on a log–log scale from which the exact equivalence point can be obtained.

Design of Internal Controls

The efficiency of PCR amplification decreases in later cycles due to depletion of reaction components, diminished enzymatic activity, and accumulation of products. This is termed the *plateau effect*. In the absence of an appropriate control, accurate quantitation of starting material can only be inferred from the amount of product produced during the early, exponential stages of the reaction. Many investigators have attempted to carry out quantitative PCR by controlling reaction conditions,[19,20] but very often the reaction has already reached a plateau when the products are barely enough for visualization following electrophoresis. Radiolabeled nucleotides or primers can be used to improve sensitivity; however, the conditions required

[19] D. A. Rappolee, D. Mark, M. J. Banda, and Z. Werb, *Science* **241,** 708 (1988).

[20] M. A. Abbott, B. J. Poiesz, B. C. Byrne, S. Kwok, J. J. Sninsky, and G. D. Ehrlich, *J. Infect. Dis.* **158,** 1158 (1988).

to obtain quantitative information by limiting PCR to the exponential phase remain difficult to optimize.[21]

These problems can be overcome by the addition of an internal control. Several reports have described quantitative PCR amplification using housekeeping genes (e.g., β-actin) as the internal controls,[22,23] and the convenience of this technique has led to its widespread use despite controversy over its validity.[24] Although this strategy normalizes tube-to-tube variations, differences in the amplification of two unrelated DNA fragments arise due to differences in primer annealing and transcription efficiencies. Such effects can become highly magnified by the exponential nature of PCR, leading to major errors in quantitation. Another strategy, PCR MIMICS,[25] involves the construction of a generic DNA fragment, which can be amplified by the same primers as the target sequence. While this yields a more truly competitive system and reduces the chances of unequal amplification efficiency, it fails to control for the efficiency of the RT step.

The internal control that we used was an *in vitro* transcribed RNA molecule generated from a modified CRABP-II cDNA containing a 42-bp deletion within the coding region. It was designed in such a way that the same primers are used for both the target and internal control RNAs throughout the RT-PCR process. The expected size of the PCR product is 214 bp for the internal control and 256 bp for CRABP-II mRNA. The consideration of how many bases should be deleted in the internal control was based on two opposing principles. Although a large deletion would increase the ease of separation of the control and natural CRABP-II PCR products, it might also lead to a significant difference in the efficiency of PCR amplification. The 42-bp deletion we selected corresponds to a 16.4% decrease in length. For targets up to 1 kb, there is no significant difference in amplification efficiency for DNA fragments with size differences of 33% or less.[23,26,27]

Since even trace amounts of the DNA template from *in vitro* transcription of the internal control RNA would result in inflated levels of RT-PCR

[21] J. Singer-Sam, M. O. Robinson, A. R. Bellve, M. I. Simon, and A. D. Riggs, *Nucleic Acids Res.* **18,** 1255 (1990).

[22] L. D. Murphy, C. E. Herzog, J. B. Rudick, A. T. Fojo, and S. E. Bates, *Biochemistry* **29,** 10351 (1990).

[23] J. Chelly, J. C. Kaplan, P. Maire, S. Gautron, and A. Kahn, *Nature* **333,** 858 (1988).

[24] P. N. Hengen, *Trends Biochem. Sci.* **20,** 476 (1995).

[25] P. D. Siebert and J. W. Larrick, *Biotechniques* **14,** 244 (1993).

[26] L. Zhou, G. Otulakowski, J. Pang, D. G. Munroe, R. J. Capetola, and C. Y. Lau, *Nucleic Acids Res.* **20,** 6215 (1992).

[27] T. E. Golde, S. Estus, M. Usiak, L. H. Younkin, and S. G. Younkin, *Neuron* **4,** 253 (1990).

products, this template is eliminated by treatment of the synthetic RNA with RNase-free DNase, followed by purification on an agarose gel. The absence of DNA template from the internal RNA control can be tested by performing PCR for 30 to 40 cycles in the absence of a prior RT reaction.

Primer Design

To avoid any possible nonspecific amplification due to contamination of RNA by human genomic DNA, the primers CRB-1 and CRB-2 were designed to match sequences located in different exons of the human CRABP-II gene.[28] The specificity of this primer pair for cDNA was confirmed by the absence of a specific amplification product from human genomic DNA after 35 cycles of PCR.

Amplification of homologous mRNAs could also interfere with the assay. Human CRABP-I and CRABP-II share an overall amino acid sequence identity of 73%. To avoid nonspecific amplification of CRABP-I cDNA, the primers were designed from regions of maximum sequence divergence. Sequencing of amplified PCR products confirmed the specificity of the two primers.

We have further extended our quantitative RT-PCR method to a human skin-grafted nude mouse model in which human CRABP-II mRNA induction is used to evaluate the pharmacological activity of topical retinoids.[29] The PCR primers CRB-1 and CRB-2 are homologous to the corresponding mouse CRABP-II sequences and are not capable of distinguishing between mouse and human cDNA under our PCR conditions. Therefore, for this application, a human-specific CRABP-II primer (CRB-4) was designed for the RT reaction to ensure that only human mRNA could serve as template.

Methods

RNA Purification

Total RNA prepared by guanidinium/CsCl or guanidinium/acid–phenol extraction is suitable for quantitative RT-PCR as described later.[30,31] RNA concentration is determined spectrophotometrically.

[28] A. Astrom, U. Pettersson, and J. J. Voorhees, *J. Biol. Chem.* **267,** 25251 (1992).

[29] G. Otulakowski, L. Zhou, W. P. Fung-Leung, G. J. Gendimenico, S. E. S. Samuel, and C. Y. Lau, *J. Invest. Dermatol.* **102,** 515 (1994).

[30] T. Maniatis, E. F. Fritsch, and J. Sambrook, "Molecular Cloning: A Laboratory Manual." Cold Spring Harbor Laboratory Press, Cold Spring Harbor, Massachusetts, 1982.

[31] P. Chomczynski and N. Sacchi, *Anal. Biochem.* **162,** 156 (1987).

Internal Control RNA

Six micrograms of human CRABP-II cDNA[17] in the phagemid vector Bluescript SK (Stratagene, La Jolla, CA) is linearized at a unique site within the coding region by digestion for 2 hr at 37° with 10 units *Xcm*I. Following phenol–chloroform extraction and ethanol precipitation, the linearized plasmid is redissolved in 15 μl distilled water and digested for 10 min at 30° with 0.25 units *BAL*-31 exonuclease (New England Biolabs, Beverley, MA) in 1× buffer (600 m*M* NaCl, 12 m*M* $CaCl_2$, 12 m*M* $MgCl_2$, 20 m*M* Tris-HCl, pH 8.0, 1 m*M* EDTA). Digested DNA ends are blunted with Klenow (30 min at room temperature) and self-ligated with T4 DNA ligase for 12 hr at 14°. The deletion and the new recombinant plasmid are confirmed by DNA sequencing. A plasmid, named CRABP-D2, containing a 42-bp deletion (corresponding to the cDNA sequence from 213 to 254 bp as described by Astrom *et al.*[17]) was picked as the template for *in vitro* transcription using T3 polymerase (Promega, Madison, WI) following the manufacturer's instructions. The RNA generated is treated with RNase-free DNase RQ1 (Promega), passed through a Sephadex G-50 spin column, and purified on a low melting temperature agarose gel.[30] The RNA is quantitated spectrophotometrically at 260 nm, diluted to a concentration of 1 attomol/μl, and stored at −70°.

Primers

(1) CRB-1: 5′-GGCAA CTGGA AAATC ATCCG ATCGG AAAAC; (2) CRB-2: 5′-CACTC TCCCA TTTCA CCAGG CTCTT ACAG. The oligonucleotide CRB-2 is also used as the primer for reverse transcription of RNA from human skin biopsies and human skin fibroblast cell lines to generate the CRABP-II-specific first-strand cDNA. A human CRABP-II-specific primer, CRB-4 (5′-CAGTG AAGCA GGGCG GTGAG CAT), is used for reverse transcription in the human skin-grafted nude mouse samples.

Reverse Transcription

The amount of total RNA required is dependent on the level of target mRNA expression, which varies from sample to sample. To allow for up to 100-fold variation in the equivalence point, set up a 1:2 serial dilution of total RNA (64 ng down to 0.5 ng), each mixed with 0.2 attomol of CRABP-D2 internal control RNA. The final volume is brought to 7 μl with nuclease-free water. The samples are heated at 65° for 4–5 min and chilled on ice. Aliquoted RT mixture is then added to the individual tubes to bring the final volume to 20 μl. The RT reaction contains 0.25 μM

CRB-2 (CRB-4 in the case of human skin-grafted mouse samples), 0.5 m*M* each dNTP, 1 unit of RNase inhibitor, 10 m*M* dithiothreitol (DTT), 200 units of MMLV (Moloney murine leukemia virus) reverse transcriptase (BRL), and 1× RT buffer (50 m*M* Tris-HCl, pH 8.3, 75 m*M* KCl, 3 m*M* $MgCl_2$). The reaction is incubated at 37° for 1 hr, heated at 90° for 5 min, and chilled on ice.

PCR

A 2-μl aliquot of the RT reaction is mixed with 48 μl of PCR master mix containing 0.4 μl*M* CRB-1 and CRB-2 primers, 50 μ*M* each dNTP, 2.5 units of *Taq* polymerase (Perkin-Elmer, Norwalk, CT), and 1× PCR buffer (10 m*M* Tris-HCl, pH 8.3, 50 m*M* KCl, 2.5 m*M* $MgCl_2$, 0.01% (w/v) gelatin). PCR is performed in the GeneAmp 9600 PCR system (Perkin-Elmer) for 30 cycles. The first 2 cycles consisted of a denaturation step at 94° for 1 min, an annealing step at 62° for 2 min and an elongation step at 72° for 3 min. The temperature and duration for the remaining 28 cycles were 94° × 30 sec, 62° × 30 sec, and 72° × 30 sec.

Quantitative Analysis

A 10-μl aliquot of PCR product is loaded onto an 8% nondenaturing polyacrylamide gel and subjected to electrophoresis. The ethidium bromide-stained gel is photographed with Type 55 Polaroid positive/negative film. The amplified DNA bands are quantitated using a Bio Image analyzer (Millipore, Bedford, MA). The ratio of the IODs obtained from the target and internal control DNA bands is plotted on a logarithm scale against the amount of total RNA present in the RT reaction. The amount of human CRABP-II mRNA in a test sample is given by the *x* ordinate at that point on the plot where the *y* ordinate equals 1.

Validation of Quantitative RT-PCR for CRABP-II

Quantitation of CRABP-II mRNA from Human Skin Fibroblast Cells

To confirm the performance of the system, we used both quantitative RT-PCR and Northern blot analysis to examine the induction of CRABP-II mRNA in cultured human skin fibroblasts treated with RA. Total RNA, purified from cells treated with vehicle alone (ethanol) or vehicle containing 10^{-6} *M* RA, was serially diluted and subjected to the RT-PCR method just described. Two DNA bands of 256 and 214 bp were observed after PAGE, consistent with the predicted sizes of target and internal control RT-PCR products, respectively (Figs. 1A and 1B).

In the vehicle-treated samples, the intensity of the two bands was almost identical (i.e., equivalence point reached) in lane 5 (Fig. 1A, upper panel) suggesting that the amount of CRABP-II mRNA in 0.8 ng of total RNA (tube 5) prior to RT-PCR was almost equal to the amount of internal control CRABP-D2 RNA (0.02 attomol). This corresponds to a CRABP-II mRNA concentration of 25 amol/μg total RNA. In RA-treated samples, the equivalence point is reached in lane 3 (Fig. 1A, lower panel), corresponding to a CRABP-II mRNA concentration of 100 amol/μg.

Figure 1B illustrates results obtained following quantitation of band intensity using a Millipore Bio Image analyzer. The ratio of the IODs of target to control bands was plotted against the amount of total RNA. Consistent with the quantitative nature of the assay, the plotted relationship was linear on a log–log scale over at least a 100-fold range. The exact equivalence point can be obtained by interpolation, yielding values of 0.20 and 0.84 ng total RNA containing 0.02 attomol CRABP-II mRNA for RA- and vehicle-treated cells, respectively. This corresponds to actual concentrations of 100 and 24 amol/μg total RNA, or a 4.2-fold increase in CRABP-II mRNA after RA treatment.

To compare results obtained from PCR with a more conventional analysis, the same RNA preparations were used for Northern blot analysis. The 20-μg aliquots of total RNA were electrophoresed on an agarose gel under denaturing conditions and transferred to nylon membrane. This blot was sequentially hybridized with cDNA probes to CRABP-II and human glycerol-3-phosphate dehydrogenase (G3PDH, control). Following quantitation of autoradiographic images on the Bio Image analyzer, CRABP-II mRNA concentration (normalized to G3PDH) was found to be induced 4.4-fold by RA treatment, in agreement with the result of quantitative PCR (data not shown).

Application to Human Skin Biopsies

Six volunteers were treated topically with 0.1% (w/w) RA cream, 2% (w/w) sodium lauryl sulfate (SLS) cream, or control treatments for 16 hr.[17,32] Total RNA isolated from keratome biopsies of treated skin (Dr. J. T. Elder, University of Michigan, Ann Arbor, MI) was analyzed by quantitative PCR as described earlier. Results from the six subjects are shown in Table I. The base level of CRABP-II mRNA in untreated skin

[32] A. Tavakkol, C. E. Griffiths, K. M. Keane, R. D. Palmer, and J. J. Voorhees, *J. Invest. Dermatol.* **99,** 146 (1992).

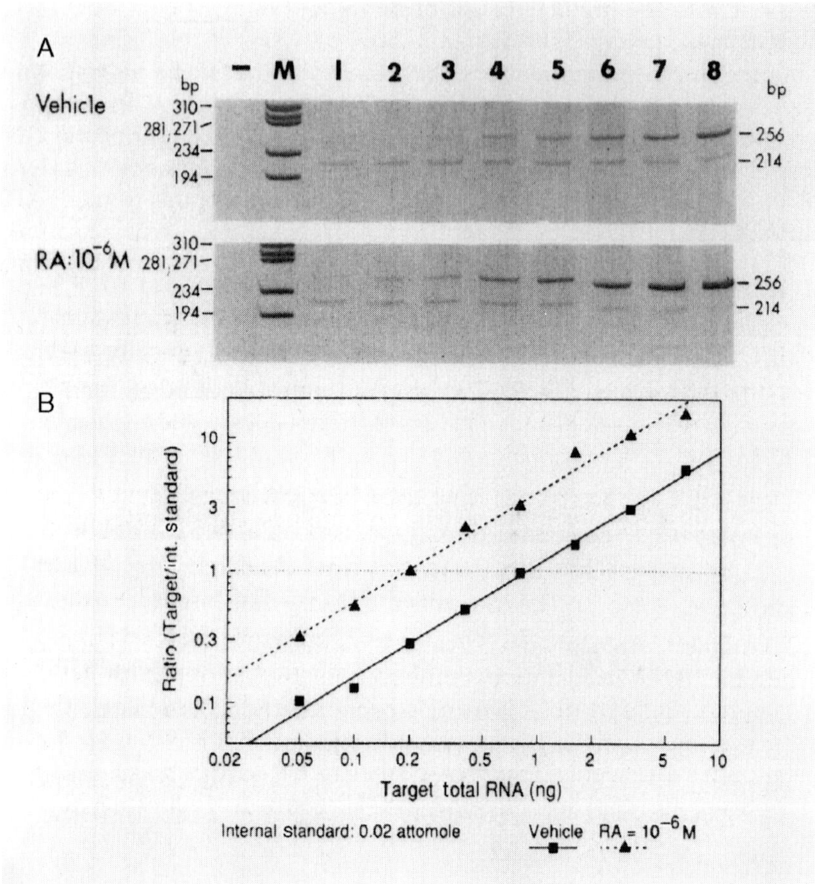

FIG. 1. Quantitation of CRABP-II RNA from human skin fibroblast cells. (A) Ethidium bromide staining of PC products separated in 8% polyacrylamide gel. *Upper:* Cells treated with vehicle (ethanol) for 24 hours; *lower:* cells treated with RA (10^{-6} *M*) for 24 hours. Note that only one-tenth of the RT reaction was used for PCR amplification so that the corresponding total RNA (ng) in lanes 1 to 8 is 0.05, 0.1, 0.2, 0.4, 0.8, 1.6, 3.2, and 6.4, respectively. The internal standard in each amplification reaction is equivalent to 0.02 attomole RNA. Lane M is DNA marker and lane – a negative control. The 256- and 214-bp PCR products are indicated to the right of the gel. (B) The DNA bands were quantitated using the Bio Image system and the ratio of the IODs obtained was plotted on a log–log scale against the amount of total target RNA. The IOD obtained from the 214-bp band was multiplied by 256/214 before plotting to correct for differences in ethidium bromide staining due to the difference in molecular weight. (Reproduced from Zhou *et al.*,[26] by permission of Oxford University Press.)

TABLE I
QUANTITATION OF CRABP-II mRNA FROM HUMAN SKIN BIOPSIES[a]

Treatment	Exp. 1	Exp. 2	Average
Subject 1			
Control	24	17	21
Vehicle	29	17	23
SLS[b]	32	18	25
RA[b]	290	240	270
Subject 2			
Control	15	17	16
Vehicle	25	26	26
SLS	22	32	27
RA	170	150	160
Subject 3			
Control	25	25	25
Vehicle	50	56	53
SLS	83	91	87
RA	180	210	195
Subject 4			
Control	20	18	19
Vehicle	41	32	37
SLS	18	14	16
RA	200	181	191
Subject 5			
Control	12	11	12
Vehicle	44	31	38
SLS	29	25	27
RA	118	111	115
Subject 6			
Control	15	17	16
Vehicle	13	13	13
SLS	13	12	13
RA	118	114	116

[a] Attomol/μg RNA.
[b] SLS, Sodium lauryl sulfate; RA, retinoic acid.

(designated "Control") ranged from 11.5 to 25 amol/μg total RNA. There was a slight increase in CRABP-II mRNA level in the samples treated with vehicle and some SLS-treated samples. Treatment with 0.1% RA cream induced statistically significant increases in CRABP-II mRNA whether compared to "Control" (7.3- to 12.8-fold), to "Vehicle" (3.1- to 11.7-fold), or to SLS (4.0- to 12-fold) treatments, respectively.

Application to a Mouse Model

The human skin-grafted nude mouse is an excellent model for dermatological studies. However, the use of such grafts to evaluate the potential clinical efficacy of pharmacological compounds has been limited by the small size of grafts and a lack of quantitative measures of skin responsiveness. Application of quantitative RT-PCR for CRABP-II overcomes this problem for compounds with retinoid activity.

The preparation of the mouse model has been described.[29] RNA was prepared from excised grafts previously treated for 4 days with topical RA. Human RNA from such preparations must be specifically quantitated to account for possible contamination with mouse RNA. Thus, 1-μl aliquots of sample RNA were reverse transcribed as described above but using a human G3PDH-specific primer and 5 μCi of [^{32}P]dCTP. Five microliters of the resulting product was added to 300 μl of cold 5% trichloroacetic acid (w/v). Precipitated cDNA was collected on a Whatman (Clifton, NJ) 934 AH glass fiber filter, washed extensively with cold 95% ethanol, and the amount of incorporated radioactivity determined by liquid scintillation counting. All reactions were carried out in triplicate, averaged, and the concentration of human RNA was calculated relative to a control RNA sample from nongrafted human skin.

For quantitative RT-PCR, a human CRABP-II-specific primer, CRB-4, was used in the RT reaction and CRB-1 and CRB-2 primers were used in the PCR step. Figure 2 shows the results of quantitative PCR from 21 skin graft samples. The 4.4-fold induction of CRABP-II mRNA in 0.1% RA-treated grafts relative to vehicle-treated grafts, while lower than that found in the keratome biopsies described earlier, falls within the range reported elsewhere in the literature.[33]

General Applications

The quantitative PCR technique described is an extremely sensitive method, capable of detecting very minute changes in CRABP-II level with relatively little starting material. Compared to the more conventional techniques of Northern blot hybridization or RNase protection, the sample size requirement can be reduced from 10–20 μg of total RNA to <200 ng. Some potential applications of this technology are discussed next.

[33] S. Hirschel-Scholz, G. Siegenthaler, and J.-H. Saurat, *Eur. J. Clin. Invest.* **19,** 220 (1989).

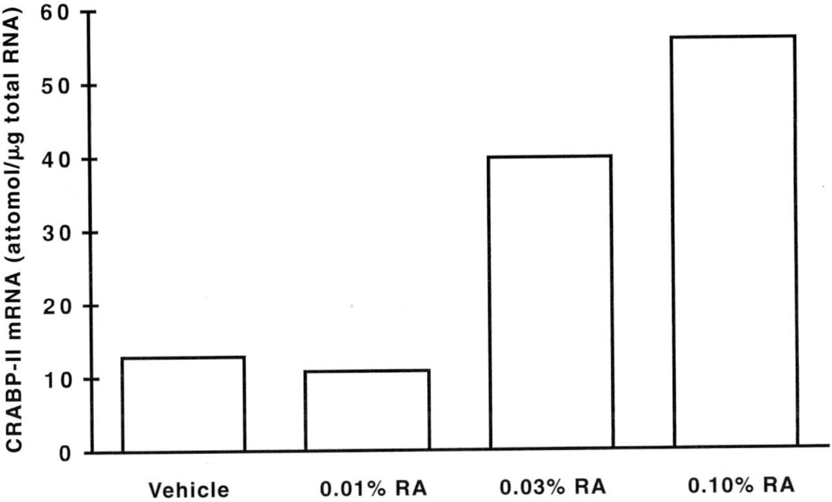

FIG. 2. Levels of CRABP-II RNA in grafted skin. Skin grafts were treated daily for 4 days with vehicle alone or with RA at the indicated concentrations. Grafts were excised following treatment and human CRABP-II RNA was quantitated within the treated graft tissue using the RT-PCR technique. CRABP-II RNA levels were significantly higher in grafts treated with 0.03% RA and 0.1% RA ($p < 0.05$ versus vehicle, one-sided Dunnett's test).

Evaluation of Pharmacological Activities of Retinoid Analogs or Retinoid Mimetics

Topical treatment with retinoids is reported to be effective in the treatment of both acne[34] and photoaging.[35,36] While the treatment is very effective, erythema and scaling have been reported as side effects.[36,37] New retinoid analogs that can treat acne and repair photodamaged skin without causing irritation will be ideal second-generation drugs. To identify active leads, several pharmaceutical and biotechnology companies are operating vigorous synthetic programs for novel retinoid analogs. One of the best ways to demonstrate that a novel chemical structure possesses retinoid activity is to evaluate its ability to induce CRABP-II both *in vitro* and *in vivo.* The quantitative RT-PCR method as described in this chapter provides a sensitive assay for induction of a selective marker for cutaneous retinoid activity *in vivo.* Future clinical trials of retinoid compounds can now make

[34] A. M. Kligman, J. E. Fulton, and G. Plewig, *Arch. Dermatol.* **99,** 469 (1969).

[35] A. M. Kligman, G. L. Grove, R. Hirose, and J. J. Leyden, *Am. Acad. Dermatol.* **15,** 836 (1986).

[36] E. A. Olsen, H. I. Katz, N. Levine, J. Shupack, M. M. Billys, S. Prawer, J. Gold, M. Stiller, L. Lufrano, and E. G. Thorne, *J. Am. Acad. Dermatol.* **26,** 215 (1992).

[37] L. H. Kligman, C. H. Duo, and A. M. Kligman, *Connect. Tissue Res.* **12,** 139 (1984).

use of punch biopsy skin specimens in place of keratome slices. Similarly, preclinical studies can make use of 1-cm^2 human skin grafts maintained on nude mice as described earlier. Such grafts mimic the responsiveness of human skin to pharmacological treatment much better than cultured cells or dermal equivalents.[29]

Diagnosis of Various Stages of Psoriatic Skin

When the relative expression levels of CRABP-I and CRABP-II were evaluated in psoriatic skin, it was found that CRABP-II mRNA levels were sixfold higher in lesional and twofold higher in nonlesional psoriatic skin as compared to normal skin.[38] The levels of CRABP-I, however, were decreased in both lesional and nonlesional psoriatic skin. A switch to high levels of CRABP-II may be relevant to the pathophysiology of the disease. The mechanism of action by which RA exerts a therapeutic effect on psoriasis is unknown. The ability to measure levels of CRABP-II and other RA-responsive genes in small biopsy samples may aid in the elucidation of the pathophysiology and therapeutic responses of this condition.

Analysis of Drug Interactions

Prolonged topical use of steroids results in negative side effects such as skin thinning, increased transparency, and telangiectasia. Preclinical studies have shown that these atrophogenic effects can be counteracted by retinoids without lessening the anti-inflammatory effect of the steroid treatment.[39] The best surrogate marker for this apparent opposing pharmacological outcome is CRABP-II.[40] Just as retinoids strongly induce CRABP-II expression in human skin, application of the potent topical steroid triamcinolone acetonide strongly suppresses its expression. Careful optimization of steroid and retinoid concentrations will be required in clinical trials to balance the opposing pharmacological actions of these drug classes and produce the most efficacious treatments for different inflammatory conditions. With the availability of quantitative PCR for CRABP-II as a sensitive and rapid assay, the molecular mechanism and pharmacological interaction of retinoid and steroid can likely be delineated.

[38] G. Siegenthaler, I. Tomatis, L. Didierjean, S. Jaconi, and J.-H. Saurat, *Dermatology* **185,** 251 (1992).

[39] R. H. Lesnik, J. A. Mezick, R. Capetola, and L. H. Kligman, *J. Am. Acad. Dermatol.* **21,** 186 (1989).

[40] P. Piletta, S. Jaconi, G. Siegenthaler, L. Didierjean, and J. H. Saurat, *Exp. Dermatol.* **3,** 23 (1994).

[7] Use of Transgenic Mice to Study Activation of Retinoic Acid-Responsive Promoters

By LESZEK WOJNOWSKI and ANDREAS ZIMMER

Introduction

Background

Retinoic acids (RAs) are vitamin A derivatives that play multiple roles during development and in the homeostasis of adult tissues. Retinoic acids have been implicated in the anterior–posterior patterning of the body axis and in limb bud development. Retinoic acids are teratogenic because exposure of the fetus to overdoses of RA leads to multiple developmental abnormalities, including neural tube defects, craniofacial deformities, and limb defects. Retinoic acids modulate transcription of RA-responsive genes via nuclear receptors, which function as ligand-dependent transcription factors and belong to the steroid/thyroid hormone receptor superfamily. Two subfamilies of RA receptors, RARs and RXRs, can be discriminated based on sequence homology and ligand-binding affinity. Each subfamily is composed of at least three unlinked genes (α, β, and γ), which in turn produce several different isoforms as a result of differential splicing and/ or promoter usage.[1–3]

Transgenic Approaches in Retinoic Acid Studies

Transgenic systems have been invaluable in studies of function of RAs. The spatial–temporal distribution of RAs within the embryo or a given tissue is difficult to assess since it involves laborious and tissue-disrupting analytical methods such as high-performance liquid chromatography (HPLC).[4] As an alternative, relative RA concentrations can be measured *in situ* using RA-inducible promoter sequences fused to a reporter gene. Conversely, transgenic approaches may be also used to identify novel RA-dependent genes.

Promoter Studies. Genes that are regulated by RAs contain RA-responsive elements (RAREs) in their promoter region. RAREs are tran-

[1] H. M. Sucov and R. M. Evans, *Mol. Neurobiol.* **10,** 169 (1995).

[2] D. Lohnes, M. Mark, C. Mendelsohn, P. Dolle, D. Decimo, M. LeMur, A. Dierich, P. Gorry, and P. Chambon, *J. Steroid Biochem. Mol. Biol.* **53,** 475 (1995).

[3] P. Kastner, M. Mark, and P. Chambon, *Cell* **83,** 859 (1995).

[4] C. Thaller and G. Eichele, *Nature* **327,** 625 (1987).

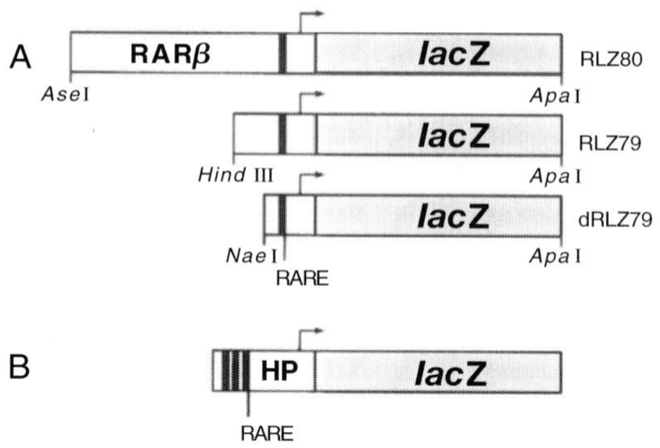

FIG. 1. (A) series of deletion constructs used to determine minimal promoter sequences of the retinoid acid receptor β2 (RARβ2).[7] Polyadenylation signal not shown. RARE, Retinoid acid-responsive element. (B) Schematic representation of a heterologous promoter (HP) construct. A few RAREs are cloned upstream of a heterologous promoter (e.g., thymidine kinase promoter or heat shock promoter). The RAREs confine the sensitivity to retinoic acid, which is reflected by the expression level of a reporter gene.

scriptional enhancers composed of a direct repeat of the GTTCAC motif with a 2–5 bp spacer. On exposure to RA, RARE enhances transcription of the adjacent gene.[5,6] This is mediated by RAR/RXR heterodimers that bind to the RARE. Transgenic animals can be engineered with RARE-containing promoter sequences controlling the expression of a reporter gene (Fig. 1). The RARE can be used in its natural context (*homologous promoters*),[7] or it can be used to enhance the expression of an unrelated, minimal promoter (*heterologous promoters*).[8,9] The first approach provides information about the expression of the gene from which the promoter is derived. The latter approach can provide information about relative tissue levels of RA. Changes in the expression pattern or level can be induced by exposing embryos to exogenous RAs (Fig. 2). This is accomplished by means of gavage of RA to pregnant animals (see later discussion).

In many cases, transgenic animal studies can be complemented by stable and transient transfection of cell lines, which are less expensive and easier to establish than mouse strains. Of course, one can also establish cell lines from transgenic mouse strains. For example, transgenic cell lines have been used to determine the spatial and temporal patterns of the RA production

[5] H. M. Sucov, K. K. Murakami, and R. M. Evans, *Proc. Natl. Acad. Sci. U.S.A.* **87,** 5392 (1990).

[6] H. de The, M. M. Vivanco-Ruiz, P. Tiollais, H. Stunnenberg, and A. Dejean, *Nature* **343,** 177 (1990).

[7] K. Reynolds, E. Mezey, and A. Zimmer, *Mech. Dev.* **36,** 15 (1991).

[8] J. Rossant, R. Zirngibl, D. Cado, M. Shago, and V. Giguere, *Genes Dev.* **5,** 1333 (1991).

[9] W. Balkan, M. Colbert, C. Bock, and E. Linney, *Proc. Natl. Acad. Sci. U.S.A.* **89,** 3347 (1992).

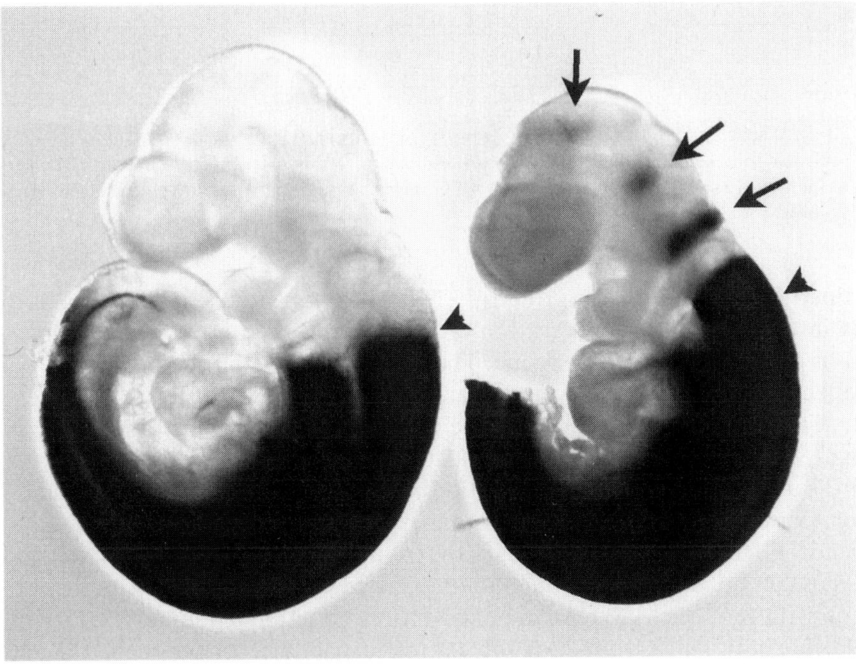

FIG. 2. Transgenic mouse embryos (9.5 p.c., construct RLZ79 from Fig. 1A) stained with X-Gal. Gavage of retinoic acid (embryo on right-hand side) extends rostrally the expression of the transgene in trunk (compare arrowheads) and induces its ectopic expression in the head region (arrows).

within embryos and tissues.[10] In this approach, RA-producing tissue is cocultured with a layer of transgenic cells. Diffusion of RA from the tissue induces local expression of the transgene in the monolayer, which can be detected by subsequent staining (see later discussion).

Gene Traps. Gene traps have been used to identify novel genes that are important for development.[11,12] In this strategy, pluripotent mouse embryonic stem (ES) cells are electroporated with a gene trap vector. The vector consists of a promoterless reporter gene (typically β-galactosidase), preceded by a splice-acceptor site. A selection marker (e.g., neomycin phosphotransferase) provides means to select clones derived from ES cells that have integrated the construct into their genomes. Alternatively, the β-galactosidase and neomycin phosphotransferase activities can be con-

[10] M. C. Colbert, E. Linney, and A. S. LaMantia, *Proc. Natl. Acad. Sci. U.S.A.* **90,** 6572 (1993).
[11] G. Friedrich and P. Soriano, *Genes Dev.* **2,** 1513 (1991).
[12] A. Gossler, A. L. Joyner, J. Rossant, and W. C. Skarnes, *Science* **244,** 463 (1989).

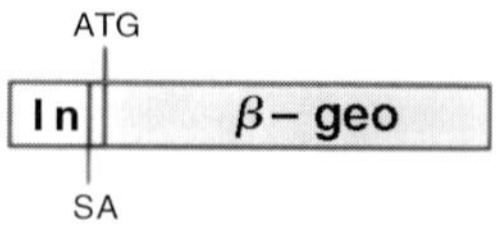

FIG. 3. Schematic representation of a gene trap construct. A promoterless cassette encoding a reporter gene/selection marker fusion protein is cloned downstream from intron sequences (In) including a splice-acceptor site (SA). ATG, Translational start site. Polyadenylation signal not shown.

tained in a single fusion protein[11] (Fig. 3). Integration of the vector into an intron leads to the generation of a fusion protein between the endogenous gene and the reporter gene. The expression level of the fusion protein reflects the activity of the promoter of the trapped gene.

The activity of the reporter gene can be visualized by histological staining. Stained clones contain β-galactosidase-tagged genes, which are expressed in the ES cells. A modification of this technique allows the investigator to identify a group of genes that is inducible by a particular soluble factor. For example, ES colonies with RA-inducible, β-galactosidase-tagged genes can be identified through their positive staining following an incubation with RA. Such clones can be used for production of chimeric embryos, which can be prescreened for the expression patterns of the tagged genes. For particularly interesting genes, the mutation can be transmitted through the germ line to establish mouse strains. Because gene trap vectors frequently act as an insertional mutagen, this strategy can also lead to a knockout mouse strain and enable the analysis of its phenotype. A detailed description of gene trap techniques can be found elsewhere in this series.[12a]

Procedures and Protocols

Constructs

Cloning of constructs for production of transgenic mice requires experience with basic techniques of molecular biology. Most important, before beginning a transgenic project, the gene of interest needs to be characterized in some detail. In most projects, the initial goal is to "replace" the 5′ coding sequences of the gene with a cassette containing a reporter gene. To accomplish this task, the investigator needs to establish the intron/exon structure of the 5′ portion of the gene and to determine the translational initiation site of the encoded protein. The possibility of alternative splicing should be also taken into consideration. Allo- or xenogeneic promoter

[12a] D. P. Hill and W. Wurst, *Methods Enzymol.* **225,** 664 (1993).

sequences of the RA-dependent gene of interest can be isolated from the respective genomic libraries. Fragments of different lengths can be used. In fact, engineering of a series of constructs with various deletions is advisable (Fig. 1). This allows the determination of "minimal promoter" sequences and the functional characterization of potential binding sites for *cis*-acting transcriptional regulators. While comparing the effect of such deletions, keep in mind that the expression levels and patterns of the reporter gene may vary considerably between transgenic strains. This variance is caused by positional effects of the transgene integration site (ectopic expression), as well as by differences in the number of transgene copies that are integrated into the genome.[13] It is therefore necessary to analyze several different transgenic strains generated with the same construct.

Promoter sequences are cloned upstream of the reporter gene to be used. The 5′ untranslated sequences and the translational start may be contained in the reporter gene cassette. Alternatively, homologous 5′ leader and translational start sites can be used. This strategy requires the cloning of an ATG-less reporter gene in frame with the translational start of the gene to be studied. This approach is usually more laborious but it has the advantage of a more physiologic transcriptional context.

Reporter Genes

The LacZ-encoded β-galactosidase[14] is an almost ideal reporter protein for studies of expression patterns in mouse embryos. This is due to (1) lack of side effects of the protein on embryonic development, (2) the autonomy of cellular expression that allows documentation of heterogeneity in cell response, (3) high sensitivity of detection and low background, and (4) the ease of embryo staining and of the quantitative measurement of the enzyme activity. β-Galactosidase cleaves the chromogenic substrate X-Gal (5-bromo-4-chloro-3-indolyl-β-D-galactopyranoside) to yield a precipitable product leading to a blue staining of the cytoplasm. The intracellular distribution of the stain can be restricted to the cell nucleus by fusing the LacZ gene to a nuclear localization signal.[15,16] This strategy makes it possible to combine X-Gal staining with other histological staining methods and immunochemistry.

To follow the spatial–temporal pattern of LacZ expression during embryogenesis, whole mouse embryos can be stained for β-gal activity until

[13] G. H. Swift, R. E. Hammer, R. J. MacDonald, and R. L. Brinster, *Cell* **38,** 639 (1984).

[14] A. Kalnins, K. Otto, U. Ruther, and B. Muller-Hill, *EMBO. J.* **2,** 593 (1983).

[15] D. Kalderon, B. L. Roberts, W. D. Richardson, and A. E. Smith, *Cell* **39,** 499 (1984).

[16] C. Bonnerot, D. Rocancourt, P. Briand, G. Grimber, and J. F. Nicolas, *Proc. Natl. Acad. Sci. U.S.A.* **84,** 6795 (1987).

E14. In older embryos and tissues of adult mice, diffusion of the X-Gal substrate may be incomplete. In this case it might be necessary to section the material prior to staining. Note also that many adult tissues (e.g., kidney, cerebellum) contain endogenous β-Gal activity, which can make the detection of the bacterial transgenic protein difficult. In such cases, a LacZ gene with a nuclear localization signal is extremely useful, as the endogenous β-galactosidase activity is mostly found in the cytoplasm.

Production of Transgenic Animals

Production of transgenic mice is an expensive and laborious procedure that requires substantial expertise in numerous manipulative and surgical techniques. A detailed description of the transgenic mice production, founder identification, breeding strategies, as well as a comprehensive discussion of these subjects, can be found in the Volume 225 of this series. It is recommended that the technique be established only when a long-term transgenic program is anticipated. Small-scale transgenic experiments can be accomplished at a much lower cost with the help of a commercial transgenic service. In both cases, the careful design of the constructs is perhaps the most important cost-saving measure.

Analysis of Transgenic Animals

Feeding of Pregnant Animals with Retinoic Acid by Gavage

1. Retinoid Acid is dissolved in corn or sesame oil and administered by gavage. The ideal volume is 300–500 μl, but up to 1 ml can be administered. Because RA is not very soluble in these oils, prepare a stock solution by dissolving RA in 95% (v/v) ethanol or dimethyl sulfoxide at a 10$\times$ concentration and then add one part of the stock solution to nine parts oil and mix by vortexing.
2. Load the suspension into a 1-ml syringe equipped with an animal feeding needle (22-gauge/25-mm length; Fig. 4A).
3. Grab the animal firmly by its neck and tail (Fig. 4B). The head needs to be fully extended.
4. Carefully insert the needle into the animal's mouth (Fig. 4C). The syringe should be parallel to the longitudinal axis of the animal. Do not use any force to insert the needle because this may lead to injuries. Slowly twist the needle until the animal swallows.
5. When the needle is completely inserted, inject the desired amount of suspension and subsequently withdraw the needle (Fig. 4D). If the gavage is done properly, the animal should not have much of the suspension in its mouth.

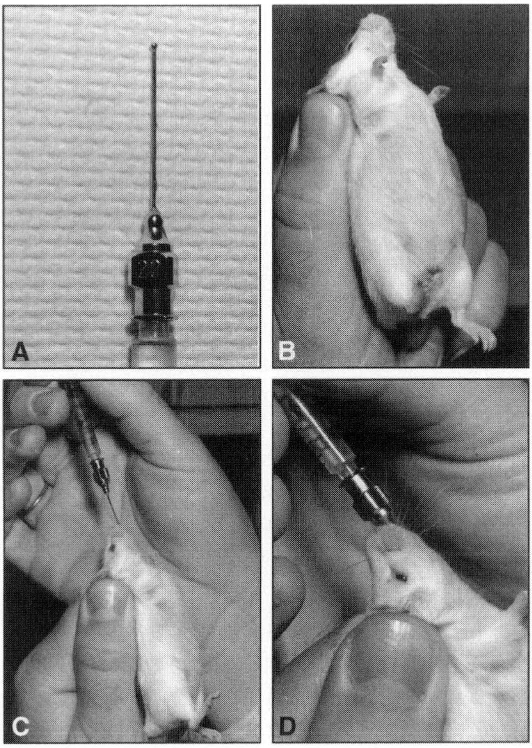

FIG. 4. Gavage in the mouse (see text for details).

Notes: The dosage is dependent on the gestation time and the RA isoform. Mouse embryos are RA-hypersensitive between embryonic days E7.0 and E8.0. Administration of 10 mg/kg all-*trans*-RA during this period is sufficient to induce severe developmental abnormalities. At later stages, the dose needs to be increased up to 100 mg/kg to be teratogenic. All-*trans*-retinoic acid is approximately five times as potent as 13-*cis*-RA.[17]

Staining of Embryos with X-Gal and Histological Analysis

1. Sacrifice a pregnant animal by means of CO_2 suffocation. Make an incision in the abdominal skin and musculature and withdraw the uterine horns. Liberate the embryos from the uterus by an incision of the uterus on the antimesometrial side and put them into cold phosphate-buffered saline (PBS).
2. Transfer the embryos into fixative (1% paraformaldehyde, 0.2% glu-

[17] M. Kessel and P. Gruss, *Cell* **67,** 89 (1991).

taraldehyde, 0.02% Nonidet P-40, in PBS). Incubate at 4° for 30 min (for embryos older than E11.5 increase the time to 60 min). Wash twice in PBS for 20 min.

3. Transfer the embryos into staining solution (1 mg/ml 5-bromo-4-chloro-3-indolyl-β-D-galactopyranoside [X-Gal], 5 m*M* $K_3Fe(CN)_6$ [potassium ferricyanide], 5 m*M* $K_4Fe(CN)_6 \cdot 3H_2O$ [potassium ferrocyanide], 2 m*M* $MgCl_2$, in PBS) and incubate at 30°. Depending on the expression levels, the incubation time varies between 2 to 16 h.
4. Stained embryos can be analyzed and photographed *in toto* or they can be sectioned and further analyzed by various histological methods. For cryosections, incubate the stained embryos overnight in cryoprotectant solution (3% polyethylene glycol (w/v), 20% sucrose (w/v), 0.1× PBS) at 4°. Completely equilibrated embryos will sink to the bottom of the solution.
5. Embed the embryos in OCT medium (Miles Inc., Elkhart, IN) and freeze in isopentane precooled on dry ice. Equilibrate to −20°, cut on a cryostat, and mount the sections on gelatinized glass slides. The sections can be used for histological and immunohistochemical analysis.

Notes: (1) Fetuses at E15 and older can be perfused to improve penetration of the fixative. For details, see Ref. 17a. (2) X-Gal stock solutions (40 mg/ml, in DMSO) should be stored at −20°. Potassium ferricyanide and potassium ferrocyanide stock solutions (0.5 *M*, in H_2O) should be protected from light and made fresh every 2 weeks. (3) Incubation time in step 3 can be extended up to 2 days. The proper pH of the staining solution (pH 7.3) and incubation temperature (≤30°) are important to avoid background due to the activity of endogenous galactosidase. (4) Cryostat sections can be restained with X-Gal. This is frequently necessary for embryos older than E11 where whole-mount staining is generally restricted to the periphery of the organs (100–200 μm in depth). (5) Stained embros can be stored almost indefinitely at room temperature in 4% paraformaldehyde (w/v)/10% methanol (v/v)/PBS.

Measurement of β-Galactosidase Activity in Tissue Extracts. The determination of β-galactosidase activity in tissue homogenates and cell lysates utilizes the chromogenic substrate for the enzyme CPRG (chlorophenol red-β-D-galactopyranoside). As CPRG is cleaved by β-galactosidase, a red product is generated. The amount of the product is proportional to the activity of the enzyme in the sample.[18]

[17a] C. Bonnerst and J.-F. Nicolas, *Methods Enzymol.* **225,** 451 (1993).

[18] D. C. Eustice, P. A. Feldman, A. M. Colberg-Poley, R. M. Buckery, and R. H. Neubauer, *BioTechniques* **11,** 739 (1991).

1a. *Cells.* Plate the cells into a 96-well, flat-bottomed microtiter plate and grow them to the desired density. Remove the medium and add 25 μl of PM2 buffer (33 mM $NaH_2PO_4 \cdot H_2O$, 66 mM $Na_2HPO_4 \cdot 7H_2O$, 2 mM $MgSO_4 \cdot 7H_2O$, 0.1 mM $MgCl_2 \cdot 4H_2O$, in H_2O; pH 8), diluted 1 : 10 with H_2O. Incubate for 10 min at room temperature. Add 100 μl of PM2 supplemented with 0.5% Triton X-100 and 40 mM 2-mercaptoethanol. Incubate for 10 min at room temperature.

1b. *Tissues.* Homogenize up to 0.5 g tissue in 1 ml of PM2. Incubate for 10 min at room temperature. Spin down cell debris in a microcentrifuge and transfer supernatant into a new tube. For each sample to be measured, transfer 50 and 100 μl of the supernatant into separate wells of a 96-well, flat-bottomed microtiter plate. Bring the volume in each well to 125 μl with PM2.

2. Set a colorimetric microplate reader to 570 nm. Start reaction by adding 25 μl of 5 mg/ml CPRG. Immediately start reading. Take readings through 1.5 OD. The OD/min value is proportional to the activity of the enzyme in a given sample.

Notes: (1) CPRG solution in PM2 has to be freshly made. For samples with very low activity of β-galactosidase, higher concentrations of CPRG (up to a final concentration of 5 mg/ml of the reaction mixture) can be used to improve the sensitivity of the assay. (2) To compare readings from separate experiments, a calibration curve with β-galactosidase (Sigma, St. Louis, MO) over the range of 0.1 to 100 mU should be included in each experiment.

[8] Use of Transgenic Mice to Eliminate Retinoic Acid Receptor Function in Specific Tissues

By MITINORI SAITOU, TOSHIHIRO TANAKA, and AKIRA KAKIZUKA

Introduction

Retinoic acid (RA) is a small lipophilic molecule and is a physiological metabolite of vitamin A (reviewed by Means and Gudas[1]). Exogenously applied RA has been shown to have a wide variety of actions on vertebrate development. These effects include induction of duplicated chicken limbs,

[1] A. L. Means and L. J. Gudas, *Annu. Rev. Biochem.* **64,** 201 (1995).

0076-6879/97 $25.00

transformations in the vertebrae, and aberrant morphogenesis of several organs. These observations have led to the prediction that endogenous RA is an important physiological regulator in vertebrate development. Therefore, the precise assignment of RA functions has long been awaited by developmental biologists.

The receptor for RA (RAR) has been identified and shown to belong to the members of the steroid/thyroid hormone receptor superfamily (reviewed by Evans[2] and Mangelsdorf and Evans[3]). Until now, three subtypes of RAR have been identified (RARα, β, and γ). However, significant functional differences among the three subtypes have not yet been identified. Accordingly, all RARs act as ligand-dependent transcription factors, which directly control the transcription of their target genes on ligand activation. These findings again emphasize the physiological importance of RA and have stimulated investigators to "knock out" the RAR genes to resolve long-unanswered questions regarding development.[4–10] Although tremendous efforts have been made to generate mice lacking RAR genes, either the functional redundancy of RARs or the embryonic lethality in the RAR-minus conditions have acted as obstacles to the clear assignment of RA functions in mammalian development as well as in adult mice.

We have chosen an alternative strategy: the targeted expression of a dominant-negative RAR. This strategy has the theoretical advantages that the dominant-negative receptor could simultaneously eliminate the functions of all RAR subtypes and that the usage of tissue- and/or stage-specific promoters could avoid early embryonic lethality. Based on this idea, we have generated a dominant-negative RAR[11] and expressed the mutant receptor by tissue-specific promoters.[12] In this chapter, using our experi-

[2] R. M. Evans, *Science* **240,** 889 (1988).

[3] D. J. Mangelsdorf and R. M. Evans, *Cell* **83,** 841 (1995).

[4] E. Li, H. M. Sucov, K.-F. Lee, R. M. Evans, and R. Jaenisch, *Proc. Natl. Acad. Sci. U.S.A.* **90,** 1590 (1993).

[5] T. Lufkin, D. Lohnes, M. Mark, A. Dierich, P. Gorry, M.-P. Gaub, M. LeMeur, and P. Chambon, *Proc. Natl. Acad. Sci. U.S.A.* **90,** 7225 (1993).

[6] D. Lohnes, P. Kastner, A. Dierich, M. Mark, M. LeMeur, and P. Chambon, *Cell* **73,** 643 (1993).

[7] D. Lohnes, M. Mark, C. Mendelsohn, P. Dollé, A. Dierich, P. Gorry, A. Gansmuller, and P. Chambon, *Development* **120,** 2723 (1994).

[8] C. Mendelsohn, D. Lohnes, D. Décimo, T. Lufkin, M. LeMeur, P. Chambon, and M. Mark, *Development* **120,** 2749 (1994).

[9] C. Mendelsohn, M. Mark, P. Dollé, A. Dierich, M.-P. Gaub, A. Krust, C. Lampron, and P. Chambon, *Dev. Biol.* **166,** 246 (1994).

[10] J. Luo, P. Pasceri, R. A. Conlon, J. Rossant, and V. Giguère, *Mech. Develop.* **53,** 61 (1995).

[11] M. Saitou, S. Narumiya, and A. Kakizuka, *J. Biol. Chem.* **269,** 19101 (1994).

[12] M. Saitou, S. Sugai, T. Tanaka, K. Shimouchi, E. Fuchs, S. Narumiya, and A. Kakizuka, *Nature* **374,** 159 (1995).

ments as examples, we discuss a strategy in which the physiological functions of RA in a specific tissue can be studied in transgenic mice. This strategy can complement the limitation of the gene knockout technique and is theoretically available to every organ, thus opening the way to define the precise roles of RA during development as well as in the adult.

Structures and Action Mechanisms of Retinoic Acid Receptors

The steroid/thyroid hormone receptor superfamily has been shown to consist of several structurally and functionally distinct domains, two of which are essential for receptor functions. The DNA-binding and ligand-binding domains of the receptor determine target gene specificity and ligand specificity, respectively. On ligand activation, the receptors act as dimeric transcription factors to regulate gene expression through binding to specific DNA sequences referred to as hormone response elements (HREs). Investigations have classified the superfamily into glucocorticoid receptor (GR) and thyroid hormone receptor (TR)/RAR subfamilies. The former makes homodimeric receptors and recognizes palindromic HREs, whereas the latter forms a heterodimeric complex and binds to a direct repeat of AGGTCA with differing lengths of spacing. The latter subfamily includes TR, RAR, vitamin D_3 receptor (VDR), and retinoid X receptor (RXR). All the members of the TR/RAR subfamily share RXR as a dimeric partner. Accordingly, an additional functional domain has been identified in the ligand-binding domain of the receptors for the dimer formation. The core amino acid sequence of the dimeric interface has been proposed as eight amino acids, forming an α-helix structure similar to a leucine zipper.[13]

Human Disease Resulting from Dominant-Negative Nuclear Receptor

Among the members of the TR/RAR subfamily, the RARs are most similar in amino acid sequence to the TRs. Furthermore, both RARs and TRs (α, β) consist of multiple subtypes, which are encoded in different genes on different chromosomes. Following these molecular findings on the nuclear receptor superfamily, the molecular basis of the syndromes of hormone resistance has been elucidated. Among them, generalized thyroid hormone resistance (GTHR)[14] is of much interest because the disease shows autosomal dominant inheritance, and thus proposal has been made that the dominant-negative TRs created by mutations on the TR genes. To date,

[13] M. Leid, P. Kastner, R. Lyons, H. Nakshatri, M. Saunders, T. Zacharewski, J.-Y. Chen, A. Staub, J.-M. Garnier, S. Mader, and P. Chambon, *Cell* **68,** 377 (1992).

[14] S. Refetoff, R. E. Weiss, and S. J. Usala, *Endocrine Rev.* **14,** 348 (1993).

more than 20 independent families with GTHR have been elucidated at the molecular level. In every case, a single amino acid deletion or substitution has been identified in the two clustered regions of the ligand-binding domain of TRβ. Interestingly, the mutations with GTHR have occurred on the amino acids, which are mostly conserved among all isoforms of TRs and RARs as well as v-*erbA,* a viral counterpart of TR,[11] and these mutations exclude the dimeric interface (Fig. 1). Note also that no mutation has been found in the DNA-binding domain of TR.

Generation of Dominant-Negative RARs

Given that the mutated amino acids are functionally similar between TR and RAR, it might be possible to transfer the dominant-negative phenotype to RAR by the same single amino acid alteration. To address this

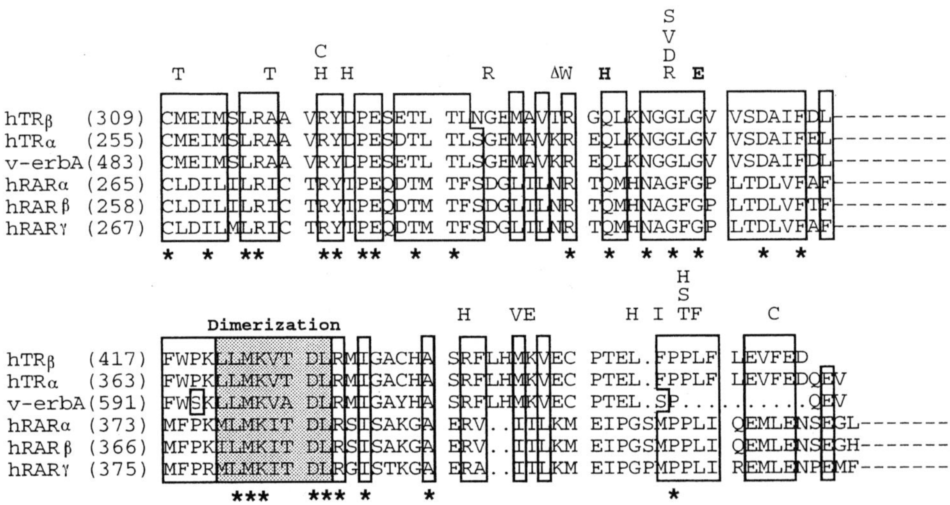

FIG. 1. Alignment of the amino acid mutations that cause thyroid hormone resistance. Partial amino acid sequences for the ligand-binding domains of TRs, v-*erbA,* and RARs are aligned. The similar amino acids conserved among all the isoforms of human TRs (α, β) and human RARs (α, β, γ), as well as v-*erbA* are boxed. Asterisks denote the amino acids identical in all the isoforms of TRs and RARs, as well as in v-*erbA.* Dots represent gaps in amino acid alignment. Dashes indicate continuous amino acids following the highlighted region. The numbers represent the positions of the first amino acids in the alignment. The mutated amino acids and deletions (displayed as Δ) responsible for thyroid hormone resistance are indicated at corresponding positions above the TRβ sequence. The two amino acid mutations used in this study are denoted in bold letters (**H** and **E**). A putative dimeric interface is shown by a shaded box.

possibility, we constructed two mutant RARs with corresponding mutations to mutated TRβs in GTHR.[11] One, RARαG303E (referred to as RAR-E), has a glycine to glutamic acid change at amino acid 303, which corresponds to TRβG347E. The other, RARαQ296H (referred to as RAR-H), has a glutamine to histidine change at amino acid 296, which corresponds to TRβQ340H. These amino acid changes were prepared by conventional site-directed mutagenesis[15,16] on the cDNA encoding RARα.

To assess the transcriptional activity of the nuclear receptor, a convenient transfection assay has become available.[2,17,18] In this assay, a cultured cell line is cotransfected with an expression vector and a reporter. The expression vector provides for the receptor protein and the reporter for an easily measurable enzyme, for example, luciferase (LUC) or chloramphenicol acetyltransferase (CAT), whose gene is placed under the control of the HRE recognized by the receptor. The application of the agonist activates the transcription of the reporter gene, resulting in the increment of the enzyme activity. Therefore, the transcriptional activity of the receptor is easily estimated by measuring the enzyme activity.

Using this assay system, we analyze the activities of the mutant RARs.[11] HepG2 cells, a well-used cell line for transient assays of the RAR activity, are transfected by the wild-type RARα expression plasmid (pCMX-hRARα) with an equal amount of plasmid for either RAR-E (pCMX-RAR-E) or RAR-H (pCMX-RAR-H). The cells are cotransfected with a luciferase reporter [TK-(βRE)$_3$-LUC] containing three copies of a retinoic acid response element (βRE) upstream of the herpes simplex virus thymidine kinase promoter. The transcriptional activities of RARs are measured by firefly luciferase activity of the cell extracts after treatment of RA in various concentrations. The LUC activities are normalized by the expression level of cotransfected β-galactosidase. These transfections are generally carried out by the calcium phosphate method[17–20] or liposome-mediated gene transfer method.[21]

The wild-type RARα shows a dose-dependent increase in the transcription activity, up to 100-fold stimulation, at 10^{-6} *M* RA (Fig. 2a). However, when the RAR-E expression vector is cotransfected with that of wild-type

[15] T. A. Kunkel, J. D. Roberts, and R. A. Zakour, *Methods Enzymol.* **154,** 367 (1987).
[16] A. Kakizuka, T. Ingi, T. Murai, and S. Nakanishi, *J. Biol. Chem.* **265,** 10102 (1990).
[17] K. Umesono, K. K. Murakami, C. C. Thompson, and R. M. Evans, *Cell* **65,** 1255 (1991).
[18] A. Kakizuka, W. H. Miller, Jr., K. Umesono, R. P. Warrell, Jr., S. R. Frankel, V. V. V. S. Murty, E. Dmitrovsky, and R. M. Evans, *Cell* **66,** 663 (1991).
[19] F. L. Graham and A. J. van der Eb, *Virology* **52,** 456 (1973).
[20] A. Kakizuka, N. Kitamura, and S. Nakanishi, *J. Biol. Chem.* **263,** 3884 (1988).
[21] H. Ikeda, M. Yamaguchi, S. Sugai, Y. Aze, S. Narumiya, and A. Kakizuka, *Nature Genet.* **13,** 196 (1996).

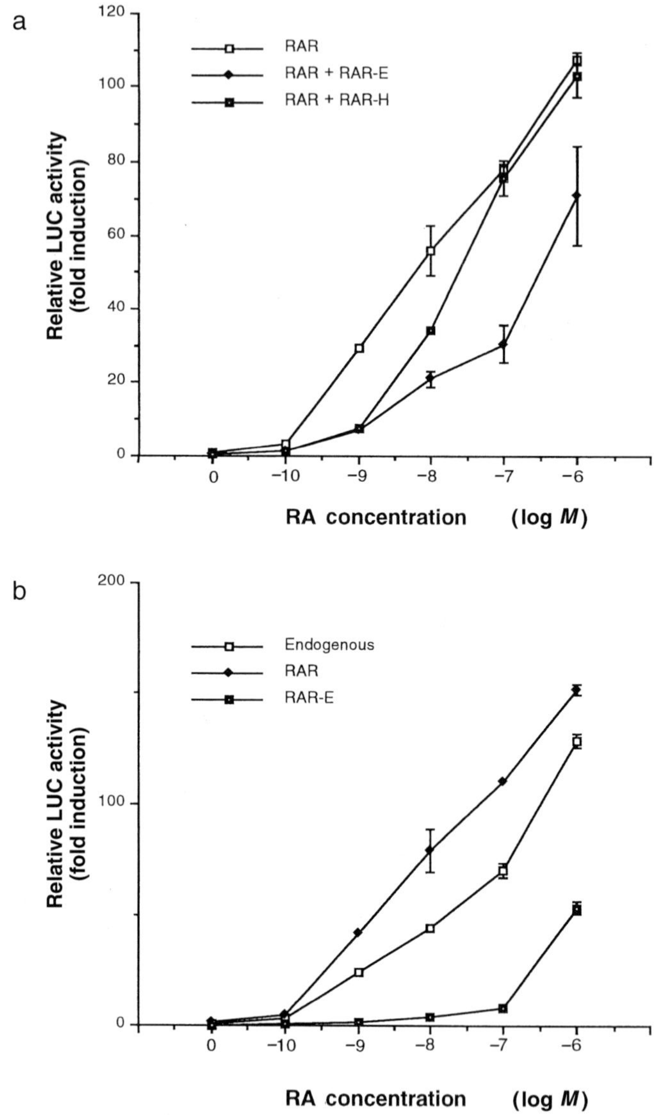

FIG. 2. Dominant-negative activities of RAR-E and RAR-H. (a) Dominant-negative activities of RAR-E and RAR-H on the wild-type RARα at various concentrations of RA. The mean activities of duplicated experiments are presented as fold inductions where the activity of wild-type RARα on the reporter gene in the absence of hormone was chosen as the reference value. (b) Dominant-negative activity of RAR-E on the endogenous RARs at various concentrations of retinoic acid. The endogenous RAR activity in the absence of hormone was chosen as the reference value. In (a) and (b), error bars are also included. Where not shown, error bars are smaller than the symbols.

RARα (1:1 ratio), the transcription activity of the wild-type RARα is significantly suppressed at all concentrations of RA examined, from 10^{-10} to 10^{-6} *M* (Fig. 2a). In contrast, RAR-H suppresses the transcriptional activity of wild-type RARα at only the physiological concentration of RA (from 10^{-10} to 10^{-8} *M*), but has no clear effect at pharmacological concentrations of RA (from 10^{-7} to 10^{-6} *M*) (Fig. 2a). In each experiment, both RAR-E and RAR-H clearly demonstrate the dominant-negative phenotype at the physiological concentration of RA on wild-type RARα.[11]

Because HepG2 cells have strong endogenous RAR activities, we next examine the transcriptional suppressive activities of the mutated RAR on the endogenous RARs. Indeed, RAR-E greatly suppresses the endogenous RAR activities (Fig. 2b). Because the amount of transfected RARs is expected to be much higher than the endogenous RARs, these results indicate that the dominant-negative activity of RAR-E is further strengthened in an excess condition against wild-type receptors. In contrast, RAR-H exhibits strengthened suppressive activities on the endogenous RAR only at the physiological concentration of RA. Additional titration experiments reveal that the suppressive effects reach a saturated level at more than 10-fold excess of the expression plasmid for the mutant receptor compared to that for the wild-type RAR, which is similar to the suppressive effects of RAR-E on the endogenous RARs.[11] These results indicate that more than 10-fold excess of the mutant receptor is expected to be expressed for the complete suppression of the endogenous wild-type RARs activities, and that the dominant-negative phenotype is functionally transferred to RAR by the single amino acid substitution so that RAR-E can function as a stronger dominant-negative RAR than RAR-H. These results also indicate that other mutations identified in GTHR can produce other dominant-negative RARs with different strengths of suppression.

Molecular Basis of Dominant-Negative Phenotype

How might the single amino acid substitution cause the dominant-negative phenotype? The exclusive presence of the mutations in the ligand-binding domain led us to speculate that the mutated receptors have reduced ligand-binding affinities. Our RA-binding experiments have demonstrated that RAR-E has a severely reduced affinity for RA,[11] which is consistent with the observation that many of the mutated TRs in GTHR show reduced affinities for the ligand. As mentioned earlier, dominant-negative receptors always contain intact dimeric interfaces and DNA-binding domains, raising the possibility that the dominant-negative phenotype requires both an intact dimeric interface and a DNA-binding domain. With additional experiments, we have shown that both intact dimeric and DNA-binding activities are indeed indispensable for the dominant-negative activities of the mutated receptors.[11]

Totally, one mechanistic explanation for the dominant-negative phenotype would be the following: the mutant RARs readily heterodimerize with RXR and make stable complexes on their responsive elements, but the mutant receptor complexes cannot respond to the ligand efficiently; this leads not only to the failure of the transactivation of the target gene but also to the interference by the mutant receptor complexes with the access of the wild-type receptors to the HREs (Fig. 3). This model has been further supported by the identification of the corepressors that suppress the transcriptional activities of the nonliganded RAR/RXR heterodimers.[3]

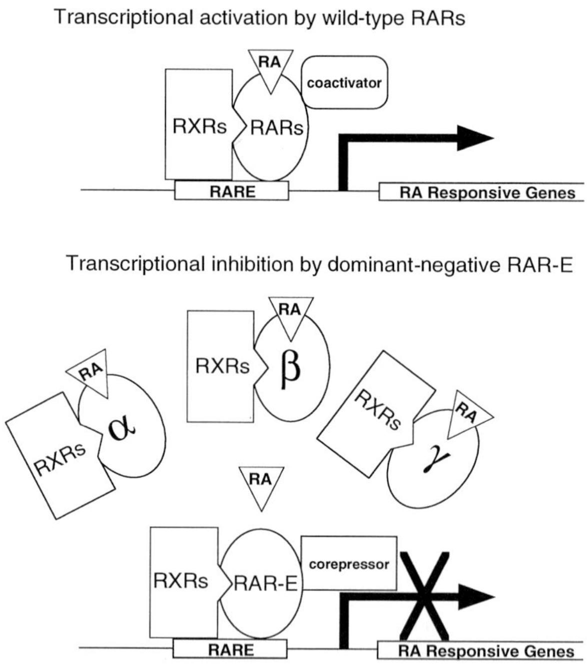

FIG. 3. A model for the dominant-negative action of RAR-E. Wild-type RAR-RXR activates the target gene transcription by binding to a RA responsive element through RA-dependent interaction with its coactivator (upper panel). In contrast, the ligand-insensitive RAR-E-RXR heterodimer makes a stable complex on an RA responsive element with a corepressor, leading not only to failure of the target gene activation but also to interference with the access of the wild-type receptor complex to the response element, thus exhibiting the dominant-negative activity (lower panel). α, RARα; β, RARβ; γ, RARγ; RARE, RA responsive element.

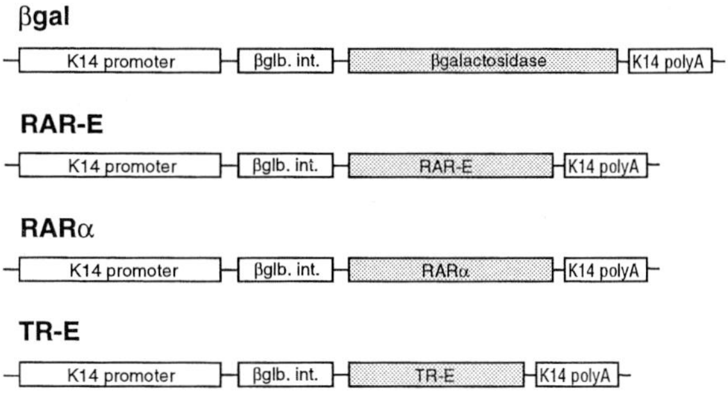

FIG. 4. Schematic structure of each transgene is presented. Open and shaded boxes represent gene or cDNA fragments common and specific in each transgene, respectively.

Targeted Expression of Dominant-Negative Retinoic Acid Receptor in Specific Tissues

Now, we have the dominant-negative RAR, which will serve as a powerful molecular tool to elucidate the physiological roles of RA; it would be possible to investigate organ-specific physiological functions of RA during embryogenesis and/or in the adult by blocking RAR-mediated signaling. This method has a theoretical advantage over gene targeting in cases like RARs where the genes consist of multiple functionally equivalent molecules or the loss of gene function leads to embryonic death. We decided to express the dominant-negative RAR (RAR-E) in the epidermis as the first experiment,[12] since the epidermis has long been considered a target organ of RA. Furthermore, we expected that any abnormal phenotypes in the epidermis would not be embryonically lethal.

The schematics of the injected transgenes are shown in Fig. 4. We use the K14 promoter, which has been reported to allow the various exogenous genes to be expressed specifically at the basal cell layer of the epidermis.[22,23] For efficient expression, the transgenes are designed to contain intronic sequences and polyadenylation sites. In this experiment, a rabbit β-globin intron and the K14 polyadenylation site are used.[12] Total length of the vector is ideally less than 10 kbp for the efficient production of the transgenic animal.[24] The DNA covering the entire transgene is separated from the plasmid portion by agarose gel electrophoresis, recovered with NA-45 mem-

[22] R. Vassar, M. Rosenberg, S. Ross, A. Tyner, and E. Fuchs, *Proc. Natl. Acad. Sci. U.S.A.* **86,** 1563 (1989).

[23] C. Byrne, M. Tainsky, and E. Fuchs, *Development* **120,** 2369 (1994).

[24] J. W. Gordon, *Methods Enzymol.* **225,** 747 (1993).

brane,[25] and suspended in TE at the concentration of 3.8 ng/μl. Female BDF1 mice are superovulated and mated to BDF1 males. The fertilized eggs are collected and their male pronuclei microinjected with the above DNA fragment by standard procedures.[24] To avoid cannibalization by the mothers,[6] we perform a cesarean operation at E19.5. For the estimation of the transgene copy number, genomic DNA is extracted and subjected to Southern blot analysis, and the signal of the transgene is compared with that of the endogenous gene.

We first express the β-galactosidase gene. The transgenic mice demonstrate clear X-Gal staining in the basal cell layer of the epidermis. No other tissues are stained. Generally, higher staining is observed in the mice with higher copies of the transgene.[12] However, some mice, even those with high copy numbers, fail to show any X-Gal staining, which is confirmed to be due to inactivation by the integrated locus.[12]

As mentioned earlier, all members of the TR/RAR subfamily share RXR as a heterodimeric partner. Therefore, we are concerned that overexpression of the dominant-negative RAR might interfere with other receptor-mediated pathways. To avoid misreading the results, we carefully set up two control experiments; we express the wild-type RARα and a dominant-negative TR (TR-E) in the same transgene context. The expression of the transgene is confirmed by Northern blot analysis in our case, which is of course replaceable by *in situ* hybridization, Western blotting, immunohistochemical analysis, etc.

The results of the targeted expression of the three constructs are summarized in Table I. All three constructions gave several transgenic mice with similar efficiency, excluding the possibility of the potential toxicity of a specific construction. Among the transgenic mice, eight offspring clearly exhibited dramatic suppression of epidermal development. Although the wild-type RAR transgene demonstrated comparable or even higher RNA levels to those of the RAR-E, wild-type RAR mice did not show any phenotypic changes.[12] This result precludes the possibility that overexpression of the mutated RAR makes the skin hypersensitive to RA and causes the phenotypic changes observed. More importantly, our results demonstrate that the phenotypic changes are completely dependent on the dominant-negative RAR but not the dominant-negative TR. Thus, we conclude that RAR-mediated pathways, and in fact RA, are absolutely necessary in normal skin development.[12]

Figure 5a demonstrates macrophenotypic changes of a severely affected animal.[12] The affected skin was tympanic and completely smooth. More-

[25] T. Maniatis, E. F. Fritsch, and J. Sambrook, *in* "Molecular Cloning: A Laboratory Manual," 2nd Ed. Cold Spring Harbor Laboratory Press, New York, 1989.

TABLE I
PHENOTYPIC CHANGES IN TRANSGENIC MICE

Gene	Copy number[a]			Phenotypes[b]			No. of examined offspring
					Affected		
	<10	10–20	>20	Normal	Moderate	Severe	
RAR-E	4	4	5	5	4	4	78
RARα	4	2	3	9	0	0	82
TR-E	1	3	2	6	0	0	56

[a] The copy number of each transgene was estimated by densitometric scanning of the autoradiographs from Southern or DNA dot blottings.
[b] When the average thickness of the epidermis was between 30 and 70% and less than 30% of the nontransgenic animal, we estimated the animals to be moderately and severely affected, respectively.

over, neither wrinkles nor hairs could be observed, although the animal had relatively normal whiskers. The affected skin seemed very thin and dried rapidly, and the animal died within a few hours probably due to dehydration. When a cesarean was performed, even the most severely affected animal could move and start breathing, which indicates that the other organs developed quite normally. This view is supported by the observation that no clear macrophenotypic changes were apparent in organs other than the epidermis. Figure 5b shows the skin sections with hematoxylin-eosin staining of the affected animal and that of the control.[12] The epidermis was severely affected and the thickness of the affected epidermis became one-fifth of that of the control. In particular, the formation of the horny and spinous layers, both of which are markers of the matured epidermis,[26] was greatly hindered. Furthermore, both the dermal–epidermal junction and the skin surface were completely flat, thus leading to loss of papillar structures and primary reliefs of wrinkles, respectively.[12] Notably, wrinkles and the spinous layer both become apparent after E15.5 in the normal back-skin epidermis.[23,26,27] The embryonic age of the affected epidermis may therefore correspond to that of around E15.5.

We next analyzed the expression of keratins, differentiation markers of the keratinocytes. Interestingly, we observed ectopic expression of K6 and K16 in the affected skin.[12] K6 and K16 have been observed in some pathological cases, for example, psoriasis and in cultured keratinocytes, and these keratins have been reported to be repressed by the administration

[26] J. Hanson, *J. Anat.* **81,** 174 (1947).
[27] H. Grüneberg, *J. Hered.* **34,** 89 (1943).

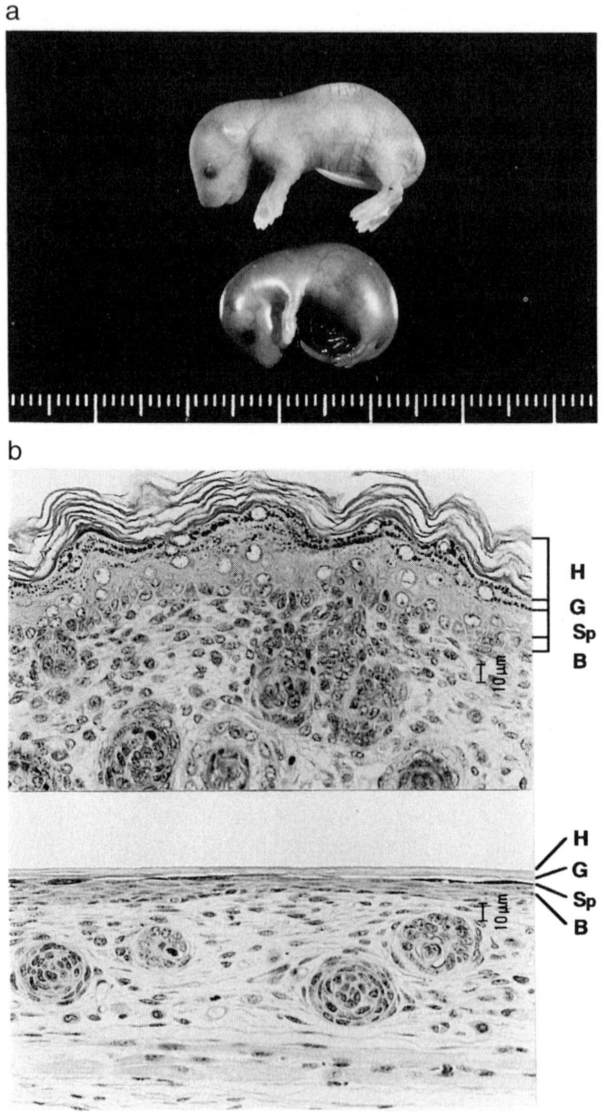

FIG. 5. Macro- and microphenotypic changes of the transgenic mice with a dominant-negative retinoic acid receptor. (a) The pictures of E19.5 embryos with normal phenotype (upper) and severely affected phenotype (lower) are demonstrated. The scale is included at the bottom. Each mark represents 1 mm. (b) The skin sections are presented from the mice shown in (a), with hematoxylin-eosin staining. The size of 10 μm is included. H, Horny layer; G, granular layer; Sp, spinous layer; B, basal cell layer.

of RA.[28] These results are consistent with the idea that the dominant-negative RAR can block the RAR-mediated signaling pathways *in vivo*.

Our additional analyses of the skin phenotype have demonstrated that RAR-E expression makes aberrant biochemical transitions from the basal cell layer to the suprabasal cell layer.[12] All of these data consistently demonstrated the absolute requirement of RA signaling in the physiological epidermal development.

Conclusion and Perspective

We have developed a novel strategy to examine the physiological functions of RA in a tissue-specific manner. This novel strategy has revealed the absolute requirement of RA in normal skin development. In addition to the epidermis, there are several other interesting targets in which to express the dominant-negative RAR, that is, lung, liver, intestine, bone, and testis. Thus, it would be possible to dissect the roles of RA systemically in these organs during embryogenesis and in adults.

Since the structure and mode of actions of the TR/RAR subfamily are quite homologous, it is possible to construct a dominant-negative molecule of other members by the same procedure, which might be used to discover unknown physiologic functions of other lipophilic molecules. Moreover, application of this strategy to so-called orphan receptors[3] would be quite challenging and might reveal the function of the orphan receptors before we know about their ligands.

Acknowledgments

This chapter was adapted from *J. Biol. Chem.* **269,** 19101 (1994), with permission of The American Society for Biochemistry and Molecular Biology.

[28] R. Kopan and E. Fuchs, *J. Cell Biol.* **109,** 295 (1989).

[9] Use of Reporter Cells to Study Endogenous Retinoid Sources in Embryonic Tissues

By MICHAEL A. WAGNER*

Introduction

Retinoids are believed to play an essential role in the biochemical and cellular processes underlying growth and development in vertebrates. Support for this hypothesis has been provided by experiments in which normal levels of retinoids in developing embryos have been perturbed. In studies designed to reduce retinoid levels *in vivo*, pregnant rats were reared on vitamin A-deficient diets resulting in fetuses that exhibited a number of morphological abnormalities, including abnormalities of the eyes, respiratory tract, heart, urogenital system, and the diaphragm.[1] Supplementing the diet of vitamin A-deficient rats with retinol reversed many of these abnormalities leading to the proposal that retinol and perhaps its metabolic end product, retinoic acid (RA), constitute the active form of vitamin A that is required for proper embryonic development and organogenesis.

Retinoids can act as intercellular signaling molecules by diffusing, due to their hydrophobicity, out of cells in which they are produced into neighboring cells. Within cells, RA binds to nuclear retinoic acid receptors (RARs) that become associated with genomic response elements in the promoter regions of retinoid responsive genes and activate their transcription. It is likely that the effects of retinoids can be attributed to the expression of these retinoic acid responsive genes and the activity of their protein products. The effects of a vitamin A-deficient diet in rats may therefore be due to the absence of the ligand vitamin A, or its metabolites, which activate the retinoid signaling pathway. Further evidence that activation of the retinoid signaling pathway mediates the effects of vitamin A *in vivo* comes from experiments in which RARs have been genetically ablated in mice.[2,3] Interestingly, these mice suffer developmental abnormalities similar to those of rats raised on vitamin A-deficient diets. These experiments,

* Present address: Department of Anatomy and Cell Biology, SUNY Health Science Center at Brooklyn, Brooklyn, New York 11203.

[1] J. G. Wilson, C. B. Roth, and J. Warkany, *Am. J. Anat.* **92,** 189 (1953).

[2] D. Lohnes, M. Mark, C. Mendelsohn, P. Dolle, A. Dierich, P. Gorry, A. Gansmuller, and P. Chambon, *Development* **120,** 2723 (1994).

[3] C. Mendelsohn, D. Lohnes, D. Decimo, T. Lufkin, M. LeMeur, P. Chambon, and M. Mark, *Development* **120,** 2749 (1994).

0076-6879/97 $25.00

together with studies analyzing the genetic, biochemical, and developmental effects of retinoids on vertebrate organisms confirm the role of retinoids, in particular RA, and the retinoid receptor-based signaling pathway as the primary means by which vitamin A exerts its biological effects.[4]

Another experimental approach that has provided insight into how vitamin A and retinoid signaling contribute to the development of the embryo involved raising retinoid concentrations in embryos above normal levels. In these studies, exogenous retinoids, such as RA, were administered to developing vertebrate embryos *in situ*, resulting in elevation of RA levels in presumably all tissues of the embryo. Embryos subjected to such treatment exhibited a number of morphological alterations, including alterations of craniofacial structures, limbs, the hindbrain of the central nervous system, and the vertebral column.[5–9] Importantly, the nature and extent of these alterations appeared to depend on the stage of development at which exogenous retinoids were administered: in general, fewer morphological alterations were evident when retinoids were administered during later stages of development. These observations suggest that for normal development to proceed, the distribution and relative levels of retinoids in early embryonic tissues must be controlled in a precise spatial and temporal manner.

To investigate the role of retinoids as important developmental signaling molecules, a necessary prerequisite is to delineate the overall "topographical" distribution of retinoids in developing embryos and to determine the local tissue source(s) of retinoids. Direct attempts at measuring retinoid levels in embryonic tissues have been impeded, in large part, by the lack of analytical methods that are sufficiently sensitive for the detection of retinoids in the minute amounts of tissue provided by embryos. Despite these limitations, the distribution and characterization of retinoids in embryonic tissues has been achieved in certain instances by organic extraction of retinoids from tissues followed by their separation and quantitation using high-performance liquid chromatography (HPLC).[10,11] To provide a more

[4] A. L. Means and L. J. Gudas, *Annu. Rev. Biochem.* **64,** 201 (1995).

[5] B. D. Abbott, M. W. Harris, and L. S. Birnbaum, *Teratology* **40,** 533 (1989).

[6] J. C. Rutledge, A. G. Shourbaji, L. A. Hughes, J. E. Polifka, Y. P. Cruz, J. B. Bishop, and W. M. Generoso, *Proc. Natl. Acad. Sci. U.S.A.* **91,** 5436 (1994).

[7] N. Papalopulu, J. D. W. Clarke, L. Bradley, D. Wilkinson, R. Krumlauf, and N. Holder, *Development* **113,** 1145 (1991).

[8] A. Ruiz i Altaba and T. Jessell, *Genes Dev.* **5,** 175 (1991).

[9] M. Kessel and P. Gruss, *Cell* **67,** 89 (1991).

[10] C. Thaller and G. Eichele, *Nature* **336,** 775 (1987).

[11] A. J. Durston, J. P. M. Timmermans, W. J. Hage, H. F. J. Hendriks, N. J. de Vries, M. Heideveld, and P. D. Nieuwkoop, *Nature* **340,** 140 (1989).

expedient method to assess relative retinoid levels in embryos, we have created a reporter cell assay that is able to detect and quantitate retinoids released from small (approximately 1 mm^3) samples of embryonic tissue maintained *in vitro*.[12]

Principle

Retinoic acid receptors are ligand-activated transcription factors present in cell nuclei. We took advantage of this property and of the retinoid signaling pathway (as described earlier) to create a reporter assay that detects and measures retinoids released from tissue explants in culture. The reporter assay consists of a RA-inducible reporter gene encoding an easily detectable protein, for example, *Escherichia coli* β-galactosidase or firefly luciferase, placed under the control of a known retinoid response element, in this case, a retinoic acid response element (RARE) taken directly from the promoter of the human β-RAR gene.[12,13] Together with a selectable drug resistance gene, the introduction of this reporter construct into cells expressing RARs allows for the establishment of a continuously dividing reporter cell line capable of detecting retinoids in the immediate microenvironment of the reporter cell.

In principle, the assay works when retinoids, either released from tissue explants or applied as a tissue extract, diffuse across the plasma membrane of the reporter cells and trigger the retinoid response pathway. Ligand-bound retinoid receptors then bind the RARE enhancer in the promoter of the reporter construct leading to induction of the reporter gene and expression of the enzymatic reporter protein. Reporter enzymatic activity indicates the presence of retinoids released from sample tissues. As described later, detection of released retinoid can be qualitative or quantitative depending on the ability either to detect or to quantitate reporter gene activity.

The retinoid reporter cell assay presented here has been successfully used to determine the distribution of retinoids in embryonic tissues.[12,14–16] Retinoid reporter cells similar to those described here have been constructed and used by other laboratories.[17,18]

[12] M. Wagner, B. Han, and T. M. Jessell, *Development* **116,** 55 (1992).
[13] H. de The, M. del Mar Vivanco-Ruiz, and P. Tiollais, *Nature* **343,** 177 (1990).
[14] M. W. Kelley, X.-M. Xu, M. A. Wagner, M. E. Warchol, and J. T. Corwin, *Development* **119,** 1041 (1993).
[15] P. McCaffery, M. Lee, M. A. Wagner, N. E. Sladek, and U. C. Drager, *Development* **115,** 371 (1992).
[16] H. L. Ang, L. Deltour, M. Zgombic-Knight, M. A. Wagner, and G. Duester, *Alcohol. Clin. Exp. Res.* **20,** 1050 (1996).
[17] M. C. Colbert, E. Linney, and A. S. LaMantia, *Proc. Natl. Acad. Sci. U.S.A.* **90,** 6572 (1993).
[18] Y. Chen, L. Huang, and M. Solursh, *Dev. Biol.* **161,** 70 (1994).

Reporter Cells

Two reporter cell lines have been developed for detecting and measuring retinoids released from tissue explants: a *lacZ* reporter cell line derived from F9 teratocarcinoma cells transfected with an *E. coli* β-galactosidase reporter gene construct and a luciferase reporter cell line derived from L cells transfected with a reporter gene construct encoding the firefly luciferase gene.[12] The F9 cells are transfected according to the method of Espeseth *et al.*[19]; L cells are transfected as described.[12] After selection in geneticin (G418)-containing media, single colonies are grown and tested for their response to RA (5×10^{-8} *M*) using an assay that measures β-galactosidase in cell lysates.[20] The highest responders are established as continuous clonal reporter cell lines and maintained under G418 selection.

The response of the F9 and L-cell reporter cell lines to varying concentrations of RA, retinol, and retinal was characterized. A dose–response curve for the L-cell luciferase reporter cell line is shown (Fig. 1). The dose–response curve of the L-cell luciferase reporter cells to RA was similar to that of the F9 *lacZ* reporter cell line, with the exception of a plateau in response beween 10^{-11} and 10^{-9} *M* RA. (The basis for this response plateau is not clear, but could be due to concentration-dependent differences in the rate of RA uptake, differential sequestration of RA, or changes in the expression and activity of RA receptors.[12]). The L luciferase reporter cell line responds to retinol at concentrations of 10^{-7} *M* and greater. The response to 10^{-7} *M* retinol is equivalent to the response to RA at approximately 10^{-9} *M*. No response was detected in F9 reporter cells to steroid hormones such as dexamethasone, β-estradiol, progesterone, testosterone, L-thyroxine, vitamin D_3, and *d*-aldosterone, or to activators of other signal transduction pathways such as forskolin or phorbol 12-myristate 13-acetate (PMA). The use of stably transfected reporter cell lines in place of transiently transfected cells provides a degree of stability and uniformity in the response of these cells that allows for a reproducible and reliable reporter assay.

Normal Passaging of Reporter Cell Lines

F9 and L-cell reporter cell lines are normally passaged in T-75 flasks with the F9 cells requiring flasks that have been gelatinized by treatment with a sterile solution of 0.1% gelatin (porcine gelatin, Sigma, St. Louis,

[19] A. S. Espeseth, S. P. Murphy, and E. Linney, *Genes Dev.* **3,** 1647 (1989).

[20] A. Reynolds and V. Lundblad, *in* "Current Protocols in Molecular Biology" (F. A. Ausubel, R. Brent, R. E. Kingston, D. D. Moore, J. G. Seidman, J. A. Smith, and K. Struhl, eds.), p. 13.6.1. Greene Publishing/Wiley-Interscience, New York, 1989.

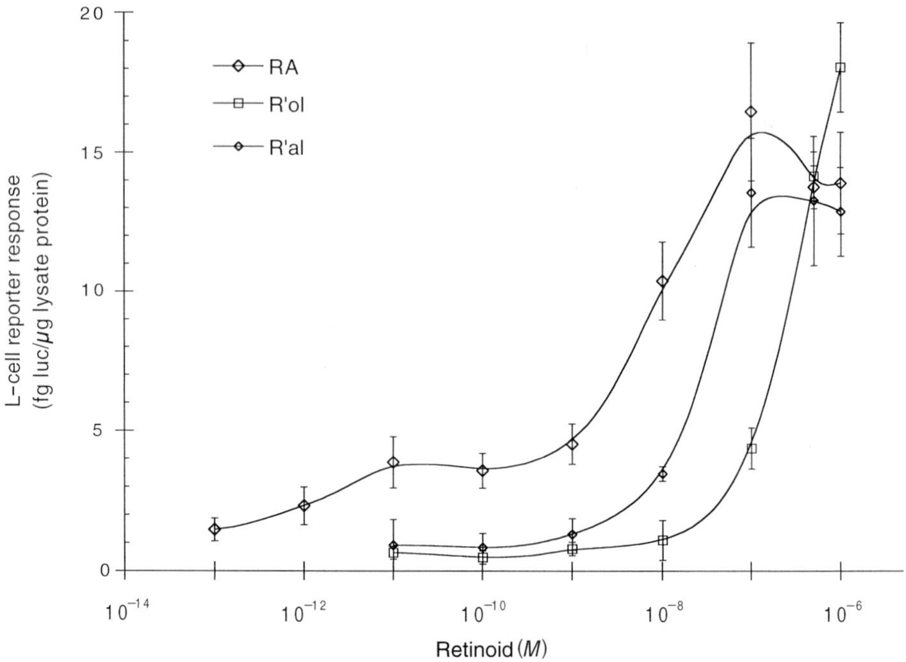

FIG. 1. L-cell–luciferase reporter cells were cultured in 24-well plates for 18–20 hr in serum-free media containing different concentrations of retinoids. Cells were harvested and assayed for luciferase activity as described.[12] Each point represents the response of the reporter cells at a given retinoid concentration with response defined as fg luciferase/μg lysate protein. (Adapted from Wagner *et al.*[12])

MO) in phosphate-buffered saline (PBS) for 1 hour. F9 reporter cells are maintained in L15-CO_2 Modification medium (Specialty Media, Lavallette, NJ) supplemented with 20% (v/v) heat-inactivated fetal calf serum and 0.8 mg/ml geneticin (G418, Gibco-BRL, Gaithersburg, MD). The relatively high concentration of fetal calf serum presumably acts to prevent the differentiation of the F9 teratocarcinoma cell line. L reporter cells are maintained in Dulbecco's modified Eagle's medium containing 10% (v/v) heat-inactivated fetal calf serum and 0.4 mg/ml G418.

For both reporter cell lines, confluent cells are passaged by trypsinizing with a 0.25% trypsin/l m*M* EDTA solution followed by trituration of the cells into suspension. Cells replated at a 1 : 10 dilution are usually confluent in approximately 3 days. For the F9 reporter cells, it is particularly important to triturate the cells well to avoid plating aggregates of cells, which can lead to the formation of patchy, nonuniform cell monolayers and poor cell growth. Care should be taken in the passaging of the reporter cell lines, particularly the F9 reporter cells, to maintain the high response of these cells to retinoids. Should the response of these reporter cells diminish with passaging, the cell line can be subcloned to regain a 95% or greater response to test concentrations of 5×10^{-8} *M* RA. This can be achieved by limiting dilution cloning of G418-resistant colonies or, if available, by fluorescence-activated cell sorting using the procedure of Nolan *et al.*[21] In general, F9 reporter cells passaged up to 10 times were used for assays before resorting to fresh cells obtained from storage in liquid N_2.

Assay Methods

F9-lacZ Reporter Assay

The F9-*lacZ* reporter assay uses *lacZ* as the reporter gene. This gene becomes activated in response to retinoids released from tissue explants maintained in short-term coculture with the F9-*lacZ* reporter cells. The product of the *lacZ* gene, *E. coli* β-galactosidase, is visualized in reporter cells using a standard histochemical procedure.[22]

A T-75 flask of F9-*lacZ* reporter cells is trypsinized and resuspended in 10 ml of selection medium containing G418 as described for normal cell passaging. One to 2 ml of this suspension is diluted into 40 ml of passaging medium and used to seed 35-mm gelatinized culture plates (approximately 20 plates at 2.0 ml per plate). The cultures are incubated at 37° in a 5% (v/v) CO_2 atmosphere until cells are just subconfluent, usually after about 2 days of culturing.

At just under subconfluency, the reporter cell monolayer is ready to be cocultured with tissue explants (approximately 1 mm^3 in size) taken from embryos or other tissues under investigation. Dissection of embryonic tissues is carried out in medium (e.g., Leibovitz's L15 medium) that maintains the viability of the tissue samples. Transfer the tissue pieces to a 35-mm culture dish containing serum-free F9-*lacZ* reporter cell assay medium (L15–CO_2 Modification medium with 1× concentrated N3 supplement[23,24] and glucose at a final concentration of 8 mg/ml) on ice. Before placing tissue explants onto the reporter cell monolayer, remove the normal passaging medium containing G418 from the cells, wash with PBS, and replace with approximately 1.0 ml of assay medium. Transfer tissue pieces onto the cell

[21] G. P. Nolan, S. Fiering, J. F. Nicolas, and L. A. Herzenberg, *Proc. Natl. Acad. Sci. U.S.A.* **85,** 2603 (1988).

[22] K. Lim and C. B. Chae, *Biotechniques* **7,** 576 (1989).

[23] H. J. Romijn, A. M. Habets, M. T. Mud, and P. S. Wolters, *Brain Res.* **254,** 583 (1981).

[24] M. Wagner, *in* "Methods in Molecular Biology" (C. Redfern, ed.). The Humana Press, Totowa, NJ, 1997.

monolayer using a fire-polished Pasteur pipette. For the tissue explants to adhere to the monolayer of reporter cells, use minimal amounts of culture medium to avoid turbulence that might otherwise dislodge tissue explants before they have had a chance to adhere. Adherence to a single spot on the monolayer is required for the reporter cells to detect locally released retinoids in the immediate area of the tissue explants. Culture the explants and F9-*lacZ* reporter cell monolayer at 37° in 5% (v/v) CO_2 for 1–2 hr to allow the tissue explants to adhere to the reporter cell monolayer. Gently add more culture medium to cover both tissue explants and cell monolayer. To monitor the responsiveness of the reporter cells, culture one monolayer in medium containing 5×10^{-8} *M* RA. Incubate all cultures overnight in a humidified atmosphere to prevent evaporation of medium.

After incubation, fix and develop the cocultures for *lacZ* expression in reporter cells.[22] Gently remove the culture medium and wash the cocultures two times with PBS. Fix the tissue and cells by incubating in a solution of 1% glutaraldehyde, 0.1 *M* sodium phosphate buffer (pH 7.0), and 1 m*M* $MgCl_2$ for 15 min at room temperature. Remove the fixative and gently wash two times with PBS. Incubate the fixed cells with a 0.2% (w/v) X-Gal solution (5-bromo-4-chloro-3-indolyl-β-galactopyranoside, Sigma, St. Louis, MO), 10 m*M* sodium phosphate buffer (pH 7.0), 150 m*M* NaCl, 1 m*M* $MgCl_2$, 3.3 m*M* $K_4Fe(CN)_6 \cdot 3H_2O$, and 3.3 m*M* $K_3Fe(CN)_6$) for a few hours to overnight at 37°. Periodically check the culture for development of the typical blue color associated with the β-galactosidase activity on the X-Gal substrate. The majority of cells in the control culture treated with RA should be blue. Remove the X-Gal solution, gently wash two times with PBS, and place a thin layer of a 50% (v/v) glycerol solution in PBS over the cocultures. Coverslip the cocultures taking care not to dislodge the tissue explants and store at 4°. (Storage at 4° often intensifies the blue reaction product.) Photograph cocultures using standard light microscopy.

L-Cell–Luciferase Reporter Assay

These reporter assays detect the fraction of total cell retinoids that are released from tissues. To determine the amount of retinoid released from tissues, levels of enzymatic reporter activity in reporter cells must be related to known concentrations of retinoids released from a point source. To achieve this, we have developed a reporter cell line using luciferase as the reporter molecule. When luciferase oxidizes its substrate luciferin, a photon is emitted that can be detected and quantified by use of a luminometer or scintillation counter. (This allows luciferase activity to be detected and measured at a much lower level than that for β-galactosidase.[25]) For quanti-

[25] J. F. Rodriguez, D. Rodriguez, J. Rodriguez, E. B. McGowan, and M. Esteban, *Proc. Natl. Acad. Sci. U.S.A.* **85,** 1667 (1988).

fying released retinoids, a standard curve relating known concentrations of retinoids to induced luciferase activity in reporter cells must be established. Known concentrations of released retinoids can be achieved by the use of ion-exchange beads loaded with RA, which act as a point source of retinoid released onto reporter cell monolayers.

Trypsinize a T-75 flask of L-cell–luciferase reporter cells and resuspend in 10 ml of L-cell passaging medium. Dilute the cells to a final concentration of 25 cells/μl and plate the cells in Terasaki microculture wells by placing 20 μl of the cell suspension into each well. Culture the cells for 3 days in a humidified atmosphere until they reach near confluency. Wash the cell monolayers with PBS and change the culture medium in the Terasaki wells to L-cell reporter assay medium [Dulbecco's modified Eagle's medium (DMEM) with 1× concentrated N3 supplement and glucose at a final concentration of 8 mg/ml].

Prepare a "standard curve" relating luciferase activity to retinoid concentration: In duplicate, set up a concentration range of [^{3}H]retinoic acid (all-*trans*-[11,12-^{3}H]retinoic acid, Dupont NEN Research Products, Wilmington, DE) in PBS from between 1 and 25 n*M*. Place a single AG1-X2 ion-exchange bead (200 μm in diameter, Bio-Rad Laboratories, Richmond, CA) in 250 μl of each serial concentration and shake for 20 min at room temperature.[26] Remove the RA solution and wash beads two times in 250 μl of PBS for 1 and 10 min, followed by a final wash in serum-free assay medium for 10 min. Place one set of beads in Terasaki wells containing only assay medium and the other set into Terasaki wells containing the L-cell reporter monolayer. Culture in a humidified atmosphere for the same period of time as the tissue cocultures.

At the end of the culture period (usually 20 h), remove the medium from the first set of Terasaki wells containing beads in culture medium alone and measure released [^{3}H]retinoic acid by scintillation counting. Convert cpms (counts per minute) to dpms (disintegrations per minute) and determine the amount of RA released using the known specific activity of the labeled RA. To assay the response of the reporter cells to RA released from the AG1-X2 beads or tissue explants, remove beads and tissue explants from the reporter cell monolayers without disturbing the monolayer and wash with PBS. Before removing the tissue explants, check each culture well under a dissecting microscope or low magnification microscope. Assay only those wells in which tissue explants successfully adhered to the cell reporter monolayer. Remove wash PBS and immediately add 10 μl of "luciferase lysis reagent" (Luciferase Assay System, Promega, Madison, WI). Let stand at room temperature for 10 min, then triturate cells with a Pipetman set at 7 μl to avoid foaming and transfer the cell lysate to an

[26] G. Eichele, C. Tickle, and B. M. Alberts, *Anal. Biochem.* **142,** 542 (1984).

Eppendorf test tube on ice. Wash the culture well again with an additional 5 μl of lysis reagent and add to the first lysate.

The luciferase assay is carried out using the Promega Luciferase Assay System. Light production can be detected using a luminometer or a scintillation counter with the coincidence circuit between photomultiplier detectors turned off, if possible.[27] For all samples, counting should proceed at a set time after cell lysate and luciferase cocktail are mixed together. Count for at least 3 min. Plot a standard curve relating luciferase activity of the L-cell reporter monolayer cocultured with [^{3}H]RA beads to RA released into the medium from the parallel set of [^{3}H]RA beads. Determine the amount of luciferase activity induced by each tissue explant and determine the amount of RA released from the tissue using the standard curve.

Comments

The relative ease and sensitivity with which this reporter cell assay can detect retinoids released from small amounts of tissue is offset by two inherent yet surmountable limitations. First, the reporter cell lines do not discern the particular isomer of retinoid being released from tissue samples. Given the recent identification of a number of naturally occurring retinoid isomers such as 3,4-didehydroretinoic acid[28] and 9-*cis*-retinoic acid,[29] and the prospect that additional isomers may be discovered, it will become increasingly necessary to discriminate between retinoid isomers in order to assess accurately the relative levels of each *in vivo*. A greater degree of isomer specificity may be achieved in the reporter assay through the use of retinoid receptor homodimers and/or heterodimers (e.g., RAR/RAR, RXR/RXR, or RAR/RXR) that specifically bind to a given retinoid isomer, together with different response elements such as RAREs or RXREs.[30,31] A second limitation of this assay is that only retinoids released from tissues are being detected by reporter cells. Retinoids that are not released, either due to sequestration by cellular retinoid-binding proteins or to intracellular turnover, will not be detected.[32,33] The assay, as presented, may suffice for

[27] V. T. Nguyen, M. Morange, and O. Bensaude, *Anal. Biochem.* **1781,** 404 (1988).

[28] C. Thaller and G. Eichele, *Nature* **345,** 815 (1990).

[29] R. A. Heyman, D. J. Mangelsdorf, J. A. Dyck, R. B. Stein, G. Eichele, R. M. Evans, and C. Thaller, *Cell* **68,** 397 (1992).

[30] D. J. Mangelsdorf, K. Umesono, S. A. Kliewer, U. Borgmeyer, E. S. Ong, and R. M. Evans, *Cell* **66,** 555 (1991).

[31] M. Leid, P. Kastner, R. Lyons, H. Nakshatri, M. Saunders, T. Zacharewski, J.-Y. Chen, A. Staub, J.-M. Garnier, S. Mader, and P. Chambon, *Cell* **68,** 377 (1992).

[32] J. F. Boylan and L. J. Gudas, *J. Cell Biol.* **112,** 965 (1991).

[33] J. F. Boylan and L. J. Gudas, *J. Biol. Chem.* **267,** 21486 (1992).

detecting and measuring only those retinoids released from cells that act as intercellular signaling molecules. If the total retinoid content of a tissue sample is to be determined, then retinoids must be "released" from the tissue by lysis and the tissue lysate assayed directly on the reporter cells.[15,24] McCaffrey *et al.*[15] have modified the F9 reporter cell assay in this way to obtain a measure of tissue retinoid content.

The reporter assay described here was originally designed to detect retinoids in tissue explants taken from embryos and maintained in short-term cultures. To use the assay in this way successfully, two important requirements must be met. The first requirement is that the development of the embryos must be timed in accordance with the plating of the reporter cells to ensure that tissue from the desired developmental stage is plated onto near-confluent reporter cell monolayers. The second requirement is that the tissue being studied must be viable under the culture conditions required for maintaining the reporter cell monolayer. The tissues of certain species require different conditions for maintaining viability *in vitro*. For instance, tissue explants from *Xenopus* are best maintained at 25°, while reporter cells must be at 37°. Cocultures of *Xenopus* tissue and reporter cells at 37° are therefore not feasible. This limitation can be circumvented by using a tissue homogenate[15] or extracting retinoids directly from amphibian tissues and assaying the homogenate or tissue extract using reporter cells maintained at 37°.[18]

The N3 supplement in the reporter cell assay medium is a substitute for serum; serum contains retinoids that may nonspecifically trigger the reporter cells and lessen the sensitivity of the assay. However, N3 can be arduous and expensive to prepare.[24] Given the short-term nature of this assay, it may be advisable first to try the assay in serum-containing medium or in medium using serum that has been stripped of retinoids by treatment with charcoal followed by sterile filtration. Alternatively, serum-free medium supplement containing insulin, transferrin, and selenite can be obtained from Boehringer Mannheim (Indianapolis, IN). Whichever medium or serum treatment is chosen, it is important that the viability of both the tissue explant and reporter cell monolayer be ensured over the duration of the assay.

Acknowledgments

I would like to thank Dr. Thomas Jessell (Howard Hughes Medical Institute, Columbia University) for his support during the generation and use of these reporter cell lines in his laboratory. I also thank Barbara Han for expert technical assistance in the production of these reporter cells.

[10] Preparation of Radiolabeled 9-*cis*- and all-*trans*-Retinoids

By PRAVEEN K. TADIKONDA and HECTOR F. DELUCA

Introduction

all-*trans*-Retinoic acid and 9-*cis*-retinoic acid have dramatically increased in importance because of the discovery of the retinoic acid receptors (RARs) and the retinoic acid X receptors (RXRs). Each of these receptors is divided into three subfamilies: α, β, and γ. all-*trans*-Retinoic acid binds to RARα, β, and γ subtypes, whereas 9-*cis*-retinoic acid binds to all the subtypes of RAR and RXR receptors. These retinoids and their other analogues are known for their role in growth, cellular differentiation, and embryonic development. These developments have kindled great interest in radiolabeled retinoids for both *in vivo* and *in vitro* studies. Their applications encompass metabolism, pharmacology, receptor binding, and cellular protein binding.

In this chapter we discuss procedures for the synthesis of radiolabeled RAR and RXR ligands, that is, all-*trans*- and 9-*cis*-retinoic acids. For the synthesis of these retinoids, three requirements have to be met: (1) the introduction of a 9-*cis* double bond and the retention of the *cis*-configuration throughout the synthesis; (2) the incorporation of radiolabel at a late stage in the synthesis; and (3) a way of introducing a label having high specific activity. The procedures discussed meet these criteria.

Synthesis of 9-*cis*-[^{3}H]Retinoic Acid

Boehm *et al.*[1] have reported a synthesis of 9-*cis*-retinoic acid with a 29 Ci/mmol specific activity in four steps starting from ethyl ester of β-ionylidine acetic acid **1** (Scheme 1). Initially, two tritium atoms are introduced by lithium aluminum tritide reduction of the ester to an alcohol followed by oxidation to an aldehyde resulting in the loss of one tritium atom, thereby reducing the specific activity of the aldehyde. Thus, (2*Z*,4*E*)-3-methyl-5-(2,6,6-trimethylcyclohex-1-enyl)penta-2,4-dienoate[2] **1** is reduced

[1] M. F. Boehm, M. R. McClurg, C. Pathirana, D. Mangelsdorf, S. K. White, J. Hebert, D. Winn, M. E. Goldman, and R. A. Heyman, *J. Med. Chem.* **37,** 408 (1994).

[2] (a) G. Cainelli, G. Cardillo, and M. Orena, *J. Chem. Soc., Perkin Trans.* **1,** 1597 (1979); (b) R. W. Dugger and C. H. Heathcock, *J. Org. Chem.* **45,** 1181 (1980); (c) U. Schwieter, C. V. Planta, R. Ruegg, and O. Isler, *Helv. Chim. Acta* **63,** 528 (1963).

0076-6879/97 $25.00

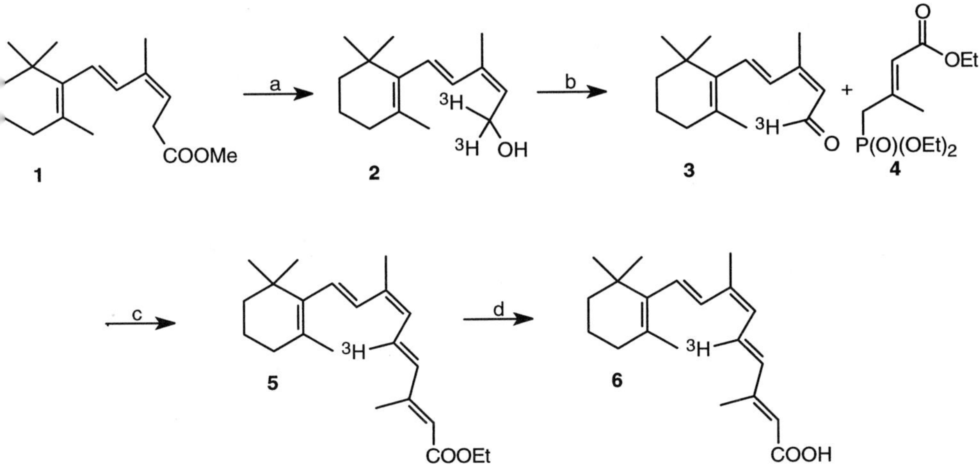

SCHEME 1. Reagents: (a) $LiAl^3H_4$, Et_2O, −78°; (b) MnO_2, dichloromethane, room temperature; (c) NaH, THF, 0–20°; (d) 5 *N* KOH, methanol, 60°, 1 hr.

to its alcohol **2** with lithium aluminum tritide at −78° followed by oxidation to its aldehyde **3** with MnO_2. This aldehyde is condensed with diethyl[3-(ethoxycarbonyl)-2-methylprop-2-enyl]phosphonate[3] in tetrahydrofuran (THF) using NaH base at 0° to obtain ethyl retinoate **5**. Finally, crude ester is hydrolyzed with 5 *N* aqueous KOH in methanol at 60° to get a mixture of retinoic acid isomers that are identified as 9-*cis*-, all-*trans*-, 9,13-di-*cis*-, and 13-*cis*-retinoic acids in a 3:3:1:1 mixture by ODS high-performance liquid chromatography (HPLC). Pure 9-*cis*-**6** and all-*trans*-retinoic acid at a 29 Ci/mmol specific activity were separated by ODS HPLC.

Synthesis of 9-*cis*-[2,3- or 3,4-$^3H_2$]Retinoic Acid

In another elegant approach,[4] taking advantage of regioselective hydrogenation of endocyclic disubstituted double bond, tritium is incorporated at either the 2,3- or 3,4-positions of the ring at the penultimate step. In this approach (Schemes 2 and 3) didehydroretinoate is synthesized, followed by tritiation of endocyclic double bond using Wilkinson's catalyst before the hydrolysis step.

[3] M. Igbal, W. G. Copan, D. D. Mucco, and G. D. Mateescu, *J. Labelled Compounds Radiopharm.* **8,** 807 (1985).
[4] Y. L. Bennani and M. F. Boehm, *J. Org. Chem.* **60,** 1195 (1995).

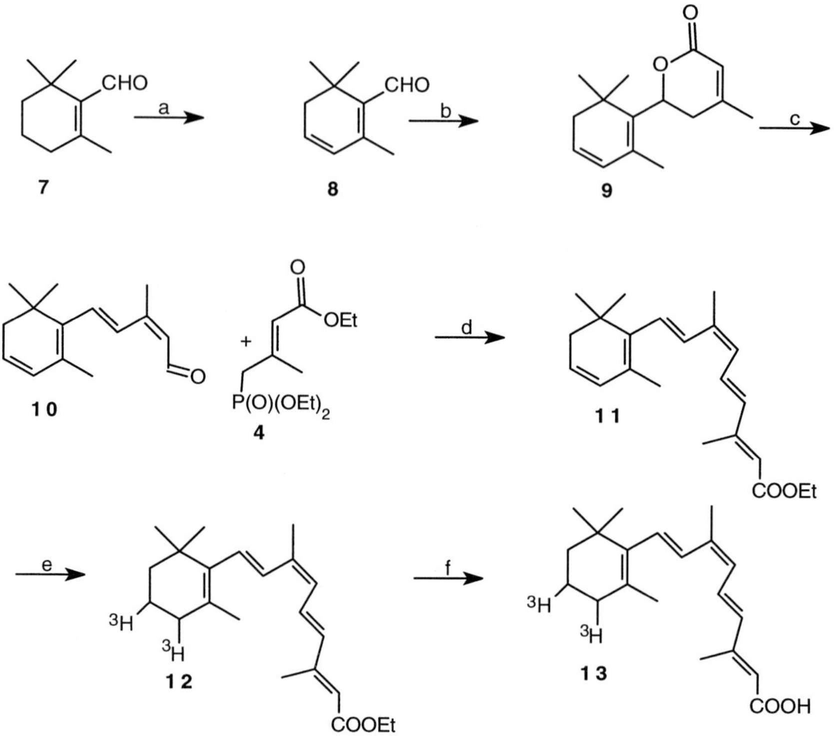

SCHEME 2. Reagents: (a) i. NBS, dichloromethane. ii. Collidine, Δ; (b) LDA, ethyl dimethylacrylate, $-78°$; (c) i. DIBAL-H, THF, $-78°$. ii. HCl, $ClCH_2CH_2Cl$; (d) *n*-BuLi, THF, DMPU, $-78°$; (e) $Rh(PPh_3)_3Cl$, benzene, $^3H_2$; (f) NaOH, ethanol.

The 9-*cis*-[3,4-$^3H_2$]retinoic acid synthesis began with bromination of β-cyclocitral **7** using *N*-bromosuccinimide (NBS), CaO, and $NaHCO_3$ at 0° followed by elimination in collidine to give α-safranal **8**.[5] α-Safranal on treatment with ethyl 3,3-dimethylacrylate using lithium diisopropylamide (LDA) in THF at $-78°$ yields the lactone **9**.[2] The lactone is reduced to the corresponding lactol using diisobutylaluminum hydride (DIBAL-H) in THF at $-78°$ followed by acid-catalyzed ring opening, afforded tetraenic aldehyde **10**. Wittig–Horner olefination of aldehyde with diethyl[3-(ethoxycarbonyl)-2-methylprop-2-enyl]phosphonate using *n*-butyllithium (*n*-BuLi), 1,3-dimethyl-3,4,5,6-tetrahydro-2(1*H*)-pyrimidone (dimethylpropy-

[5] W. M. B. Koenst, L. M. Van der Linde, and H. Boelens, *Tetrahedron Lett.* **15,** 3175 (1974).

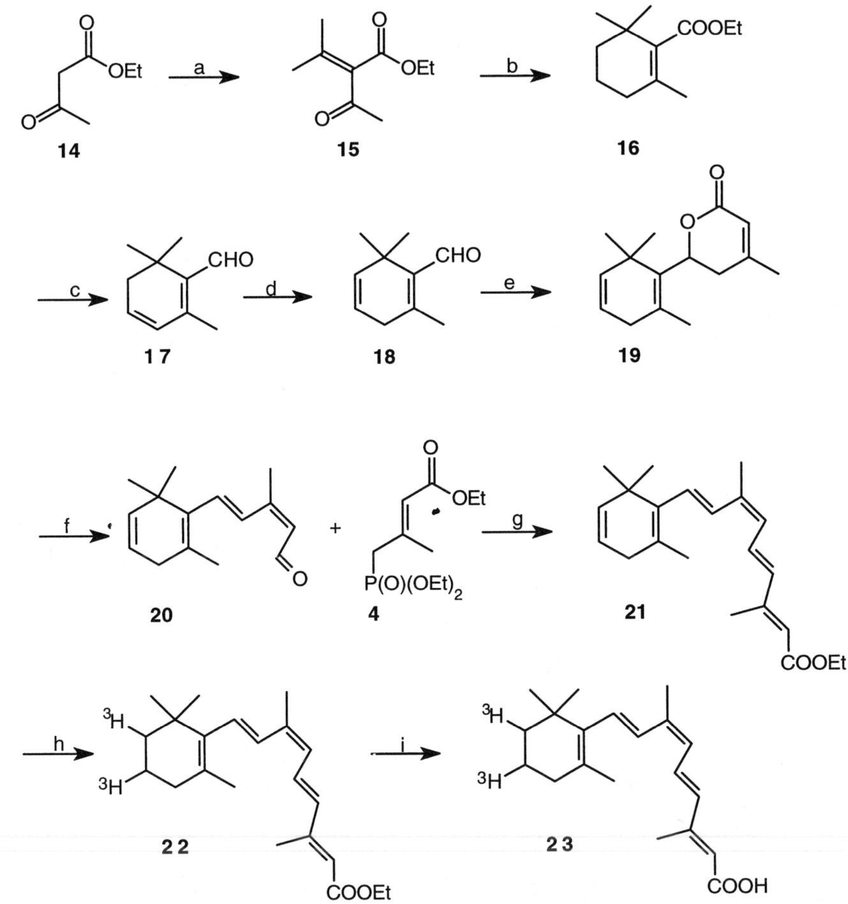

SCHEME 3. Reagents: (a) Acetone, acetic anhydride, $ZnCl_2$; (b) allyltriphenylphosphonium chloride, *n*-BuLi, Et_2O; (c) i. LAH, Et_2O. ii. DMSO, $(COCl)_2$, Et_3N, THF; (d) DBU, CH_2Cl_2; (e) LDA, ethyl dimethylacrylate, −78°; (f) i. DIBAL-H, THF, −78°. ii. HCl, $ClCH_2CH_2Cl$; (g) *n*-BuLi, THF, DMPU, −78°; (h) $Rh(PPh_3)_3Cl$, benzene, $^3H_2$; (i) NaOH, ethanol.

leneurea) (DMPU), in THF at −78° gives the required didehydroretinoate **11**, which is tritiated using Wilkinson's catalyst and tritium gas in oxygen-free benzene to afford ethyl [3,4-$^3H_2$]retinoate **12**. The ester is refluxed with 5 *N* aqueous NaOH in ethanol to obtain a mixture of retinoic acid, which is purified on ODS HPLC to obtain 9-*cis*-[3,4-$^3H_2$]retinoic acid with 58 Ci/mmol specific activity. 9-*cis*-[3,4-$^3H_2$]retinoic acid **13** was useful for competitive binding studies but this material could not be used for metabolic studies in animals because position 4 in the parent retinoid is highly prone to oxidation, which can result in partial loss of radioactivity. To avoid the possibility of loss in radiolabel during the metabolic work, the synthesis of 9-*cis*-[2,3-$^3H_2$]retinoic acid using a methodology similar to that described earlier was developed.

Synthesis of 9-*cis*-[2,3-$^3H_2$]Retinoic Acid

The required 1,4-cyclohexadiene-1-carboxaldehyde **18** for the synthesis of 9-*cis*-[2,3-$^3H_2$]retinoic acid (Scheme 3) is synthesized first by condensation of ethylacetoacetate **14** with acetone in the presence of acetic anhydride and fused zinc chloride to afford 2-isopropylidene ketoester[6] **15**, which is treated with allyltriphenylphosphoranyl ylide in a Michael–Wittig fashion to give cyclic ester **16**.[7] The cyclic ester is reduced to its alcohol with lithium aluminum hydride (LAH) in diethyl ether. Further, the alcohol is oxidized to give the corresponding aldehyde **17** under Swern oxidation conditions.[8] The aldehyde is isomerized using 1,8-diazabicyclo[4.5.0]undec-7-ene (DBU) at room temperature to obtain 1,4-cyclohexadiene-1-carboxaldehyde **18**. This carboxaldehyde is transformed into 9-*cis*-[2,3-$^3H_2$]retinoic acid **23** with 60 Ci/mmol specific activity by performing similar chemical manipulations used for the synthesis of 9-*cis*-[3,4-$^3H_2$]retinoic acid.

Synthesis of C-20-[*methyl*-^3H]-9-*cis*-Retinoic Acid

Because we required 9-*cis*-retinoic acid of higher specific activities for biological studies, we have developed a methodology[9] for the synthesis of 9-*cis*-retinoic acid of high specific activity (73–79 Ci/mmol).

Synthesis of C-20-labeled 9-*cis*-retinoic acid is outlined in Scheme 4, and the commercially available *cis*-3-methylpent-2-en-4-ynol (*E/Z* ratio 5/95) is utilized to fix the *cis* configuration at the 9-position in the parent retinoid.

The bromomagnesium acetylide is added to 2,2,6-trimethylcyclohexanone **24** to obtain the diastereomeric diols **25**. The acetylenic bond in diol **25** is reduced to a *trans*-double bond with lithium aluminum hydride in THF. The primary hydroxyl group is oxidized to an aldehyde **27** and the side chain is extended by Wittig olefination to an ester **28** keeping the

[6] I. Alkonji and D. Szabo, *Chem. Ber.* **100,** 2773 (1967).
[7] H. Wuest and G. Buchi, *Helv. Chim. Acta* **54,** 1765 (1971).
[8] A. J. Manscu, S. L. Huang, and D. Swem, *J. Org. Chem.* **43,** 2480 (1978).
[9] U.S. patent applied for.

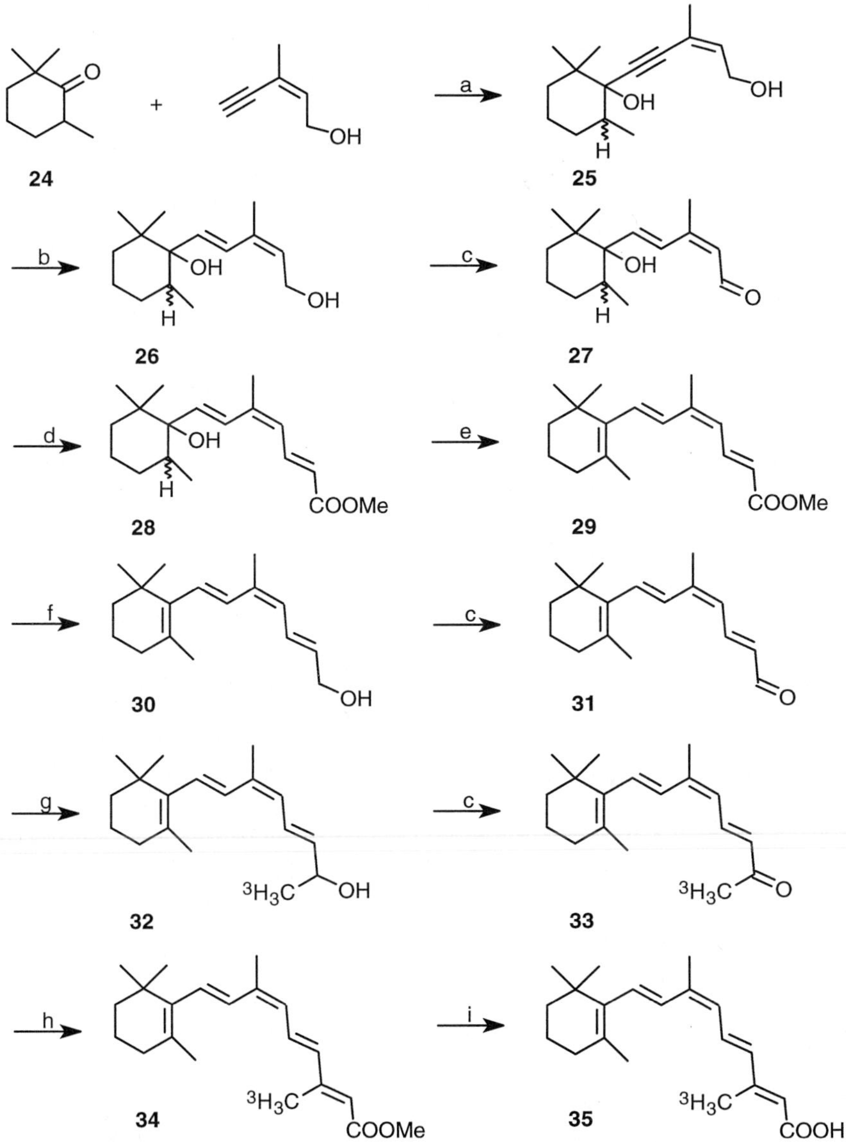

SCHEME 4. Reagents: (a) EtMgBr, THF; (b) LAH, THF; (c) MnO_2, dichloromethane; (d) $Ph_3P{=}CHCOOMe$, benzene; (e) HCOOH, hexane; (f) DIBAL-H, dichloromethane, −78°; (g) $^3H_3CMgBr$, THF; (h) $(EtO)_2P(O)CH_2COOOMe$, NaH, THF; (i) 5 *N* KOH, methanol, 50°.

tertiary hydroxyl at position 6 intact. Dehydration of the tertiary hydroxyl group was effected smoothly by treating with 80% (w/w) formic acid in hexane. DIBAL-H reduction of the ester **29** yielded alcohol **30**, which is oxidized to the corresponding aldehyde **31**. Tritium incorporation is accomplished by the addition of [$^3H_3$]methyl Grignard reagent to the aldehyde. The resulting secondary alcohol **32** is oxidized to ketone **33** and condensed with methyl diethylphosphonoacetate (NaH, THF) to obtain methyl retinoate **34**.[10] Finally the ester is hydrolyzed using methanolic KOH at 50° to afford a mixture of retinoic acid isomers.

Reversed-phase HPLC analysis of the mixture showed three peaks in a 5:11:1 ratio. The compounds were identified as 13-*cis*-, 9-*cis*- and all-*trans*-retinoic acids by coeluting them with the authentic samples. The required 9-*cis*-retinoic acid was purified on an ODS HPLC column using a solvent system of methanol/2-propanol/water/acetonitrile/acetic acid (25/15/30/30/1.2) (v/v). The specific activity of the purified C-20-labeled 9-*cis*-retinoic acid **35** was 79 Ci/mmol.

Preparation of all-*trans*-[11,12-$^3H_2$]Retinoic Acid

Liebman *et al.*[11] have utilized oxenin[12] **36** for the synthesis of all-*trans*-[11,12-$^3H_2$]retinoic acid **41** (Scheme 5). Pure recrystallized diol **36** is partially reduced with carrier-free tritium gas over a Lindlar catalyst in the presence of quinoline to obtain hydroxenine **37** at specific activities of 40–50 Ci/mmol after workup. The primary hydroxyl group of dihydroxenin is selectively acetylated and then rearranged to retinyl acetate **39** using the hydrobromic acid and the phase transfer reagent cetyltrimethylammonium bomide (CETAB). The retinyl acetate is either hydrolyzed directly to retinol **40** or simultaneously hydrolyzed and oxidized by MnO_2 and silver oxide in methanolic sodium hydroxide solution to [11,12-$^3H_2$]retinoic acid **41**, which is purified by crystallization to 98% radiochemical purity.

The specific activities obtained were in the range of 25–40 Ci/mmol. In this procedure, a small amount of scrambling of the radioisotope into other positions in addition to positions 11 and 12 is expected.

Preparation of C-20-[*methyl*-^{3}H]-all-*trans*-Retinoic Acid and Retinol

The all-*trans*-[11,12-$^3H_2$]retinol, when used for metabolic work, lost a significant amount of radioactivity. This may be because of the oxidation

[10] J. D. Bu Lock, S. A. Quarsie, and D. A. Taylor, *J. Labelled Compounds* **9,** 311 (1973).

[11] A. A. Liebman, W. Burger, D. H. Malerek, L. Serico, R. R. Muccino, C. W. Perry and S. C. Choudhry, *J. Labelled Compounds Radiopharm.* **5,** 525 (1990).

[12] O. Isler, W. Huber, A. Ronco, and M. Kofler, *Helv. Chim. Acta* **30,** 1911 (1947).

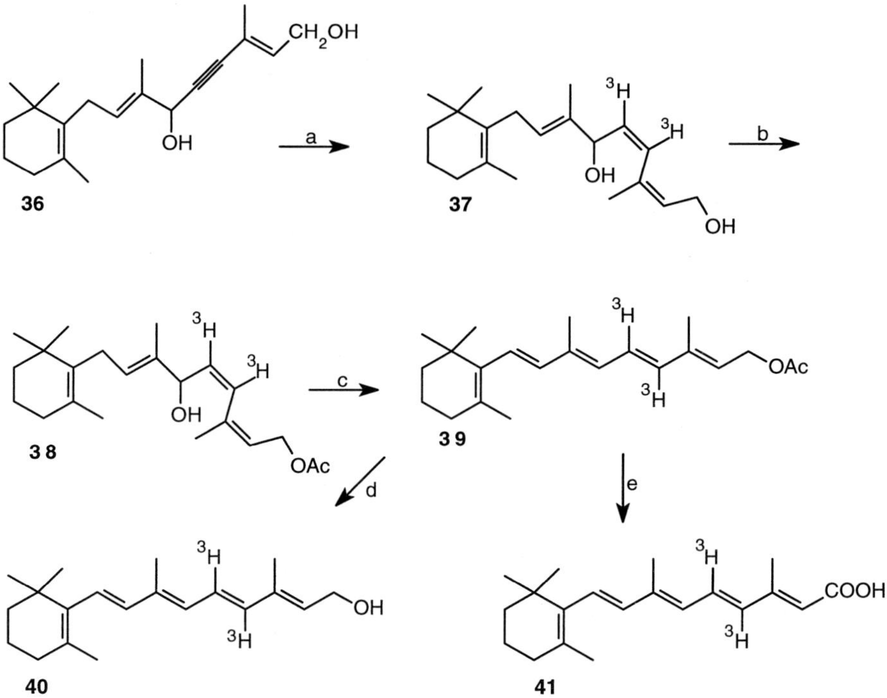

SCHEME 5. Reagents: (a) Lindlar catalyst, $^3H_2$; (b) acetic anhydride, triethylamine, toluene, 40°; (c) 60% HBr, CETAB, Py, dichloromethane; (d) 10 *N* NaOH, toluene, ethanol; (e) 1 *N* NaOH, methanol, MnO_2, Ag_2O, water.

at positions 11 or 12 during the studies. To avoid this, a label at another position was needed. We, therefore, developed a methodology for the synthesis of C-20-laeled all-*trans*-retinoic acid based on the synthesis of C-20-[*methyl*-^{3}H]-9-*cis*-retinoic acid and this acid was reduced to the C-20-labeled alcohol via its ester.

Scheme 6 outlines the strategy for the synthesis of all-*trans*-retinoic acid. β-Ionone **42** is condensed with triethyl phosphonoacetate using NaH as base in THF to obtain ester **43**. The ester without the separation of *cis*-*trans*-isomers proceeds further and is reduced to its alcohol **44** using two equivalents of DIBAL-H in dichloromethane at −78°. The trans-isomer is separated by silica gel column chromatography and oxidized with MnO_2 to its aldehyde **45**. The aldehyde on Wittig olefination with two carbon stable ylide in benzene gives tetraenic ester **46**. Reduction of the ester with

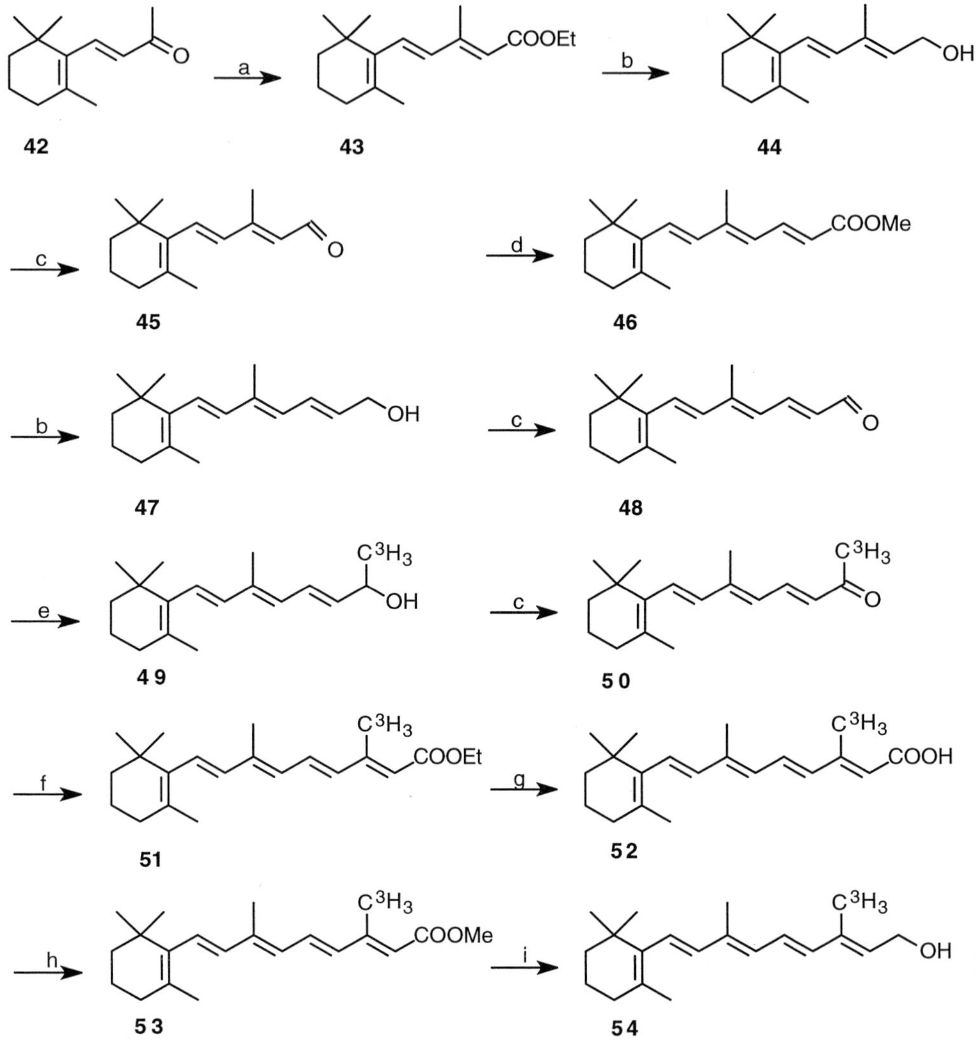

SCHEME 6. Reagents: (a) $(EtO)_2P(O)CH_2COOEt$, NaH, THF, 0–20°; (b) DIBAL-H, dichloromethane, −78°; (c) MnO_2, dichloromethane, room temperature; (d) $Ph_3P{=}CHCOOMe$, benzene, room temperature; (e) $^3H_3CMgI$, THF, 0°; (f) $(EtO)_2P(O)CH_2COOEt$, NaH, THF, 0–20°; (g) 5 *N* KOH, methanol, 50°; (h) Diazomethane, ether, 0°; (i) LAH (1.0 *M* soln. in THF), THF, −20°.

DIBAL-H in dichloromethane at −78° followed by the oxidation with MnO_2 gives the much required precursor for the incorporation of the tritium label. The [$^3H_3$]-methyl Grignard reagent is added to the aldehyde **48** to obtain the secondary alcohol **49**, which was oxidized to ketone **50** using MnO_2 oxidation. The ketone is condensed with triethyl phosphonoacetate using NaH in THF to yield ethyl retinoate **51**, which is saponified using 5 *N* KOH in methanol at 50° to obtain C-20-labeled all-*trans*-retinoic acid **52** and isomers. all-*trans*-Retinoic acid is purified on an ODS HPLC. The specific activity of the purified all-*trans*-retinoic acid is determined to be 67 Ci/mmol. Finally, the purified C-20-[*methyl*-^{3}H]-all-*trans*-retinoic acid is esterified to its methyl ester **53** using diazomethane in ether and is subsequently reduced to its alcohol **54** using lithium aluminum hydride.

Acknowledgments

This work was supported in part by program project grant DK14881, from the National Institutes of Health, a fund from the National Foundation for Cancer Research, and a fund from the Wisconsin Alumni Research Foundation.

[11] Identification and Quantification of Retinoic Acid and Other Metabolites from β-Carotene Excentric Cleavage in Human Intestine *in Vitro* and Ferret Intestine *In Vivo*

By XIANG-DONG WANG and NORMAN I. KRINSKY

Introduction

Formation of intestinal metabolites of β-carotene appears to involve both a central and an excentric cleavage pathway (Fig. 1). The central cleavage pathway uses a 15,15′-dioxygenase,[1] which is localized in the cytosolic fraction of intestinal mucosa[2] and yields two molecules of retinal (retinaldehyde).[3,4] The excentric cleavage mechanism yields a series of β-apocarotenals of different chain lengths (reviewed in Ref. 5). β-Apocaro-

[1] D. S. Goodman and J. A. Olson, *Methods Enzymol.* **15,** 462 (1969).

[2] J. A. Olson and M. R. Lakshman, *Methods Enzymol.* **189,** 425 (1990).

[3] J. Devery and B. V. Milborrow, *Br. J. Nutr.* **72,** 397 (1994).

[4] A. Nagao, A. During, C. Hoshino, J. Terao, and J. A. Olson, *Arch. Biochem. Biophys.* **328,** 57 (1996).

[5] X.-D. Wang, *J. Am. Coll. Nutr.,* **13,** 314 (1994).

0076-6879/97 $25.00

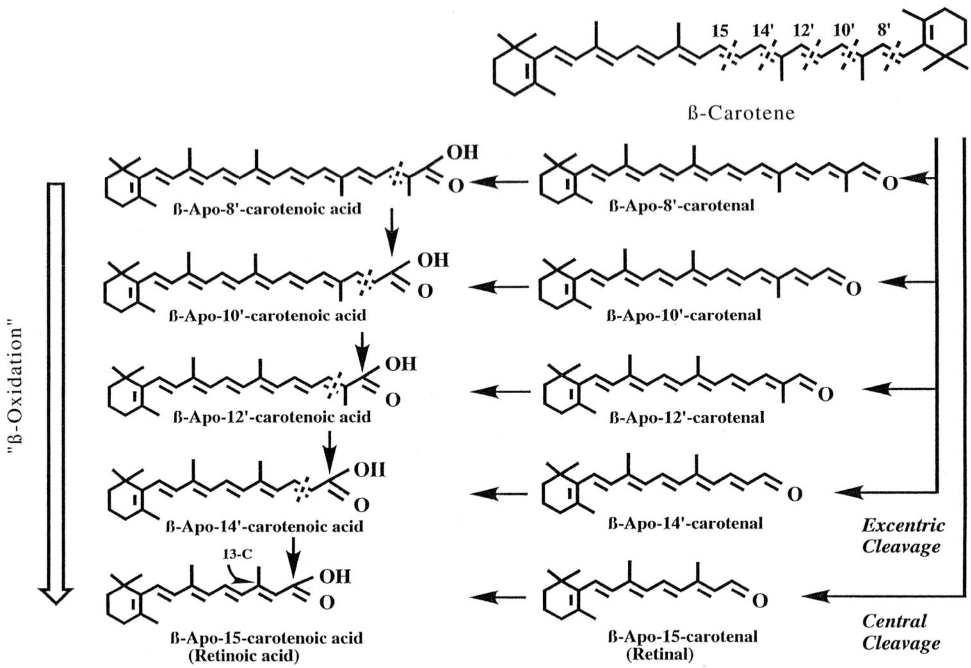

FIG. 1. Structures of some metabolites in the metabolic pathway of β-carotene.

tenals are subsequently oxidized to their corresponding β-apocarotenoic acids. By a process analogous to fatty acid β-oxidation, the β-apocarotenoic acids are converted stepwise to retinoic acid, where the β-oxidation-type process is stopped by the presence of the methyl substituent on carbon 13 (Fig. 1). These results are in agreement with earlier work by Napli and Race,[6] who demonstrated that β-carotene may be a significant precursor of retinoic acid *in vitro* in a process in which retinal is probably not involved as an intermediate.

Of particular interest is the observation that the retinoic acid produced by excentric cleavage pathway of β-carotene represents at least 50% of the retinoic acid formed both *in vitro*[7] and *in vivo*,[8] even when citral (an inhibitor of retinal oxidation) is added to the incubation mixture or the perfusate.

[6] J. L. Napoli and K. R. Race, *J. Biol. Chem.* **263,** 17372 (1988).

[7] X.-D. Wang, N. I. Krinsky, G. Tang, and R. M. Russell, *Arch. Biochem. Biophys.* **293,** 293 (1992).

[8] X. Hebuterne, X.-D. Wang, D. Smith, G. Tang, N. I. Krinsky, and R. M. Russell, *J. Lipid Res.* **37,** 482 (1996).

The assay for excentric cleavage products of β-carotene metabolism can be carried out either by the incubation of the postnuclear fraction of intestinal mucosa *in vitro* or by perfusion of the ferret intestine *in vivo*. We have found these two models to be particularly useful for assaying the production of retinoic acid and other excentric cleavage products. The use of human intestinal homogenates has been shown to be particularly useful, inasmuch as the conversion of β-carotene to vitamin A in intestinal mucosa in different animal species *in vitro* is different, usually lower than in comparison to humans *in vitro*.[5] The postnuclear fraction of human intestinal homogenates consistently formed central and excentric cleavage products when incubated with either all-*trans*- or 9-*cis*-β-carotene.[9]

The ferret has been useful as an animal model for β-carotene absorption and metabolism because these animals have many anatomic and physiological features that are similar to those of humans,[10] and we have shown that the ferret is appropriate for studying the intricacies of β-carotene metabolism *in vivo*.[11]

Chemicals: Sources, Preparation, and Purification

all-*trans*-β-Carotene, all-*trans*-retinoic acid, all-*trans*-retinol, all-*trans*-retinal, retinyl acetate, oleic acid, sodium taurocholate, Krebs phosphate buffer, dimethyl sulfoxide (DMSO), and other chemicals were purchased from Sigma Chemical Co. (St. Louis, MO). 9-*cis*-β-Carotene, 9-*cis*-retinoic acid, 13-*cis*-retinoic acid, β-apo-14′-, 12′, 10′-, and 8′-carotenal, β-apo-12′-carotenoate, and β-apo-14′, 12′, and 8′-carotenoate esters were gifts from Hoffmann-La Roche, Inc. (Nutley, NJ, and Basel, Switzerland).

Substrates are purified by passage through a 7% water-weakened alumina column[12] or by high-performance liquid chromatography (HPLC), as described later. All carotenoids and retinoids are stored under N_2 at −70° and handled under red light. β-Apo-10′-carotenoic acid is prepared by oxidation of the β-apo-10′-carotenal, using a procedure modified from Barua *et al.*[13] β-Apo-14′- and 8′-carotenoic acids are prepared by saponification of β-apo-14′- and 8′-carotenoate esters as follows: the methyl and ethyl ester of β-apo-14′- and 8′-carotenoate are saponified with 10% etha-

[9] X.-D. Wang, N. I. Krinsky, P. N. Benotti, and R. M. Russell, *Arch. Biochem. Biophys.* **313,** 150 (1994).

[10] J. G. Fox, "The Biology and Disease of the Ferret." pp. 1–65. Lea & Febiger, Philadelphia, 1992.

[11] X.-D. Wang, R. M. Russell, R. Marini, G. Tang, G. Dolnikowski, J. G. Fox, and N. I. Krinsky, *Biochim. Biophys. Acta* **1167,** 159 (1993).

[12] N. I. Krinsky, D. G. Cornwell, and J. L. Oncley, *Arch. Biochem. Biophys.* **73,** 233 (1958).

[13] A. B. Barua, M. C. Ghosh, and K. Goswami, *Biochem. J.* **113,** 447 (1969).

nolic KOH (pH > 12) overnight at room temperature, then acidified with 6 *N* HCl (pH 3) and the products extracted with hexane. The extracts containing β-apocarotenoic acid are dried under N_2. Further purification of β-apocarotenoic acids is carried out by using the HPLC method described in the HPLC section. The peak containing the β-apocarotenoid is collected for use as a substrate.

In Vitro Analyses

Enzyme Purification

Samples of fresh human small intestine, obtained immediately from discarded surgical procedures, are flushed with ice-cold HEPES buffer (pH 7.4, 20 m*M*) and cut lengthwise. The mucosa of the intestine is removed by scraping with a glass slide. These procedures, as well as the centrifugations, are carried out at 4°. The mucosal scrapings are homogenized on ice in a Brinkmann Polytron homogenizer (Westbury, NY) for 10 sec at speed 10 with 20 m*M* cold HEPES buffer (pH 7.4; w : v, 1 : 5). The homogenates are centrifuged in a Sorval RT6000 refrigerated centrifuge (DuPont Co., Newton, CT) at 2100*g* at 4° for 30 min. The resulting supernatant represents the postnuclear fraction and is used as the enzyme source.

Incubation Conditions

The postnuclear fraction of human intestinal mucosa (0.5–1.0 mg protein) is incubated under red light with the following additions in glass vials at 37° in a shaking waterbath for up to 30 min: 20 m*M* HEPES buffer (pH 7.4), 150 m*M* KCl, 2 m*M* NAD^+, 2 m*M* dithiothreitol (DTT), and substrates (2 to 10 μ*M* β-carotene dissolved in 10 μl DMSO), in a final volume of 1 ml. Control vials lack either substrate or the tissue homogenate. In addition, in some experiments, the homogenate is boiled for 5 min before incubation with substrate. The vials are left uncovered and therefore exposed to room air as the gas phase. Protein concentration is determined using the BCA (bicinchroninic acid) protein assay (Pierce Co., Rockford, IL).

Extraction Procedures

After incubation, 100 μl of 0.5 *N* KOH in ethanol is added to the 1-ml reaction mixture to stop the reaction, followed by the addition of the internal standards, retinyl acetate, and γ-carotene, each in 100 μl of ethanol and 2 μ*M* α-tocopherol in 100 μl of ethanol as an antioxidant. The metabolites are extracted by adding 2 ml hexane, vortexed, and the mixture then centrifuged for 3 min at 320*g* at 4°. The hexane layer is removed and the

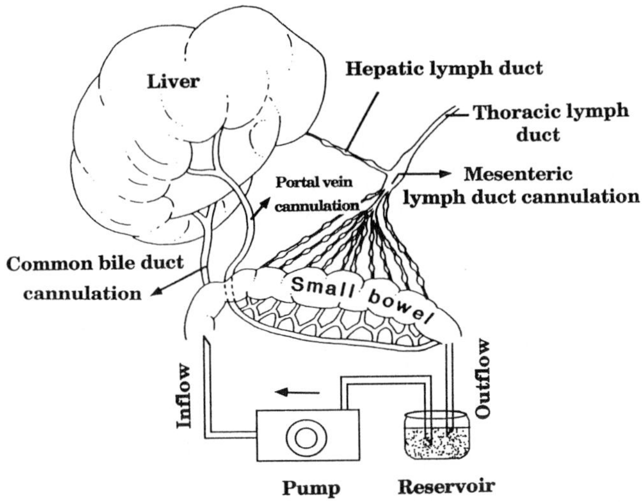

FIG. 2. Presentation of sites of cannulation and sampling in the intestinal perfusion model used for studies of β-carotene absorption and metabolism in ferrets.[14]

residue is acidified (pH 3.0) by adding 50 μl 6 *N* HCl. A second extraction is performed with 2 ml hexane to remove acidic metabolites, such as retinoic acid. The two extractions were pooled, dried under N_2, and resuspended in 50 μl ethanol for injection in the HPLC system described later.

In Vivo Intestinal Perfusion Experiments

Male ferrets (*Mustela putorius furo*) from Marshall Farms (North Rose, NY) were used to study the intestinal absorption and metabolism of β-carotene. They were fed dry ferret food (Win-Hy Foods, Tulsa, OK). The dry ferret food contained β-carotene 0.54 μg/g and retinyl esters (17.7 μg/g, as determined by HPLC in our laboratory. Maintenance and husbandry of adult ferrets (900–1600 g), as well as the procedure of cannulation and sampling, have been described in previous papers.[11,14]

Figure 2 demonstrates the sites of cannulation and sampling for the intestinal perfusion experiments. Following an overnight fast, 3.0 ml of corn oil is administered orally to ferrets to dilate the intestinal lymphatics. Thirty minutes later, anesthesia is induced with ketamine hydrochloride (30 mg/kg) and xylazine (3 mg/kg) administered intramuscularly. Ferrets are intu-

[14] X. D. Wang, N. I. Krinsky, R. Marini, G. Tang, J. Yu, J. G. Fix, and R. M. Russell, *Am. J. Physiol.* **263,** G480 (1992).

bated with 3.0-mm-i.d. endotracheal tubes, and anesthesia is maintained with 2–3% isoflurane in 100% oxygen. Anesthetized ferrets are kept on circulating hot water blankets at 38° and the body temperature is monitored throughout the entire perfusion.

Through a midline abdominal incision, the proximal inflow catheter (0.64-cm o.d. 0.32-cm i.d.; Tygon flexing plastic tubing, Norton, Akron, OH) is inserted 5 cm distal to the ligament of Treitz of the small intestine, and the proximal outflow catheter is introduced 60 cm distal to the inflow catheter. To prevent the perfusate from washing back into the stomach or continuing into the large intestine, encircling ligatures are tied immediately proximal to the inflow catheter and distal to the outflow catheter. The intestinal segment is flushed with normal saline to remove intestinal contents. The common bile duct as well as the mesenteric lymph duct are cannulated using polyethylene catheters (1.27-mm o.d., 0.86-mm i.d., PE-90, Clay Adams, Becton Dickinson and Co., Parsippany, NJ). Bile and lymph are collected into heparinized tubes wrapped in aluminum foil.

The portal vein is identified and then cannulated, allowing heparinized polyethylene tubing (1.22-mm o.d., 0.76-mm i.d., PE-60, Clay Adams, Becton Dickinson and Co.) to be passed into the portal vein without obstructing portal blood flow. This cannula is secured with surgical glue. Dextrose (5%) is perfused through the intestinal segment at 0.5 ml/min until the perfusion experiment begins. The lymph, bile, and portal blood collected in the hour prior to experimental perfusion are saved and analyzed for baseline measurements.

The perfusion experiments are carried out under red light to avoid isomerization or degradation of β-carotene. A syringe pump (Model 22, Harvard Apparatus, South Natick, MA) is used to perfuse a micellar solution through the intestinal segment at a flow rate of 2.0 ml/min. all-*trans*-β-Carotene or 9-*cis*-β-carotene (2 to 10 μ*M*) dissolved in 0.5 ml DMSO is prepared under red light in a mixed micellar solution containing 0.4 μ*M* α-tocopherol, 2.5 m*M* oleic acid, and 10 m*M* sodium taurocholate in Krebs phosphate buffer at pH 7.0. The micellar solution is formed by sonication for 15 min at 80 W of power before the perfusion. The stability of β-carotene after sonication and after 8 hr of storage at room temperature is checked by HPLC. No oxidative products were detected. The extent of micellar incorporation of β-carotene in the perfusate is examined by filtration through a 0.2-μm filter (UNIFLOTM, Schleicher and Schuell, Inc., Keene, NH). β-Carotene was not detected by HPLC in the solution after passage through the filter, which indicates that the extent of micellar incorporation of β-carotene in the perfusate was close to 100%.

For the next 2–4 hr, the β-carotene-containing micellar solution is perfused continuously. The portal vein cannula is sampled every hour: a 1.0-

ml sample is withdrawn by syringe within a 3-min period and the same volume of normal saline is simultaneously injected into the portal vein. After perfusion the animals are killed by puncturing the abdominal aorta under deep isoflurane anesthesia. The perfused intestinal segment is removed, freed of its mesentery and serosal fat, and weighed. The intestinal mucosa are scraped with a glass slide, and homogenized in a Brinkmann Polytron homogenizer with ice-cold HEPES buffer and methanol (v:v, 2:1). After the intestinal scrapings are collected, the segments are suspended with a 5-g weight tied to one end for 24 hr of drying to ensure a constant degree of stretching. At the end of the drying period, the length of each segment is recorded.

Extraction of *in Vivo* Products

The samples (lymph, serum, bile, intestinal mucosal scrapings) are extracted as follows: 100 μl of an ethanolic solution of 0.5 *N* KOH is added to either 0.8–2.0 ml of lymph or serum or 0.5 g of intestinal mucosal scrapings, followed by the addition of the internal standards, retinyl acetate, tocol, and γ-carotene, each in 100 μl of ethanol. The metabolites are extracted by adding 2 ml hexane, and the mixture is then centrifuged for 3 min at 320 g at 4°. The hexane layer is removed and the residue acidified by adding 50 μl 6 *N* HCl. A second extraction is performed with 2 ml hexane. The two extractions are pooled, dried under N_2, and resuspended in 50 μl ethanol for injection in the HPLC system described next.

High-Performance Liquid Chromatography

A gradient reversed-phase HPLC system described earlier[15,16] for the analysis of retinoids and carotenoids is used with minor modifications. The HPLC system consists of two Waters 510 pumps (Waters Chromatography Division of Millipore Corp., Milford, MA); a Waters 490E multiwavelength spectrophotometer detector is set at 340 nm (retinol and retinoic acid), 380 nm (retinal and β-apo-14′-carotenoic acids), 400 nm for β-apocarotenoids, and 450 nm for β-carotene and β-apocarotenoids (Table I). A Pecosphere-3 C_{18} 0.46- × 8.3-cm cartridge column (Perkin-Elmer, Norwalk, CT) and Waters 840-Digital 350 data station are used. A Waters 715 Ultra Wisp autosampler is used for sample injection.

[15] X.-D. Wang, G.-W. Tang, J. G. Fox, N. I. Krinsky, and R. M. Russell, *Arch. Biochem. Biophys.* **285,** 8 (1991).

[16] G. Tang and N. I. Krinsky, *Methods Enzymol.* **214,** 69 (1993).

TABLE I
CHARACTERISTICS OF VARIOUS β-APOCAROTENOIDS AND RETINOIDS[a]

Compounds	Retention time (min)	Absorption maxima (nm)	Detection wavelength[b] (nm)	$E^{1\%}_{1\,cm}$ [c]	Molecular weight
β-Apo-8′-carotenal	11.8	465	450	2640	416
β-Apo-8′-carotenoic acid	9.7	441	450	2516	432
β-Apo-10′-carotenal	10.8	444	450	2190	376
β-Apo-10′-carotenoic acid	8.5	424	400	2066	392
β-Apo-12′-carotenal	10.6	428	400	2160	350
β-Apo-12′-carotenoic acid	8.3	408	400	2036	366
β-Apo-14′-carotenal	8.8	404	400	1664	310
β-Apo-14′-carotenoic acid	6.1	378	380	1601	326
β-Apo-15-carotenal (retinal)	7.3	375	380	1548	284
β-Apo-15-carotenol (retinol)	6.5	325	340	1835	286
β-Apo-15-carotenoic acid (retinoic acid)	5.7	340	340	1485	300

[a] The retention times and absorption maxima are those detected in the HPLC separation described in the high-performance liquid chromatography section.
[b] The detection wavelength indicates which channel is used to detect these compounds.
[c] The $E^{1\%}_{1\,cm}$ value is adjusted for each detection wavelength.

The gradient procedure at a flow rate of 1 ml/min is as follows: 100% solvent A [acetonitrile (CH_3CN) : tetrahydrofuran (THF) : water, 50 : 20 : 30, v/v/v, with 0.35% acetic acid and 1% ammonium acetate in water] for 3 min, followed by a 6-min linear gradient to 40% solvent A and 60% solvent B (CH_3CN : THF : water, 50 : 44 : 6, v/v/v, with 0.35% acetic acid and 1% ammonium acetate in water), a 12-min hold at 40% solvent A/60% solvent B, then a 7-min gradient back to 100% solvent A. Individual carotenoids and retinoids are identified by coelution with standards, and quantified relative to the internal standards (retinyl acetate and γ-carotene), by determining peak areas calibrated against known amounts of standards. In this HPLC system, the retinoid and carotenoid we are interested in is eluted as shown in Table I. Figure 3 illustrates the HPLC separation of various intermediates of β-carotene metabolism observed in the human intestinal homogenate system.[7] For analysis of retinoic acid isomers, we used the same HPLC system, which could separate 9-*cis*-, 13-*cis*-, and all-*trans*-retinoic acid (Fig. 4). We also use another gradient reversed-phase HPLC system[17,18] for the analysis of retinoic acid isomers. In this method, retinoic acid isomers

[17] G. Tang and R. M. Russell, *J. Lipid Res.* **31,** 175 (1990).
[18] X. Hebuterne, X.-D. Wang, E. J. Johnson, N. I. Krinsky, and R. M. Russell, *J. Lipid Res.* **36,** 1264 (1995).

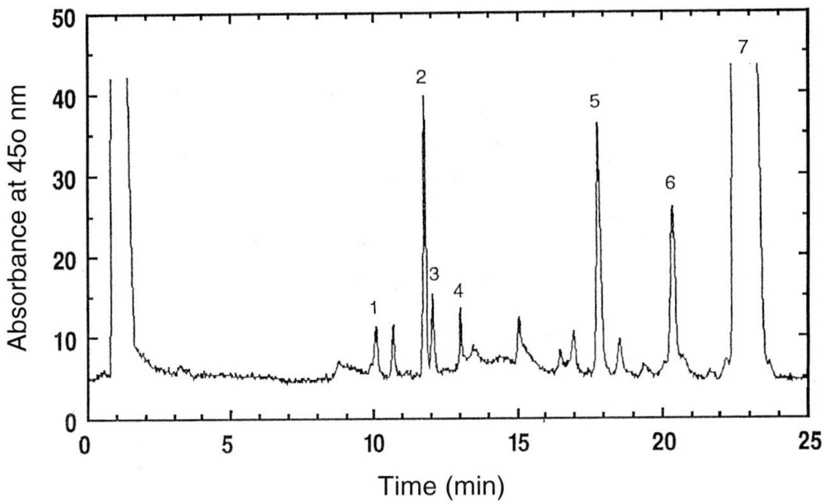

FIG. 3. The HPLC separation of various intermediates of β-carotene metabolism observed in the human intestinal homogenate system, using the HPLC described, and monitoring the metabolites at 450 nm.[7] Peak identification: 1, β-apo-14′-carotenal; 2, β-apo-12′-carotenal; 3, β-apo-10′-carotenal; 4, β-apo-8′-carotenal; 5, 5,6-epoxy-β-carotene; 6, γ-carotene (internal standard); 7, β-carotene.

are analyzed by reversed-phase HPLC on two Pecosphere 3 × 3CR ODS cartridge columns (Perkin-Elmer) using CH_3OH–H_2O [solvent A, 75:25 (v/v) 1% ammonium acetate in H_2O] and 100% CH_3OH (solvent B) as previously described.[17] A normal phase HPLC system for separation of retinal and retinoic acid was described earlier.[19]

Identification of Metabolites

The identification of metabolites from β-carotene is based on HPLC, UV spectrum, chemical derivatization, and GC/MS. Individual carotenoids and retinoids are initially identified by coelution with standards, and quantified relative to the internal standard (γ-carotene for carotenoids and retinyl acetate for retinoids) by determining peak areas calibrated against known amounts of standards. An additional Waters 994 programmable photodiode array detector is used for measurement of absorption spectra. To obtain a more detailed analysis of the retinoic acid isomers formed during these incubations, the retinoic acid fraction from the HPLC chromatogram is collected and dried under the N_2. The residue is dissolved in 3 ml peroxide-

[19] J. L. Napoli, *Methods Enzymol.* **189,** 470 (1990).

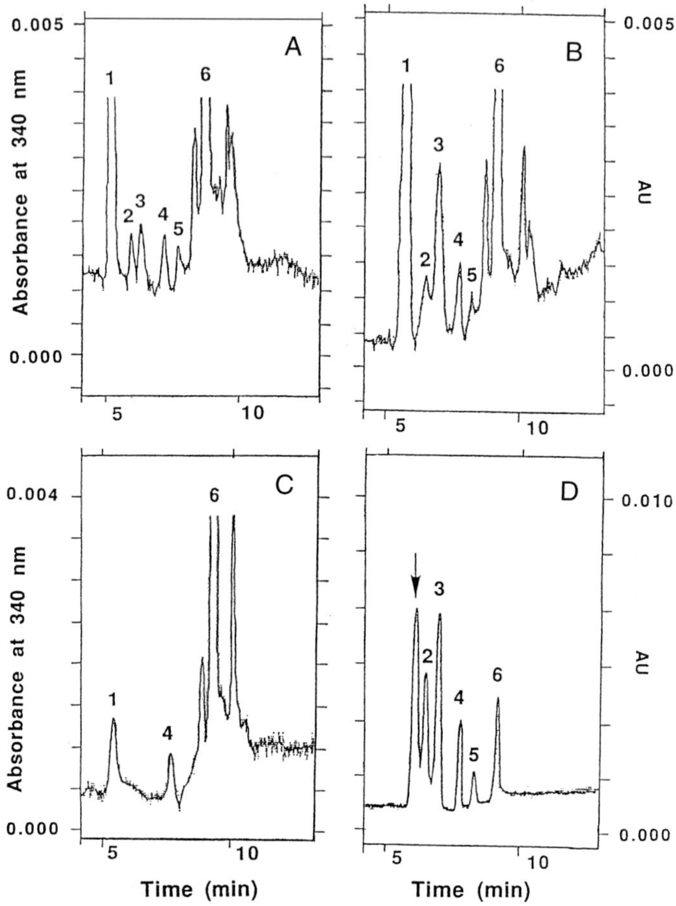

FIG. 4. HPLC profile of extracts of a human intestinal postnuclear fraction after incubation with either 9-*cis*-β-carotene or all-*trans*-β-carotene. (A) After incubation with 6 μ*M* 9-*cis*-β-carotene. (B) After incubation with 6 μ*M* 9-*cis*-β-carotene plus 4 μ*M* all-*trans*-β-carotene. (C) Control incubation using tissue boiled for 5 min. (D) Elution pattern of various standard retinoids. Peak identification: 1, β-apo-13-carotenone; 2, 9-*cis*-retinoic acid; 3, all-*trans*-retinoic acid; 4, retinol; 5, retinal; 6, retinyl acetate (internal standard). The peak indicated by the arrow is a 13-*cis*-retinoic acid standard. (Reprinted with permission from Wang *et al.*[9])

free diethyl ether and derivatized with diazomethane to form the methyl retinoates.[9] Authentic 9-*cis*-retinoic acid and all-*trans*-retinoic acid, as well as the retinoic acid fraction from either the incubation of 9-*cis*-β-carotene with human intestinal mucosa fraction or the perfusion of β-carotene with ferret intestine, are reanalyzed using the same HPLC system, as shown in

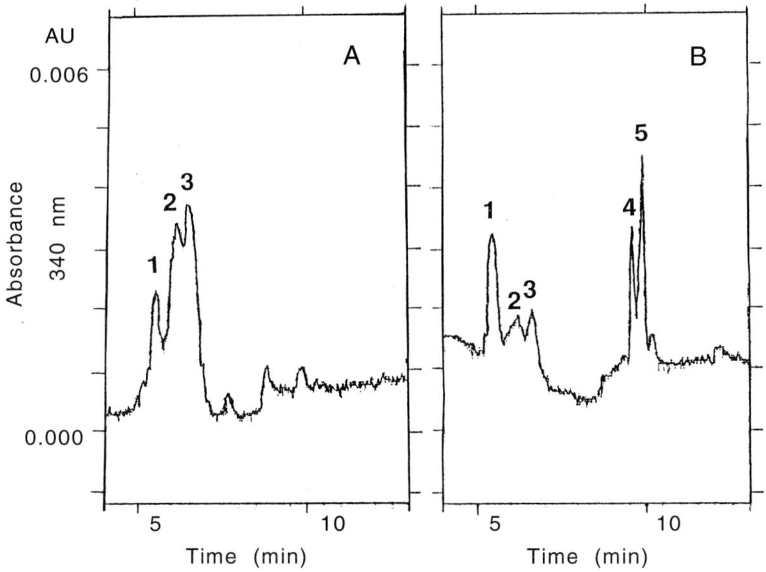

FIG. 5. HPLC profile of the polar retinoid fraction obtained from extracts of human intestinal postnuclear fractions after incubation with 9-*cis*-β-carotene. (A) Original HPLC profile. (B) After derivatization with ethereal diazomethane and an additional HPLC separation. Peak identification: 1, β-apo-13-carotenone; 2, 9-*cis*-retinoic acid; 3, all-*trans*-retinoic acid; 4, 9-*cis*-metnyl retinoate; 5, all-*trans*-methyl retinoate. (Reprinted with permission from Wang *et al.*[9])

Fig. 5. Methyl 9-*cis*-retinoate and methyl all-*trans*-retinoate from authentic samples of retinoic acid elute at 9.5 and 9.8 min, respectively. To confirm the identification of the retinoic acids formed, the methyl retinoate formed is analyzed using negative ion chemical ionization (NCI)-GC/MS.[11] The NCI-GC/MS consists of a Hewlett Packard Model 5988A mass spectrometer and a Hewlett Packard Model 5890 series II GC. The gas chromatography (GC) column is an aluminum-clad, fused-silica capillary column coated with a high-temperature HT-5 stationary phase. The on-column injection port temperature is initially set at 53° and the column oven at 50°. Both are programmed to increase by 20°/min to 350°. Helium is used as the GC carrier gas and methane is employed as the reaction gas for electron capture-NCI mass spectrometry. The ion source temperature is set at 150° and the ion source pressure is 0.5 torr of methane. The retention time and molecular mass in the GC/MS output are compared with standard methyl retinoate. The detection limit for this method is as low as 1 pg.

TABLE II

SYNTHESIS OF RETINOIC ACID FROM 9-*cis*-β-CAROTENE AND/OR ALL-*TRANS*-β-CAROTENE[a,b]

Substrate	Concentration (μ*M*)	Retinoic acid (pmol/h/mg protein) 9-*cis*-	all-*trans*-	13-*cis*-	Total	Ratio of 9-*cis*/total
9-*cis*-β-Carotene	4	16 ± 1	18 ± 2	ND	34	0.47
all-*trans*-β-Carotene	2	ND	18 ± 4	ND	18	0.00
9-*cis*-β-Carotene	4					
+ all-*trans*-β-Carotene	2	16 ± 2	38 ± 6	ND	54	0.30

[a] In the human intestinal postnuclear fraction.

[b] Data are expressed as the mean ± SEM (at least three determinations). ND, not detected. (Reprinted with permission from Wang *et al.*[9])

Results of *in Vitro* Investigation

Our previous studies demonstrated that retinoic acid can be produced from β-carotene and β-apocarotenals.[7] The incubation of varying concentrations of either β-apo-8′-carotenal or β-apo-12′-carotenal with human intestinal homogenate produced 6.6 or 2.7 times more retinoic acid than retinal, respectively; when β-carotene was used as the substrate, only one-third as much retinal as retinoic acid was produced. We have shown that only retinoic acid was produced when β-apo-14′-carotenoic acid was the substrate in the mitochondrial fraction of rabbit liver.[20] To ascertain if the production of retinoic acid occurred via the central cleavage pathway, involving direct oxidation of retinal to retinoic acid, or via the excentric pathway, involving oxidation of β-apocarotenals to retinoic acid, we used the inhibitor, citral, which could block the oxidation of retinal in human intestinal mucosa in these experiments *in vitro*. We demonstrated that 1 m*M* citral did not block the production of various β-apocarotenals and retinoic acid from the metabolism of β-carotene.

The identification of 9-*cis*-retinoic acid as the specific ligand for the RXR class of nuclear receptors raises the question as to the source of 9-*cis*-retinoic acid in the body. There is little information available about the formation of 9-*cis*-retinoic acid, although it is known that it is not formed from all-*trans*-retinoic acid during a normal extraction procedure. 9-*cis*-Retinoic acid may arise via isomerization of all-*trans*-retinoic acid in the body, by the oxidation of 9-*cis*-retinal or through the conversion of 9-*cis*-β-carotene to this biologically active retinoid. In our recent study, we focused on the enzymatic conversion of either-all-*trans*-β-carotene or 9-*cis*-

[20] X.-D. Wang, R. M. Russell, D. E. Smith, and N. I. Krinsky, *J. Biol. Chem.* **271,** 26490 (1996).

TABLE III
CONCENTRATION OF RETINOIDS AND β-APO-12′-CAROTENAL IN FERRET INTESTINAL MUCOSA[a]

	Perfusion with			
	β-Carotene		Retinal	
Substance	− Citral	+ Citral	− Citral	+ Citral
Retinoic acid	36 ± 3	19 ± 5[b]	30 ± 2	ND
β-Apo-12′-carotenal	79 ± 10	68 ± 19	ND	ND
Retinol	251 ± 3	437 ± 68[b]	994 ± 337	2084 ± 185[b]
Retinyl esters	7310 ± 2882	6847 ± 2978	10440 ± 2176	5446 ± 897

[a] After a 2-hr perfusion of 10 μM β-carotene or 1 μM retinal with or without citral. Values are means ± SEM (n = 3). (Modified with permission from Hebuterne *et al.*[8])
[b] Significantly different at $P < 0.05$.

β-carotene into all-*trans*-retinoic acid and 9-*cis*-retinoic acid by incubation of homogenates of human intestinal mucosa. We demonstrate[9] that the intestinal cleavage of dietary 9-*cis*-β-carotene can provide a source of 9-*cis*-retinoic acid for the human body (Table II). This observation is based on the identification of 9-*cis*-retinoic acid as a product of 9-*cis*-β-carotene metabolism by HPLC comigration with the authentic material (Fig. 4), and the formation of a methyl retinoate that comigrated with authentic 9-*cis*-methylretinoic acid (Fig. 5).

Results of *in Vivo* Experiments

We have extended our *in vitro* observation to the *in vivo* ferret model.[18] In the intestinal perfusion of β-carotene in ferret model, retinoic acid was

TABLE IV
SYNTHESIS OF RETINOIC ACID ISOMERS FROM MICELLAR SOLUTIONS OF 9-*cis*-β-CAROTENE OR ALL-*TRANS*-β-CAROTENE[a]

Substrate (10 μM)	Retinoic acid (pmol/h/mg protein)		
	9-*cis*	all-*trans*-	13-*cis*-
Micellar solution control	1.1 ± 0.1	5.8 ± 0.7	0.9 ± 0.3
9-*cis*-β-Carotene	4.0 ± 0.5*	9.0 ± 1.0	1.1 ± 0.3
all-*trans* β-Carotene	1.0 ± 0.3	14.0 ± 0.9*	1.1 ± 0.2

[a] During intestinal perfusion of the ferret. Data are expressed as the mean ± SEM (at least three ferrets in each group). (Modified with permission from Hebuterne *et al.*[18])

identified by comparing the retention times in HPLC, by UV spectrum, by methylation, and by subsequent GC/MS analysis.[8,11,18] Retinoic acid formation was completely inhibited when retinal was perfused with citral through the ferret intestine (Table III). However, retinoic acid and β-apocarotenal were both formed from perfusion of ferret intestine with β-carotene in both the presence and absence of citral (Table III), which proves the existence of an excentric cleavage pathway of β-carotene for retinoic acid synthesis in living body.

The *in vitro* results were further demonstrated in the ferret model. The ferret intestinal perfusion of β-carotene isomers have shown that, after the perfusion of all-*trans*-β-carotene, all the retinoic acid formed was in the all-*trans* form, whereas the perfusion of 9-*cis*-β-carotene results in the biosynthesis of about 50% of the total retinoic acid as the 9-*cis* isomer (Table IV).[18]

Conclusion

β-Carotene is an important precursor of retinoic acid in the intestinal mucosa both *in vitro* and *in vivo.* The intestinal cleavage of dietary 9-*cis*-β-carotene can provide a source of 9-*cis*-retinoic acid for the human body. The conversion of β-carotene to retinoic acid involves at least two pathways, namely, a central cleavage pathway and an excentric cleavage pathway.

Acknowledgments

Much of the work reported here has been supported by National Institutes of Health grant CA49195 and U.S. Department of Agriculture grant 94-37200-0444.

[12] Assessing Metabolism of β-[^{13}C]Carotene Using High-Precision Isotope Ratio Mass Spectrometry

By ROBERT S. PARKER, J. THOMAS BRENNA, JOY E. SWANSON, KEITH J. GOODMAN, and BONNIE MARMOR

Introduction

Many fundamental aspects of the metabolism of β-carotene in the human remain unresolved, including the range of efficiency of absorption, extent and stoichiometry of conversion of β-carotene to vitamin A, extent of postabsorptive conversion to vitamin A, and rate of plasma turnover.

0076-6879/97 $25.00

Some of these parameters, for example, postabsorptive conversion to vitamin A, are difficult or impossible to assess directly, and must be studied indirectly using modeling approaches, ideally under steady-state conditions. Most (perhaps all) of the above issues are best addressed using tracer approaches, since tracers offer many advantages over use of unlabeled β-carotene. Chief among these advantages are the ability to distinguish between administered and endogenous β-carotene and the ability to measure or model metabolic events under steady-state conditions. Tracer methods using ^{3}H- or ^{14}C-labeled β-carotene were used in the mid-1960s,[1,2] but cannot now be applied to health individuals. A stable tracer method based on the use of octadeuterated β-carotene has been published.[3] In this chapter we describe a stable isotope tracer method based on the use of highly enriched β-[U-^{13}C]carotene and high-precision isotope ratio mass spectrometry (IRMS). A preliminary report of this approach has appeared,[4] and details and improvements in the method are described later. Biological applications of IRMS have been reviewed.[5]

IRMS instruments are highly specialized mass spectrometers designed for the high-precision determination of isotope ratios for C, H, N, O, or S. Unlike organic mass spectrometers, samples must be converted to one of several gases prior to introduction to the IRMS. For C, samples are usually combusted to yield CO_2, with an isotope ratio representative of the material of interest. Carbon dioxide is admitted to a tight electron impact (IE) ion source and produces molecular ions at m/z 44, 45, and 46, which are monitored continuously using three Faraday cup detectors. The beams are comprised primarily of $^{12}C^{16}O_2$, $^{13}CO_2$ + $^{12}C^{17}O^{16}O$, and $^{12}C^{18}O^{16}O$. The m/z 46 signal is used to adjust the m/z 45 for the contribution of ^{17}O, yielding a ratio of $^{13}C/^{12}C$. For tracer applications employing a baseline correction, the ^{17}O correction is negligible and can be ignored. Since 1990, gas chromatography (GC) interfaced to IRMS by means of an inline microcombustion furnace (GCC-IRMS) has been available commercially and has facilitated high-precision determination of $^{13}C/^{12}C$ and $^{15}N/^{14}N$ from mixtures separated by GC. Advances toward high-precision organic D/H have been described.[6]

[1] R. Blomstrand and B. Werner, *Scand. J. Clin. Lab. Invest.* **19,** 339 (1967).

[2] D. S. Goodman, R. Blomstrand, B. Werner, H. S. Huang, and T. Shiratori, *J. Clin. Invest.* **45,** 1615 (1966).

[3] S. R. Dueker, A. D. Jones, G. M. Smith, and A. J. Clifford, *Anal. Chem.* **66,** 4177 (1994).

[4] R. S. Parker, J. E. Swanson, B. Marmor, K. J. Goodman, A. B. Spielman, J. T. Brenna, S. M. Viereck, and W. K. Cranfield, *Ann. N.Y. Acad. Sci.* **691,** 86 (1993).

[5] J. T. Brenna, *Acc. Chem. Res.* **27,** 340 (1994).

[6] H. Tobias and J. T. Brenna, *Anal. Chem.* **68,** 3002 (1996).

Procedures

Purification of all-trans-β[^{13}C]Carotene from Algal Extracts

Crude hexane–acetone extracts of *Dunaliella* sp. grown in a closed system with $^{13}CO_2$ as sole carbon source were obtained from Martek (Columbia, MD). Algal lipids are subjected to potassium hydroxide (KOH) saponification to remove glycerides and chlorophylls. Analysis of ^{13}C enrichment in perhydro-β-carotene (see below) using organic mass spectrometers, or IRMS following serial dilution with unlabeled β-carotene, indicated ^{13}C substitution of >98%. all-*trans*-β-[U-^{13}C]carotene can be purified to >98% by repeated crystallization from petroleum ether, the remainder being primarily α-carotene.

Preparation and Administration of β-[^{13}C]Carotene

β-[U-^{13}C]Carotene (1–2 mg) is dissolved in 1 ml dichloromethane, and 1 g high oleic acid safflower oil is added. The solvent is removed under vacuum until the expected weight of oil plus β-carotene is achieved. The oil solution of β-carotene is diluted with 19 ml additional safflower oil and emulsified into 70 ml non-vitamin-fortified skim milk plus 30 g banana (for emulsion stability and taste) using a hand-held homogenizer. This emulsion containing labeled β-carotene is consumed with a small standardized meal, plus an additional 100 ml of non-vitamin-fortified skim milk to rinse the container and palate.

To standardize conditions in the upper gastrointestinal tract and allow clearance of previously consumed carotenoids, subjects are placed on a low carotenoid diet from 48 hr prior to the β-[U-^{13}C]carotene dose through 36 hr postdose. Standard lunch and evening meals are consumed 3 and 9 hr postdose, respectively. Standardization is particularly important with subjects undergoing repeated testing.

On the morning of dosing, subjects are fitted with an indwelling catheter with a three-way Luer stopcock in a forearm vein using sterile technique. The stopcock assembly is convenient for frequent blood sampling. Blood samples are collected in a plastic syringe and placed in heparinized culture tubes for plasma separation. Between blood draws the stopcock assembly is flushed with sterile saline containing 10 U/ml heparin to maintain patency. Blood samples are ideally collected hourly over the initial 15 hr if pharmacokinetic data during the absorption period is required, as illustrated in Fig. 2. Less frequent sampling can be performed after this period.

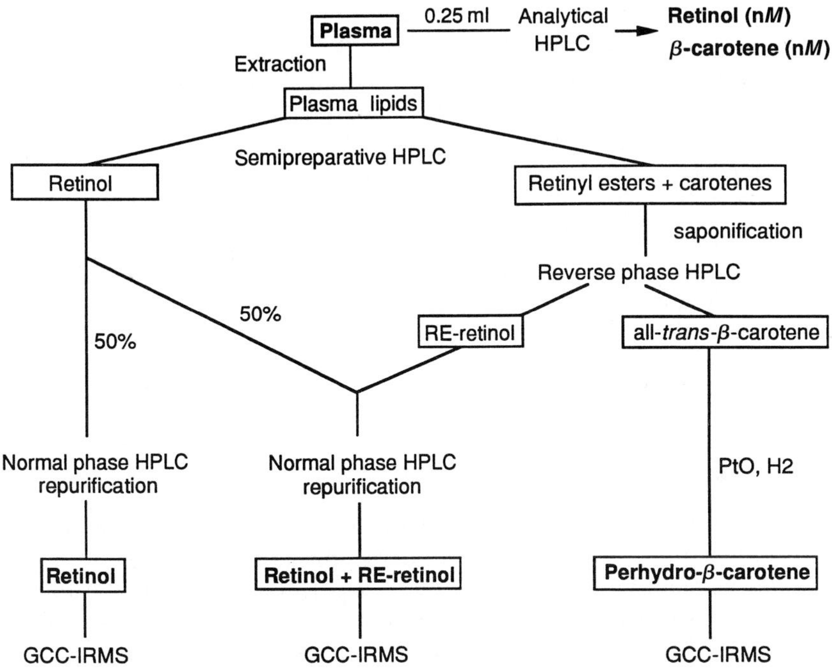

FIG. 1. Sample processing scheme for determination of plasma concentration of retinol and β-carotene by HPLC, and carbon isotope ratio in unesterified retinol, total vitamin A, and perhydro-β-carotene fractions.

High-Pressure Liquid Chromatography Quantification of β-Carotene and Retinol Concentrations in Plasma

The overall analytical scheme is illustrated in Fig. 1. The general approach involves measurement of (1) plasma retinol and β-carotene concentration by HPLC, and (2) $^{13}C/^{12}C$ by GCC-IRMS. The plasma concentration of labeled β-carotene, retinol, and retinyl ester is then calculated using both HPLC and GCC-IRMS data as described later.

Plasma concentrations of β-carotene and retinol are determined by a modification of the method of Thurnham *et al.*[7] Duplicate plasma portions (0.25 ml) were deproteinized with one volume ethanol containing internal standard (retinyl acetate) and extracted twice with two volumes hexane. Combined hexane extracts are dissolved in 40 μl dimethylformamide, diluted with 210 μl mobile phase [acetonitrile–methanol–chloroform,

[7] C. I. Thurnham, E. Smith, and P. S. Flora, *Clin. Chem.* **34,** 377 (1988).

47 : 47 : 6 (v/v), containing 0.05 *M* ammonium acetate and 1% triethylamine (v/v)], sonicated, and subjected to HPLC analysis.

The HPLC conditions consist of a 4.6-mm × 15-cm Spherisorb ODS-2 column (LKB Instruments Ltd., Surrey, UK) maintained at 26°, a flow rate of 1.2 ml/min, and a photodiode array detector (Waters 996, Millipore Corp., Milford, MA). The retention times of retinol, retinyl acetate (internal standard), and β-carotene were 2.1, 2.6, and 16.9 min, respectively. Plasma concentrations of retinol and β-carotene are calculated using calibration curves and corrected for volume recovery using the internal standard. The method is validated against plasma samples of known β-carotene and retinol concentrations obtained from the National Institute of Standards and Technology. The coefficients of variation for retinol and β-carotene are about 3% and 7%, respectively.

Preparation of Fractions for GCC-IRMS Analysis

Duplicate plasma samples (1.5–2.2 g samples are most convenient) are deproteinized with one volume ethanol and extracted twice with three volumes hexane. Unesterified retinol is separated from retinyl esters and β-carotene using reversed-phase semipreparative HPLC on a Vydac TP201 column (10 mm × 25 cm, Separations Group, Hesperia, CA), using methanol–dichloromethane (76 : 24, v/v) at a flow rate of 1.2 ml/min and a temperature of 35°. Retinol elutes at 9.7 min, and the fraction containing β-carotene and retinyl esters (and other carotenes) between 16.5 and 18 min.

The retinyl ester–carotene fraction is saponified in 2% ethanolic KOH at 45° for 25 min and extracted with hexane. The hexane phase, containing carotenes plus retinol derived from retinyl ester, is evaporated and redissolved in methanol-dichloromethane (90 : 10) and subjected to analytical reversed phase HPLC using a Vydac TP201 column (4.6 mm × 15 cm, Separations Group) and a mobile phase of methanol-dichloromethane (90 : 10) at 0.8 ml/min. Retinol and all-*trans*-β-carotene elute at 2.2 and 8 min, respectively, and are collected in glass screw-cap vials.

The β-carotene fraction is evaporated to dryness, redissolved in chloroform, and hydrogenated to the thermally stable perhydro-β-carotene analog using platinum oxide under hydrogen gas, overnight at room temperature in the dark. The hydrogenated β-carotene samples are filtered, redissolved in 10 μl hexane, and subjected to GCC-IRMS as described in the next section.

The plasma unesterified retinol fraction is divided into two equal portions, and one portion combined with the retinyl ester–retinol fraction to yield a *total vitamin A* fraction. The total vitamin A fraction and the remaining half of the unesterified retinol fraction is then subjected to further purification by normal phase HPLC using a 15-cm nitrile column and a mobile phase of hexane–2-propanol (95:5, v/v) at 0.8 ml/min. The retinol fraction is then subjected to IRMS as described later. Unlike β-carotene, retinol can be subjected to gas chromatography without thermal degradation.

GCC-IRMS Analyses of Retinol and Perhydro-β-Carotene Fractions

The carbon isotopic ratio of retinol and perhydro-βC is determined using a Finnigan MAT (San Jose, CA) model 252 high-precision isotope ratio mass spectrometer interfaced to a 5880A Hewlett Packard gas chromatograph via a ceramic combustion furnace maintained at 850°. For both retinol and perhydro-β-carotene analysis, hexane solutions (1 μl) are injected onto a DB-1 capillary GC column (0.32-μm i.d. × 15 m, J&W Scientific, Folsom, CA) using an on-column injector (J&W Scientific). The linear velocity of the high-purity helium carrier gas is 20 cm/sec. For perhydro-β-carotene, the GC is programmed from 60 to 265° at 25°/min, 265 to 300° at 10°/min, held at 300° for 4 min, and then increased to 325° at 30°/min and held for 10 min to ensure elution of any remaining compounds. perhydro-β-carotene elutes at about 13 min (approximately 300°). For retinol analysis, the GC was programmed from 60 to 200° at 25°/min, 200 to 235° at 6°/min, and 235 to 315° at 30°/min to elute any remaining materials. Retinol elutes at approximately 10 min.

The column eluant is continuously and quantitatively combusted to CO_2 by the combustion furnace, the water of combustion removed by a continuous-flow water trap, and CO_2 swept into the ion source. Carbon dioxide of precisely known isotope ratio (OzTech Trading Corp., Fremont, CA), calibrated relative to an international carbonate standard, PeeDee Belemnite (PDB), is admitted directly into the ion source at preprogrammed times, usually at the beginning of the run and about 1 min following elution or perhydro-β-carotene. As discussed earlier, masses 44, 45, and 46 are simultaneously monitored. The coefficient of variation of isotope ratio in retinol or perhydro-β-carotene is typically less than 6% from plasma samples.

Calculation of Plasma Concentration of ¹³C-Labeled Retinol, Retinyl Ester, and β-Carotene

The observed $^{13}C/^{12}C$ ratio, standardized against the calibrated external CO_2 standard, is initially reported in the $\delta^{13}C$ notation ["δ", or ^{13}C concentration in parts per thousand (per mil) relative to PDB]. Such values can be converted to a more convenient form, atom percent ^{13}C (AP), according to the following equation:

$$\text{Atom \% }^{13}\text{C} = \frac{(100 \times R_{\text{PDB}})(\delta^{13}\text{C}/1000 + 1)}{(1 + R_{\text{PDB}})(\delta^{13}\text{C}/1000 + 1)} \tag{1}$$

where δ represents the reported delta value of the sample (e.g., retinol), and R_{PDB} represents the $^{13}C/^{12}C$ intensity ratio for the international CO_2 standard, PDB (0.0112372). Baseline (predose) AP values are subtracted from all subsequent values to yield atom percent excess (APE) ^{13}C in each fraction (retinol, total vitamin A, perhydro-β-carotene). APE is proportional to the percentage of total analyte molecules, which are present as the ^{13}C-enriched isotopomer at any point in time after baseline, and controls for both natural abundance ^{13}C and ^{13}C enrichment persisting from previous doses of β-[^{13}C]carotene. APE can also be expressed in terms of atom fraction excess ^{13}C (F), where F = APE/100.

The plasma concentration of [^{13}C]retinol and β-[^{13}C]carotene is calculated as the product of the atom fraction excess ^{13}C (F) and the mean plasma concentration (determined by HPLC), corrected for the atom fraction excess ^{13}C in the dose, for example:

$$\beta\text{-}[^{13}\text{C}]\text{Carotene (n}M) = \frac{(F_{\beta\text{C}}) \times \beta\text{C(n}M)}{F_{\text{DOSE}}} \quad (2)$$

Because plasma retinyl esters levels are typically too low for convenient direct analysis by GCC-IRMS, the plasma concentration of [^{13}C]retinyl ester is calculated as the concentration of [^{13}C]retinol in the combined retinol plus retinyl ester fraction minus the concentration of [^{13}C]retinol in the unesterified retinol fraction. For this calculation, the HPLC concentration term [numerator in Eq. (2)] is represented by one-half that of plasma unesterified retinol, and the mass contribution of retinyl ester–retinol to the total retinol fraction is ignored, because it typically represents less than 1% of total plasma retinol, particularly with tracer oral doses of β-carotene. Standard errors associated with the plasma concentration of labeled analytes are calculated, taking into account the error associated with both HPLC and GCC-IRMS assays.

Results

Examples of the short-term (0–50 hr) and long-term kinetics of plasma ^{13}C-labeled β-carotene, retinyl ester, and unesterified retinol in a subject after a single oral dose of approximately 2 mg β-[U-^{13}C]carotene is illustrated in Figs. 2 and 3. The concentration peak is labeled β-carotene and retinyl ester at 5 hr (Fig. 2) corresponds to the known kinetics of chylomicrons, and represents absorption of unmetabolized ^{13}C-labeled β-carotene and its chief intestinal metabolite, [^{13}C]retinyl ester. Labeled β-carotene exhibits a second broad peak between 24 and 48 hr, reflecting hepatic secretion of β-carotene in very low density lipoproteins (VLDLs) and subsequent lipolysis of VLDL to lower density lipoproteins. [^{13}C]Retinol, a minor

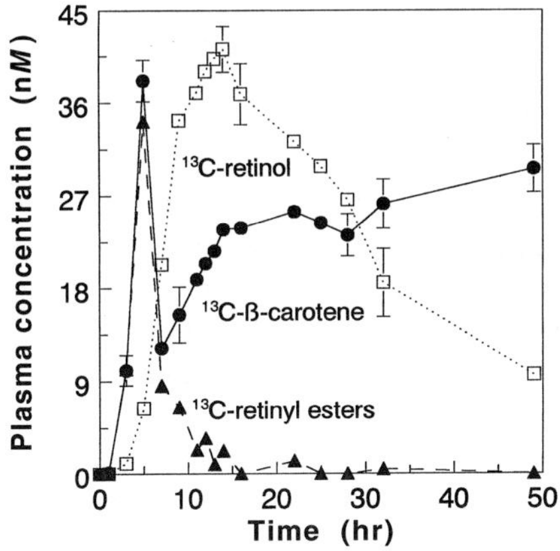

FIG. 2. Kinetics of ^{13}C-labeled β-carotene, retinyl esters, and retinol in human plasma over 50 hr following a single oral dose of 2 mg all-*trans*-β-[U-^{13}C]carotene. Data are means and standard deviations of duplicate determinations at each time point.

intestinal metabolite of β-[^{13}C]carotene, exhibits a single peak at about 12 hr, reflecting hepatic secretion of the retinol–retinol-binding protein–transthyretin complex. Longer term plasma kinetics of ^{13}C-labeled retinol and β-carotene are illustrated in Fig. 3.

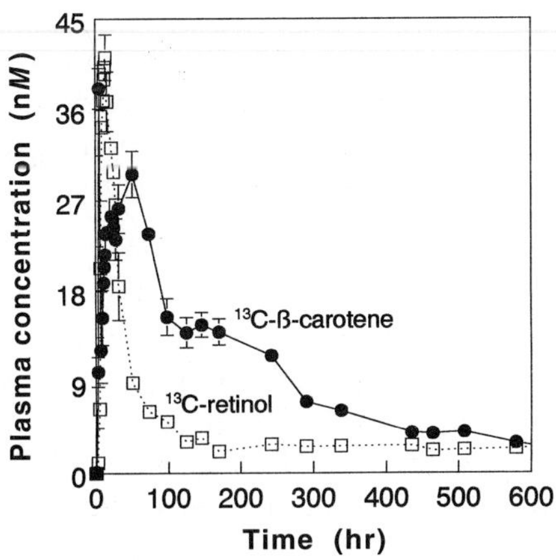

FIG. 3. Long-term kinetics of ^{13}C-labeled β-carotene and retinol in human plasma following a single oral dose of 2 mg all-*trans*-β-[U-^{13}C]carotene. Data are means and standard deviation of duplicate determinations at each time point. (Data from the same subject as that of Fig. 2.)

Discussion

Advantages of Approach

In GCC-IRMS, combusted analytes are detected as the CO_2^+ produced in a high-sensitivity tight El ion source. Each carbon atom of the analyte molecule has an independent chance of ionization and therefore detection limits are related to moles of analyte carbon, rather than to moles of analyte as in GC/MS. Further, the ionization probability and molecular ion stability of CO_2 is constant for all analytes. For these reasons, compounds of low ionization efficiency, low stability, and high molecular weight tend to be detected with high sensitivity by GCC-IRMS compared to conventional organic GC/MS. β-Carotene is one of the better examples; as a hydrocarbon it tends to be poorly ionized and its molecular ion is unstable. On combustion it yields 40 mol of CO_2 per mole β-carotene.

GCC-IRMS permits use of low doses typical of daily β-carotene intake, which do not perturb endogenous pool sizes of β-carotene or retinol. As illustrated in Fig. 2, β-[^{13}C]carotene is clearly evident in plasma by 3 hr after a dose of 2 mg β-[U-^{13}C]carotene. In contrast, no detectable increase in concentration of labeled βC could be observed prior to 5 hr postdose with 40 mg β-carotene-d_8 uring organic mass spectrometry.[3] We have used this approach with oral doses of β-[^{13}C]carotene as low as 5 μg.[8] Use of such low doses is aided by the relatively low total body pool of β-carotene,[9] and the high precision of $^{13}C/^{12}C$ measurement afforded by IRMS. Tracer doses are valuable for modeling purposes where steady-state conditions are necessary, and for studying metabolic interconversions well below saturation kinetics.

The described approach is also valuable for obtaining data on terminal elimination kinetics of β-carotene or retinol in humans, as illustrated in Fig. 3. Even after 400 hr, changes in the carbon isotope ratio in plasma β-carotene can easily be observed when measured at 100-hr intervals.

Cautionary Notes

A concern inherent to IRMS analysis is isotopic fractionation during sample preparation. Certain isotopomers may fractionate during HPLC, such that isotopomers of differing ^{13}C or deuterium enrichment may be partially or completely resolved. Reversed-phase HPLC has been reported to resolve isotopomers of deuterated β-carotene completely or partially.[3]

[8] C.-S. You, R. S. Parker, K. J. Goodman, J. E. Swanson, and T. N. Corso, *Am. J. Clin. Nutr.* **64,** 177 (1996).

[9] R. S. Parker, *Am. J. Clin. Nutr.* **47,** 33 (1988).

Such phenomena may be problematic if HPLC fractions are being collected for isotope ratio analysis. In the case of β-[^{13}C]carotene, we have been unable to achieve even partial resolution of natural abundance and ^{13}C-enriched β-carotene using the highest resolution system employed in the sample preparation (semipreparative reversed-phase HPLC). However, evidence of minor fractionation was observed in mixtures of natural abundance and highly ^{13}C-enriched β-carotene, as the front half of the β-carotene peak consistently exhibited $^{13}C/^{12}C$ ratios slightly higher than the rear half of the peak.[10] Detailed examination of fatty acid methyl ester HPLC fractionation has shown that the leading edge of natural abundance fatty acid peaks tends to be enriched in ^{13}C while the tails tend to be modestly ^{13}C depleted.[11] It is generally recommended that entire HPLC peaks be collected to avoid problems stemming from fractionation.

Unlike organic MS, GCC-IRMS cannot distinguish analyte from contaminant, because all are converted to CO_2. The isotope ratio of retinol or perhydro-β-carotene can be altered by coeluting or partially resolved compounds. It has been shown that conventional peak integration methods applied to overlapping GCC-IRMS peaks can produce inaccurate isotope ratios, even for pairs of compounds of well-matched isotope ratio.[12] Because baseline separation is critical, substantial purification of analytes is recommended when mixtures are formidable.

High-precision IRMS are designed to measure small differences in isotope ratio very close to natural abundance. Even with low oral doses (1 mg) of β-[^{13}C]carotene, ^{13}C enrichment in plasma β-carotene can reach 10 AP or more since the plasma pool size is low and the clearance rate is relatively slow. Measurement accuracy of isotope ratios at levels higher than about 10 AP must be carefully assessed with isotopically calibrated standards. At these higher levels, organic MS becomes a better choice although its precision is limited. Recently, GCC-IRMS has been combined with organic MS in a single instrument by splitting the GC effluent 90% to the RMS and 10% to an ion MS.[13] This approach optimally handles enrichments from natural abundance to 100% ^{13}C and is available commercially.

Last, precision of the isotope ratio measurement is limited by the mass of analyte injected. Low concentrations of β-carotene may require

[10] R. S. Parker, unpublished data (1997).
[11] R. Caimi and J. T. Brenna, *J. Chromatogr. A* **757,** 307 (1997).
[12] K. J. Goodman and J. T. Brenna, *Anal. Chem.* **66,** 1294 (1994).
[13] W. Meier-Augenstein, W. Brand, and D. Rating, *Biol. Mass Spectrom.* **23,** 376 (1994).

larger blood samples. Alternatively, curve-fitting algorithms may produce satisfactory precision and accuracy for samples of low signal intensity.[14]

Potential Improvements to the Method

With adequate sample purification, retinol and perhydro-β-carotene fractions could be combined and their isotope ratios determined within the same GCC-IRMS run. If separate retinol and retinyl ester data are required, the number of GC injections would be reduced by one-third. If only total plasma retinol and perhydro-β-carotene data are needed, such a combination would result in a 50% reduction in the number of GC injections.

The approach described in this chapter involves calculation of plasma concentration of labeled analyte by using a combination of HPLC and GCC-IRMS data. Ideally, quantification of both analyte concentration and isotope ratio could be performed simultaneously by GCC-IRMS. Development of a battery of appropriate internal standards would be required for this approach to control for losses of both retinol and β-carotene during extraction, HPLC purification, and hydrogenation. Internal standard choice must entail not only similar behavior during fraction preparation (or hydrogenation), but also elution near retinol and perhydro-β-carotene during GC. The latter requirement is important because GC oven temperature is inversely related to carrier flow, and changes in flow may cause changes in the split ratio at the open split upstream of the mass spectrometer.

[14] K. J. Goodman and J. T. Brenna, *J. Chromatogr. A.* **689,** 63 (1995).

[13] Atmospheric Pressure Chemical Ionization and Electron Capture Negative Chemical Ionization Mass Spectrometry in Studying β-Carotene Conversion to Retinol in Humans

By GUANGWEN TANG, BRUCE A. ANDRIEN, GREGORY G. DOLNIKOWSKI, and ROBERT M. RUSSELL

Introduction

The nutritional importance of β-carotene (β-C) was first established in 1930, when it was discovered to be a precursor of vitamin A.[1] Epidemiologi-

[1] T. Moore, *Biochem. J.* **24,** 692 (1930).

0076-6879/97 $25.00

cal and laboratory studies have now suggested that β-C may have biological activities itself as an intact molecule.[2–7] Therefore, both β-C and its metabolites possess important physiological functions that need further study. Investigators have utilized synthetic β-C (unlabeled), radioactive isotope labeled β-C or stable isotope labeled β-C to study β-C metabolism in humans. When a human subject is given a large dose of synthetic β-C, the blood β-C concentration may be elevated substantially. However, blood retinol concentrations are homeostatically controlled. Therefore, supplementation with β-C does not result in increased concentration of retinol in blood. The conversion of β-C to vitamin A cannot be investigated in humans by feeding them unlabeled β-C because it is impossible to distinguish newly administered β-C and its metabolites from body reserves.

Two studies by Goodman *et al.*[8] and Blomstrand and Werner[9] used radioactive β-C (^{14}C or ^{3}H). These studies have provided most of our knowledge about how humans absorb and metabolize β-C. Isotopically labeled compounds, such as [^{13}C]β-C[10] and [$^{2}H_8$]β-C (β-C-d_8)[11] have been used to study β-C in humans. In a study conducted by Parker *et al.*,[10] 1 mg of perlabeled [^{13}C]β-C was given to a middle-age male subject in the form of a milk beverage. High-performance liquid chromatography (HPLC) was used to separate the all-*trans*-β-C, retinol, and retinyl ester fractions from plasma samples. Isotope ratios of all fractions were determined by gas chromatography–combustion–gas isotope ratio mass spectrometry (GC-C-IRMS). These studies indicated that small doses of [^{13}C]β-C, typical of the daily dietary intake, can be traced from plasma β-C, retinol, and retinyl ester pools for up to 24 days postdose. Data from Marmor *et al.*[12] demonstrated that it is possible to trace plasma pools of β-C and their biokinetics

[2] R. Peto, R. J. Doll, J. D. Buckley, and M. B. Sporn, *Nature* **290,** 201 (1981).

[3] G. W. Burton and K. U. Ingold, *Science* **244,** 569 (1984).

[4] N. I. Krinsky, *Clin. Nutr.* **7,** 107 (1988).

[5] A. Bendich, *J. Nutr.* **119,** 112 (1989).

[6] L. X. Zhang, R. V. Cooney, and J. S. Bertram, *Carcinogenesis* **12,** 2309 (1991).

[7] E. B. Rimm, M. J. Stampfer, A. Ascherio, E. Giovannucci, G. Colditz, and W. C. Willett, *N. Engl. J. Med.* **328,** 1450 (1993).

[8] D. S. Goodman, H. S. Huang, and T. Shiratori, *J. Biol. Chem.* **241,** 1929 (1966).

[9] R. Blomstrand and B. Werner, *Scand. J. Clin. Lab. Invest.* **19,** 339 (1967).

[10] R. S. Parker, J. E. Swanson, B. Marmor, K. J. Goodman, A. B. Spielman, J. T. Brenna, S. M. Viereck, and W. K. Canfield, *in* "Carotenoids in Human Health" (L. M. Canfield, N. I. Krinsky, and J. A. Olson, Eds.), Vol. 691, pp. 86–95. New York Academy of Sciences, New York, 1993.

[11] S. R. Dueker, A. D. Jones, G. M. Smith, and A. J. Clifford, *Anal. Chem.* **66,** 4177 (1994).

[12] B. Marmor, R. S. Parker, J. E. Swanson, C.-S. You, Y. Wang, K. Goodman, J. T. Brenna, and W. Canfield, *FASEB* **8,** A192 (1994).

up to 5 weeks postdose. However, the high cost of per labeled [^{13}C]β-C limits the size of sample groups.

In another study conducted by Dueker *et al.*,[11] 73 μmol of β-C-d_8 was given to a male subject and plasma samples were drawn for a 24-day period. The β-C fraction and its metabolite, retinol, were isolated from the plasma using a solid-phase extraction protocol. The isotope ratio of β-C-d_8/β-C-d_0 in the isolated plasma β-C was determined either by reversed-phase HPLC or tandem mass spectrometry (MS/MS) with electron ionization. They reported a strong correlation between the ratios of β-C-d_8/β-C-d_0 in the plasma determined by HPLC and MS/MS methods. The β-C-d_8/β-C-d_0 ratio peaked at 7 hr. The retinol sample was derivatized and analyzed for the ratio of retinol-d_4/retinol by GC-MS. The retinol-d_4/retinol ratio peaked at 24 hr. Although the MS/MS method was able to detect 100 pmol, the HPLC method was able to detect as little as 1.87 pmol of β-C-d_8.

A method for detecting carotenoids employed liquid chromatography/electrospray ionization–mass spectrometry (LC/ESI-MS).[13] The detection limit of this method for β-C was between 1 and 2 pmol. However, this method still needs to be tested using biological samples.

In this study we have developed a method that uses flow injection atmospheric pressure chemical ionization–mass spectrometry (FI/APCI-MS) to measure the enrichment of β-C-d_8 in the serum of a subject who was orally supplemented with β-C-d_8. We also investigated the kinetics of the metabolism of β-C-d_8 to retinol-d_4, using gas chromatography–mass spectrometry with electron capture negative chemical ionization (GC/EC-NCI-MS) to measure the enrichment of derivatized retinol-d_4 in the serum.

Experimental

Standards

Crystalline all-*trans*-β-C-d_8 (11, 11′, 19, 19, 19, 19′, 19′, 19′-$^2H_8$-β-C, 82.0% in all-*trans* form, 8.0% in 13-*cis* form, 4.2% in 9-*cis* form, and 3.4% in 15-*cis* form) in a sealed amber ampoule was provided by BASF (Ludwigshafen, Germany). The purity of β-C-d_8 was checked by HPLC and was 97.5% spectroscopically pure, but it contained β-C-d_7 (15.7%), β-C-d_0 (2.9%) and β-C-d_6 (0.3%) as measured by APCI-MS described in this chapter.

Sample Preparation

Blood Sample Collection. Blood sampling followed the regulations of the Human Investigation Committee at Tufts University. After an overnight

[13] R. B. van Breemen, *Anal. Chem.* **67,** 2004 (1995).

fast, a 47-year-old female volunteer weighing 67 kg consumes gelatin capsules containing 235 μmol of β-C-d_8 in 2 g corn oil with a high fat breakfast (30 g fat from two plain bagels and 40 g almond butter). For 21 days, the subject consumed regular meals. Serum samples are collected at 0, 1, 3, and 6 hr and at 1, 2, 4, 8, 14, and 21 days after the β-C-d_8 dose and stored at $-70°$.

Extraction and Separation of β-C. Three milliliters of $CHCl_3/CH_3OH$ (2:1, v/v) is added to a 1-ml serum sample. The mixture is vortexed and centrifuged for 15 min at 4° and 800 *g*. The $CHCl_3$ layer is collected. Two milliliters of hexane is added to the aqueous layer to reextract fat-soluble nutrients. The hexane layer and the $CHCl_3$ layer are combined and evaporated. The residue is dissolved in 200 μl of ethanol and injected onto an HPLC equipped with a C_{30} carotenoid column (YMC, Inc., Wilmington, NC).[14] The gradient procedure employs solvents based on combinations of methanol, methyl terbutyl ether, and water. Solvent A is 83:15:2 (v/v/v) and solvent B is 8:90:2 (v/v/v). Eight-three percent solvent A and 17% solvent B are used for 5 min followed by a 12-min linear gradient to 45% solvent B, a 5-min linear gradient to 100% B, a 5-min hold at 100% solvent B, and finally a 2-min gradient back to 83% solvent A and 17% solvent B. In this HPLC system, the retinol peak elutes from 3.2–5.0 min and β-C elutes from 21–23 min. The β-C fraction is rechromatographed through the same system to exclude contamination from α-carotene. The prepurified β-C fraction was finally chromatographed on a new Nova-pak C_{18} column (3.9 × 150 mm and 4-μm particle size, from Waters, Milford, MA) using 90% solvent B and 10% methanol and collected from 5.5–7.5 min. The purified β-C fraction is evaporated under N_2. The residue is redissolved in 50 μl of absolute ethanol and kept at $-20°$ until analyzed by FI/APCI-MS.

Extraction, Separation, and Derivatization of Retinol. The retinol HPLC fraction from the serum extract as described previously was dried under N_2. Forty microliters of (*N-tert*-butyldimethylsily)trifluoroacetamide (MTBSTFA) is added to the residue in the test tube. The test tube is capped with a ground glass stopper and heated at 130° for 50 min.[15] The reaction mixture is transferred by a glass pipette to a brown vial with a conical-shaped inner wall, evaporated to 20 μl under N_2, and then kept at $-20°$ until GC/MS analysis.

Sample Analysis

HPLC Analysis of Serum Samples. Concentrations of β-C and retinol in a 100-μl aliquot of serum are measured by HPLC equipped with a

[14] L. C. Sander, K. E. Sharpless, N. E. Craft, and S. A. Wise, *Anal. Chem.* **66,** 1667 (1994).
[15] G. J. Handelman, M. J. Haskell, A. D. Jones, and A. J. Clifford, *Anal. Chem.* **65,** 2024 (1993).

Pecosphere-3C C_{18} column (Perkin-Elmer, Nowalk, CT) and a Waters 994 programmable photodiode array detector with the wavelength set at 450 nm for carotenoids and 340 nm for retinoids.[16] The concentration of β-C and retinol in serum and the percentage isotopic enrichment of β-C-d_8 and retinol-d_4 are used to calculate the molar enrichment of β-C-d_8 and retinol-d_4.

FI/APCI-MS Analysis of β-C in Human Serum. A FI/APCI-MS is used for β-C analysis. The mobile phase of absolute ethanol is pumped through a 0.005-inch-i.d. PEEK tube by a Waters 6000A LC pump at a flow rate of 1 ml/min. A manual injection valve (Rheodyne 8125, Rheodyne, Cotati, CA) with a 5-μl loop (Rheodyne 8020) is installed between the LC pump and the APCI source. Each of the extracted β-C samples is manually injected three times approximately a half minute apart and had a concentration of approximately 2–3 ng/μl in ethanol. In addition, there are two blank injections of ethanol between sample sets to flush and clean the system. A quadrupole mass spectrometer (Hewlett Packard 5988) is fitted with an electrospray interface (Analytica of Branford 102506) and an APCI source cover (Analytica of Branford 103590).[17] The APCI nebulizer is operated with a pressure of 60 psi of nitrogen supplied from boiling off a liquid nitrogen cylinder. The temperature of the APCI heater and the voltage of the capillary exit are optimized at 325° and 140 V to minimize the fragmentation and thermal degradation of the β-C. The countercurrent drying gas is also supplied from boiling off of a liquid nitrogen cylinder and it is optimized with an operating temperature of 200° at a flow rate of 2.2 liter/min. This gas is filtered just prior to the countercurrent drying gas heater with a charcoal filter (Supelco, Supelpure HC 2-2446). This APCI source used a corona discharge from a needle generated by the following potentials: V(corona needle) = +1350 V, V(end plate) = −2250 V, and V(capillary) = −2650 V. These potentials produced ion currents of I(corona needle) = 2.5 μA, I(end plate) = 2 μA, and I(capillary) = 0.4 μA. The MS scan range was 535–550 Da in 0.1-Da steps with a scan speed of 5.25 scan/sec and a half-maximum peak width of 0.6 Da.

GC/MS Analysis of Retinol in Human Serum. One microliter of derivatized retinol is injected by an HP 7673A autosampler into an HP 5890 GC. The GC employs a cool on-column injector. The on-column injector was fitted with a 1-m deactivated fused silica retention gap. The

[16] G. Tang, G. G. Donikowski, M. C. Blanco, J. G. Fox, and R. M. Russell, *J. Nutr. Biochem.* **4,** 58 (1993).

[17] B. A. Andrien, J. P. Quinn, and C. M. Whitehouse, *Atmospheric Pressure Chemical Ionization by Corona Discharge with Pneumatic Nebulization.* Paper presented at the 42nd ASMS Conference on Mass Spectrometry and Allied Topics, Chicago, Illinois (1994).

retention gap is connected with an zero dead volume metal connector (HP 5061-5801) with ferrule (HP 5061-5804) to a 15-m × 0.25-mm-i.d. fused-silica capillary column, coated with a DB-1 stationary phase of 0.1-μm film thickness from J. W. Scientific (Folsom, CA). The temperature of the column oven and the on-column injector is programmed from 50 to 285° at 15°/min. The GC/MS interface temperature is set at 285°. Helium is used as the carrier gas. The *tert*-butylmethylsilyl derivative of retinol elutes at ~7 min. The GC eluate is detected by an HP 5988A quadrupole mass spectrometer, using 0.5-torr methane negative ion chemical ionization. The temperature of the ion source is 150°. The mass spectrometer is scanned repetitively from 260 to 280 Da. The data are collected and analyzed with an HP 1000 minicomputer running Rev. F of the RTE data system.

When the labeled retinol enrichment is below 1%, it is difficult to obtain accurate isotope ratios, due to dynamic range limitations of the GC/MS detection system. In these cases, we increase the sample injection volume from 1 to 3 μl, increase the gain of the electron multiplier, and reinject the sample. This approach causes the unlabeled peak at *m/z* 268 to go off scale and the peak at *m/z* 272 to rise out of the background. We then measure the ratio of the labeled retinol ion at *m/z* 272 to the natural abundance ^{13}C-retinol ion at *m/z* 269. Previous measurement of the *m/z* 268 to *m/z* 269 ratio allows us to calculate the *m/z* 268 to *m/z* 272 ratio. In this way we are able to measure percent enrichments to 0.1%.

Results

Initially we attempted to analyze β-C by GC/MS, but were unable to do so because β-C isomerizes and decomposes on the GC column.[18] Therefore, we chose to bring the β-C directly to the mass spectrometer via the APCI interface.

The choice of solvent in APCI is important because the solvent plays a critical role in the ionization process. Under APCI conditions, β-C is readily ionized by proton transfer from protonated ethanol to a protonated molecule with a molecular mass of 537 Da. Figure 1 shows that β-C analyzed by APCI-MS exhibits a linear correlation between concentration and signal response. The APCI analysis for β-C extracted from a blood sample is shown as replicate flow injection profiles in Fig. 2A. The detection limit was 50 pg of standard β-C (Fig. 2B).

A β-C-d_8 standard synthesized by BASF was analyzed using FI/APCI-MS. We prepared standard solutions of β-C-d_8 and β-C-d_0 and measured

[18] R. B. van Breemen, *Anal. Chem.* **68,** 299A (1996).

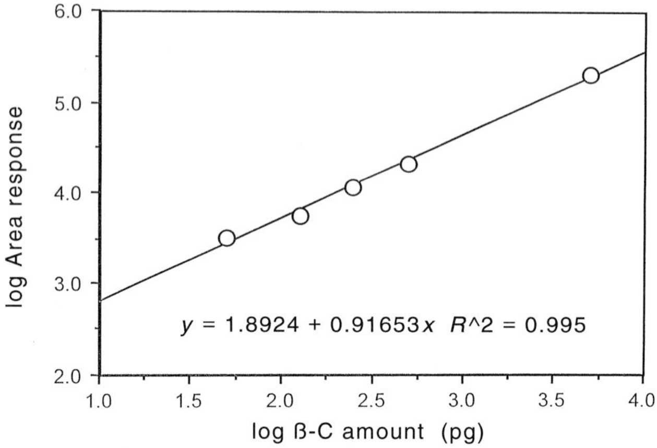

FIG. 1. The correlation between β-carotene concentration and signal response obtained from flow injection atmosphere pressure chemical ionization mass spectrometry.

the concentration separately using a UV/VIS spectrophotometer. We then prepared mixtures of the β-C-d_8 and β-C-d_0 standards in different ratios and analyzed them by APCI-MS. The results are shown in Table I. The difference between the UV/VIS spectrum absorbance and the APCI/MS measurement is within measurement error.

As one can readily observe from Fig. 2, the heights of the flow injection peaks are highly variable. The typical coefficient of variation (*CV*) for three replicate flow injection peak heights is 25%. The peak heights depend to some extent on the speed with which the instrument operator opens the sample loop valve. Fortunately, the ratios of peak areas are more reproducible. Table II shows the coefficient variations of three replicate injection peak areas versus the percent enrichment of the biological samples. The typical *CV* is about 10% in the range of >10% enrichment. Below 5% enrichment, the data are highly variable. Because labeled β-C percent enrichments of >10% are easily achieved, the lack of reproducibility at low enrichment is not a limiting factor in this study.

We measured the naturally occurring isotopes of β-C by integrating the reconstructed ion profiles of m/z values of 537 ($M + H^+$ of β-C-d_0), 538 ($M + H^+$ of [^{13}C]β-C-d_0, and 539 ($M + H^+$ of [$^{13}C_2$]β-C-d_0) (Fig. 3). Because we did not have an isotopically pure β-C-d_8 standard, we measured the total isotopic enrichment of the labeled β-C in human serum by integrating the reconstructed ion profiles at m/z values of 544 ($M + H^+$ of β-C-d_7), 545 ($M + H^+$ of ^{13}C-β-C-d_7 and $M + H^+$ of β-C-d_8), and 546 ($M + H^+$ of ^{13}C-β-C-d_8 and $^{13}C_2$-β-C-d_7) (Fig. 3).

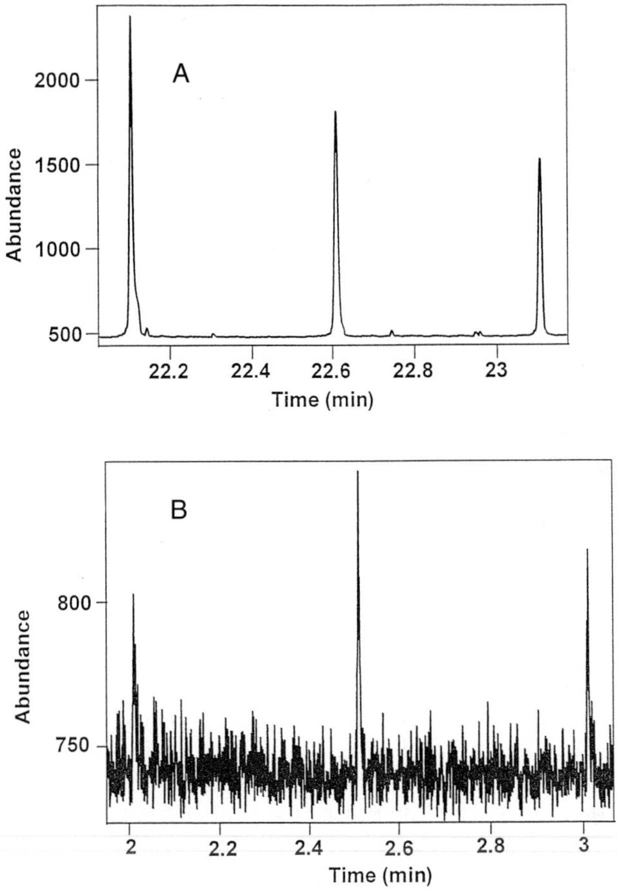

FIG. 2. Replicate injection profiles for the analysis of β-carotene by flow injection atmosphere pressure chemical ionization mass spectrometry. (A) Replicate injection profile for the extract from a human serum. (B) Replicate injection profile for the 50 pg (0.09 pmol) of standard β-carotene.

The β-C blood response curve for labeled β-C is shown in Fig. 4. The absolute concentration of labeled β-C-d_8 in serum reached 329 n*M* 2 days after the administration of β-C-d_8 (Fig. 4) while unlabeled β-C in serum peaked at 1 day after the β-C-d_8 dose. The highest enrichment of labeled β-C was 30.9%, which occurred at 2 days after the β-C-d_8 dose (Fig. 5).

We have measured retinol by both GC/MS and APCI-MS methods. For GC/MS, retinol collected from serum was derivatized with MTBSTFA and the subsequent retinyl MTBS ether was analyzed by GC/electron cap-

TABLE I
MIXTURE OF LABELED β-CAROTENE ($d_8 + d_7 + d_6$) AND STANDARD β-CAROTENE-d_0 MEASURED BY UV/VIS SPECTRUM AND APCI/MS[a]

Ratio of labeled to standard	Labeled β-carotene ($d_8 + d_7 + d_6$) in total β-carotene (%)	
	UV/VIS absorbance	APCI-MS
1:1	49.6 ± 0.3	46.1 ± 3.8
1:3	24.8 ± 0.2	23.0 ± 1.1

[a] Data were shown as mean ± SEM.

ture negative chemical ionization–MS (GC/ECNCI-MS) (Fig. 6A). The GC/ECNCI-MS of retinyl MTBS ether shows no molecular ion, but does show a major fragment ion at m/z 268 (m/z 272 for retinol-d_4) (Fig. 6). For APCI/MS, retinol extracted from serum was purified by the same HPLC method used for LC/MS. The detection limit for APCI-MS of underivatized retinol at m/z 269 (MH^+-H_2O) was 500 pg by flow injection in methanol with 1% formic acid. Because the detection limit by GC/MS was better than by LC/APCI-MS (see later discussion) and because there were some coeluting impurities that overlapped with the retinol isotope peaks we decided to rely on the GC/MS data.

The signal response to various concentrations of the retinol derivative showed a linear correlation (Fig. 7). The detection limit was 6 pg of standard retinol MTBSTFA derivative (Fig. 6B). The total enrichment of labeled retinol was determined by integrating the peak area under the reconstructed mass chromatograms of the negative ions at m/z 271 (d_3), 272 (d_4 + ^{13}C-d_3), and 273 (^{13}C-d_4) (Fig. 8). The natural abundance of retinol was determined by integrating the peak area under the reconstructed mass chromatograms of the negative ions at m/z 268 (d_0), m/z 269 (^{13}C-d_0), and m/z 270

TABLE II
DETECTION PRECISION OF β-C USING APCI-MS METHOD IN THREE REPLICATES

Enrichment	Standard deviation	*CV*
0.266	0.031	11.7
0.309	0.029	9.4
0.258	0.025	11.6
0.227	0.025	11.0
0.192	0.019	9.9
0.184	0.014	7.6

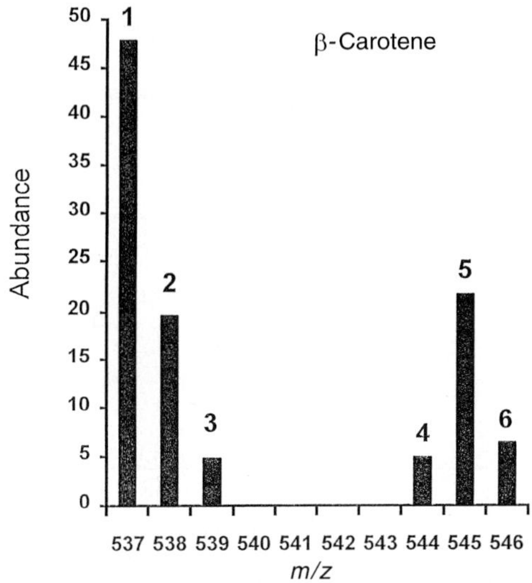

FIG. 3. Deuterium enrichment for a human serum extract is calculated from the isotopic profile obtained rom flow injection atmosphere pressure chemical ionization mass spectrometry. In bar graph, 1 is M + H$^+$ of β-C-d_0 at m/z 537; 2 is for M + H$^+$ of [^{13}C]β-C-d_0 at m/z 538; 3 is M + H$^+$ of [$^{13}C_2$]β-C-d_0 at m/z 539; 4 is M + H$^+$ of β-C-d_7 at m/z 544; 5 is M + H$^+$ of [^{13}C]β-C-d_7 and M + H$^+$ of β-C-d_8 at m/z 545; and 6 is M + H$^+$ of [^{13}C]β-C-d_8 at m/z 546.

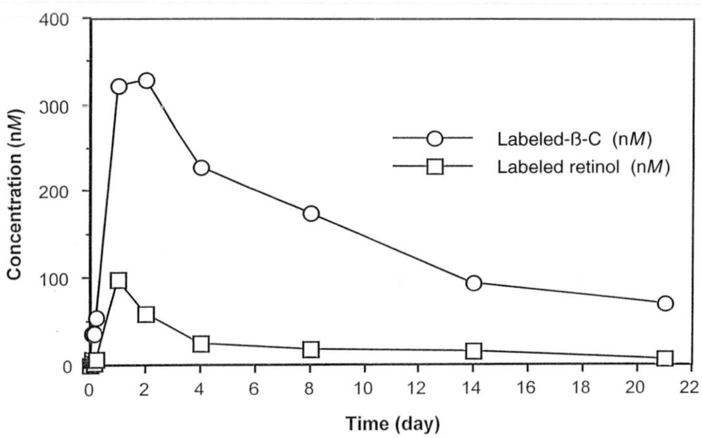

FIG. 4. Concentration of β-carotene-d_8 and retinol-d_4 in the serum collected at 0, 3, and 6 hr and at 1, 2, 4, 8, 14, and 21 days following a 126-mg β-carotene-d_8 supplementation.

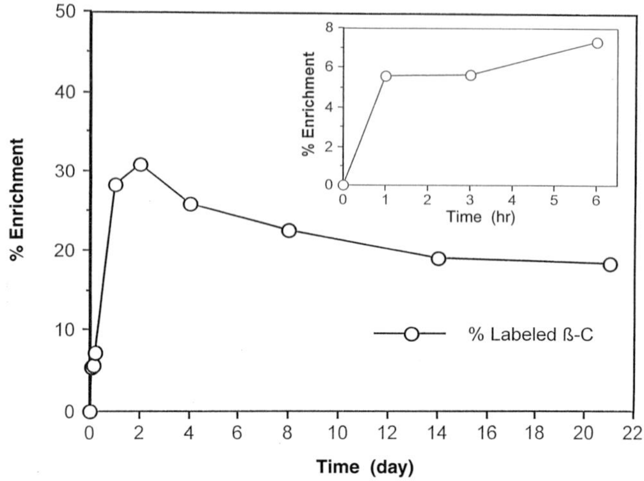

FIG. 5. Percent enrichment of β-carotene-d_8 in human serum following a 126-mg β-carotene-d_8 supplementation.

($^{13}C_2$-d_0) (Fig. 8). The analysis indicated that the enrichment of retinol-d_3 was less than 1%. Therefore, the contribution of [^{13}C]retinol-d_3 to m/z 272 (<0.2%) was ignored. Deuterated retinol enrichment in serum after β-C-d_8 supplementation showed an increase starting at 6 hr and reached a peak at 24 hr with 5.8% enrichment and absolute concentration of 92 nM (Figs. 4 and 9).

Discussion

After extraction and cleanup by LC, FI/APCI-MS and GC/ECNCI-MS can measure the β-C-d_8 and retinol-d_4 enrichment in human blood samples with a detection limit that is at least 10 times more sensitive than any other published mass spectrometry method. From a 1-ml serum sample, the labeled β-C enrichment can be measured as low as 5% and the labeled retinol enrichment can be as low as 0.1%. We used a pharmacological dose of labeled β-C in these experiments. However, our ultimate goal is to measure β-C absorbance from food under physiological conditions (6 mg β-C per meal). From our current data, it appears that this goal is attainable.

Because the dose contained β-C-d_7 along with its natural abundance ^{13}C isotopes, there are isotopic impurities that overlap with the β-C-d_8 peak in our samples. Therefore, it was necessary in our calculations to use all of the isotopic peaks at m/z 544–546 for labeled β-C and m/z 271–273 for labeled retinol. Including all of the isotopes instead of just β-C-d_8 and

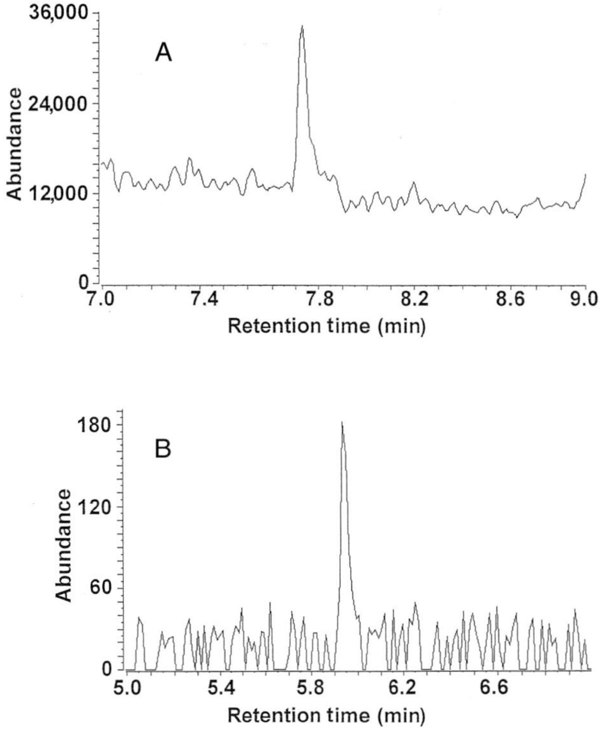

FIG. 6. A representative GC/ECNCI-MS chromatogram from the analysis of retinol deuterium enrichment in a human serum. (A) Profile for the extract from a human serum. (B) Profile for a 6 pg (0.02 pmol) of standard retinol.

retinol-d_4 in the analysis increases the accuracy of the isotope ratio analysis by up to 8% based on analysis of standards of known isotopic composition. The 2.9% β-C-d_0 in β-C-d_8 dose was not corrected because its contribution to the accuracy of the analysis is trivial.

β-C blood response kinetics show that blood enrichment of β-C-d_8 is highest at 48 hr. However, the concentration of total β-C (unlabeled plus labeled) peaks at 24 hr. Because these are fasting blood samples, this observation may indicate hepatic release of unlabeled β-C from the liver where it had been deposited previously. Because we did not collect any blood samples between 6 and 24 hr, we could not see the valley in the blood response curve between 6 and 24 hr observed by Dueker *et al.*[11]

Forty-eight hours after the β-C-d_8 dose, 0.5 mg β-C-d_8 was circulating in the body. This figure was calculated based on the highest serum concentration of β-C-d_8 and total blood volume in the subject (4% of body weight).

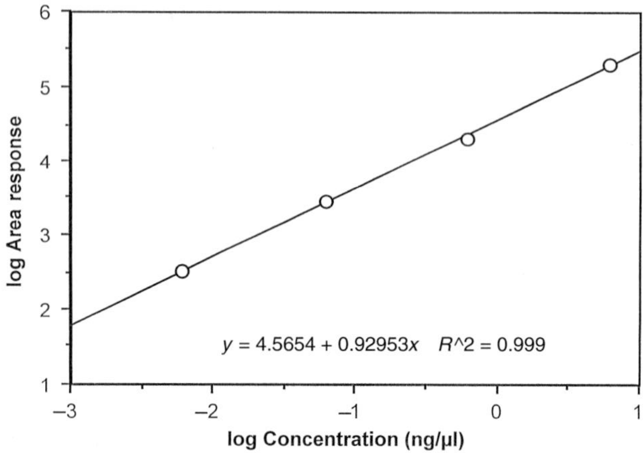

FIG. 7. The correlation between concentration of retinyl-*tert*-butyldimethylsilyl ether and the signal area response obtained from the gas chromatography electron capture negative chemical ionization mass spectrometry.

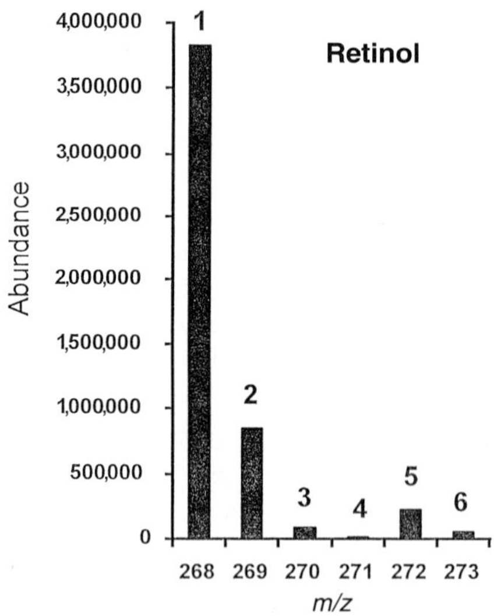

FIG. 8. A representative isotopic profile obtained from the GC/ECNCI-MS analysis of retinyl-*tert*-butyl dimethyl ether. Bar graph 1 is *m/z* 268 for retinol-d_0, 2 is *m/z* 269 for [^{13}C]retinol-d_0, 3 is *m/z* 270 for [$^{13}C_2$]retinol-d_0, 4 is *m/z* 271 for retinol-d_3, 5 is *m/z* 272 for retinol-d_4, and 6 is *m/z* 273 for [^{13}C]retinol-d_4.

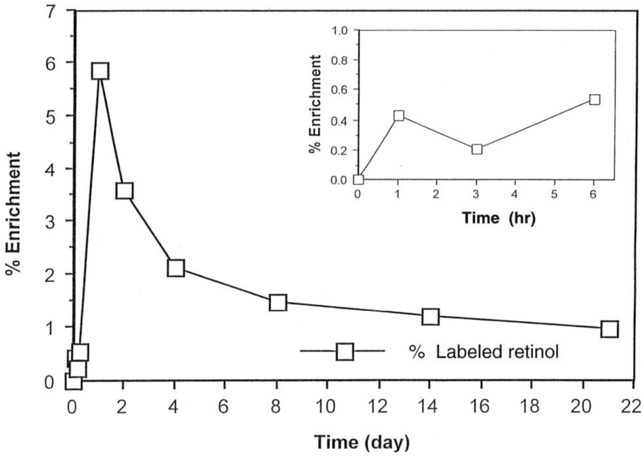

FIG. 9. Percent enrichment of retinol-d_4 in human serum following a 126-mg β-carotene-d_8 supplementation.

However, unlabeled β-C peaked at 24 hr after β-C-d_8 with 1.2 mg unlabeled β-C circulating in the body. This observation indicated that β-C-d_8 is retained in the body, most likely in liver or fat tissues, and is released gradually into the bloodstream.

When a human subject is given a high dose of β-C (235 μmol), as was the case in this experiment, intestinal absorption of β-C-d_8 is incomplete and inefficient.[15] Before 24 hr, the serum response curves mainly represent intestinally derived β-C-d_8 and retinol-d_4. The retinol-d_4 concentration in serum samples increases at 6 hr after the β-C-d_8 dose and reaches its highest concentration at 24 hr after the dose. The retinol-d_4 serum response peak is earlier in retention time, sharper in elution curve, and lower in concentration than the β-C-d_8 serum response peak (Fig. 4).

Isotope effects have been observed that are associated with steric, inductive, and hyperconjugative effects.[19] We do not expect to see any detectable isotope effects in absorption and metabolism of β-C-d_8 to retinol-d_4 because the labeling is at the C-19 (three deuteriums) and C-11 (one deuterium) positions of β-C molecule, which are too trivial to change the affinity of β-C or the strength of C–C double bonds in β-C.

These stable isotope/mass spectrometry methods have provided direct evidence that β-C is converted to retinol in humans. Having developed the FI/APCI-MS and GC/ECNCI-MS methods, we can learn more about

[19] A. F. Thomas, "Deuterium Labeling in Organic Chemistry." Meredith Corp., New York, 1971.

the kinetics of blood response to a β-C-d_8 dose and the kinetics of its metabolite retinol-d_4 in humans. Through further development and application of these MS methods, we will be able to answer many questions about *in vivo* β-C metabolism.

Acknowledgments

The authors thank Dr. Paust from BASF for supplying us β-carotene-d_8. The authors also thank Bonnie Marmor for editorial assistance and helpful suggestions. This work was supported in part by USDA contract 53-3K06-5-10 and by NIH grant R01CA49195.

Section II

Vitamin D

[14] Synthesis of [3α-^{3}H]Vitamin D_3 and 1α,25-Dihydroxy[1β-^{3}H]Vitamin D_3

By RAHUL RAY and MICHAEL F. HOLICK

Synthesis of [3α-^{3}H]Vitamin D_3

[3α-^{3}H]Vitamin D_3 is synthesized (Fig. 1, structures **I–VII**) from 7-dehydrocholesterol by a procedure previously described.[1]

*Preparation of Cholesta-4,6-diene-3-one (**II**)*

A solution of 7-dehydrocholesterol (**I**) (250 mg) in 25 ml of toluene is refluxed in a flask fitted with Dean–Stark trap until 5 ml of toleune is collected in the trap. The flask is cooled to 25° and cyclohexanone (1.5 ml) added. Refluxing continues until 2 ml of toluene is collected in the trap. The reaction mixture is cooled to 25° and aluminum isopropoxide (68 mg) added quickly. The resulting solution is refluxed for an additional 30 min, cooled to 25° and 20 ml of a 1 : 1 (v/v) mixture of ether and water added. The contents of the flask is transferred to a separatory funnel and the bottom aqueous layer discarded. The organic layer is washed with 0.1 *N* HCl (5 ml) and water (10 ml). The bottom aqueous layer is discarded and the organic solution dried over anhydrous magnesium sulfate. The organic layer is concentrated *in vacuo* and the crude reaction mixture is purified by preparative thin-layer chromatography (TLC) on silica (1000-μm plates, Analtech, Vineland, NJ). The yield of the desired product cholesta-4,6-diene-3-one (**II**) is 182 mg (73.2%).

*Preparation of Cholesta-3,5,7-trien-3-ol Acetate (**III**)*

Cholesta-4,6-diene-3-one (**II**) is dissolved in 2.4 ml of anhydrous acetic anhydride and 0.6 ml of anhydrous pyridine and gently refluxed in a nitrogen atmosphere for 4 hr. Crystals of the enol acetate form on cooling the reaction mixture to 25°. The mixture is kept at 4° for 12 hr and the crystals of the enol acetate are collected by filtration. On recrystallization from a minimum amount of ethanol, the desired enol acetate (**III**) is obtained in approximately 60% yield. The UV spectrum of the enol acetate (in methanol) had peaks at 302, 314, and 329 nm with a maximum at 314 nm.

[1] S. A. Holick, M. F. Holick, J. E. Frommer, J. W. Henley, and J. A. Lenz, *Biochemistry* **19,** 3933 (1980).

METHODS IN ENZYMOLOGY, VOL. 282 0076-6879/97 $25.00

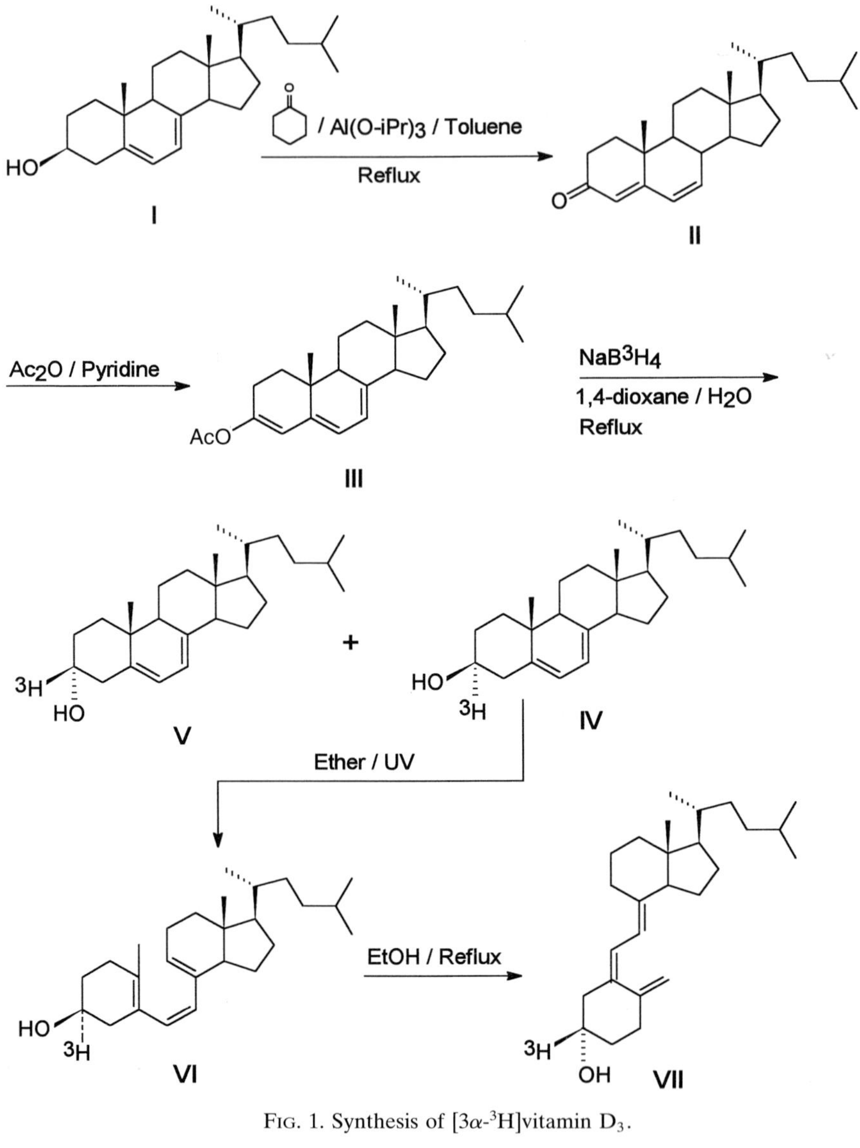

FIG. 1. Synthesis of [3α-^{3}H]vitamin D_3.

Preparation of 7-Dehydrocholesterol and 3-Epi-7-Dehydrocholesterol

A mixture of enol acetate (**III**, 3 mg) and sodium borohydride (3 mg) in 1,4-dioxane (0.5 ml) and water (0.03 ml) is refluxed gently with gentle magnetic stirring under nitrogen for 3 hr. After the reflux, the solution is cooled to 25° and diluted with 5 ml of anhydrous ether and 2 ml of 5% aqueous HCl. After brief stirring, the ether layer is withdrawn. The aqueous solution is extracted twice with two 5-ml portions of ether. The combined organic layer is washed with water (3 ml) and dried over anhydrous magnesium sulfate and concentrated under a stream of N_2 to produce a white solid. The reaction mixture is dissolved in a small volume of ethyl acetate and applied to a silica plate (1000 μm) that is eluted with 20% ethyl acetate in *n*-hexane. The two UV-active bands are isolated. Yields of 7-dehydrocholesterol (more polar band) and 3-epi-7-dehydrocholesterol (less polar band) are 1.29 and 0.23 mg, respectively, as determined by UV spectroscopy utilizing an extinction coefficient of 10,500 at 281 nm.

*Preparation of 7-Dehydro[3α-^{3}H]Cholesterol (**IV**) and 7-Dehydro-3-epi[3β-^{3}H]Cholesterol (**V**)*

One curie of sodium boro[^{3}H]hydride (specific activity 11 Ci/mmol, Amersham International, Springfield IL) is carefully added to a 10-ml round-bottom flask containing 3 mg of cholesta-3,5,7-trien-3-ol acetate and a magnetic stirring bar. The flask is closed with a rubber septum, and 0.5 ml of freshly distilled 1,4-dioxane and 0.03 ml of water are added into the reaction flask with syringes fitted with hypodermic needles. A reflux condenser is fitted into the flask, and a N_2 inlet line introduced into the reaction system. The reaction mixture is gently refluxed (oil bath temperature 110°) for 3 hr. After the reflux, the solution is cooled to 25°, and 5 ml of ether and 2 ml of 5% HCl are cautiously added to the reaction flask via the septum at the top of the reflux condenser with syringes fitted with hypodermic needles. Gas is evolved at this step so the entire preparation should be carried out in a well-ventilated fume hood. After stirring for 15 min, the reflux condenser is removed and the top organic layer very cautiously withdrawn with a Pasteur pipette. The aqueous solution is carefully extracted with two additional 5-ml portions of ether.

The combined organic solution is washed with water (3 ml). Again, the top organic layer is carefully withdrawn with a Pasteur pipette, and anhydrous magnesium sulfate (0.5 g) is added to the organic solution (in a 25-ml Erlenmeyer flask), which is allowed to stand for 30 min with occasional gentle shaking. The end of a Pasteur pipette is blocked with glass wool, and organic solution added (caution) through the open end to filter off the solids. The solids are washed twice with 2-ml portions of ether, and the combined filtrate dried under a gentle stream of N_2 to produce a white solid residue. This residue is dissolved in a minimum volume of ethyl acetate and applied (caution) to a silica TLC plate. Standard samples of 7-dehydrocholesterol and 7-dehydro-3-epicholesterol are also applied to plate, which is eluted with 20% ethyl acetate in hexane. After the elution, the two UV-active bands corresponding to 7-dehydrocholesterol and 7-dehydro-3-epicholesterol are scraped off the plate and extracted with ethyl acetate. Solutions of 7-dehydro[3α-^{3}H]cholesterol (**IV**) and 7-dehydro-3-epi[3β-^{3}H]cholesterol (**V**) are dried in a gentle stream of N_2 and redissolved in freshly distilled toluene (10 ml). Aliquotes of this solution are mixed with scintillation liquid and counted for radioactivity in a liquid scintillation counter. The yields of 7-dehydro[^{3}H]cholesterol and 7-dehydro-3-epi[^{3}H]-cholesterol were 16.2 and 2 mCi, respectively with a specific activity of 2.75 Ci/mmol.

*Preparation of [3α-^{3}H]Previtamin D_3 (**VI**)*

A solution of 7-dehydro[^{3}H]cholesterol (**IV**, 5.6 mCi) in 5 ml of anhydrous ether is pipetted in a quartz test tube. A gentle stream of N_2 is passed on the solution until the volume is reduced to 3 ml. The test tube is immediately covered with a rubber septum and a balloon filled with N_2 is attached to the test tube. The test tube is taped on the side of the outer jacket of the lamp and irradiated for 10 min using a medium-pressure Hanovia Hg arc lamp. The lamp has an outer cooling jacket, and cold water is passed through the jacket during irradiation. Alternatively, the test tube can be clamped 2.5 cm away from the lamp without the water jacket. The irradiated sample is concentrated under N_2 and applied to a TLC plate (1000 μm) that is eluted with 14% ethyl acetate in hexane. After 2 elutions, bands (UV-active) corresponding to [^{3}H]previtamin D_3 (**VI**, least polar, 1.44 mCi), a mixture of [^{3}H]lumisterol$_3$ and [^{3}H]tachysterol$_3$ (intermediate polarity, did not separate very well, 0.8 mCi) and 7-dehydro[^{3}H]cholesterol (**IV**, most polar, 2 mCi) are isolated.

*Preparation of [3α-^{3}H]Vitamin D_3 (**VII**)*

The sample of [^{3}H]previtamin D_3 (**VI**) from the previous step is dissolved in 2 ml of freshly distilled ethanol and gently refluxed in a N_2 atmosphere. After 5 hr of refluxing, the solution is concentrated under N_2 and applied to a silica TLC plate. A standard sample of vitamin D_3 is also applied to the plate. The plate is eluted twice with 14% (v/v) ethyl acetate in hexane and the UV-active band corresponding to vitamin D_3 is isolated with ethyl acetate. The yield of the product is 1.47 mCi. Radioactive homogeneity of the [^{3}H]vitamin D_3 (**VII**) is determined by high-performance liquid chromatography (HPLC) of a radioactive sample spiked with a standard sample of vitamin D_3 using a silica column (Altech Associates, State College, PA) and 8% (v/v) ethyl acetate in hexane as eluant. The radioactive sample produced a single radioactive peak corresponding to the UV peak of vitamin D_3.

The UV spectrum of [^{3}H]vitamin D_3 (in ethanol) has a typical maximum at 265 nm and a minimum at 228 nm. The specific activity of the radioactivity samples obtained by UV spectroscopy matches well with the calculated value of 2.75 Ci/mmol.

Synthesis of 1α,25-Dihydroxy[1β-^{3}H]Vitamin D_3 and 1β,25-Dihydroxy[1α-^{3}H]Vitamin D_3

The two specific radiolabeled isomers of 1α,25-dihydroxyvitamin D_3 are synthesized (Fig. 2, structures **VIII–XIII**) by a method previously described.[2]

*Preparation of 1-Keto-25-Hydroxyprevitamin D_3 (**IX**)*

1α,25-Dihydroxyvitamin D_3 [1,25-$(OH)_2D_3$, **VIII**][3] (22 mg) is dissolved in 2 ml of freshly distilled dichloromethane that is kept over 4-Å molecular sieves overnight before use. Activated manganese dioxide (150 mg) (Aldrich Chemical Co., Milwaukee, WI) is added to it. The mixture is stirred magnetically in an argon atmosphere (argon-filled balloon) for 20 hr followed by filtration of the reaction mixture over a bed of Celite. The solid is washed thoroughly with dichloromethane and the combined organic solution concentrated in a stream of argon. The desired product (1-keto-25-hydroxyprevitamin D_3, **IX**) is obtained as an oil (18.4 mg) by preparative TLC purification (silica plate, 33% ethyl acetate in hexane, R_f 0.5).

*Preparation of 1α, 25-Dihydroxy[1β-^{3}H]Previtamin D_3 (**X**) and 1β,25-Dihydroxy[1α-^{3}H]Previtamin D_3 (**XI**)*

Sodium boro[^{3}H]hydride [500 mCi (DuPont Co., Boston, MA), specific activity 78 Ci/mmol] is added to an ice-cold solution of 1-keto-25-hydroxyprevitamin D_3, **IX** (4 mg) in 1 ml of freshly distilled methanol and 0.01 ml of freshly distilled water. The solution is stirred magnetically at 0° for 30 min followed by the addition of 1 ml of acetone. The resulting solution is stirred on ice for an additional 30 min and then dried to a solid with a gentle stream of argon. This step should be carried out with caution, and the entire operation must be conducted in a well-ventilated fume hood. An aqueous solution of saturated NaCl (1 ml) is added to the solid residue and the aqueous solution is extracted with five 1-ml portions of ethyl acetate.

[2] S. A. Holick, M. F. Holick, and J. A. McLaughlin, *Biochem. Biophys. Res. Commun.* **97,** 1031 (1980).

[3] 1,25-$(OH)_2D_3$ was a kind gift from Dr. Milan Uskokovic, Hoffmann La-Roche, Nutley, NJ. Alternatively, 1,25-$(OH)_2D_3$ can be purchased from Calbiochem, San Diego, CA.

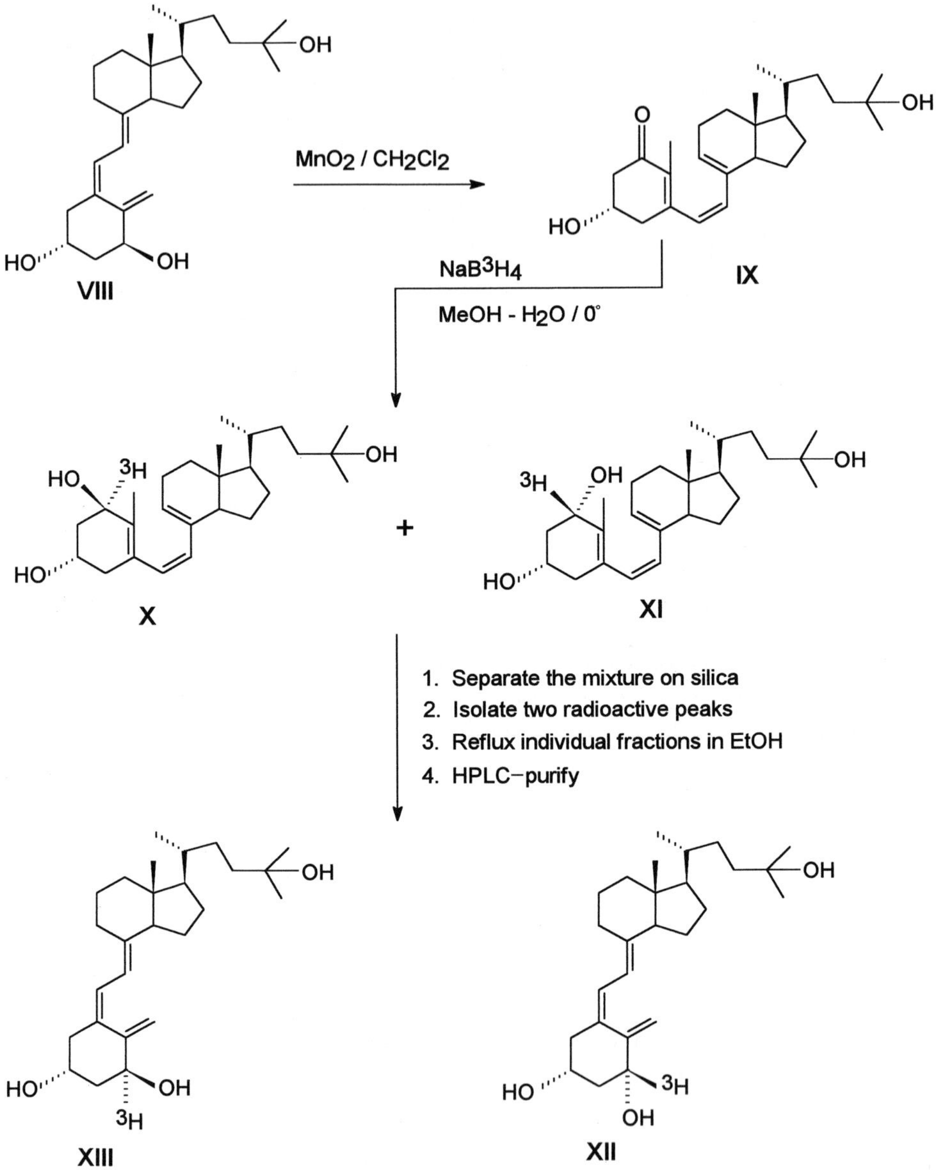

FIG. 2. Synthesis of radiolabeled isomers of 1α,25-dihydroxyvitamin D_3.

The reaction mixture (342 mCi) is concentrated with a gentle stream of argon and charged at the top of a silica column [2 g of silica for flash chromatography (Aldrich)]. After applying the sample at the top of the silica bed, the rest of the column is filled with 20% ethyl acetate in hexane, and the end of the column is closed with a rubber septum. A balloon filled with argon is attached to the column through the septum, and elution continues under mild argon pressure from the balloon. Fifty 1-ml fractions are collected and each fraction is assayed for radioactivity. Two radioactive peaks center around fraction 8 (1β,25-dihydroxy[1α-^{3}H]previtamin D_3) (**XI**, 7.4 mCi), and fraction 15 (1α,25-dihydroxy[1β-^{3}H]previtamin D_3) (**X**, 10.3 mCi) pooled separately.

*Preparation of 1β,25-Dihydroxy[1α-^{3}H]Vitamin D_3 (**XII**)*

A sample of 1β,25-dihydroxy[1α-^{3}H]previtamin D_3 (**XI**, 5.5 mCi) is dissolved in freshly distilled ethanol (1 ml) and the solution gently refluxed for 6 hr in an argon atmosphere. The reaction mixture is dried in a gentle stream of argon and the residue is redissolved in a minimum volume of 10% (v/v) 2-propanol in hexane. The sample is injected into a Waters HPLC system (Waters Associates, Milford, MA) consisting of a M6000A pump, an U6K injector and a model 441 UV detector (set at 254 nm) fitted with an ECONOSIL silica column (5 μm, Altech Associates). The HPLC is run with the same solvent at a flow of 1.5 ml/min and fractions are collected at 1-min intervals. Aliquots from the collected fractions are counted for radioactivity, and the fractions collected between 11 through 14 min (1β,25-dihydroxy[1α-^{3}H]previtamin D_3, **XI**) and 15 through 19 min (1β,25-dihydroxy[1α-^{3}H]vitamin D_3, **XII**) are pooled. The yield of compound (**XII**) is 2.3 mCi.

*Preparation of 1α,25-Dihydroxy[1β-^{3}H]Vitamin D_3 (**XIII**)*

A 6.5-mCi sample of 1α,25-dihydroxy[1β-^{3}H]previtamin D_3 (**X**) is treated the same way as in the case of 1β-dihydroxy[1α-^{3}H]previtamin D_3 (**XI**) and the product is chromatographed via HPLC using a solvent system consisting of methanol (4%) and 2-propanol (4%, v/v) in hexane[4] at a flow rate of 2.2 ml/min. Fractions at 1-ml intervals are collected and aliquots are counted for radioactivity. Fractions at 14 and 15 min (1α,25-hydroxy[1β-^{3}H]previtamin D_3, **X**) and 16 through 19 min (1α,25-dihydroxy[1β-^{3}H] vitamin D_3, **XIII**) are pooled. The yield of the desired product is 2.8 mCi.

[4] When 10% 2-propanol in hexane is used as eluant, there is no separation between 1α,25-dihydroxyprevitamin D_3 and 1α,25-dihydroxyvitamin D_3.

UV spectra of 1α,25-dihydroxy[1β-^{3}H]vitamin D_3 (**XIII**) and 1β,25-dihydroxy[1α-^{3}H]vitamin D_3 (**XII**) in methanol are identical with those of their unlabeled counterparts. Specific activities of the radiolabeled materials are calculated to be 15 Ci/mmol based on an extinction coefficient of 18,400 at 265 nm.

[15] Assay of 1,25-Dihydroxyvitamin D_3 from Serum Samples: Use of Receptor-Binding or Enzyme-Coupled Reporter Analysis

By MATTHEW J. BECKMAN and HECTOR F. DELUCA

Introduction

The measurement of 1,25-dihydroxyvitamin D_3 [1,25-$(OH)_2D_3$] in biological samples is of great importance from a medical point of view and as a research tool. An assessment of serum 1,25-$(OH)_2D_3$ concentration can be helpful in the diagnosis and treatment of a variety of bone diseases: parathyroid gland dysfunction, malignancies, renal failure, end-organ resistance diseases, and metabolic disorders. Further, it is useful as a research tool in medicine, physiology, and metabolism. Traditional methodologies for 1,25-$(OH)_2D_3$ measurement include bioassay,[1] radioimmunoassay,[2,3] cytoreceptor assay,[4] or radioreceptor assay.[5] Methods for serum metabolite isolation and preparation of stable 1,25-$(OH)_2D_3$ receptor (VDR) have greatly improved the use of radioreceptor technology.[6,7] Also, there have been improvements in radioimmunoassay technology for 1,25-$(OH)_2D_3$.[8] Each of these methods has played a valuable role in the characterization of the aforementioned disorders, however, they rely on complicated preassay workup strategies to isolate and purify 1,25-$(OH)_2D_3$. Advances in the

[1] P. H. Stern, A. J. Hamstra, H. F. DeLuca, and N. H. Bell, *J. Clin. Endocrinol. Metab.* **46,** 891 (1978).

[2] R. Bouillon, *J. Ster. Biochem.* **19,** 921 (1983).

[3] E. Mawer, J. Berry, J. Cundall, P. Still, and A. White, *Clin. Chem. Acta* **190,** 199 (1990).

[4] S. C. Manolagas, F. L. Culler, J. E. Howard, and A. S. Brickman, and L. J. Deftos, *J. Clin. Endocrinol. Metab.* **56,** 751 (1983).

[5] P. F. Brumbaugh and M. R. Haussler, *Life Sci.* **13,** 1737 (1973).

[6] R. L. Horst, T. A. Reinhardt, and B. W. Hollis, *Kid. Int.* **29,** S28 (1990).

[7] T. A. Reinhardt, R. L. Horst, J. W. Orf, and B. W. Hollis, *J. Clin. Endocrinol. Metab.* **58,** 91 (1984).

[8] B. W. Hollis, J. Q. Kamerud, A. Kurkowski, H. Beauieu, and J. L. Napoli, *Clin. Chem.* **42,** 586 (1996).

METHODS IN ENZYMOLOGY, VOL. 282
0076-6879/97 $25.00

development of monoclonal antibodies (mAb) to VDR[9] and gene reporter systems for the assessment of 1,25-$(OH)_2D_3$ action[10] can now be utilized as powerful tools for rapid, accurate, and sensitive measurements of 1,25-$(OH)_2D_3$. This chapter describes two of the newer approaches for determining 1,25-$(OH)_2D_3$ concentrations from small volumes of serum. Neither of these assays requires extensive extraction, bound/free protein separation, or high-performance liquid chromatography because of the highly specific nature of each assay.

The first technique is an application of mAb against the VDR in a radioreceptor assay. We have typically used pig intestinal nuclear extract (PINE) as an abundant source of VDR[11] and mAb XVIE6 or XVIE10,[9] which specifically detect pig VDR, but newer recombinant methods for expressing and purifying VDR now allow for a broader use of this assay among interested users. The technique was first described by Koyama *et al.*[12] and their results using PINE are summarized here. The second technique describes a sensitive and rapid molecular method for measuring 1,25-$(OH)_2D_3$ and vitamin D analogs from biological samples. This method utilizes a reporter gene (luciferase) under the control of a highly inducible vitamin D target gene promoter of 1,25-$(OH)_2D_3$ 24-hydroxylase (24-OHase).[13]

Extraction, Separation, and Purification of Serum 1,25-$(OH)_2D_3$

Sample preparation for the radioreceptor assay requires that the 1,25-$(OH)_2D_3$ be isolated from serum proteins and interfering metabolites such as 25-OH-D_3 or 24,25-$(OH)_2D_3$. To 0.5 ml of serum, 800 cpm of [26,27-^{3}H]-1,25-$(OH)_2D_3$ (Dupont–New England Nuclear, Boston, MA) in 10 μl of ethanol is added to monitor extraction recovery. A lipid hydrolysis step is incorporated by adding 50 μl 1 *M* KOH and incubating the sample at room temperature for 1 hr. This step is followed by adding 0.5 ml phosphate-buffered saline (PBS) and three volumes of dichloromethane and vortexing for 3 min. The lower organic phase is transferred to a new tube, while the upper phase is extracted twice more with three volumes of dichloromethane. The dichloromethane washes are combined, dried under nitrogen, and resuspended in 96:4 (hexane:2-propanol). Meanwhile, Sep-Pak silica col-

[9] M. C. Dame, E. A. Pierce, J. M. Prahl, C. W. Hayes, and H. F. DeLuca, *Biochemistry* **25,** 4523 (1986).

[10] H. Darwish and H. F. DeLuca, *Crit. Rev. Euk. Gene Expr.* **3,** 89 (1993).

[11] M. C. Dame, E. A. Pierce, and H. F. DeLuca, *Proc. Natl. Acad. Sci. U.S.A.* **82,** 7825 (1985).

[12] H. Koyama, J. M. Prahl, A. Uhland, M. Nanjo, M. Inaba, Y. Nishizawa, H. Morii, Y. Nishii, and H. F. DeLuca, *Analyt. Biochem.* **205,** 213 (1992).

[13] N. Arbour, T. Ross, J. Prahl, and H. F. DeLuca, *Ph. D. diss.* **6,** 152 (1996).

umns (Waters Co., Milford, MA) are preactivated[7] by serial treatment with 5 ml each of methanol, chloroform, hexane, and 96:4 hexane:2-propanol. The samples are applied to the columns and washed serially with 10 ml of 96:4 (hexane:2-propanol) to remove 25-OH-D_3 and other compounds of similar polarity, and 8 ml of 94:6 (hexane:2-propanol) to remove 24,25-$(OH)_2D_3$ from the sample. The 1,25-$(OH)_2D_3$ is eluted in 10 ml of 85:15 (hexane:2-propanol). The samples are dried under nitrogen and redissolved in 70 μl of ethanol, 20 μl of which is used to determine the recovery of tracer, and duplicate 20-μl aliquotes of each sample can be applied to the radioreceptor assay.

Sample preparation for the luciferase assay does not require that 1,25-$(OH)_2D_3$ be separated from 25-OH-D_3 or other metabolites because these metabolites do not interfere at the volumes used to run this assay. Also the detection limit of this assay is 0.1 pg, which like our radioreceptor assay is about an order of magnitude more sensitive than many of the traditional methodologies for 1,25-$(OH)_2D_3$ assay. Therefore, we perform a streamlined extraction procedure using ethyl acetate. To 0.1 ml of serum add 0.3 ml of ethyl acetate. Mix by gentle vortexing at 5-min intervals for 30 min. Spin at 500*g* for 10 min. Transfer the upper layer to a new tube. Repeat the extraction step twice more with brief vortexing and combine all organic layers. Dry down under nitrogen at 37°. Resuspend the sample in 50 μl of ethanol and use 2 μl in each luciferase assay. Recoveries are monitored by adding 10,000 cpm [26,27-^{3}H]-1,25-$(OH)_2D_3$ to 0.1 ml of control serum and extracting in the same way as sample serum. The recovery extracts are transferred to a scintillation vial and dried down and then the radioactivity is measured.

Radioreceptor Assay

Assay Principle

The assay is run in disequilibrium. Standard 1,25-$(OH)_2D_3$ (0.3–40 pg) or samples in 20 μl of ethanol are incubated for 1 hr at room temperature with 270 ml of $TEDK_{300}$ solution (50 m*M* Tris-HCl, 1.5 m*M* EDTA, 5 m*M* dithiothreitol, 300 m*M* KCl, pH 7.4), containing 20 fmol of PINE VDR, 1 μl (6.7 nmol) of biotinylated mAb, and 0.75 mg bovine serum albumin (BSA). The mixture is incubated for another hour at room temperature with 5000 cpm of [26,27-^{3}H]-1,25-$(OH)_2D_3$ in 10 μl of ethanol and then maintained overnight on ice. The immunoprecipitation step is done using a saturating amount of avidin-Sepharose. The Sepharose is washed three times with 750 μl of Tris/ENTA-Triton and transferred to scintillation vials, to which 4 ml of Biosafe-II (Research Products International Corp., Mount

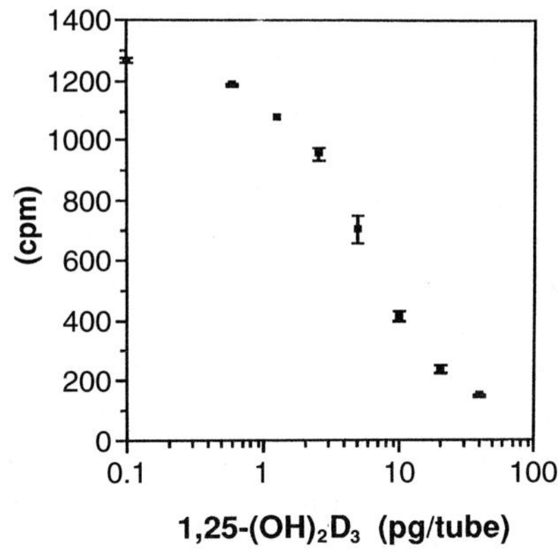

FIG. 1. Standard curve for 1,25-$(OH)_2D_3$ using 20 fmol of porcine intestinal VDR, 1 mg of biotinylated mAb, 0.25% BSA, and 5000 cpm of [26,27-^{3}H]-1,25-$(OH)_2D_3$ under nonequilibrium conditions. Each plot represents mean $\pm$ SD of triplicates.

Prospect, IL) cocktail is added and the radioactivity measured scintillation counting. The 1,25-$(OH)_2D_3$ concentrations (pg/ml) are determined from a standard curve and by correcting the values to 100% recovery.

Technical Aspects

The procedures for preparing PINE and mAb XVIE6 and XVIE10 have been previously described.[9] Monoclonal antibodies are dialyzed against 0.1 *M* sodium bicarbonate buffer pH 8.3, and are biotinylated by incubating 1 mg/ml of antibody solution with 60 mg of N-hydroxysulfo-succinamidobiotin (NHS-biotin) (Pierce, Rockford, IL) at room temperature for 4 hr. Free biotin is removed by dialysis against PBS. Avidin-Sepharose is synthesized as reported,[14] in which avidin is coupled to Sepharose CL-4B (Pharmacia, Piscataway, NJ) using a cyanogen bromide activation. The VDR content of PINE is determined by a modified[15] hydroxyapatite assay.[16] Protein concentrations are determined by Bradford assay.[17]

Results

Percent recovery in this assay is excellent, yielding an average of 95.0 $\pm$ 9.5% of added 1,25-$(OH)_2D_3$ when 2.5, 5, 10, 20, and 40 pg of 1,25-$(OH)_2D_3$ were added to the 0.5-ml serum aliquotes, whereas there was no detection of added 24,25-$(OH)_2D_3$ or 25-OH-D_3. Figure 1 represents a

[14] J. Kohn and M. Wilcheck, *Biochem. Biophys. Res. Commun.* **107,** 878 (1982).

[15] M. Inaba and H. F. DeLuca, *Biochim. Biophys. Acta* **1010,** 20 (1989).

[16] W. R. Wecksler and A. W. Norman, *Anal. Biochem.* **92,** 314 (1979).

[17] M. M. Bradford, *Anal. Biochem.* **72,** 248 (1976).

typical standard curve for 1,25-$(OH)_2D_3$ using 20 fmol of VDR, 1 μg of mAb, and 0.75 mg of BSA in 300 μl of reaction mixture. The sensitivity of this assay system defined as 2 SD of the BO tubes was 0.3–0.6 pg/tube. Fifty percent displacement of bound tracer was obtained at approximately 5.8 pg/tube. The cross-reactivities of 25-OH-D_3 and 24,25-$(OH)_2D_3$ compared to that of 1,25-$(OH)_2D_3$ were 0.05 and 0.017%, respectively (Fig. 2). The coefficient of correlation was 0.958 with a slope of 0.937 and an intercept of 3.7 pg/ml. The intra-assay coefficient of variation (*CV*), when the same sample was serially measured from human blood, averaged 26.1 ± 3.87 pg/ml, for a 14.6% *CV*. By application of this system to clinical samples (Fig. 3), the average 1,25-$(OH)_2D_3$ concentration in serum from 14 normal subjects was 36.6 ± 10.5 pg/ml. Patients with primary hyperparathyroidism gave values significantly higher than control, while significantly lower values were observed in patients with chronic renal failure ($P < 0.05$, multiple comparison, Scheffe's type). Figure 4 shows a gradual increase in serum 1,25-$(OH)_2D_3$ of normal human subjects administered 2 μg 1α-OH-D_3 per day for 7 days.

Suggestions

The concentration of BSA (0.25–0.5%) used in crucial for the stabilization of the VDR and for eliminating the influence of some lipids on ligand-receptor binding. Other means of extraction can be used in place of dichlo-

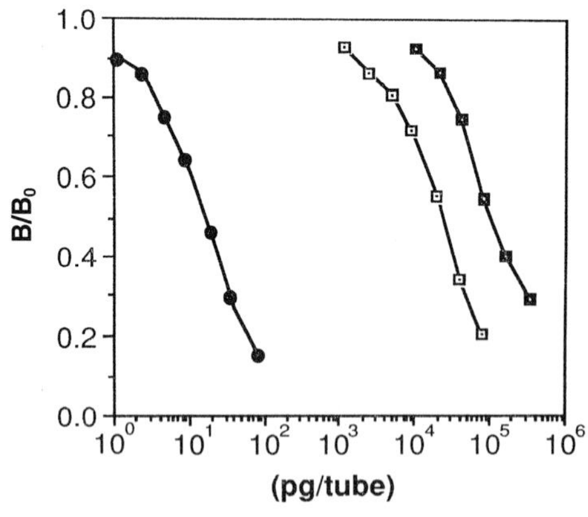

FIG. 2. Competition of [26,27-^{3}H]-1,25-$(OH)_2D_3$ binding to VDR by 1,25-$(OH)_2D_3$ (filled circles), 25-OH-D_3 (squares), and 24,25-$(OH)_2D_3$ (filled squares). Cross reactivity was calculated at the point of 50% competition.

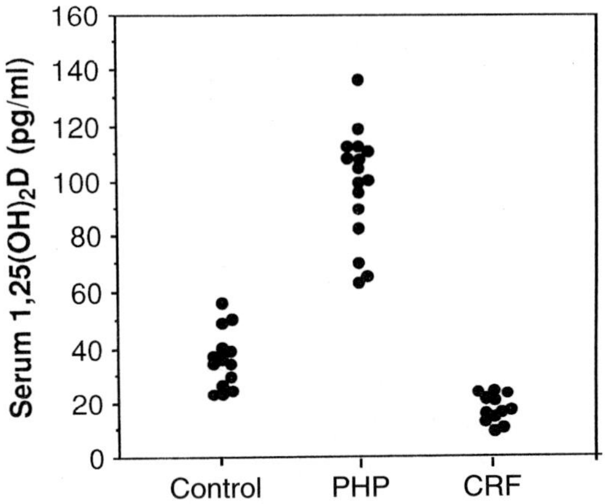

FIG. 3. Concentrations of 1,25-$(OH)_2D_3$ levels in sera from normal (n = 14) patients with primary hyperparathyroidism (PHP; n = 16) and patients with chronic renal failure (CRF; n = 12).

romethane, such as ethyl acetate. Also, if time is limiting, a 2–3 hr incubation is sufficient to accommodate a full VDR–1,25-$(OH)_2D_3$ interaction; however, the overnight incubation on ice is recommended for optimal sensitivity and complete immunoprecipitation of VDR. The wide availability of puri-

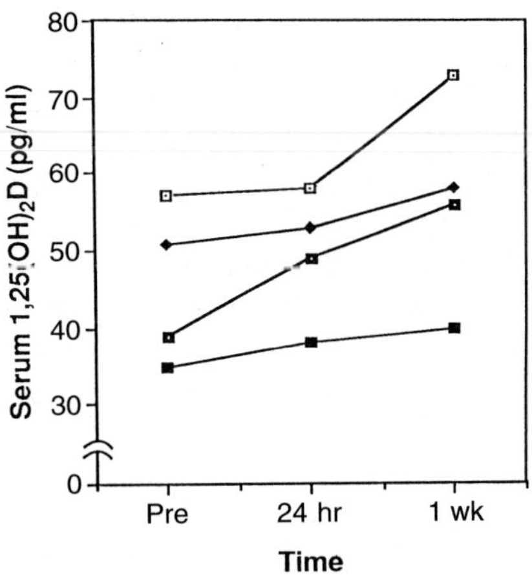

FIG. 4. Concentrations of 1,25-$(OH)_2D_3$ of sera from normal volunteers after oral administration of 1α-OH-D_3. Each volunteer was given 2 mg of 1α-OH-D_3 every day for 7 days. Blood was obtained 1 and 7 days after treatment and 1,25-$(OH)_2D_3$ was assayed. Each curve represents the data for one individual.

fied recombinant VDR[18,19] makes this assay feasible in any laboratory. The only restriction here is that the antibody used must react with the specific type of VDR being employed. The two mAb XVIE6 and SVIE10, are pig specific, but there are others that cross-react with the VDR of several species.[20,21]

Luciferase-Coupled Assay

Assay Principle

The 24-OHase has two VDREs in tandem separated by 93 nucleotides and is a highly responsive control element of the vitamin D system.[22] We have used a 1.5-kb *Stu*I fragment of the rat 24-OHase gene promoter, −1399 to +76 nucleotides of the published sequence to subclone into the *Nhe*I site of the expression vector, $pMAM_{neo}$-LUC (Clontech Laboratories, Palo Alto, CA). This vector contains a luciferase reporter gene and an SV40 (simian virus 40) promoter-driven neomycin resistance gene, which allows for selection of stably transfected cells. The insertion of the 24-OHase promoter into the multiple cloning site of $pMAM_{neo}$-LUC confers 24-OHase-directed luciferase gene expression. In addition, there is an MMTV-LTR promoter that allows for dexamethasone induction of luciferase gene expression (Fig. 5). The construct (designated p23) yields maximal levels of 1,25-$(OH)_2D_3$-inducible luciferase activity without interference of endogenous MMTV-LTV promoter-driven transcription. Stable cell transfection of the recombinant plasmid containing p23 is performed in rat osteosarcoma (ROS 17/2.8) cells (American Type Culture Collection, Rockville, MD) using a lipofectin-based procedure.[23] Viable cells maintained in the presence of the aminoglycoside G418 (Gibco-BRL, Gaithersburg, MD) for 2 weeks are replated in complete medium without G418, and are then cloned by limiting dilution. A clone was selected based on it having a low background and giving a high signal in response to 1,25-$(OH)_2D_3$. An assay consists of seeding cells into wells of a 98-well microtiter plate and treating the cells with the appropriate concentration of 1,25-$(OH)_2D_3$ or sample in 2 μl ethanol for 16 hr.

[18] T. K. Ross, J. M. Prahl, and H. F. DeLuca, *Proc. Natl. Acad. Sci. U.S.A.* **88,** 6555 (1991).
[19] C. L. Smith, G. L. Hager, J. W. Pike, and S. J. Marx, *Mol. Endocrinol.* **5,** 867 (1991).
[20] E. A. Pierce, M. C. Dame, R. Bouillon, H. Van Baelen, and H. F. DeLuca, *Proc. Natl. Acad. Sci. U.S.A.* **82,** 8429 (1985).
[21] E. A. Pierce and H. F. DeLuca, *Arch. Biochem. Biophys.* **261,** 241 (1988).
[22] C. Zierold, H. M. Darwish, and H. F. DeLuca, *J. Biol. Chem.* **270,** 1675 (1995).
[23] L. Lu, P. L. Zeitlin, W. B. Guggino, and R. W. Craig, *Pflugers Arch.-Eur. J. Physiol.* **415,** 198 (1989).

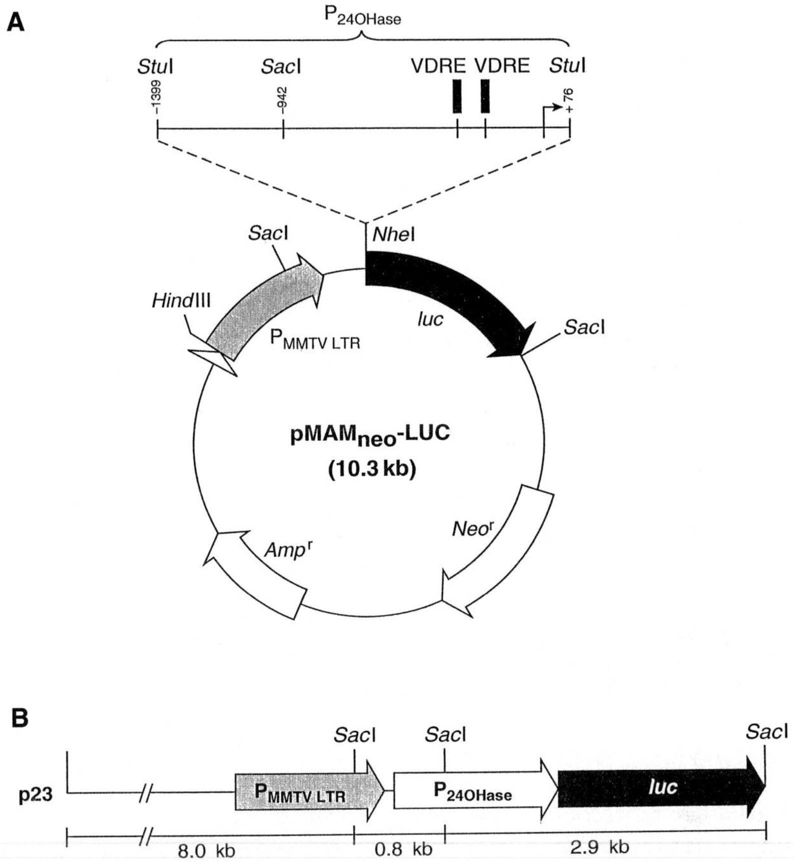

FIG. 5. Construction of the rat 24-hydroxylase promoter/luciferase gene reporter. (A) A 1.5-kb *Stu*I fragment of the 24-hydroxylase gene promoter containing tandem VDREs was subcloned into an expression vector, pMAM$_{neo}$-LUC. The luciferase gene is under the control of the 24-hydroxylase promoter. The vector also has SV40 early promoters, which permit selection of stable transfectants that are antibiotic resistant. (B) Orientation of the recombinant p23. Plasmid p23 was digested with *Sac*I, generating three fragments of 8.0, 2.9, and 0.8 kb as shown. The 24-hydroxylase and MMTV-LTR promoters are positioned upstream of the luciferase gene and direct luciferase expression in tandem.

Luciferase Assay

Cell are lysated in the wells, washed with PBS, and suspended in 50 μl luciferase buffer, diluted according to the supplier (Promega, Madison, WI). Ten microliters of lysate is added directly to luciferin substrate solution (Promega). Luciferase activity is detected using a monolight 2010 luminometer (Analytical Luminescence Laboratory, San Diego, CA).

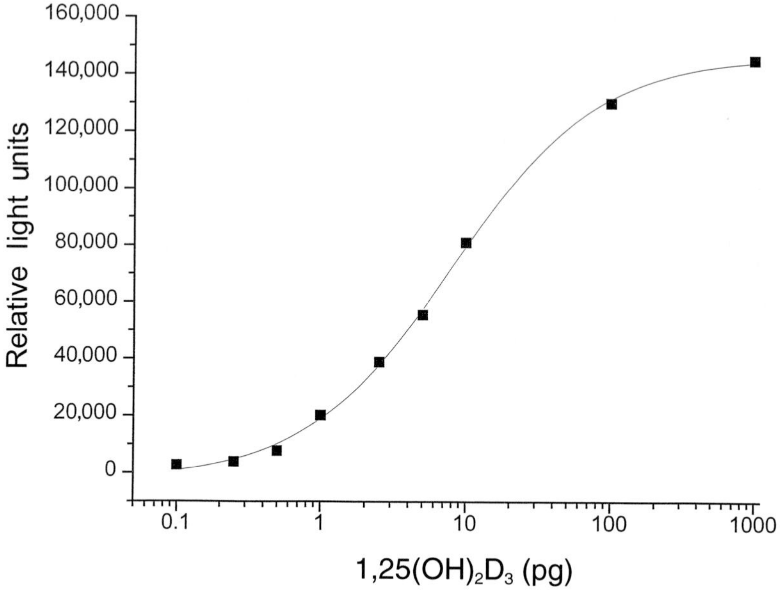

FIG. 6. Standard curve (0.1–1000 pg) of 1,25-$(OH)_2D_3$ used in luciferase assay and expressed as relative light units. Values were fitted to a sigmoidal curve fit using Origin 4.0 for Windows (Microcal, Northampton, MA).

Technical Aspects

This assay represents a new concept in the detection and measurement of 1,25-$(OH)_2D_3$ and can be applied to the measurement of vitamin D analogs as well, provided that the analog has known 24-OHase induction activity. The stable transfection must be performed in a cell line with adequate levels of VDR, such as ROS 17/2.8 and T-47D cells. In addition, the luciferin substrate is unstable and therefore must be stored at $-70°$ and in the dark between uses. For nonquantitative studies, luminescence measurements can be done using a scintillation counter[24]; however, we recommend that luminometer detection, if available, be used for enhanced sensitivity and precision of luminescence measurements.

Results

In Fig. 6, a typical luciferase assay standard curve (range, 0.1–10 pg) is presented. Luminescence is expressed as relative light units (RLUs). The

[24] R. Fulton and B. Van Ness, *Biotechniques* **14,** 762 (1993).

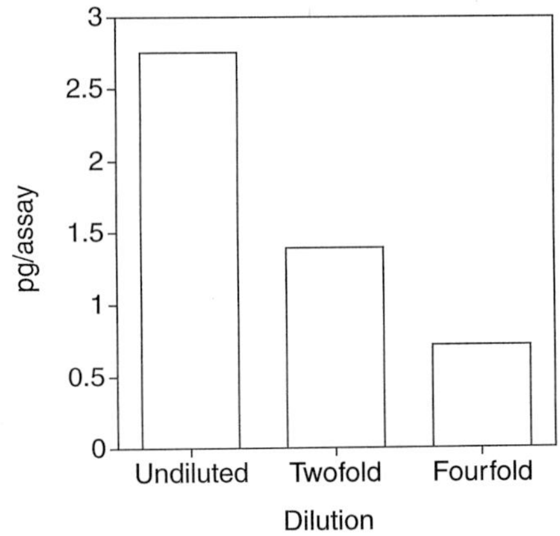

FIG. 7. 1,25-$(OH)_2D_3$ (pg/assay) from rat serum as measured using the luciferase assay. Two hundred microliters of serum was extracted with ethyl acetate as described in the text. The samples were dissolved in 20, 40, or 80 μl of ethanol and 2 μl of each was used for the assay. To determine pg/ml of serum, the values obtained for the diluted samples were multiplied by 50, 100, or 200, respectively.

relationship between 1,25-$(OH)_2D_3$ and luciferase activity is sigmoidal, which might be due to cooperativity associated with the two VDREs in the 24-OHase promoter sequence. The assay is accurate within the range of 0.5 and 10 pg, and can be plotted on logarithmic graph paper. Samples are diluted so that they fall within this range, and the RLU values are read off the curve as pg/well and then back-calculated to pg/ml by multiplying by the dilution factor. As little as 50 μl can be successfully extracted for an accurate determination of 1,25-$(OH)_2D_3$ concentration. Figure 7 shows an example from rat serum serially diluted by one-half each time, undiluted twofold, and fourfold, respectively. With each dilution, the 1,25-$(OH)_2D_3$ concentration was halved.

The advantages of this assay are its excellent sensitivity, very limited sample handling time, and that it is the first 1,25-$(OH)_2D_3$ assay that does not require the use of radioactivity. A disadvantage for some users is that cell culture is required to maintain the stable-transfected cell line and to run the luciferase assay. However, for most research laboratories cell culture is routine, making the luciferase assay easily adaptable to this setting. Running the assay in 98-well microtiter plates makes it feasible to run numerous samples simultaneously, and a luminometer equipped with a plate reader can analyze the entire set of samples in a matter of minutes.

Acknowledgments

This work was supported in part by grants DK14881 and DK0 7665 from the National Institutes of Health, a fund from the National Foundation for Cancer Research, and a fund from the Wisconsin Alumni Research Foundation.

[16] Quantitation of 25-Hydroxyvitamin D and 1,25-Dihydroxyvitamin D by Radioimmunoassay Using Radioiodinated Tracers

By BRUCE W. HOLLIS

Introduction

Vitamin D occurs in two distinct forms: vitamin D_2 (ergocalciferol) and vitamin D_3 (cholecalciferol). Vitamin D_3 is a 27-carbon derivative of cholesterol: vitamin D_2 is a 28-carbon molecule derived from the plant sterol ergosterol. Besides containing an extra methyl group, vitamin D_2 differs from vitamin D_3 in that it contains a double bond between carbons 22 and 23. It is important to note that in humans, vitamins D_2 and D_3 provide equal potency. Metabolic activation of vitamin D is achieved through hydroxylation reactions at both the carbon-25 of the side chain and subsequently the carbon-1 of the A ring.

Since the mid-1970s, methods have been developed that allow accurate measurement of circulating levels of 25-hydroxyvitamin D (25-OH-D) and 1,25-dihydroxyvitamin D [1,25-$(OH)_2$D]. These assays are based on procedures involving high-performance liquid chromatography (HPLC),[1,2] competitive protein-binding assay (CPBA),[3] radioreceptor assay (RRA),[4,5] and radioimmunoassay (RIA).[6–10] The assay of 25-OH-D is useful in detecting

[1] J. A. Eisman, R. M. Shepard, and H. F. DeLuca, *Anal. Biochem.* **80,** 298 (1977).
[2] B. W. Hollis and N. E. Frank, *J. Chromatogr.* **343,** 43 (1985).
[3] J. G. Haddad and K. J. Chyu, *J. Clin. Endocrinol. Metab.* **33,** 992 (1971).
[4] J. A. Eisman, A. J. Hamstra, B. E. Kream, and H. F. DeLuca, *Arch. Biochem. Biophys.* **176,** 235 (1976).
[5] B. W. Hollis, *Clin. Chem.* **32,** 2060 (1986).
[6] B. W. Hollis and J. L. Napoli, *Clin. Chem.* **31,** 1815 (1985).
[7] B. W. Hollis, J. Q. Kamerud, S. R. Selvaag, J. D. Lorenz, and J. L. Napoli, *Clin. Chem.* **39,** 529 (1993).
[8] B. W. Hollis, J. Q. Kamerud, A. Kurkowski, . Beaulieu, and J. L. Napoli, *Clin. Chem.* **42,** 586 (1996).
[9] R. Bouillon, P. DeMoor, E. R. Baggiolini, and M. R. Uskokovic, *Clin. Chem.* **26,** 562 (1980).

0076-6879/97 $25.00

states of vitamin D deficiency and excess. The assay of 1,25-$(OH)_2D$ is useful in diagnosing states of inadequate (such as pseudohypoparathyroidism and hypoparathyroidism) or excessive activation (sarcoidosis, tuberculosis, Hodgkin's disease). As the clinical demand for 25-OH-D and 1,25-$(OH)_2D$ analysis increases, simplification and streamlining of the analytical methods must progress. This has meant moving away from cumbersome procedures such as HPLC and radioassays that involve the use of tritium toward methodology involving RIA coupled with ^{125}I detection. We describe a synopsis of previously published procedures that incorporates the use of radioiodine into RIAs for both 25-OH-D and 1,25-$(OH)_2D$.[7,8]

Materials and Reagents

25-OH-D Radioimmunoassay

Crystalline 25-$(OH)D_3$ is from Hoffmann-LaRoche (Nutley, NJ). 23,24,25,26,27-Pentanor-C(22)-carboxylic acid of vitamin D is synthesized as previously described.[6] Primary 25-OHO-D antiserum is prepared as previously described[7] or purchased from INCSTAR Corp. (Stillwater, MN). Radioiodinated 25-OH-D is prepared as previously described[7] using Bolton–Hunter reagent to a specific activity of 2000 Ci/mmol or purchased from INCSTAR Corp. Swine-skin gelatin is from Sigma Chemical Co. (St. Louis, MO). Donkey anti-goat second antibody is from INCSTAR Corp. HPLC-grade acetonitrile was from Fisher Chemical Co. (Pittsburgh, PA). Unless otherwise noted, all other reagents are reagent grade.

1.25-$(OH)_2D$ Radioimmunoassay

Crystalline 1,25-$(OH)_2D_3$ and 1,25-$(OH)_2$-24,25,26,27-tetranor-C(23)-carboxylic acid are from Hoffmann-LaRoche. Primary 1,25 $(OH)_2D$ antiserum is raised in a sheep immunized with 1.25-$(OH)_2D_3$, 25-hemisuccinate conjugated to bovine serum albumin (BSA).[10] Radioiodinated 1,25-$(OH)_2D$ is prepared as previously described[8] using Bolton–Hunter reagent to a specific activity of 2000 Ci/mmol or purchased from INCSTAR Corp. Bond-Elut C_{18}–OH silica and silica cartridges (500 mg) and the Vac-Elut cartridge rack are from Varian Instruments (Harbor City, CA). N-Evap evaporator, Model 112, is from Organomation (Northborough, MA). Donkey anti-goat second antibody is from INCSTAR Corp. HPLC-grade hexane, acetonitrile,

[10] L. J. Fraher, S. Adami, T. L. Clemens, G. Jones, and J. L. H. O'Riordan, *Clin. Endocrinol.* **18,** 151 (1983).

dichloromethane, methanol, and 2-propanol are from Fisher Chemical Co. Unless otherwise noted, all other reagents are reagent grade.

Preparation of Assay Calibrators

25-OH-D Radioimmunoassay

Human serum is "stripped" free of vitamin D metabolites by treatment with activated charcoal. Absence of 25-OH-D in the stripped sera is confirmed by direct ultraviolet detection of 25-OH-D in serum following HPLC.[2] Subsequently, crystalline 25-OH-D_3 dissolved in absolute ethanol is added to the stripped sera to yield calibrators at concentrations of 0, 5, 12, 40, and 100 ng/ml.

1.25-$(OH)_2D$ Radioimmunoassay

Human serum is again "stripped" free of vitamin D metabolites using activated charcoal. Absence of 1,25-$(OH)_2D$ in the stripped sera is confirmed by RRA.[5] Subsequently, crystalline 1,25-$(OH)_2D$ dissolved in absolute ethanol is added to the stripped sera to final concentrations of 0, 5, 12, 20, 40, 100, and 200 pg/ml.

Preassay Sample Preparation

25-OH-D Radioimmunoassay

25-(OH)-D is extracted from calibrators and samples as follows: 0.5 ml of acetonitrile is placed into a 12- × 75-mm borosilicate glass tube after which 50 μl of sample or calibrator is dropped through the acetonitrile. After vortex mixing, the tubes were centrifuged (2000 *g*, 4°, 5 min) and 25 μl of supernatant transferred to 12- × 75-mm borosilicate glass tubes and placed on ice.

1,25-$(OH)_2D$ Radioimmunoassay

0.5–0.75 ml of sample or calibrator is placed into 12- × 75-mm glass tubes, an equal volume of acetonitrile is added. The mixture is vigorously vortex mixed and the solution centrifuged (2000*g*, 20°, 10 min). The supernate is removed into 12- × 75-mm borosilicate glass tubes to which one sample volume of 25 mg/ml sodium metaperiodate is added and incubated for 30–60 min at ambient temperature. Apply the supernate–sodium periodate mix to a C_{18}–OH cartridge that has been prewashed successively with 5 ml of 2-propanol and 5-ml methanol. Wash the cartridge successively with 5 ml of methanol/water (70/30, v/v), 5 ml of hexane/dichloromethane (90/10, v/v), and 5 ml of hexane/2-propanol (99/1, v/v). The C_{18}–OH cartridge is now placed *into* a silica cartridge previously washed successively with 5 ml methanol, 5 ml 2-propanol, and 5 ml hexane/2-propanol (80/20, v/v). Elute 1,25-$(OH)_2D$ onto the silica cartridge using 5 ml of hexane/2-propanol (92/8, v/v). Remove the C_{18}–OH cartridge and wash the silica cartridge with an additional 2 ml of 92/8 mixture. Finally, elute 1,25-$(OH)_2D$ from the silica cartridge using 5 ml of hexane/2-propanol (80/20, v/v). Regenerate each C_{18}–OH cartridge for reuse by washing with 2 ml of methanol. The silica cartridges can be reused without any further washing steps. Evaporate the fraction containing 1,25-$(OH)_2D$ under nitrogen at up to 55°, cool the tubes, and reconstitute the residues in 50 μl of absolute ethanol. Gap and mix each sample and store at −20° until the RIA is to be performed.

Radioimmunoassay

25-OH-D Radioimmunoassay

The assay tubes are 12- × 75-mm borosilicate glass tubes containing 25 μl of acetonitrile-extracted calibrators or samples. To each tube add ^{125}I-labeled 25-(OH)D derivative (50,000 cpm in 50 μl (v/v) 1:1 ethanol, 0.01 *M* phosphate buffer, pH 7.4). Then add to each tube 1.0 ml of primary antibody diluted 1:15,000-fold in sodium phosphate buffer (50 m*M*, pH 7.4) containing 0.1% swine-skin gelatin. Nonspecific binding was estimated using the above buffer minus the antibody. Vortex mix the contents of the tubes and incubate them for 90 min at 20–25°. Then add 0.5 ml of the second antibody-precipitating complex to each tube, vortex mix, incubate at 20 25° for 20 min, and centrifuge (2000*g*, 20°, 20 min). Discard the supernate and determine radioactivity in a gamma well counting system. The 25-OH-D values are calculated directly from the standard curve by the counting system using a smooth-spline method of calculation. The entire 25-OH-D assay procedure is displayed in Fig. 1.

1,25-$(OH)_2D$ Radioimmunoassay

The assay tubes are 12- × 75-mm borosilicate glass tubes containing 20 μl of the ethanol-reconstituted extracted calibrators or samples. To each tube add ^{125}I-labeled 1,25-$(OH)_2D$ derivative (50,000 cpm in 50 μl (v/v) 1:1 ethanol, 0.01 *M* phosphate buffer, pH 7.4). Then add to each tube 0.25 ml of primary antibody diluted 150,000-fold in sodium phosphate buffer

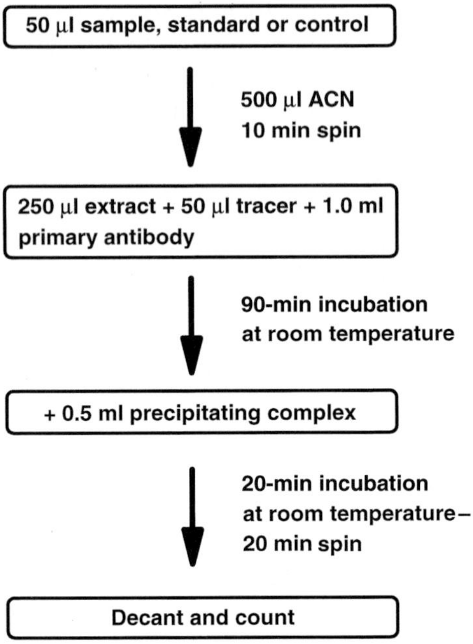

FIG. 1. Flow diagram of ^{125}I-based 25-OH-D RIA.

(50 m*M*, pH 6.2 containing swine-skin gelatin and polyvinyl alcohol (molecular weight 13,000–23,000) at concentrations of 0.1 and 0.35%, respectively. Nonspecific binding was estimated using the above buffer minus the antibody. Vortex mix the contents of the tubes and incubate them for 2 hr at 20–25°. Then add 0.5 ml of the second antibody-precipitating complex to each tube, vortex mix, incubate at 20–25 ° for 20 min, and centrifuge 2000*g*, 20°, 20 min). Discard the supernate and determine radioactivity in a gamma well counting system. The 1,25-$(OH)_2$D values are calculated directly from the standard curve by the counting system using a smooth-spline method of calculation. The entire 1,25$(OH)_2$D assay procedure is displayed in Fig. 2.

Results

25-OH-D Radioimmunoassay

Table I depicts the cross-reactivity of vitamin D and several of its metabolites with the goat antiserum generated against the 23,24,25,26,27-pentanor-C(22)-carboxylic acid calciferol. Several vitamin D metabolites could equally displace ^{125}I-labeled 25-OH-D derivative from the antibody.

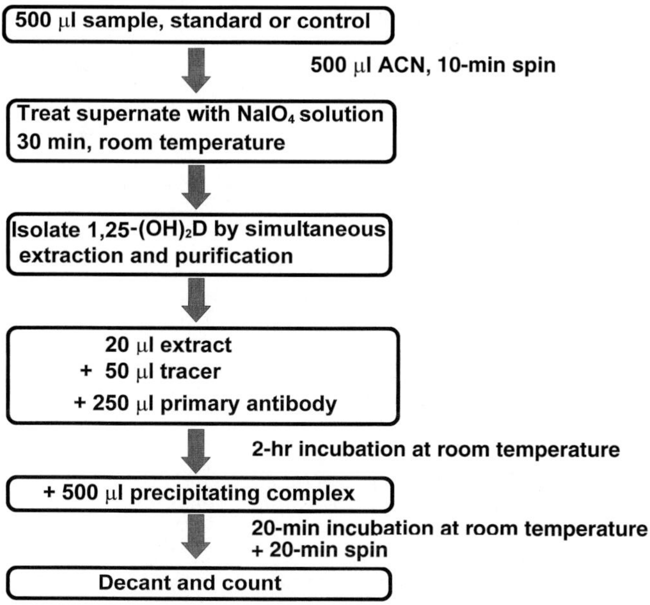

FIG. 2. Flow diagram of ^{125}I-based 1,25-$(OH)_2D$ RIA.

TABLE I
CROSS-REACTIVITY OF VARIOUS VITAMIN D COMPOUNDS WITH 25-OH-D ANTISERUM AND ^{125}I-LABELED VITAMIN D DERIVATIVE[a]

Steroid	Cross-reactivity (%)[b]
Vitamin D_2	0.8
Vitamin D_3	0.8
Dihydrotachysterol	<0.1
25-OH-D_2	100
25-OH-D_3	100
25-OH-D_3-26,23-lactone	100
24,25-$(OH)_2D_2$	100
24,25-$(OH)_2D_3$	100
25,26-$(OH)_2D_2$	100
25,26-$(OH)_2D_3$	100
1,25-$(OH)_2D_2$	2.5
1,25-$(OH)_2D_3$	2.5

[a] From Hollis *et al.*,[7] with permission.
[b] Ability to displace 50% of the ^{125}I tracer from the 25-OH-D antiserum diluted 15,000-fold.

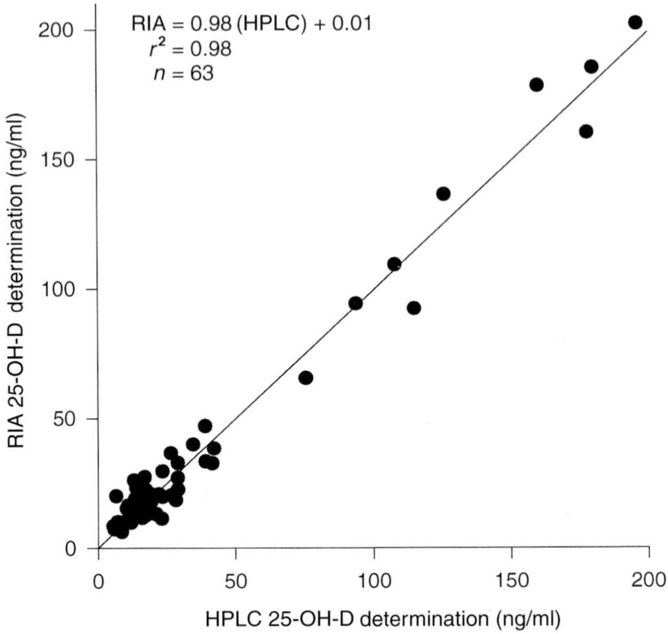

FIG. 3. 25-OH-D values obtained in various human subjects by direct ^{125}I-based RIA (y) and by direct ultraviolet quantification of 25-OH-D following HPLC (x). (From Hollis *et al.*,[7] with permission.)

Exceptions to this were vitamins D_2, D_3, 1,25-$(OH)_2D$, and dihydrotachysterol, which were far less effective in displacement. The detection limit of this RIA, defined as 3 standard deviations from the mean for data on the zero sample, was 2 ng/ml 25-$(OH)D_3$. Bound tracer was displaced 50% by 25-OH-D_3 at approximately 40 ng/ml.

The estimated analytical recovery of 25$(OH)D_3$ in this RIA from four separate human serum samples was 97.3 ± 10.4% ($\bar{x} \pm$ SD). Assay precision (% *CV*), both within- and between-assay variation, was determined at three different points of the standard curve. At all points, the assay variation was nearly identical.[7] Within-assay variation was <6.0%; between-assay variation was <15%. Assay parallelism was assessed in four different serum samples by serial dilution of the original plasma in vitamin D-free serum and all demonstrated linearity.[7]

For further method validation, results obtained by the RIA for 63 human subjects and 20 vitamin D-deficient samples were compared with those obtained by a conventional assay,[2] in which 25-OH-D is measured in serum by direct quantification of ultraviolet absorbance after liquid chromatogra-

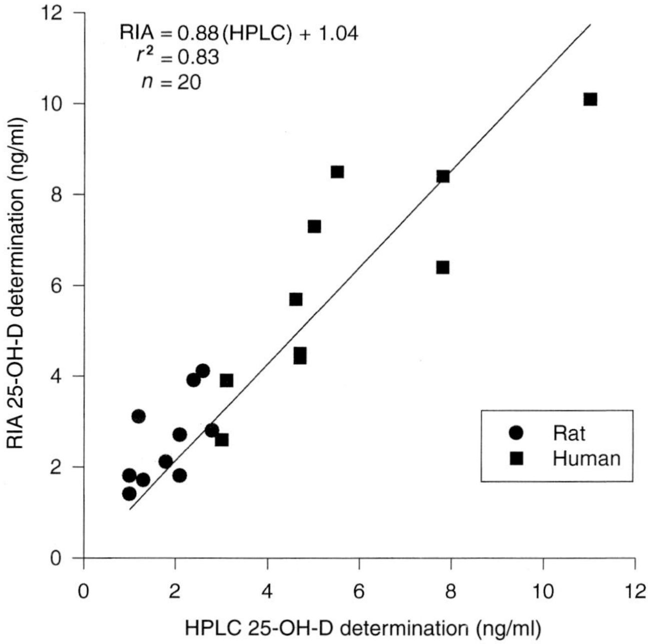

FIG. 4. 25-OH-D values obtained in vitamin D-deficient humans and rats by direct ^{125}I-based RIA (y) and by direct ultraviolet quantification of 25-OH-D following HPLC (x).

phy. Assessment of vitamin D status as determined by the two assay methods was shown to be quite similar (Figs. 3 and 4). The respective values from 36 normal subjects, 12 biliary atresia patients, and 8 vitamin D therapy patients are listed in Table II.

1,25-(OH)$_2$D Radioimmunoassay

Table III depicts the cross-reactivity of vitamin D_3 and several of its metabolites with antibody S-11. Note that these cross-reactivities are based on the entire extraction and purification system, including sample pretreatment with sodium metaperiodate. Any vitamin D metabolite not possessing a 1-hydroxyl function is virtually eliminated from cross-reacting in this assay system. However, several vitamin D metabolites containing a 1-hydroxyl function other than 1,25-(OH)$_2$D can cross-react in this RIA. 1,25-(OH)$_2$D$_2$ has approximately 70% potency as compared to 1,25(OH)$_2$D$_3$, while other 1-hydroxylated metabolites demonstrate a fourfold or less ability to compete in this RIA. The respective circulating levels of 1,25-(OH)$_2$D from 50 normal subjects obtained by the RIA with or without sample pretreatment

TABLE II
CONCENTRATION OF 25-OH-D AS DETERMINED BY ^{125}I RADIOIMMUNOASSAY IN VARIOUS PHYSIOLOGICAL STATES[a]

Subject type	*n*	Concentration (ng/ml)	
		Mean	Range
Normal[b]	36	25.7	9.9–41.5
Biliary atresia	12	6.3	4.3–8.3
Vitamin D therapy[c]	8	145	92–202

[a] From Hollis *et al.*,[7] with permission.
[b] Samples from subjects in Minnesota in October.
[c] Samples from subjects with hypoparathyroidism or pseudohypoparathyroidism receiving pharmacologic doses of vitamin D_2.

by sodium metaperidoate are displayed in Fig. 5. Untreated samples clearly exhibited elevated "apparent" 1,25-$(OH)_2$ levels as compared to sodium metaperidoate-treated samples by regression analysis. There is a clear relationship between these two assays as displayed in Fig. 5. The data clearly demonstrate that without sample pretreatment, circulating 1,25-$(OH)_2D$ levels are overestimated by an average of 41%.

TABLE III
CROSS-REACTIVITY OF VARIOUS VITAMIN D COMPOUNDS WITH 1,25-$(OH)_2D$ ANTISERUM S-11 AND ^{125}I-LABELED 1-HYDROXYLATED TRACER[a]

Steroid	Cross-reactivity (%)[b]
Vitamin D_3	<0.001
25-OH-D_3	<0.001
24,25-$(OH)_2D_3$	0.005
25,26-$(OH)_2D_3$	<0.001
1,25-$(OH)_2D_2$	73.1
1,25-$(OH)_2D_3$	100
1,25-$(OH)_2D_3$-26,23-lactone	13.1
1,24,25-$(OH)_3D_3$	25.1
1,25,26-$(OH)_3D_3$	23.2

[a] Following sample treatment and purification. From Hollis *et al.*,[8] with permission.
[b] Ability to displace 50% of the ^{125}I tracer from the 1,25-$(OH)_2D$ antiserum diluted 1:150,000-fold.

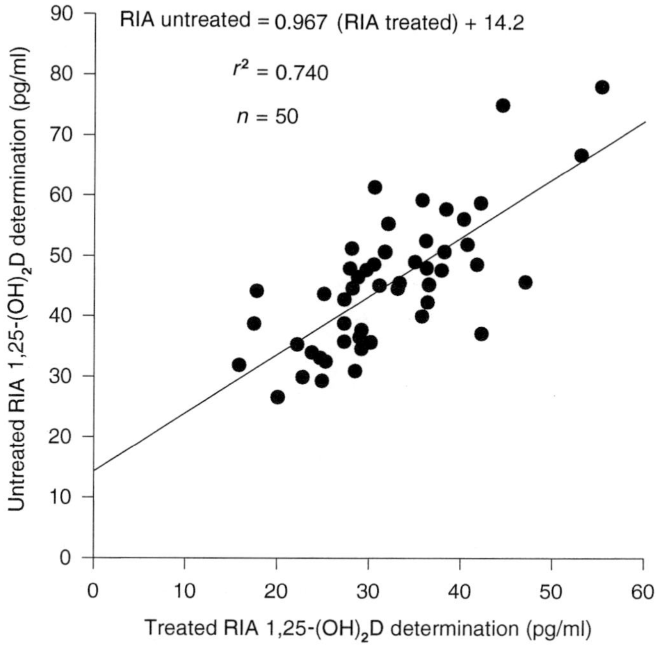

FIG. 5. Comparison of circulating 1,25-$(OH)_2D$ obtained by the ^{125}I-based RIA with and without sample pretreatment with sodium metaperiodate. (From Hollis *et al.*,[8] with permission.)

The detection limit of this RIA was 2.4 pg/ml. Bound tracer was displaced 50% by 1,25-$(OH)_2D_3$ at approximately 20 pg/ml. The estimated analytical recovery of 1,25-$(OH)_2D_3$ and 1,25-$(OH)_2D_2$ in the assay at two levels in a pooled serum sample was 95 and 68%, respectively. Assay precision (% *CV*), both within- and between-assay variation, was determined at three different points on the standard curve. Assay variation was greatest at the lowest levels, but in all cases within acceptable limits ($\leq$15%) for this type of RIA. Assay parallelism was assessed in four different samples by serial dilution of the original plasma in vitamin D-free serum and all demonstrated linearity.[8]

Final method validation was achieved by comparison of the RIA with a standard RRA procedure for 1,25-$(OH)_2D$ assessment.[5] Results obtained by the RIA for 50 human subjects were compared with values obtained by the RRA (Fig. 6) and the data from the two procedures are highly related. The clinical utility of this RIA is demonstrated in Fig. 7 by its ability to differentiate normal from pathological samples.

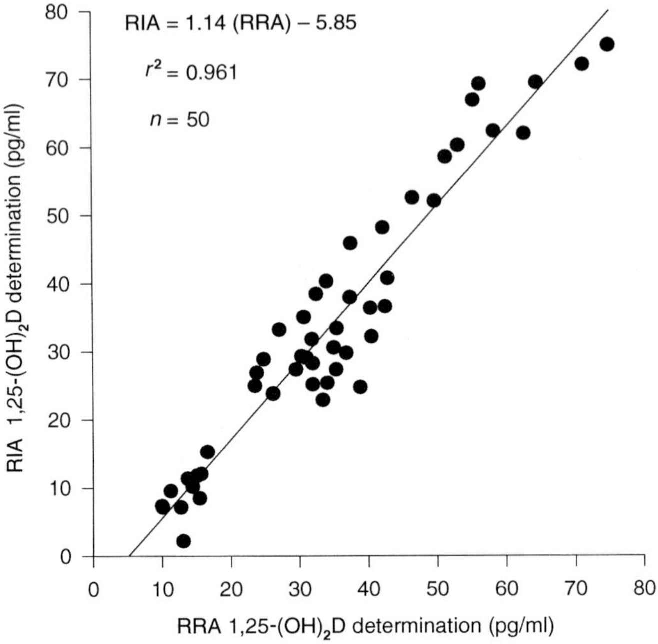

FIG. 6. Comparison of circulating 1,25-$(OH)_2D$ obtained by the ^{125}I-based RIA and by traditional RRA. (From Hollis *et al.*,[8] with permission.)

Discussion

Many methods have been developed for the determination of vitamin D nutritional (25-OH-D) and hormonal [1,25-$(OH)_2D$] status.[1–10] One major deficiency in most of these procedures is that they utilize tritium-labeled tracers. We have developed RIAs based on ^{125}I tracers that assess the circulating levels of 25-OH-D and 1,25-$(OH)_2D$.[7,8] In the past, RIAs for 25-OH-D and 1,25-$(OH)_2D$ had been developed but were dependent on ^3H tracers and substantial sample prepurification prior to final quantitation. The methods presented here not only have the advantage of an ^{125}I tracer, but also require a minimum of sample handling prior to assay.

For the 25-OH-D RIA, we raised an antisera against a synthetic vitamin D analog, 23,24,25,26,27-pentanor-C(22)-carboxylic acid.[6] Coupling this compound to BSA allowed us to obtain antibodies that cross-reacted equally with most vitamin D_2 and D_3 metabolites (Table I). Although many vitamin D metabolites other than 25-OH-D are present in the circulation, they contribute by only a small percentage (1–7%) to the overall assessment

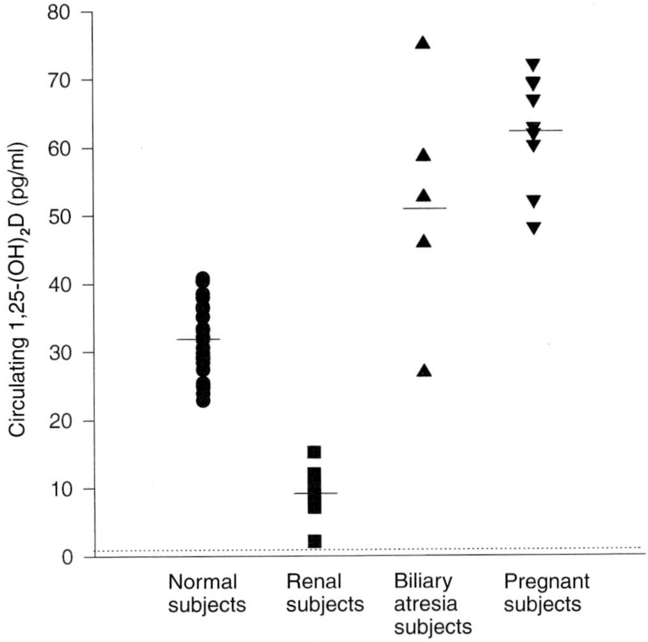

FIG. 7. Circulating 1,25-$(OH)_2D$ as determined by the ^{125}I-based RIA from 50 normal subjects, 11 chronic renal failure subjects, 5 biliary atresia subjects, and 9 pregnant subjects. (From Hollis *et al.*,[8] with permission.)

of nutritional vitamin D status as compared with 25-OH-D.[11] The parent compounds, vitamins D_2 and D_3, are unreliable indicators of overall vitamin D status and usually are present in the circulation in very low concentrations.[11] The validation of this RIA in predicting nutritional vitamin D status at both normal and extremely deficient levels is shown in Figs. 3 and 4, respectively.

In contrast to the 25-OH-D RIA, the RIA for 1,25-$(OH)_2D$ requires much greater specificity and sensitivity due to its extremely low concentrations in the circulation. For this reason, the RIA for 1,25-$(OH)_2D$ is more technically demanding than is the RIA for 25-OH-D. These excess demands for the 1,25-$(OH)_2D$ RIA reside almost totally in sample preparation and purification prior to assay. To date, the best antibodies toward 1,25-$(OH)_2D$ have, at best, a cross-reactivity with the non-1-hydroxylated metabolites of vitamin D of ~1%.[10] Given that these non-1-hydroxylated metabolites circulate at concentrations 1000 times greater than that of 1,25-$(OH)_2D$,

[11] B. W. Hollis and W. B. Pittard, *J. Clin. Endocrinol. Metab.* **59,** 652 (1984).

the magnitude of the problem becomes apparent. In the past, these unwanted metabolites were removed prior to RIA by HPLC sample prepurification,[9] which is extraordinarily cumbersome and not suited to routine clinical analysis. We therefore chose a simplified chromatographic procedure that had been incorporated into an RRA for 1,25-$(OH)_2D$ a decade ago.[5] However, when we utilized this purification procedure with the RIA, the "apparent" circulating 1,25-$(OH)_2D$ concentrations were elevated. Further investigation revealed that the vitamin D metabolite most responsible for this overestimation was 25,26-$(OH)_2D$. Efforts to resolve this metabolite by chromatographic means short of HPLC failed. We therefore instituted a novel sample pretreatment with sodium metaperiodate. This pretreatment step converted all of the 24,25-$(OH)_2D$ and 25,26-$(OH)_2D$ into their respective aldehyde and ketone forms, which are easily removed by the current chromatographic scheme. The effectiveness of this treatment is demonstrated in Fig. 5. Further, the concentrations of 1,25-$(OH)_2D$ as determined in serum from various groups of healthy and pathological subjects agree well with values known to occur under these conditions (Fig. 7).

In summary, the ^{125}I-based RIAs for 25-OH-D and 1,25-$(OH)_2D$ described here simplify the quantitation for these two important vitamin D metabolites. These RIAs have several unique features, including ^{125}I-based tracers, calibrators that allow direct quantification from a calibration curve without the need to estimate recovery losses and, most important, total elimination of sample prepurification (25-OH-D) or elimination of prepurification by HPLC [1,25-$(OH)_2D$]. Thus, these new RIAs represent a major advance for assessing circulating 25-OH-D and 1,25-$(OH)_2D$ in terms of both time savings and cost effectiveness.

[17] Assay of Vitamin D Derivatives and Purification of Vitamin D Hydroxylases

By YOSHIHIKO OHYAMA, SHIN-ICHI HAYASHI, EMIKO USUI, MITSUHIDE NOSHIRO, and KYU-ICHIRO OKUDA

Introduction

Vitamin D_3 is hydroxylated at position 25 by liver microsomal and mitochondrial vitamin D_3 25-hydroxylases (25-hydroxylase).[1,2] The micro-

[1] J. L. Omdahl and H. F. DeLuca, *Physiol. Rev.* **53,** 327 (1973).
[2] K.-I. Okuda, E. Usui, and Y. Ohyama, *J. Lipid Res.* **36,** 1641 (1995).

0076-6879/97 $25.00

somal enzyme purified from male rats is the same as CYP2C11, a male-specific P450. The mitochondrial enzyme is identical with 5β-cholestane-3α,7α,12α-triol (an intermediate in the conversion of cholesterol to cholic acid) 27-hydroxylase, and is categorized as CYP27.[2,3] The product, 25-hydroxyvitamin D_3 (25-OH-D_3), is transported to the kidney through the bloodstream, where it is hydroxylated at the 1α-position to give 1α,25-dihydroxyvitamin D_3 [1,25-$(OH)_2D_3$] in calcium or 1,25-$(OH)_2D_3$ depleted animals. It is hydroxylated at position 24 to give 24*R*,25-dihydroxyvitamin D_3 [24,25-$(OH)_2D_3$] in calcium-repleted animals by a 25-hydroxyvitamin D_3 24-hydroxylase (24-hydroxylase). This enzyme also hydroxylates 1α,25-$(OH)_2D_3$ to 1α,24*R*,25-trihydroxyvitamin D_3. We have purified both 25- and 24-hydroxylases to homogeneity and isolated their cDNA clones. These hydroxylases are members of the P450 superfamily that are *b*-type hemoproteins functioning as terminal monooxygenase in the oxygenation of various organic compounds.

In this chapter the assay and purification of both 25- and 24-hydroxylases are described. Molecular cloning of these enzymes is described elsewhere in this volume.[4]

Enzyme Assay

The P450 enzymes catalyze a monooxygenation that involves the uptake of two electrons from NADPH with the reduction of one atom of O_2 to water and insertion of the other atom of O_2 into the substrate. In mitochondria, both a flavoprotein containing FAD as a prosthetic group (NADPH–ferredoxin reductase) and an iron-sulfur protein (ferredoxin) participate in the electron transfer:

$$\text{NADPH} \rightarrow \text{NADPH–ferredoxin reductase} \rightarrow \text{ferredoxin} \rightarrow \text{P450}$$

In microsomes, a flavoprotein containing both FAD and FMN (NADPH–cytochrome P450 reductase) transfers electrons:

$$\text{NADPH} \rightarrow \text{NADPH–P450 reductase} \rightarrow \text{P450}$$

The specificity of these proteins is generally low. For reconstitution, the electron-transferring proteins of various tissues can be mixed to form active complexes.

[3] D. R. Nelson, T. Kamataki, D. J. Waxman, F. P. Guengerich, R. W. Estabrook, R. Foreseen, F. J. Gonzalez, M. J. Coon, I. C. Gunsalus, O. Gotoh, K.-I. Okuda, and D. W. Nebert, *DNA Cell Biol.* **12,** 1 (1993).

[4] M. Noshiro, Y. Ohyama, E. Usui, M. Akiyoshi-Shibata, Y. Yabusaki, and K.-I. Okuda, *Methods Enzymol.* **282,** [19], (1997) (this volume).

Enzyme activity is measured by either of the following two methods depending on whether the enzyme is membrane bound or in a solubilized state. In the latter case the incubation mixture should be fortified with an electron transfer system.

In these methods, NADPH–adrenodoxin reductase is used as a NADPH–ferredoxin reductase and adrenodoxin is used as ferredoxin. Both are prepared from bovine adrenals. NADPH-P450 reductase is prepared from rat liver microsomes. These proteins are commercially available.

Materials

1α-Hydroxyvitamin D_3 (1α-OH-D_3), 25-hydroxyvitamin D_3, and 1α,25-dihydroxyvitamin D_3 are the products of Duphar (Weesp, The Netherlands). NADPH–adrenodoxin reductase and adrenodoxin are prepared according to the methods of Suhara *et al.*[5,6] NADPH–cytochrome P450 reductase is prepared as described by Yasukochi and Masters.[7]

Assay of Liver Microsomal Vitamin D_3 25-Hydroxylase

Buffer I: 200 m*M* potassium phosphate buffer (pH 7.0) containing 1 m*M* EDTA

Assay for Microsomal Enzyme

Incubation mixture:

Microsomes	0.2–1.0 mg of protein
Buffer I	500 μl
1α-OH-D_3 (10 nmol/μl of ethanol)	10 μl
NADPH (50 m*M*)	20 μl
Water	To a final volume of 1 ml

The incubation mixture containing all of the components except the NADPH is preincubated for 2 min at 37°, and the reaction started by adding NADPH. Incubation is continued with shaking for 10–20 min at the same temperature. At the end of the reaction, 5 ml of benzene is added and the reaction mixture is vortexed and centrifuged at 1000*g* for 10 min to separate the organic layer. A 4-ml aliquot of the benzene extract is removed by pipette and the solvent is evaporated by flushing with N_2 gas. The residue is dissolved in a small amount of chloroform, and an aliquot is subjected to high-performance liquid chromatography (HPLC) analysis.[8]

[5] K. Suhara, Y. Ikeda, S. Takemori, and M. Katagiri, *FEBS Lett.* **28,** 45 (1972).

[6] K. Suhara, S. Takemori, and M. Katagiri, *Biochim. Biophys. Acta* **263,** 272 (1972).

[7] Y. Yasukochi and B. S. S. Masters, *J. Biol. Chem.* **251,** 5337 (1976).

[8] S. Hayashi, M. Noshiro, and K.-I. Okuda, *J. Biochem.* (*Tokyo*) **99,** 1753 (1986).

Assay for Solubilized Enzyme

Incubation mixture:

P450	0.1–0.5 nmol
NADPH–cytochrome P450 reductase (100 U/ml)	20 μl
Dilauroylglyceryl-3-phosphorylcholine (1 mg/ml)	10 μl
Buffer I	250 μl
1α-OH-D_3 (4 nmol/μl of ethanol)	5 μl
NADPH (50 m*M*)	10 μl
Water	To a final volume of 0.5 ml

Incubation and analysis are carried out as described earlier.

Assay of Liver Mitochondrial Vitamin D_3 25-Hydroxylase

Buffer II: 80 m*M* potassium phosphate buffer (pH 7.4) containing 2 m*M* $MgCl_2$

Buffer III: 160 m*M* potassium phosphate buffer (pH 7.4) containing 0.5 m*M* EDTA

Assay for Mitochondrial Enzyme

Incubation mixture:

Mitochondrial suspension	2–10 mg of protein
Buffer II	250 μl
1α-OH-D_3 (10 nmol/μl of ethanol)	5 μl
DL-isocitrate (1 *M*)	12.5 μl
Water	To a final volume of 0.5 ml

The incubation mixture is preincubated for 2 min at 37° and the reaction is started by adding DL-isocitrate. Incubation is continued for 5–20 min and terminated by adding 5 ml of benzene. Analysis is carried out as described earlier.[9]

Assay for Solubilized Enzyme

Incubation mixture:

P450	10–50 pmol
Adrenodoxin (400 nmol/ml)	10 μl
NADPH–adrenodoxin reductase (10 U/ml)	10 μl
Buffer III	250 μl

[9] O. Matsumoto, Y. Ohyama, and K.-I. Okuda, *J. Biol. Chem.* **263,** 14256 (1988).

1α-OH-D_3 (4 nmol/μl of ethanol)	5 μl
NADPH (50 m*M*)	10 μl
Water	To a final volume of 0.5 ml

Comments. The 25-hydroxylase has a severalfold higher turnover number toward 1α-hydroxyvitamin D_3 as a substrate than that toward vitamin D_3. Therefore, the former substrate is used for the routine assay of the enzyme. The enzyme activity toward vitamin D_3 can be assayed similarly if a sufficient amount of the substrate is used (100 nmol/ml of the final volume). In this case the reaction is terminated by adding 1 ml of ethanol and the products are extracted with 5 ml of petroleum ether. Isocitrate is a functional electron donor only in the *in vitro* assay of mitochondrial vitamin D_3 25-hydroxylases.[10,11] Other members of the tricarboxylic acid cycle do not work as well except that citric acid is a slightly less effective electron donor than isocitrate. When the 25-hydroxylase activity is assayed using isocitrate as electron donor only the mitochondrial enzyme activity is measured. The NADPH generating system is preferred to NADPH in incubation of the crude enzyme, since NADPH is also consumed by other enzymes present.

Assay of Renal 25-Hydroxyvitamin D_3 24-Hydroxylase

Buffer IV	15 m*M* Tris–acetate (pH 7.4) containing 0.19 *M* sucrose, 2 m*M* magnesium acetate, and 1 m*M* EDTA
Buffer V	200 m*M* Tris-HCl (pH 7.7) containing 1 m*M* EDTA

Assay for Mitochondrial Enzyme

Kidneys are homogenized with buffer IV and mitochondria are prepared by differential centrifugation. Mitochondria are suspended in the same buffer.

Incubation mixture:

Kidney homogenate or mitochondrial suspension	5–15 mg of protein
25-OH-D_3 (10 nmol/μl of ethanol)	10 μl
Sodium succinate (1 *M*)	25 μl
Buffer IV	To a final volume of 1 ml

After a 2-min preincubation, the reaction is initiated by adding sodium succinate. Incubation is continued for 10–30 min at 37° and terminated by adding 5 ml of benzene. Extraction and analysis are performed as described earlier.[12,13]

[10] S. Taniguchi, N. Hoshita, and K.-I. Okuda, *Eur. J. Biochem.* **40,** 607 (1973).

[11] K.-I. Okuda, P. Weber, and V. Ullrich, *Biochem. Biophys. Res. Commun.* **74,** 1071 (1977).

[12] Y. Ohyama, S. Hayashi, and K.-I. Okuda, *FEBS Lett.* **255,** 405 (1989).

[13] Y. Ohyama and K.-I. Okuda, *J. Biol. Chem.* **266,** 8690 (1991).

Assay for Solubilized Enzyme

Incubation mixture:

P450	5–50 pmol
Adrenodoxin (400 nmol/ml)	5 μl
NADPH–adrenodoxin reductase (10 U/ml)	5 μl
Buffer V	250 μl
25-OH-D_3 (2 nmol/μl of ethanol)	5 μl
NADPH (50 m*M*)	10 μl
Water	To a final volume of 0.5 ml

Comments. Because high concentrations of detergents are inhibitory to the reaction, no more than 50 μl of the detergent-solubilized sample should be included in the total incubation mixture of 0.5 ml. When 24,25-$(OH)_2D_3$ is treated with periodate, the compound is cleaved at the *vic*-diol[14] and is not detected. This serves as an additional confirmation of the identity of the compound.

Analysis of Metabolites by High-Performance Liquid Chromatography

The metabolites of vitamin D_3 are routinely separated by either normal- or reversed-phase HPLC and the effluents are monitored at 265 nm.[15] Identification of the products by HPLC using only a single elution system may lead to an erroneous conclusion.[16] For confirmation of product identification, it is desirable to perform HPLC using at least two different straight phase systems that differ in elution properties and one reversed-phase system. For further confirmation, mass spectrum and/or nuclear magnetic resonance spectra may be required. For conventional assays in the process of enzyme purification normal phase HPLC is preferable owing to its rapidity. Retention times of some vitamin D_3 metabolites in various HPLC systems are shown in Table I.

Purification of Vitamin D Hydroxylases

Both the 25- and 24-hydroxylases are labile and are easily inactivated by unfavorable treatment during the course of purification. It is therefore important to purify the enzymes based on assay of enzyme activity, because inactivation in the course of purification cannot be detected by the conventional spectroscopic assay method that is based on the CO-difference spectrum of P450.

[14] M. Burgos-Trinidad, A. J. Brown, and H. F. DeLuca, *Biochemistry* **25,** 2692 (1986).
[15] J. L. Napoli, N. J. Koszewski, and R. L. Horst, *Methods Enzymol.* **123,** 127 (1986).
[16] A. J. Brown and H. F. DeLuca, *J. Biol. Chem.* **260,** 14132 (1985).

TABLE I
ELUTION POSITIONS OF VITAMIN D_3 METABOLITES FROM HPLC COLUMNS

Product	Column	Solvent for injection	Mobile phase (v/v)	Flow rate (ml/min)	Retention time (min)
25-OH-D_3	Silica gel[a]	2-Propanol	Hexane/2-propanol (93.5:6.5)	1.4	8
	ODS[b]	Methanol	Methanol/H_2O (93:7)	1.0	9
1,25-$(OH)_2D_3$	Silica gel[a]	Chloroform	Hexane/2-propanol/methanol (88:6:6)	1.4	12
	ODS[b]	Methanol	Methanol/H_2O (80:20)	1.0	18
24,25-$(OH)_2D_3$	Silica gel[a]	Chloroform/ethyl acetate (4:1) (v/v)	Hexane/2-propanol/methanol (88:6:6)	1.4	9
	ODS[b]	Methanol	Methanol/H_2O (80:20)	1.0	16

[a] Data were obtained with a Finepak SIL-5 (4.6 × 250 mm, JASCO Co., Tokyo).
[b] Data were obtained with a Finepak SIL C18 (4.6 × 250 mm, JASCO Co., Tokyo).

Generally P450s are purified by the following procedures: solubilization by use of suitable detergent(s), fractionation by polyethylene glycol, hydrophobic chromatography, and further chromatographies using hydroxyapatite and ion-exchange columns. Chromatographic properties of both 25-hydroxylase and 24-hydroxylase are summarized in Table II.

General cautions for purification of vitamin D_3 hydroxylase are as follows: (1) Exposure of the enzyme to low ionic strength results in deterioration. Therefore, the ionic strength of buffers for equilibration and dialysis is kept higher than 0.038 [corresponding to 20 m*M* phosphate buffer (pH 7.4) or to 29 m*M* Tris-HCl buffer (pH 7.4)] in our methods. (2) The ratio

TABLE II
PROPERTIES OF VITAMIN D_3 HYDROXYLASE PROTEINS

Property	Microsomal 25-hydroxylase	Mitochondrial 25-hydroxylase	Mitochondrial 24-hydroxylase
Suitable hydrophobic column	ω-Aminooctyl-Sepharose	ω-Aminohexyl-Sepharose	Pentyl-Sepharose
Elution buffer from hydroxyapatite column	80 m*M* KP_i[a]	90 m*M* KP_i	120 m*M* KP_i
Elution buffer from DEAE column	Nonbound	60 m*M* NaCl	40 m*M* NaCl
Elution buffer from CM column	60 m*M* KP_i	NA[b]	NA[b]
Suitable detergents	Emulgen 913	Lubrol PX	Lubrol PX, Tween 20
Molecular weight[c]	50,000	52,500	53,000

[a] Potassium phosphate buffer.
[b] NA, not applicable.
[c] Estimated by SDS–PAGE.

of protein/cholate is selected to solubilize the purifying enzyme effectively while keeping solubilization of other proteins at a minimum. (3) It is important to collect the fractions showing high specific enzyme activity in the ω-aminohexyl-Sepharose column chromatography step. Otherwise the final preparation is contaminated with other proteins that are difficult to remove later. (4) It is also important to select the proper detergents for purification of these enzymes. The detergents described are those found to be most effective. Use of inadequate detergents causes an unrecoverable loss of the enzyme activities. (5) Proteinase inhibitors, leupeptin, and pepstatin are included in buffers used in early stages of purification so as to avoid possible degradation of protein by proteinases.

Purification of Liver Microsomal Vitamin D_3 25-Hydroxylase

Livers from 73 male rats of the Wistar strain weighing about 200 g are excised, thoroughly perfused with chilled 0.9% NaCl, and then homogenized with nine volumes of 20 m*M* Tris-HCl buffer (pH 7.5) containing 0.25 *M* sucrose, 1 m*M* EDTA, and 1 μg/ml each of leupeptin and pepstatin. The homogenate is centrifuged at 10,000*g* for 15 min. The supernatant is centrifuged at 100,000*g* for 60 min. The packed pellets of microsomes are washed with 0.15 *M* KCl containing 1 m*M* EDTA and 1 μg/ml each of leupeptin and pepstatin. All buffer solutions used in the following purification procedures contained 20% glycerol and 1 μg/ml each of leupeptin and pepstatin unless otherwise stated.

The washed microsomes are suspended in 100 m*M* Tris-HCl buffer (pH 7.4) that contained 0.1 *M* KCl, 1 m*M* EDTA, and 1 m*M* 1,4-dithiothreitol (DTT). To solubilize microsomes 10% sodium cholate is added slowly with mild stirring to give a final cholate/protein ratio of 3:1 (w/w) and gentle stirring is continued for 60 min. The mixture is centrifuged at 100,000*g* for 60 min and the supernatant is fractionated with polyethylene glycol.[17] The fraction precipitated with 8–12% of polyethylene glycol is collected by centrifugation, suspended in 100 m*M* potassium phosphate buffer (pH 7.4) containing 1 m*M* EDTA and 1 m*M* DTT, and dialyzed overnight against the suspension buffer.

The dialyzed sample is then resolubilized with 10% cholate adjusting the final concentration of protein 4 mg/ml and that of cholate 1%. The mixture is centrifuged at 100,000*g* for 60 min. The supernatant is applied to an ω-aminooctyl-Sepharose 4B column (4 × 16 cm) that has been equilibrated with 100 m*M* phosphate buffer containing 0.5% cholate, 1 m*M*

[17] M. J. Coon, T. A. van der Hoeven, S. B. Dahl, and D. A. Haugen, *Methods Enzymol.* **52,** 109 (1978).

EDTA, and 1 mM DTT. The column is washed with the equilibration buffer and proteins are eluted with 100 mM potassium phosphate buffer containing 0.4% cholate, 1 mM EDTA, 1 mM DTT, and 0.08% Emulgen 913. Fractions showing high 25-hydroxylation activity are pooled and dialyzed against 30 mM phosphate buffer (pH 7.4) containing 0.5 mM EDTA, 0.5 mM DTT, 0.2% Emulgen 913, and 0.2% cholate. The dialyzed sample is applied to a hydroxypapatite column (2.5 × 20 cm) equilibrated with the same buffer as that used for dialysis. The column is washed with 40 mM phosphate buffer containing 0.5 mM EDTA, 0.5 mM DTT, 0.2% Emulgen 913, and 0.2% cholate. 25-Hydroxylase is eluted with 80 mM potassium phosphate buffer. The fractions showing high 25-hydroxylation activity are pooled and dialyzed against 20 mM phosphate buffer containing 1 mM EDTA, 0.5 mM DTT, 0.2% Emulgen 913, and 0.2% cholate. The dialyzed sample is applied to a DEAE-Sepharose CL-6B column (1.2 × 6.5 cm) equilibrated with the dialyzing buffer; the column is washed with the same buffer. The 25-hydroxylase is not bound to the column but passes through the column. No enzyme activity is observed in the bound fractions that are eluted with 50 and 150 mM phosphate buffers.

The pass-through fraction showing 25-hydroxylation activity is applied to the second hydroxyapatite column (1.4 × 12 cm) equilibrated with 20 mM phosphate buffer containing 1 mM EDTA, 0.5 mM DTT, 0.2% Emulgen 913, and 0.2% cholate. The column is washed with 50 mM phosphate buffer containing 0.5 mM EDTA, 0.5 mM DTT, 0.1% Emulgen 913, and 0.2% CHAPS. Proteins are eluted with 150 mM phosphate buffer containing 0.5 mM DTT, 0.5 mM EDTA, 0.1% Emulgen 913, and 0.2% CHAPS. Fractions showing high 25-hdyroxylation activity are pooled and dialyzed against 20 mM phosphate buffer containing 1 mM EDTA, 0.5 mM DTT, 0.1% Emulgen 913, and 0.2% CHAPS. The dialyzed sample is applied to a CM-Sepharose CL-6B column (1.2 × 12 cm) equilibrated with the same buffer used for dialysis. The column is washed with 30 mM phosphate buffer containing 1 mM EDTA, 0.5 mM DTT, 0.1% Emulgen 913, and 0.2% CHAPS, and bound proteins are eluted with 60 mM potassium phosphate buffers. Fractions enriched in 25-hydroxylase activity are pooled, dialyzed, and rechromatographed on the second CM-Sepharose CL-6B column (1.2 × 12 cm) under the same conditions. The purified sample is applied to the third hydroxyapatite column (1.2 × 3 cm) to remove the neutral detergents. Overall purification is as shown in Table III.[8]

The final preparation shows a single band on sodium dodecyl sulfate–polyacrylamide gel electrophoresis (SDS–PAGE). It exhibits not only vitamin D_3 25-hydroxylation activity but also hydroxylates testosterone at the 2α- or 16α-position. It is well known that a male-specific P450, P450 2C11, hydroxylates testosterone at the 2α- or 16α-position. The NH_2-terminal

TABLE III
PURIFICATION OF MICROSOMAL VITAMIN D_3 25-HYDROXYLASE

Purification step	Protein (mg)	Total P450[a] (nmol)	Total activity (nmol/min)	Specific activity[b] (pmol/min/mg of protein)	Purification[c] (-fold)	Yield[d] (%)
Microsomes	14,087	11,516	12.4	0.9	—	—
Polyethylene glycol fractionation (8–12%)	4,057	9,281	65.3	16.1	1	100
ω-Aminooctyl-Sepharose	426	2,840	112	263	16	172
Hydroxyapatite I	156	1,285	73.1	468	29	112
DEAE-Sepharose	113	1,134	71.0	626	39	109
Hydroxyapatite II	67.2	790	59.0	881	55	90
CM-Sepharose I	29.1	426	47.3	1,620	101	72
CM-Sepharose II	10.7	180	25.9	2,419	150	40
Hydroxyapatite III	5.6	96	20.9	3,706	230	32

[a] P450 content is estimated by the CO difference spectra [T. Omura and R. Sato, *J. Biol. Chem.* **239,** 2370 (1964)].
[b] Enzyme activity is measured by using vitamin D_3 as a substrate.
[c,d] Purification (-fold) and yield are calculated after the solubilization step.

amino acid sequence of the rat microsomal 25-hydroxylase is the same as that of P450 2C11.

Comments. This P450 is present in male, but not female, rat liver microsomes. Microsomal 25-hydroxylase activity in female rat liver is not identical with this P450 and is less well characterized.[18]

Purification of Liver Mitochondrial Vitamin D_3 25-Hydroxylase

Livers from 12 female rats weighing about 200 g are excised, thoroughly perfused with chilled 0.9% NaCl solution, and homogenized with nine volumes of 20 m*M* Tris-HCl buffer (pH 7.4) containing 0.25 *M* sucrose, 1 m*M* EDTA, and 1 μg/ml each of pepstatin and leupeptin. The homogenate is centrifuged at 600*g* for 12 min, and the supernatant is again centrifuged at 8000*g* for 15 min. The precipitated pellets are washed twice with the same buffer and then suspended in 60 m*M* Tris-HCl buffer (pH 7.4) containing 20% glycerol, 0.5 m*M* EDTA, 0.5 m*M* DTT, 1 μg/ml each of leupeptin and pepstatin. To this mitochondrial suspension, 10% cholate is added slowly under mild stirring to give a final cholate/protein ratio of 2:1 (w/w) and gentle stirring is continued for 60 min at 4°. All buffer solutions used in the following purification procedures contained 20% glycerol and 1 μg/ml each of leupeptin and pepstatin unless otherwise stated. The solubilized mitochondrial suspension is centrifuged at 100,000*g* for 60 min, and the supernatant is subjected to polyethylene glycol fractionation.[17]

A fraction precipitated by 4–12% of polyethylene glycol 6000 is collected by centrifugation and suspended in 100 m*M* potassium buffer (pH 7.4)

[18] S. Hayashi, E. Usui, and K.-I. Okuda, *J. Biochem.* (*Tokyo*) **103,** 863 (1988).

containing 1 m*M* EDTA, and 0.5 m*M* DTT. The suspension is dialyzed against the suspension buffer. The dialyzate is resolubilized with 1% sodium cholate and centrifuged at 100,000*g* for 60 min. The supernatant is applied to an ω-aminohexyl-Sepharose column (2.1 × 16 cm) equilibrated with 100 m*M* potassium phosphate buffer (pH 7.4) containing 0.5% sodium cholate, 1 m*M* EDTA, and 0.5 m*M* DTT and then eluted with 100 m*M* potassium phosphate buffer containing 0.4% sodium cholate and 0.2% Lubrol PX, 1 m*M* EDTA, and 0.5 m*M* DTT. A typical chromatogram is shown in Fig. 1. Fractions with 25-hydroxylase activity are combined. The combined fractions are dialyzed against 20 m*M* potassium phosphate buffer (pH 7.4) containing 0.5 m*M* EDTA, 0.4 m*M* DTT, 0.1% sodium cholate, and 0.08% Lubrol PX, and applied to a hydroxyapatite column (1.2 × 10.6 cm) equilibrated with the same buffer as that used for dialysis. The column is washed with 40 m*M* potassium phosphate buffer (pH 7.4) containing 0.5 m*M* EDTA, 0.4 m*M* DTT, 0.1% sodium cholate, and 0.08% Lubrol PX, and then eluted by increasing phosphate buffer concentration to 90 m*M*. Fractions with enzyme activity are pooled and dialyzed against 30 m*M* Tris-HCl buffer (pH 7.4) containing 0.4 m*M* EDTA, 0.4 m*M* DTT, 0.4% Lubrol PX, and 0.1% sodium cholate. From this step on, protease inhibitors are omitted and the next column chromatography is performed at room temperature.

Aliquots are injected into a TSK gel DEAE-5PW HPLC column (Tosoh Co., 7.5 × 75 mm, Tokyo, Japan) equilibrated with 30 m*M* Tris-HCl buffer (pH 7.4) containing 0.4 m*M* EDTA, 0.4 m*M* DTT, 0.4% Lubrol PX, and

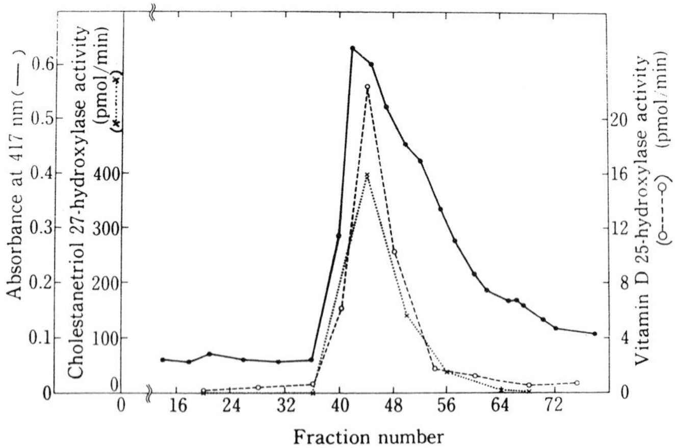

FIG. 1. Chromatography of liver mitochondrial vitamin D_3 25-hydroxylase on an ω-aminohexyl-Sepharose column. The enzyme activity toward 5 β-cholestane-3α,7α,12α-triol (27-hydroxylase) is also presented.

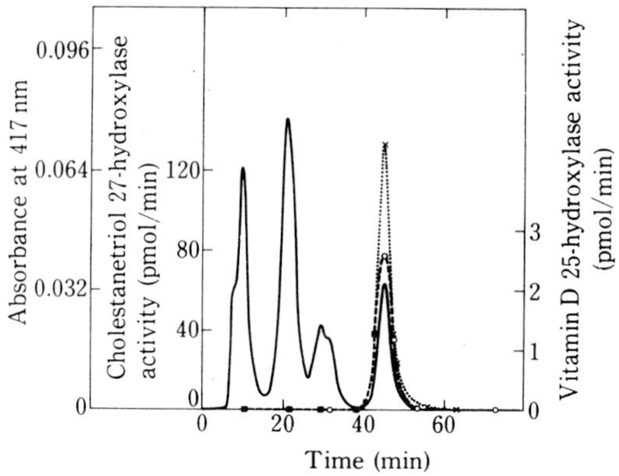

FIG. 2. Elution profiles of the liver mitochondrial 25-hydroxylase on a TSK gel DEAE-5PW column. Protein was eluted with a linear gradient of NaCl (0–0.1 *M* from 0–70 min, 0.1–0.5 *M* from 70–100 min). Solid line, absorbance at 417 nm; dotted line, 27-hydroxylase activity; dashed line, 25-hydroxylase activity.

0.1% sodium cholate. Proteins are eluted with linear gradients of sodium chloride (0–0.1 *M* from 0–70 min, 0.1–0.5 *M* from 70–100 min) (Fig. 2). The flow rate is 0.4 ml/min and the effluent monitored at 417 nm to detect hemoproteins. Fractions showing enzyme activity are collected and stored at −80° without significant loss of activity at least for 4 weeks. Overall purification is shown in Table IV.[9]

Purification of 25-Hydroxyvitamin D_3 24-Hydroxylase

Kidneys are isolated from 20 rats weighing about 200–300 g that received vitamin D_3 (50,000 IU dissolved in 0.1 ml of corn oil/day/animal) for 1 week, stored at −80°, and thawed before preparation. They are homogenized with nine volumes of 15 m*M* Tris-HCl (pH 7.4) containing 0.19 *M* sucrose,

TABLE IV
PURIFICATION OF MITOCHONDRIAL VITAMIN D_3 25-HYDROXYLASE

Purification step	Protein (mg)	Total P450 (nmol)	Total activity[a] (nmol/min)	Specific activity (pmol/min/mg of protein)	Purification[b] (-fold)	Yield[c] (%)
ʻlitochondria	3,940	559	7.32	1.86	—	—
·olyethylene glycol fractionation (4–12%)	755	207	35.5	47.0	1	100
›-Aminohexyl-Sepharose	65.1	59.8	14.5	223	4.7	41
ʼydroxyapatite	24.1	25.1	10.1	419	8.9	28
›EAE-5PW	0.12	1.4	1.91	15,917	339	5.4

[a] Enzyme activity is measured by using 1α-OH-D_3 as a substrate.
[b,c] Purification (-fold) and yield are calculated after the solubilization step.

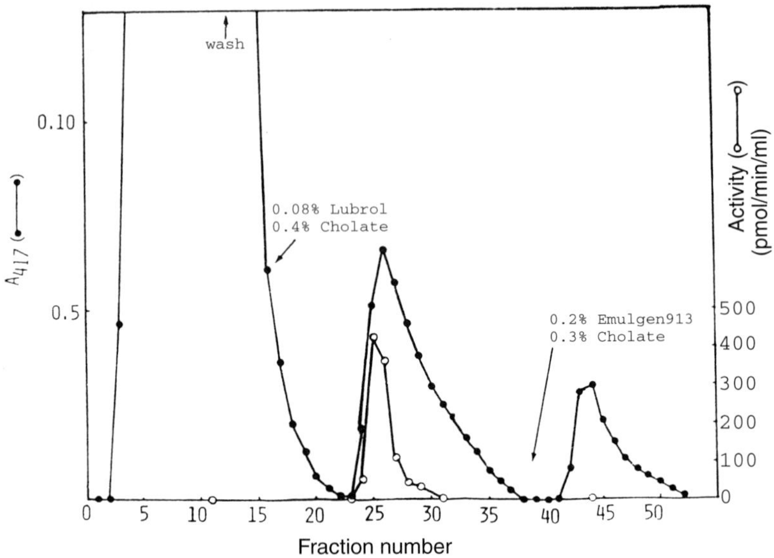

FIG. 3. Chromatography of kidney mitochondrial 25-hydroxyvitamin D_3 24-hydroxylase on a pentyl-Sepharose column.

1 m*M* DTT, 1 m*M* EDTA, and 1 μg/ml each of pepstatin and leupeptin. The homogenate is centrifuged at 200*g* for 12 min, and the supernatant is again centrifuged at 6000*g* for 12 min. The precipitate is washed twice with the same buffer and suspended in the same buffer. The suspension is sonicated at 0° for a total period of 1 min with 30-sec intervals at 20 kHz and 50 W. The suspension is centrifuged at 100,000*g* for 60 min. The precipitate is suspended in 100 m*M* potassium phosphate buffer (pH 7.4) containing 20% glycerol, 0.5 m*M* EDTA, 1 m*M* DTT, and 1 μg/ml each of leupeptin and pepstatin. (All buffer solutions used in the following purification procedures contained 20% glycerol, and 1 μg/ml each of leupep-

TABLE V
PURIFICATION OF 25-HYDROXYVITAMIN D_3 24-HYDROXYLASE

Purification step	Protein (mg)	Total P450 (nmol)	Total activity (nmol/min)	Specific activity (nmol/min/mg of protein)	Purification (-fold)	Yield (%)
Mitochondria	1104	ND[a]	ND	—	—	—
Solubilized fraction	469	16.5	81	0.173	1	100
Pentyl-Sepharose	23.4	5.0	43	1.85	11	53
Hydroxyapatite	5.6	2.1	35	6.32	37	43
DEAE-5PW	0.11	0.29	5.7	54.6	316	7

[a] ND, not determined.

tin and pepstatin unless otherwise stated.) To this suspension 10% cholate containing 0.08% Lubrol PX is added slowly to give a final cholate/protein ratio of 3:5 (w/w). The solubilized suspension (about 10 mg of protein/ml) is centrifuged at 100,000*g* for 60 min.

The supernatant is applied to a pentyl-Sepharose column (2 × 13 cm) equilibrated with 100 m*M* potassium phosphate buffer (pH 7.4) containing 0.4% sodium cholate, 0.5 m*M* EDTA, and 0.5 m*M* DTT. The column is washed with the equilibration buffer and 24-hydroxylase is eluted with 100 m*M* phosphate buffer (pH 7.4) containing 0.4% sodium cholate, 0.08% Lubrol, 0.5 m*M* EDTA, and 0.5 m*M* DTT (Fig. 3). Fractions showing the enzyme activity are pooled and dialyzed against 30 m*M* phosphate buffer (pH 7.4) containing 0.1 m*M* EDTA, 1 m*M* DTT, 0.4% sodium cholate, and 0.03% Tween 20. The dialyzate is then applied to a hydroxyapatite column (1.2 × 6.5 cm) equilibrated with the same buffer as that used for dialysis. The column is washed with 60 m*M* phosphate buffer (pH 7.4) containing 0.1 m*M* EDTA, 1 m*M* DTT, 0.4% sodium cholate, and 0.05% Tween 20, and then 24-hydroxylase is eluted with 150 m*M* phosphate buffer. Fractions with the enzyme activity are collected and dialyzed against 35 m*M* Tris-HCl buffer (pH 7.4) containing 0.4 m*M* EDTA, 0.5 m*M* DTT, 0.1% Lubrol, 0.3% Tween 20, and 0.05% sodium cholate. From this step on protease inhibitors are omitted and HPLC is performed at room temperature.

Aliquots of the combined fractions are injected into a TSK-gel DEAE-5PW column equilibrated with the dialyzing buffer. Proteins are eluted with linear gradients of sodium cholate (0–0.1 *M* from 0–70 min). The flow rate is 0.4 ml/min, and the effluents are monitored at 417 nm. Fractions showing enzyme activity are collected and stored at −80° without deterioration at least for 4 weeks. Overall purification is shown in Table V.[12,13]

Comments. Pretreatment of rats with vitamin D_3 [or 1,25-$(OH)_2D_3$] will increase the 25-hydroxyvitamin D_3 24-hydroxylase activity about 8- to 10-fold. The enzyme can be extracted more effectively from frozen kidneys than from fresh.

[18] Assay of 25-Hydroxyvitamin D 1α-Hydroxylase and 24-Hydroxylase

By MATTHEW J. BECKMAN and HECTOR F. DELUCA

Introduction

The hormonal form of vitamin D, that is, 1,25-dihydroxyvitamin D_3 [1,25-$(OH)_2D_3$], is formed largely if not exclusively from 25-OH-D_3 catalyzed by the 25-OH-D_3 1α-hydroxylase (P450cc1α).[1] As a hormone, 1,25-$(OH)_2D_3$ is secreted from renal proximal tubules and has many functions in calcium homeostasis and skeletal maintenance.[2] A compensatory response to 1,25-$(OH)_2D_3$ action in its target sites is the induction and expression of 1,25-$(OH)_2D_3$ 24-hydroxylase (P450cc24) (Fig. 1). 24-Hydroxylation is the initial step in the catabolism of 1,25-$(OH)_2D_3$ and plays a role in limiting the life of cellular 1,25-$(OH)_2D_3$.

Because P450cc1α and P450cc24 serve important roles in regulating both the formation and distribution of 1,25-$(OH)_2D_3$, several assays for these enzymes have been developed. The classic assays for P450cc24 and P450cc1α in either whole homogenate or mitochondria have relied on a measurement of the product largely by high-performance liquid chromatography (HPLC) or by radioreceptor assay.[3,4] The P450cc1α is of extremely low abundance and is found exclusively in renal proximal tubules.[5] Despite the severalfold induction of P450cc1α activity by vitamin D deficiency, the relative amount of P450cc1α per total protein is low enough that the purification of P450cc1α has also been problematic.[6]

The inducibility of P450cc24 by 1,25-$(OH)_2D_3$ to high abundance in rat kidneys has provided the means by which it can be enriched and subsequently purified by a very laborious procedure that employed extraction, separation, and HPLC to monitor P450cc24 activity by product formation.[7] A more simple assay of P450cc24 activity was recently developed[8] that can be used in conjunction with either purification steps or for general studies

[1] R. Brommage and H. F. DeLuca, *Endocrine Rev.* **6,** 491 (1985).
[2] H. F. DeLuca, *FASEB J.* **2,** 224 (1988).
[3] G. Jones, *Methods Enzymol.* **123,** 141 (1986).
[4] J. Napoli, N. Koszewski, and R. Horst, *Methods Enzymol.* **123,** 127 (1986).
[5] H. Kawashima, S. Torikai, and K. Kurokawa, *Proc. Natl. Acad. Sci. U.S.A.* **78,** 1199 (1981).
[6] R. A. Ettinger and H. F. DeLuca, *Arch. Biochem. Biophys.* **316,** 14 (1995).
[7] Y. Ohyama, S. Hayashi, and K. Okuda, *FEBS Lett.* **255,** 405 (1989).
[8] M. Burgos-Trinidad, A. J. Brown, and H. F. DeLuca, *Anal. Biochem.* **190,** 102 (1990).

METHODS IN ENZYMOLOGY, VOL. 282
0076-6879/97 $25.00

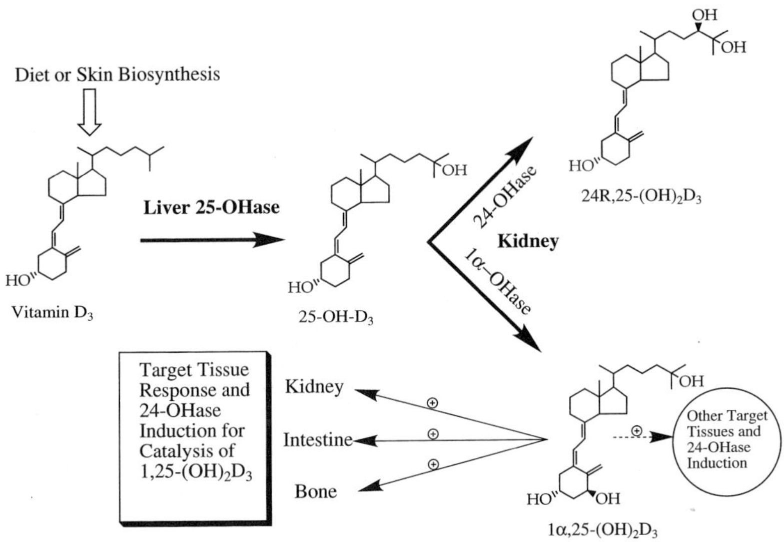

FIG. 1. Schematic of the functional metabolism of vitamin D_3. Vitamin D_3 is converted to 25-OH-D_3 in the liver via a microsomal P450cc25, whereas 25-OH-D_3 is further metabolized in the kidney to either 24,25-$(OH)_2D_3$ via the mitochondrial P450cc24 for deactivation, or 1,25-$(OH)_2D_3$ via the P450cc1α for biological activation. Target tissue P450cc24 is induced by 1,25-$(OH)_2D_3$ activating the catabolism of 1,25-$(OH)_2D_3$ at these sitse. The + symbols indicate induction of P450cc24.

of P450cc24 kinetics or regulation. This method derives from the fact that periodate cleaves the C–C bond of carbons bearing vicinal hydroxyls. In the case of a vitamin D_3 side chain, periodate cleaves the C-24 and C-25 bond of 24,25-$(OH)_2D_3$ or 1,24,25-$(OH)_3D_3$ of the known vitamin D metabolites.

In the case of P450cc1α, Brown *et al.*[9] have synthesized a specialized 1α-^{3}H-radiolabeled substrate that, when hydroxylated at the 1α-position, releases ^{3}H into water. The objective for synthesizing such a substrate is to simplify the P450cc1α assay step to enable purification of P450cc1α.

Properties of P450cc24 and P450cc1α

P450cc24 and P450cc1α are members of the cytochrome P450 superfamily, and are further classified as a subgrouping of P450 proteins known as mitochondrial mixed function monooxygenases. Both enzymes share the requirement of NADPH, ferredoxin, and ferredoxin reductase for func-

[9] A. J. Brown, K. Perlman, H. K. Schnoes, and H. F. DeLuca, *Anal. Biochem.* **164,** 424 (1987).

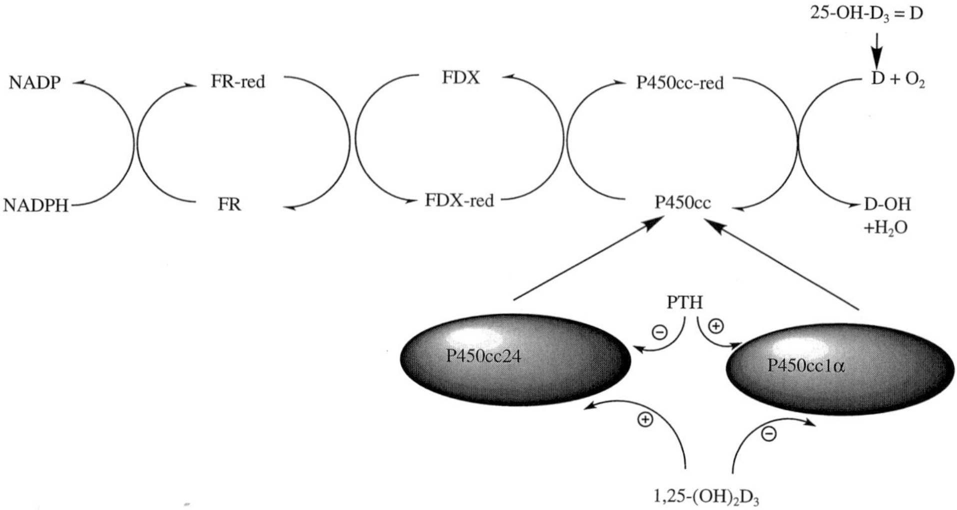

FIG. 2. Schematic model of the mechanism of mitochondrial P450cc24 and P450cc1α activation. Reducing equivalents are supplied via NADPH to a flavoprotein (ferredoxin reductase) then to an iron-sulfur protein (ferredoxin) then to the P450cc monooxygenase, which catalyzes the hydroxylation by utilizing one oxygen atom from O_2; the other atom forms H_2O. Parathyroid hormone and 1,25-$(OH)_2D_3$ reciprocally regulate renal P450cc24 and P45cc1α activities. In the case of P450cc24, 1,25-$(OH)_2D_3$ induces its gene expression. The + and − symbols indicate either positive or negative regulation.

tional enzymatic activity (Fig. 2). Rat kidney P450cc24 was purified to homogeneity by Ohyama *et al.*[7] and was found to have a molecular mass of 55 kDa. Several mammalian P450cc24 cDNAs have now been cloned and sequenced[10,11] and alignment of these sequences shows that there is >90% homology among them. Much less is known of the molecular properties of P450cc1α because the isolation of this enzyme has been thwarted by very low abundance and instability. Further, reliable and rapid assays for measuring P450cc1α activity have not been available. The details of both the periodate cleavage assay for the 24-hydroxylase and the tritium release assay for the 1α-hydroxylase are provided along with data illustrating the versatility with which these sensitive and rapid assays can be applied to the study of P450cc24 and P450cc1α.

[10] Y. Ohyama, M. Noshiro, and K. Okuda, *FEBS Lett.* **278,** 195 (1991).
[11] K. S. Chen, J. M. Prahl, and H. F. DeLuca, *Proc. Natl. Acad. Sci. U.S.A.* **90,** 4543 (1993).

Periodate Cleavage Assay for Measuring P450cc24 Activity

Preparation of 25-OH-[26,27-^{3}H]D$_3$ and 1,25-(OH)$_2$-[26,27-^{3}H]D$_3$ Substrates

The substrate used for the periodate cleavage assay can be either 25-OH-[26,27-^{3}H]D$_3$ (160 Ci/mmol) or 1,25-(OH)$_2$-[26,27-^{3}H]D$_3$ (160 Ci/mmol), both purchased from DuPont–New England Nuclear (Boston, MA). The basic requirements of the substrate are C-25 hydroxylation, [26,27-^{3}H]-side-chain label of high specific activity and high radiopurity >98%. Radiopurity can be established easily by employing a routine HPLC step prior to use. For 25-OH-D$_3$, a silica column is equilibrated in 98:2 (hexane/2-propanol) and gives a retention time of about 12–15 min at 2 ml/min flow rate. For 1,25-(OH)$_2$D$_3$, the solvent system used is 90:10 (hexane/isopropanol) for a retention time of 15–17 minutes at 2 ml/min flow rate. Then samples require further purification over a C_{18} reversed-phase column HPLC using 10% water in methanol for 25-OH-D$_3$ and 30% water in methanol for 1,25-(OH)$_2$D$_3$, at flow rates of 2 ml/min, respectively. The periodate cleavage assay is quantified by counting the tritium found in the aqueous phase or [^{3}H]acetone.

Preparation of Cofactors

There are three circumstances in which P450cc24 activity is measured: (1) whole cell, (2) cell/tissue homogenates or mitochondria, and (3) semipurified or purified P450cc24. In whole-cell experiments, the P450cc24 machinery for electron transport and cofactors are intact and the only component required to elicit P450cc24 activity is substrate. This assay does not measure P450cc24 activity per se but rate of production of 24,25-(OH)$_2$D$_3$ or 1,24,25-(OH)$_3$D$_3$. When homogenates or mitochondria are employed as the source of P450cc24, it must be kept in mind that although most of the cofactors are still present and functional, an energy source is required for replenishing NADPH. For this purpose, succinic acid or other tricarboxylic acid substrates can be used. In all cases where semipurified or purified P450cc24 is used, including solubilized mitochondria, membrane fractions from recombinant expression systems or P450cc24 at various stages of purification, NADPH, ferredoxin, and ferredoxin reductase are required. We have traditionally purified adrenodoxin (ADX) and adrenodoxin reductase (AR) from bovine adrenal glands.[12,13] However, we are currently puri-

[12] T. Sugiyama and T. Yamano, *FEBS Lett.* **52,** 145 (1975).

[13] K. Suhara, S. Takemori, and M. Katagiri, *Biochim. Biophys. Acta* **263,** 272 (1972).

fying recombinantly expressed bovine ADX and AR cofactors.[14,15] An NADPH regenerating system is also required and consists of NADP, glucose 6-phosphate and glucose-6-phosphate dehydrogenase (available from Sigma, St. Louis, MO).

Solubilization of P450cc24 or P450cc1α is accomplished after isolation of kidney mitochondria. Mitochondria are washed and then resuspended in solubilization buffer [20% (v/v) glycerol, 160 mM KPO_4, pH 7.4, and 0.1 mM EDTA]. Prior to solubilization, phenylmethylsulfonyl fluoride and dithiothreitol are added to final concentrations of 0.25 and 1.0 mM, respectively. A 15% (w/v) sodium cholate (recrystallized) solution in 10 mM potassium phosphate, pH 7.4, is then added dropwise to a final concentration of 0.6%. After stirring at 0° for 1 hr, the mixture is centrifuged at 100,000g for 60 min. The supernatant, containing soluble P450cc24 or P45cc1α, can be stored at −80° until use.

P450cc24 Assay Methods

The assay for P450cc24 enzyme activity can be generally accomplished by one of the following three methods described here.

Whole-Cell Experiments: Sf21 Insect Cells Expressing Recombinant P450cc24. Spodopters frugiperds (Sf)21 cells are first washed with phosphate-buffered saline (PBS), pH 6.4, and then resuspended in TC-100 medium [10% bovine serum albumin (BSA)]. In these experiments we recommend using a 96-well microtiter plate. Assays are run in triplicate for cells (1×10^6 cells/well) infected with either wild-type or recombinant virus in 95 μl medium. The substrate, 100 pmol of either 25-OH-[26,27-^{3}H]D_3 or 1,25-$(OH)_2$-[26,27-^{3}H]D_3 (specific activity 1500–2000 dpm/pmol), is added in 5 μl of 95% ethanol, and the mixture is incubated for 10 min in a 37° incubator with the lid of the titer plate off. Background measurements are determined from the wells containing wild-type Sf21 cells, which contain no recombinant P450cc24. The reaction is stopped by the addition of 20 μl 1 N acetic acid.

Homogenate Experiments Prepared from Tissues, Washed Cells, or Isolated Mitochondria. The buffer used for this assay is 0.25 M sucrose Tris–acetate, pH 7.4 (10 mM Tris neutralized with acetic acid). If using a tissue sample or washed cells, a 20% homogenate is prepared in the buffered sucrose. If using mitochondria, adjust the protein concentration between 5–10 mg/ml in sucrose buffer. Samples are pipetted as 200-μl aliquots into

[14] M. F. Palin, L. Berthiaume, J. G. Lehoux, M. R. Waterman, and J. Sygusch, *Arch. Biochem. Biophys.* **295,** 126 (1992).

[15] Y. Sagara, A. Wada, Y. Takata, M. R. Waterman, K. Sekimizu, and T. Horiuchi, *Biol. Pharm. Bull.* **16,** 627 (1993).

1.5-ml Eppendorf tubes containing 100 μl of either succinate or sucrose buffer. Reactions are run in triplicate with three samples containing succinate and three samples without succinate to measure background. P450cc24 activity is initiated by addition of 200 pmol of either 25-OH-[26,27-^{3}H]D_3 or 1,25-$(OH)_2$-[26,27-^{3}H]D_3 (specific activity 1500–2000 dpm/pmol) in 5 μl of 95% ethanol and incubation in a 37° waterbath for 10 min with the caps open. Reactions are stopped by the addition of 50 μl 1 *N* acetic acid.

Measurements from Semipurified or Purified Sources of P450cc24. This assay is more complicated because of the many components that must be added. The reaction mixture contains the solubilized P450cc24 fraction, with or without the bovine adrenal reconstitution system in 10 m*M* potassium phosphate, pH 7.4, in a final volume of 190 μl. Again these assays can be run in a 1.5-ml Eppendorf tube. The reconstitution system consists of 1.6 nmol adrenodoxin, 0.1 nmol adrenodoxin reductase, 100 nmol NADP, 1.0 mmol glucose 6-phosphate, and 0.2 units glucose-6-phosphate dehydrogenase. The reaction is initiated by adding 200 pmol of either 25-OH-[26,27-^{3}H]D_3 or 1,25-$(OH)_2$-[26,27-^{3}H]D_3 (specific activity 1500–2000 dpm/pmol) in 10 μl 95% ethanol and incubated for 10 min at 37° in a shaking waterbath. Background is monitored by measuring reaction mixtures in which the bovine adrenal reconstitution system is replaced by 10 m*M* potassium phosphate, pH 7.4. The reaction is stopped by adding 40 μl 1 *N* acetic acid.

Periodate Cleavage Step

A saturated (8% w/v) solution of aqueous sodium periodate ($NaIO_4$) is mixed into each well of the microtiter plate (150 μl) or Eppendorf tube (500 μl) and incubated on ice for 30 min. If doing the whole cell or homogenate assay, first centrifuge down the solid components of the reaction mixture. The [^{3}H]acetone is separated from the labeled substrate and other reaction components by aspirating the reaction mixture through a 1-ml Superclean LC-18 solid phase extraction (SPE) cartridge (Supelco, Bellefonte, PA) mounted on a VacuElute manifold system. Up to 10 reactions can be processed simultaneously using this system. The reaction tube is rinsed twice with 0.5 ml ice-cold water followed by addition of the rinses to the cartridge. The cartridge eluate is collected in a 20-ml vial, mixed with 15 ml of Bio-Safe II, and radioactivity determined using a liquid scintillation analyzer. The amount of product is calculated from the specific activity of the substrate.

Uses of Periodate Cleavage Assay

The versatility and speed (1–3 hr) of the periodate cleavage assay affords the opportunity to process several more samples at once and in a shorter time than with classical HPLC methodologies, even if automated. The major emphasis of our work is the study of P450cc24 regulation and the involvement of P450cc24 in the regulation of vitamin D metabolism. Our laboratory has cloned and expressed human and chicken P450cc24 cDNAs.[11,16] The periodate cleavage assay provides a fast and reliable means of testing for positive clones of recombinant P450cc24 activity (Fig. 3). Burgos-Trinidad *et al.*[8] compared the periodate cleavage assay to the traditional HPLC method and demonstrated the periodate cleavage method to be as sensitive and reproducible (Fig. 4). Inaba *et al.*[17] has studied the characteristics of the human P450cc24 from HL-60 cells with the periodate cleavage assay. Reciprocal plots for human 25-OH-D_3–P450cc24 and 1,25-$(OH)_2D_3$–P450cc24 gave apparent K_m values for 25-OH-D_3 as 0.52 μM and for 1,25-$(OH)_2D_3$ as 0.02 μM (Fig. 5). In the same way Burgos-Trinidad and DeLuca[18] established these values for chicken P450cc24 as 1.47 μM for 25-OH-D_3 and 0.14 μM for 1,25-$(OH)_2D_3$. Also, the periodate cleavage assay is being used in our laboratory to study the downregulation of P450cc24 induced by parathyroid hormone (PTH) in AOK-B50 cells. These cells are derived from LLC-PK1 cells that have the PTH-receptor stably transfected into them to form the AOK-B50 strain.[19] In these cells, P450cc24 is potently induced by 1,25-$(OH)_2D_3$ whereas the upregulation is blocked by the presence of PTH (Fig. 6).

Tritium Release Assay for Measuring P450cc1α Activity

Preparation of Substrate

The development of the release assay was contingent on the synthesis of 25-OH-[1α-^{3}H]D_3 (specific activity ~11 Ci/mmol) as substrate.[9] As shown in Fig. 7 (structures **I–VI**), the starting material was 1,25$(OH)_2$-3,5-cyclovitamin D_3 (**I**) prepared from 25-OH-D_3.[20] Allylic oxidation of **I** with manganese dioxide gave 1-oxo-25-OH-3,5-cyclovitamin D_3 (**II**). Compound **II** was reduced with sodium borotritide yielding 1α,25-$(OH)_2$-[1β-^{3}H]-3,5-cyclovitamin D_3 (**IIIa**) and 1β,25-$(OH)_2$-[1α-^{3}H]-3,5-cyclovitamin D_3 (**IIIb**). Compounds **IIIa** and **IIIb** were purified by normal phase HPLC[9] and further treated with mesochloride and lithium aluminum hydride followed by cycloreversion to form 3β-acetates of 25-OH-[1α-^{3}H]D_3 (**VIa**) and 25-OH-

[16] F. Jehan, R. Ismail, Z. Lu, and H. F. DeLuca, *J. Bone Min. Res.* **11,** 7482 (1996).
[17] M. Inaba, M. Burgos-Trinidad, and H. F. DeLuca, *Arch. Biochem. Biophys.* **284,** 257 (1991).
[18] M. Burgos-Trinidad and H. F. DeLuca, *Biochim. Biophys. Acta* **1078,** 226 (1991).
[19] F. R. Bringhurst, H. Juppner, J. Guo, P. Urena, J. T. Potts, Jr., H. M. Kronenberg, A. B. Abou-Samra, and G. V. Segre, *Endocrinology* **132,** 2090 (1993).
[20] H. Paaren, H. Schnoes, and H. DeLuca, *J. Org. Chem.* **45,** 3253 (1980).

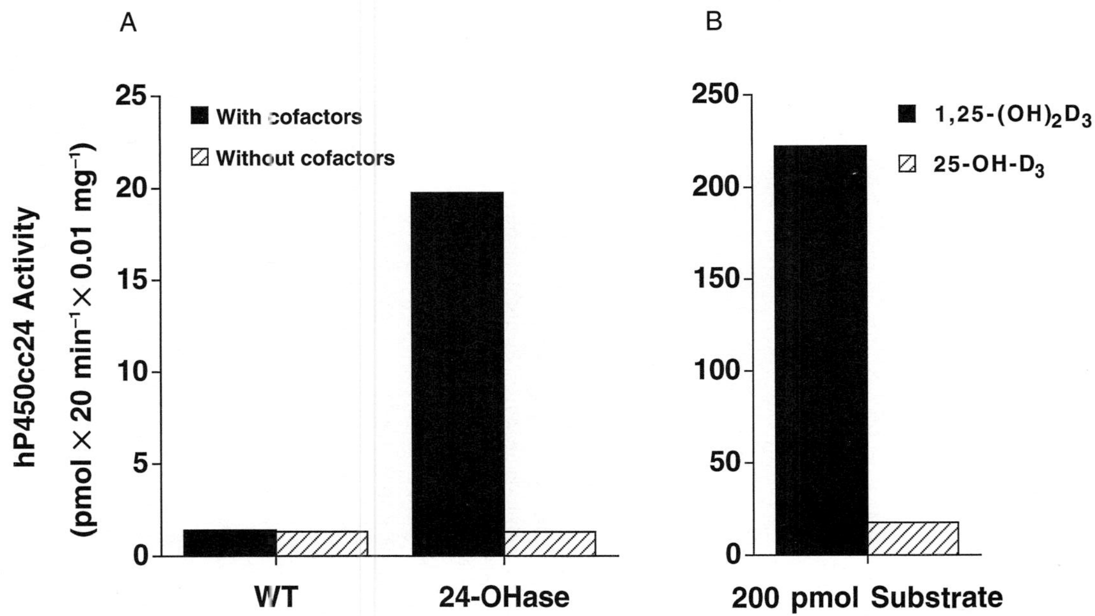

FIG. 3. (A) Dependence of recombinant h-P450cc24 for its cofactors adrenodoxin, adrenodoxin reductase, and NADPH. P450cc24 activity was determined by a periodate cleavage assay, using 10 mg of solubilized *Sf*21 cell mitochondria obtained from either wild-type AcNPV (WT) or AcNPV-P450cc24 (24-OHase) infected *Sf*21 cells. (B) The selectivity of h-P450cc24 for its two natural substrates [1,25-$(OH)_2D_3$ and 25-OH-D_3]. Reactions were initiated by addition of 200 pmol of appropriate 26,27-^{3}H-labeled substrate; specific activity 800 dpm/pmol.

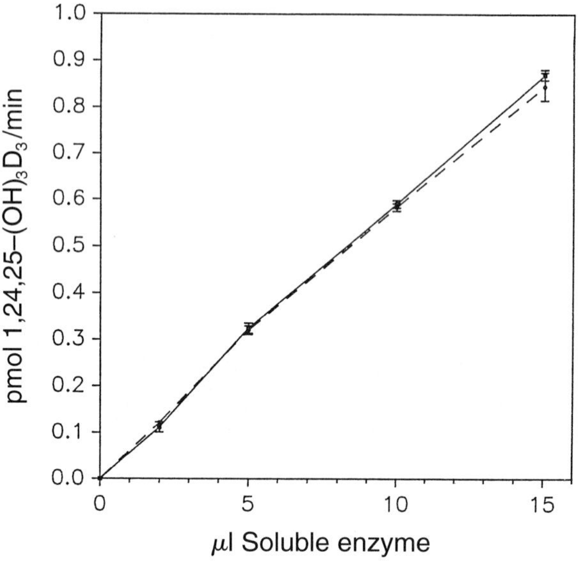

FIG. 4. Rate of 1,25-$(OH)_2D_3$ production as a function of solubilized mitochondrial protein. The incubations contained increasing amounts of the soluble fraction (7.3 μg/μl). All the reactions were corrected for sodium cholate to a final concentration of 0.045% (w/v). Picomoles of product were calculated after subtracting controls in the absence of the reconstitution system at each protein concentration measured. Identical samples were analyzed by using the periodate assay (dashed line) or HPLC assay (solid line). Values are expressed as the mean ± SD of triplicate determinations.

[1β-^{3}H]D_3 (**VIb**). These compounds were purified by normal phase HPLC and then saponified and repurified[9] by normal phase HPLC yielding a mixture 25-OH-[1α-^{3}H]D_3 and 25-OH-[1β-^{3}H]D_3, compounds **VIa** and **VIb**, respectively.

P450cc1α Assay Methods

Reaction mixtures contain 1.25 μ*M* NADPH, 12.5 m*M* glucose-6-phosphate, 7.5 unit glucose-6-phosphate dehydrogenase, 12.5 μ*M* ADX, 0.5 μ*M* AR, 30 mg protein from the solubilized mitochondrial supernatant, and 100 pmol 25-OH-[1α-^{3}H]D_3 (4000 dpm/pmol) in a 200 μl total volume of 100 m*M* potassium phosphate, pH 7.4, and 0.1 μ*M* EDTA. Reactions are initiated by addition of substrate [in 10 μl 95% (v/v) ethanol] and incubated at 37° for 30 min. The incubation time and concentration of P450cc1α are in the linear range for 1,25-$(OH)_2D_3$ production. Reactions are stopped by addition of 30 μl 1 *N* acetic acid, and the mixture passed through a Superclean LC-18 SPE cartridge (Supelco) followed by two 1-ml washed with

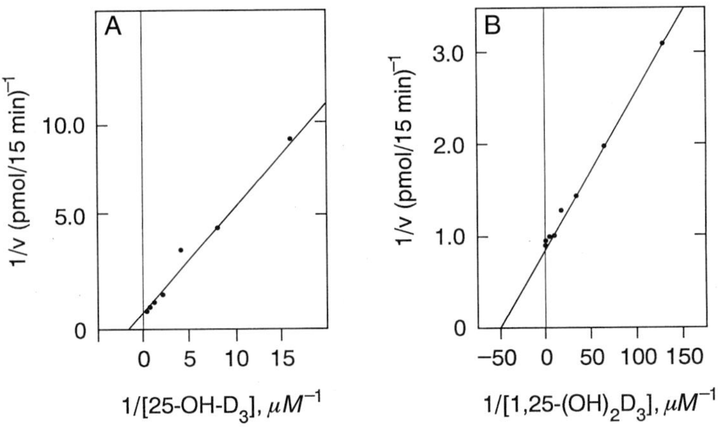

FIG. 5. Reciprocal plots for (A) 25-OH-D_3–P450cc24 and (B) 1,25-$(OH)_2D_3$–P450cc24 from HL-60 cells. HL-60 cells were treated with 10^{-7} *M* 1,25-$(OH)_2D_3$ for 24 hr. HL-60 mitochondria were incubated with increasing concentrations of 25-OH-[26,27-^{3}H]D_3 or 1,25-$(OH)_2$-[26,27-^{3}H]D_3 at 37° for 15 min. The enzyme activity was determined by the periodate method. The apparent K_m for 25-OH-D_3 was 5.2×10^{-7} *M* and for 1,25-$(OH)_2D_3$ was 2.0×10^{-8} *M*. Each point represents the average of two determinations.

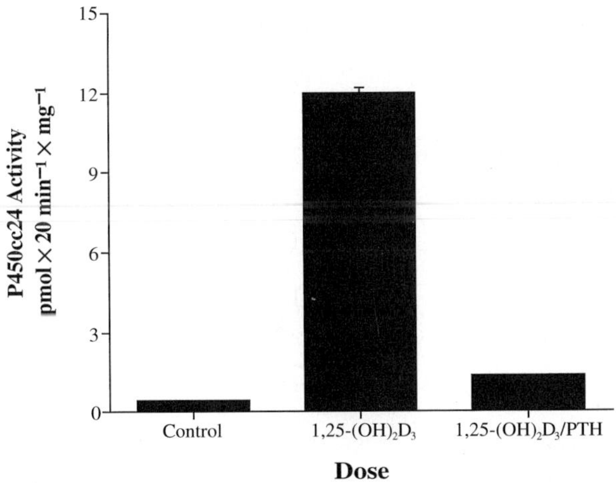

FIG. 6. The regulation of P450cc24 activity in AOK-B50 proximal kidney tubule cells (derived from an LLC-PK1 line) that contain stably transfected receptors (100,000/cell) for PTH. Activities were expressed as pmol × 20 min^{-1} × mg protein^{-1} from the periodate assay used on AOK-B50 homogenates. Ethanol solvent was used as a control, compared with 1,25-$(OH)_2D_3$ (13 n*M*) in ethanol and with 1,25-$(OH)_2D_3$ in ethanol and 90 n*M* PTH. The cells were incubated for 8 hr, then analyzed. PTH was dosed at the 4-hr time point. Cells were washed in ice-cold PBS to remove medium extracellular dose compounds. SDs were determined from three independent experiments.

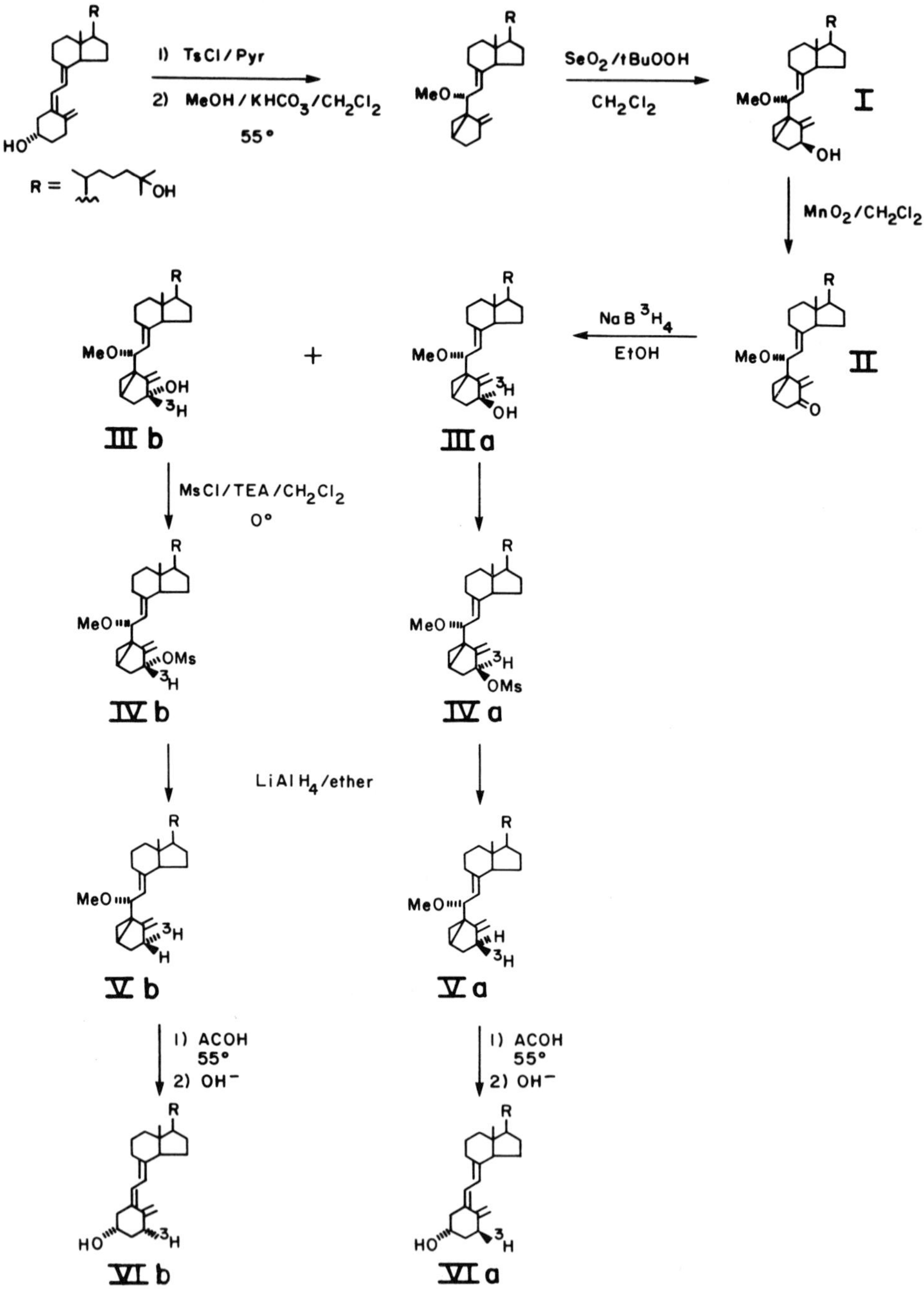

FIG. 7. Reaction scheme for the synthesis of 25-OH-[1α-^{3}H]D_3 and 25-OH-[1β-^{3}H]D_3.

water. The radiolabeled substrate remains bound to the cartridge while ^3H_2O released by 1α-hydroxylation passes through and is mixed with 15 ml of scintillation fluid (Bio-Safe II, Research Products International Corp., Mount Prospect, IL) prior to determining the amount of radioactivity in a liquid scintillation spectrometer.

Uses of Tritium Release Assay

Reconstitution of solubilized chicken kidney mitochondria with NADPH, bovine ADX, and bovine AR restored the 1α-hydroxylation activity of P450cc1α. P450cc1α activity assessed by tritium release agrees well with production of 1,25-$(OH)_2D_3$ as measured by HPLC.[9] Therefore, like the periodate cleavage assay for P450cc24, the tritium release assay offers a fast and reliable alternative to the standard HPLC method for determining P450cc1α activity. The advantage of the tritium release assay over HPLC is the rapid workup, stable product formation (^3H_2O), and lack of interference with secondary product formation by P450cc24 or other enzymatic reactions that would promote underestimation of P450cc1α activity by further metabolizing 1,25-$(OH)_2D_3$.

The major practical use of the tritium release assay is as a fast assay step in the progression of P450cc1α purification (Fig. 8). Our laboratory (R. Ettinger, M. Phelps, and H. F. DeLuca, unpublished results, 1993) found that P450cc1α is unstable in a time-dependent manner even in the presence of a broad spectrum of protease inhibitors and running all experiments at 4°. Because of this, the use of HPLC assays that require a great deal of time to complete the running of numerous samples prohibits advances in P450cc1α purification.

Perspective for the Future

Three important areas for future consideration are (1) to complete the purification of P450cc1α, (2) gain further insight into the reciprocal nature of renal P450cc24/P450cc1α regulation, and (3) examine more extensively the pathways of vitamin D metabolism. Work in the first two areas has been ongoing, but there is still very little information available as to other enzymes involved in the metabolism of vitamin D, such as for 26-hydroxylation and lactone formation. It is conceivable that a 26-hydroxylase tritium release assay could be developed that would be a useful tool in the exploration of this important side pathway of vitamin D metabolism. Also, the importance of P450cc24 in the metabolism of vitamin D_3 has received

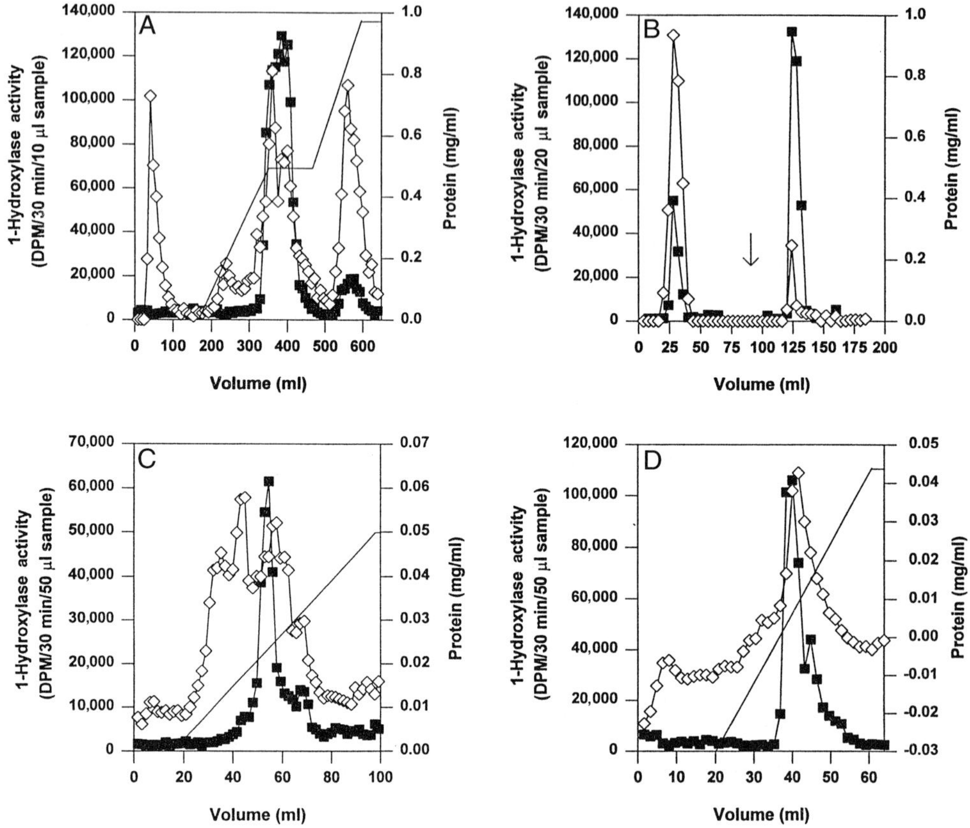

FIG. 8. Separation of the P450cc1α from chicken kidney mitochondrial proteins by (A) sequential, Q-Sepharose, (B) adrenodoxin-Sepharose, (C) Mono S, and (D) Mono P chromatography. In all profiles, 1-hydroxylase activity (solid squares) was determined by use of the tritium release assay, and protein concentrations (empty diamonds) were measured by the Bradford method. In the Q-Sepharose chromatographic profile, soluble chicken mitochondrial protein was precipitated with 35% saturation ammonium sulfate and applied to a Q-Sepharose column. Bound proteins were eluted with a NaCl gradient (solid line) from 0 to 0.23 *M*, 0.23 to 0.5 *M*, and 0.5 *M*. Q-Sepharose fractions containing 1-hydroxylase activity were pooled and applied to an adrenodoxin-Sepharose column. Bound proteins were eluted with 300 m*M* NaCl and 0.2% sodium cholate. An arrow indicates when this buffer was applied. Adrenodoxin-Sepharose fractions containing 1-hydroxylase activity were pooled and applied to a Mono S column. Bound proteins were eluted with a NaCl gradient (solid line) from 0 to 0.5 *M*. In the Mono P chromatographic profile (D), Mono S fractions containing 1-hydroxylase activity were pooled and applied to a Mono P column. Bound proteins were eluted with a NaCl gradient (solid line) from 0 to 0.5 *M*. (From R. Ettinger and H. F. DeLuca, manuscript submitted.)

renewed attention with the discovery of Akiyoshi-Shibata *et al.*[21] that recombinant rat P450cc24 is multicatalytic, catalyzing C-24 and C-23 hydroxylations and C-24 keto-oxidation.

Acknowledgments

This work was supported in part by grants DK14881 and DK07665 from the National Institutes of Health, a fund from the National Foundation for Cancer Research, and a fund from the Wisconsin Alumni Research Foundation.

[21] M. Akiyoshi-Shibata, T. Sakaki, Y. Ohyama, M. Noshiro, K. Okuda, and Y. Yabusaki, *Eur. J. Biochem.* **224,** 335 (1994).

[19] Molecular Cloning of Vitamin D_3 Hydroxylases

By MITSUHIDE NOSHIRO, YOSHIHIKO OHYAMA, EMIKO USUI, MEGUMI AKIYOSHI-SHIBATA, YOSHIYASU YABUSAKI, and KYU-ICHIRO OKUDA

Introduction

Vitamin D_3 is hydroxylated to 25-hydroxyvitamin D_3 by a vitamin D_3 25-hydroxylase (25-hydroxylase) located in liver mitochondria or endoplasmic reticulum.[1,2] 25-Hydroxyvitamin D_3 is then transported to the kidney through the bloodstream. In kidney mitochondria, it is hydroxylated at the 1α-position by a 25-hydroxyvitamin D_3 1α-hydroxylase (1α-hydroxylase) to 1α,25-dihydroxyvitamin D_3 in calcium-depleted animals or hydroxylated at the 24R-position to 24R,25-dihydroxyvitamin D_3 by a 25-hydroxyvitamin D_3 24-hydroxylase (24-hydroxylase) in animals that are calcium or 1,25-dihydroxyvitamin D_3 repleted. We have purified both 25- and 24-hydroxylases to homogeneity[3,4] and isolated their cDNA clones.[5,6] These hydroxylases constitute novel families in the P450 super family,[7] that is,

[1] J. L. Omdahl and H. F. DeLuca, *Physiol. Rev.* **53,** 327 (1973).
[2] K.-I. Okuda, E. Usui, and Y. Ohyama, *J. Lipid Res.* **36,** 1641 (1995).
[3] O. Masumoto, Y. Ohyama, and K.-I. Okuda, *J. Biol. Chem.* **263,** 14256 (1988).
[4] Y. Ohyama and K.-I. Okuda, *J. Biol. Chem.* **266,** 8690 (1991).
[5] E. Usui, M. Noshiro, and K.-I. Okuda, *FEBS Lett.* **262,** 135 (1990).
[6] Y. Ohyama, M. Noshiro, and K.-I. Okuda, *FEBS Lett.* **278,** 195 (1991).
[7] D. R. Nelson, T. Kamataki, D. J. Waxman, F. P. Guengerich, R. W. Estabrook, R. Feyereisen, F. Z. Gonzalez, M. J. Coon, I. C. Gunsalus, O. Gotoh, K.-I. Okuda, and D. W. Nebert, *DNA Cell Biol.* **12,** 1 (1993).

0076-6879/97 $25.00

CYP27 for 25-hydroxylase and CYP24 for 24-hydroxylase. In this chapter, molecular cloning of these hydroxylases is described. Assay and purification of the enzymes are described elsewhere in this volume.[8]

Preparation of Antibodies against 25- and 24-Hydroxylases

Screening of cDNA libraries is performed by nucleic acid hybridization or by an immunological detection method. In this experiment, the latter method is described.

Specific polyclonal antibodies against the purified hydroxylases are prepared from 8-week-old BALB/c female mice. The purified enzyme is mixed with Ribi adjuvant (Ribi, Inc., Hamilton, MT) by vortexing for 2–3 min, and 0.2–0.5 ml of the mixture containing 2–10 μg of antigen is injected into the abdomen of mice. The same injection is repeated two or three times every 2–4 weeks. Two to four weeks after the last injection, a booster [2–5 μg of the antigen in 10 m*M* phosphate-buffered saline (PBS) per mouse] is injected into the tail vein. After 3–7 days mice are sacrificed and their serum is pooled as polyclonal antibodies. The titer and specificity of the pooled antibodies are determined by Western blotting[9] and/or an enzyme-linked immunosorbent assay (ELISA).[10] The serum can be used to screen the cDNA library without further purification. For preparation of monoclonal antibody, spleen cells are isolated and fused with myeloma cells to produce monoclonal antibodies according to the method described by Köhler and Milstein.[11]

Comment. Because Ribi adjuvant contains defined ingredients (cell wall skeleton, monophosphoryl lipid A, and trehalose dimycolate of tubercle bacilli) instead of the whole bacterial extract, it enhances the immune response substantially. Lesser amounts of antigen are therefore required.

Cloning of cDNAs for Vitamin D_3 Hydroxylases

Construction of cDNA libraries, isolation of cDNA clones, and nucleotide sequencing are performed according to established methods.[12]

[8] Y. Ohyama, S. Hayashi, E. Usui, M. Noshiro, and K.-I. Okuda, *Methods Enzymol.* **282,** [17], 1997 (this volume).

[9] H. Towbin, T. Staehelin, and J. Gordon, *Proc. Natl. Acad. Sci. U.S.A.* **76,** 4350 (1979).

[10] E. Engvall and P. Perlmann, *Immunochemistry* **8,** 871 (1971).

[11] G. Köhler and C. Milstein, *Nature* (*London*) **256,** 495 (1975).

[12] J. Sambrook, E. F. Fritsch, and T. Maniatis, "Molecular Cloning: A Laboratory Manual," Cold Spring Harbor Press, New York, 1989.

Isolation of cDNA Clone for Vitamin D_3 25-Hydroxylase

Total RNA of normal rat liver (1–2 g) is extracted by the guanidine hydrochloride method.[13] Poly(A)$^+$ RNA is prepared from the total RNA using oligo(dT)-cellulose column chromatography and used to produce the λgt11 cDNA library.[12] The library is plated with *Escherichia coli* Y1090 and screened using the specific polyclonal antibodies prepared earlier as a probe (first antibody). ^{125}I-Labeled anti-mouse IgG antibodies (Amersham International plc, Buckinghamshire) are used as the second antibodies to detect the first antibodies.[14] Out of about 1.4×10^6 clones, positive clones showing immunoreactive signals are isolated. Whether the clones obtained are long enough to contain the total coding region is judged from the size of the mRNA of 25-hydroxylase determined by Northern hybridization. The clone that is considered to include the total coding regions is subcloned into a pUC19 plasmid and then subjected to restriction enzyme mapping. Nucleotide sequence is determined by the dideoxynucleotide chain termination method[15] using 7-deaza-dGTP[16] and Sequenase (Amersham International).[17] The open reading frame is 1599 bp long and encodes 533 amino acids.

When the amino-terminal sequence (Ala–Ile–Pro–Ala–Ala–) of the purified protein is compared with that of the protein sequence deduced from the cDNA sequence (EMBL/GenBank/DDBJ database accession No. Y07534), the former is found to be identical with the deduced amino acid sequence from residues 33 through 37 of the latter. This establishes that the peptide removed by processing comprised amino acid residues 1 through 32. The cleaved peptide contains 6 basic amino acid residues together with 18 hydrophobic amino acid residues characteristic of the presequence of a mitochondrial precursor protein. The mature enzyme consists of 501 amino acid residues corresponding to a molecular weight of 57,182.[5]

Isolation of cDNA Clone for 25-Hydroxyvitamin D_3 24-Hydroxylase

Total kidney RNA is prepared by the guanidine hydrochloride method[13] from rats that were injected intraperitoneally with vitamin D_3 (50,000 IU in corn oil) for 5 days, and poly(A)$^+$ RNA is isolated using the oligo(dT)-cellulose column. λgt11 and λZAP cDNA libraries are prepared from the poly(A)$^+$ RNA, and screened by the use of the specific polyclonal antibod-

[13] R. J. MacDonald, G. H. Swift, A. E. Przybyla, and J. M. Chirgwin, *Methods Enzymol.* **152,** 219 (1987).

[14] R. A. Young and R. W. Davis, *Proc. Natl. Acad. Sci. U.S.A.* **80,** 1194 (1983).

[15] F. Sanger, S. Nicklen, and A. R. Coulson, *Proc. Natl. Acad. Sci. U.S.A.* **74,** 5463 (1977).

[16] S. Mizusawa, S. Nishimura, and F. Seela, *Nucleic Acids Res.* **14,** 1319 (1986).

[17] S. Tabor and C. C. Richardson, *Proc. Natl. Acad. Sci. U.S.A.* **84,** 4767 (1987).

ies. The clones isolated from the λZAP cDNA library are subcloned into pBluescript by the *in vivo* excision method.[18] The clone that seemed to contain the total coding region is sequenced. Deletion mutants of the clone are prepared by using an *Exo*III/mung bean nuclease deletion kit (Takara Shuzo Co., Otsu, Japan) and the nucleotide sequence is determined as described earlier (EMBL/GenBank/DDBJ database accession No. X59506). It reveals a 1542-bp open reading frame encoding 514 amino acid residues. The amino-terminal amino acid sequence (Arg–Ala–Pro–Lys–Glu–Val–Pro–Leu–) of the purified 24-hydroxylase determined by the automated Edman degradation method agrees perfectly with the deduced amino acid sequence from residue 36 to 43, establishing that processing of the precursor for the mitochondrial protein occurs between residue 35 and 36. The cleaved 35-residues that contain 6 arginine and 2 lysine residues together with 16 hydrophobic amino acid residues are characteristic of the presequence of mitochondrial precursor protein. The mature protein comprises 479 amino acid residues corresponding to a molecular weight of 55,535, which is close to 53,000 estimated from sodium dodecyl sulfate–polyacrylamide gel electrophoresis (SDS–PAGE) of the purified enzyme.[6]

Comments. The cDNA libraries for 24-hydroxylase are prepared from rats that were injected with vitamin D_3. The injection of vitamin D_3 increases the 24-hydroxylase activity about 8 to 10-fold in 10 days, reflecting the increase in the concentration of 24-hydroxylase mRNA. This induction may be due to 1,25-dihydroxyvitamin D_3 that is formed from vitamin D_3 *in vivo.*[19]

Expression of Vitamin D_3 Hydroxylases

Mammalian, yeast, insect, and bacterial cells are used as the heterologous expression system for identification of the isolated cDNA. Although COS cells are a convenient system for identification, it is difficult to obtain the amount of expressed proteins necessary for the study of enzyme characteristics. For this purpose, other expression systems such as yeast, insect, and bacterial cells are preferred. The basic experimental conditions for such expression systems have been described elsewhere in another volume of this series.[20] In this section the expression of vitamin D_3 hydroxylases in COS, yeast, and bacterial cells is described.

[18] L. G. Davis, M. D. Dibner, and J. F. Batty, "Basic Method in Molecular Biology," Elsevier, New York, 1986.

[19] T. Shinki, C. H. Jin, A. Nishimura, Y. Nagai, Y. Ohyama, M. Noshiro, K.-I. Okuda, and T. Suda, *J. Biol. Chem.* **267,** 13757 (1992).

[20] D. V. Goeddel, ed., *Methods Enzymol.* **185** (1990).

TABLE I

Expression of Liver Mitochondrial Vitamin D_3 25-Hydroxylase in COS 7 Cells

Expression	1α,25-Dihydroxyvitamin D_3 formed[a] (pmol/min/mg protein)	5β-Cholestane-3α,7α,12α,27-tetrol formed[b] (pmol/min/mg protein)
Transfected COS cells	2.33	114
Control COS cells	0.17	6.7

[a] 1α-Hydroxyvitamin D_3 is used as substrate.

[b] 5β-Cholestane-3α,7α,12α-triol is used as substrate.

Expression of Vitamin D_3 25-Hydroxylase (CYP27) in COS-7 Cells

The 1.9-kbp insert of the isolated cDNA containing the total coding region is ligated into the *Sma*I site of expression vector pSVL (Pharmacia, Piscataway, NJ). Simian COS-7 cells are transfected by the constructed plasmid according to the DEAE-dextran method,[12] 5 μg of expression plasmid being used per 1×10^6 cells. Cells are incubated for 72 hr at 37° under 5% CO_2 in 10 ml of Dulbecco's modified Eagle's medium containing 10% fetal calf serum. To assay 25-hydroxylase activity, cells are harvested 72 hr after transfection, washed with PBS, frozen at −80°, thawed, and homogenized with an equal volume of 0.25 *M* sucrose containing 50 m*M* phosphate buffer (pH 7.4) and 1 m*M* EDTA. Mitochondria are obtained by differential centrifugation. The precipitate obtained by centrifugation at 10,000*g* for 15 min is suspended and solubilized by sodium cholate at the final concentration of 0.5%. The 25-hydroxylase activity is measured by the reconstitution method, using NADPH, adrenodoxin, and NADPH–adrenodoxin reductase as the electron donors as described elsewhere in this volume.[8] The typical reconstitution system revealed 25-hydroxylase activity toward 1α-hydroxyvitamin D_3 with a specific activity of 2.33 pmol/min/mg of mitochondrial protein that is more than 10-fold higher than that of the control (Table I).[21]

Expression of 25-Hydroxylase (CYP27) in Yeast Cells

The expression plasmid pAC25 for vitamin D_3 25-hydroxylase in yeast system is constructed as shown in Fig. 1. Two *Eco*RI–*Sac*I fragments (0.32 and 1.85 kb) are prepared from the cDNA clone described earlier and then inserted into the *Eco*RI–*Sac*I site of pUC19 to yield pUC25N and pUC25C, respectively. The *Eco*RI–*Nco*I fragment of pUC25N is replaced by the

[21] E. Usui, M. Noshiro, Y. Ohyama, and K.-I. Okuda, *FEBS Lett.* **274,** 175 (1990).

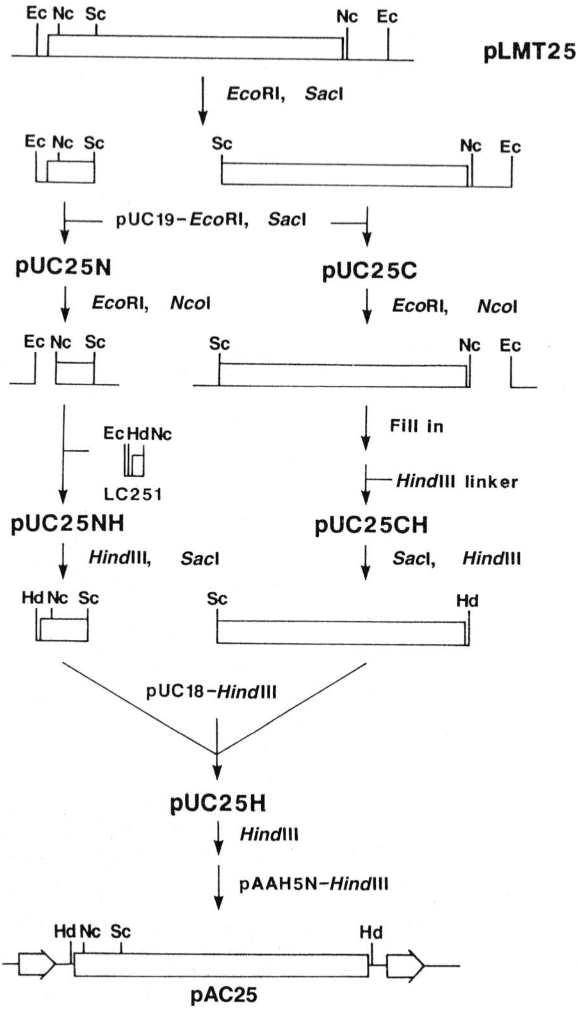

FIG. 1. Construction of the expression plasmid pAC25 for the precursor protein of rat mitochondrial vitamin D_3 25-hydroxylase. Open boxes indicate the protein-coding region. Open arrows indicate yeast alcohol dehydrogenase I (ADH) promoter and terminator regions. Restriction sites indicated are Ec, *Eco*RI; Hd, *Hind*III; Nc, *Nco*I; Sc, *Sac*I. The synthesized linker, LC252, is:

```
5'-AATTCAAGCTTAAAAAAATGGCTGTGTTGAGCCGCATGAGACTGAGATGGGCGCTTCTGGACACTCGTGTGATGGGC-3'
    3'-GTTCGAATTTTTTTACCGACACAACTCGGCGTACTCTGACTCTACCCGCGAAGACCTGTGAGCACACTACCCGGTAC-5
```

(Reprinted from *FEBS Letters*, **280**(2), M. Akiyoshi-Shibata, E. Usui, T. Sakaki, Y. Yabusaki, M. Nosho, K. Okuda, and H. Ohkawa, "Expression of Rat Liver Vitamin D3 25-Hydroxylase cDNA in *Saccharomyces cerevisiae*," pp. 367–370. Copyright 1991 with kind permission of Elsevier Science–NL, Sara Burgerhartstraat 25, 1055 KV Amsterdam, The Netherlands.)

synthetic *Eco*RI–*Nco*I linker DNA (LC252) to construct pUC25NH. The original *Nco*I site in the 3′-untranslational region of pUC25C is filled in and ligated to a *Hind*III linker to yield pUC25CH. Both *Hind*III–*Sac*I fragments (0.27 and 1.37 kb) from pUC25NH and pUC25CH, respectively, are generated and doubly inserted into the *Hind*III site of pAAH5N to produce the expression plasmid pAC25. *Saccharomyces cerevisiae* AH22 [ϱ^+] cells are transformed with the expression plasmid pAC25. Yeast spheroplast is prepared and then subjected to subcellular fractionation.

The expressed protein is detected in the mitochondrial fraction of the recombinant yeast by Western blotting. The typical reconstitution system shows the 25-hydroxylase activity with a turnover number of 0.14 mol/min/mol of cytochrome P450.[22]

Comments. Vitamin D_3 25-hydroxylase has been shown to be identical with 5β-cholestane-3α,7α,12α-triol 27-hydroxylase. The enzyme expressed in transformed COS or yeast cells by recombinant expression plasmids[23,24] as well as the purified enzyme[2,25] possesses both activities. For this reason the enzyme catalyzing 25-hydroxylase is designated CYP27. There is also an inherited disease, cerebrotendinous xanthomatosis (CTX), that is caused by an anomaly of the gene structure of CYP27.[26] Berginer *et al.*[27] reported that osteoporosis and an increased risk of bone fractures observed in some CTX patients may be explained by low or null activity of the liver mitochondrial vitamin D_3 25-hydroxylation due to an abnormal enzyme formed from the defective gene.

Expression of 24-Hydroxylase (CYP24) in COS-7

A DNA fragment containing the complete coding region is produced by digestion with restriction enzymes of *Dde*I and *Pmac*I and inserted into the *Sma*I site of the mammalian expression vector, pSVL (Pharmacia, Piscataway, NJ). COS-7 cells are transfected with the constructed plasmid, cultured, and harvested as described above. A mitochondrial fraction is prepared from the transformed COS cells, solubilized, and used for reconstitution as described elsewhere in this volume.[8] The typical result shows the 24-hydroxylase activity with a specific activity of 14.2 pmol/min/mg of mitochondrial protein that is 10-fold higher than that of the control.[6]

[22] M. Akiyoshi-Shibata, E. Usui, T. Sakaki, Y. Yabusaki, M. Noshiro, K.-I. Okuda, and H. Ohkawa, *FEBS Lett.* **280,** 367 (1991).

[23] K. Oeda, T. Sakaki, and H. Ohkawa, *DNA* **4,** 203 (1985).

[24] G. Daum, P. C. Böhni, and G. Schatz, *J. Biol. Chem.* **257,** 13028 (1982).

[25] Y. Ohyama, O. Masumotro, E. Usui, and K.-I. Okuda, *J. Biochem.* (*Tokyo*) **109,** 389 (1988).

[26] J. J. Cali, C.-L. Hsieh, V. Frank, and D. W. Russell, *J. Biol. Chem.* **266,** 7779 (1991).

[27] V. M. Berginer, S. Shany, D. Alkalay, J. Berginer, S. Dekel, G. Salen, S. Tint, and D. Gazit, *Metab. Clin. Exp.* **42,** 69 (1992).

Comments. COS cells are a cell line derived from African green monkey kidney cells (CV-1),[28] and retain at least a part of the regulatory mechanism for metabolism of vitamin D_3. When the substrate, 25-hydroxyvitamin D_3 is directly added into the culture medium for assay, it stimulates the cell's own regulatory system and markedly enhances the biosynthesis of endogenous 24-hydroxylase.[29] The effect of the externally introduced 24-hydroxylase gene is thereby obscured. To obtain the clearer effect of the introduced cDNA, the transfected cells should be cultured in the absence of vitamin D_3 and the enzyme activity should be measured by the reconstitution method employing the proper electron transport system.

Expression of 24-Hydroxylase (CYP24) in Escherichia coli

Expression in bacteria is often employed to obtain a sufficient amount of recombinant proteins. However, often the product is not completely identical with the native enzyme and may have different properties. There is no guarantee that posttranslational processing such as cleavage of presequence, glycosylation, and incorporation of prosthetic groups is carried out properly. This is because artificial modification such as deletion or substitution is usually carried out in the coding region of the enzyme to sustain expression in bacterial system. Without such modification, mammalian proteins do not express well in bacterial cells.[31] The present method produces a protein that is identical with the native enzyme.

An expression vector is constructed by the method shown in Fig. 2. The cloning vector pUC19 is modified to yield pUC19Ac3 by inserting the synthesized AC3 linker into *Hinc*II site. The *Acc*II–*Acc*III fragment coding for the N-terminal region of the mature 24-hydroxylase is prepared from the full-length cDNA clone pCC24-8[6] and doubly inserted, together with the synthesized *Hind*III–*Acc*II linker (LKC24R2), which contains the ribosome-binding site and ATG initiation codon, between the *Hind*III and *Acc*III sites of pUC19Ac3, to give pUKC24R2N. The *Nae*I–*Pst*I fragment coding for the C-terminal region of 24-hydroxylase is prepared from pCC24-8 and inserted between the *Nae*I and *Pst*I sites of pBluescript KS(+) (pBC24CH). The *Nae*I–*Hind*III fragment (320 bp) for the C-terminal region and the *Acc*I–*Nae*I fragment (1.2 kb) for the central region of the 24-

[28] Y. Gluzman, *Cell* **23,** 175 (1981).

[29] Y. Ohyama, K. Ozono, M. Uchida, T. Shinki, S. Kato, T. Suda, O. Yamamoto, M. Noshiro, and Y. Kato, *J. Biol. Chem.* **269,** 10545 (1994).

[30] M. Akiyoshi-Shibata, T. Sakaki, Y. Ohyama, M. Noshiro, K.-I. Okuda, and Y. Yabusaki, *Eur. J. Biochem.* **224,** 335 (1994).

[31] H. J. Barnes, M. P. Arlotto, and M. R. Waterman, *Proc. Natl. Acad. Sci. U.S.A.* **88,** 5597 (1991).

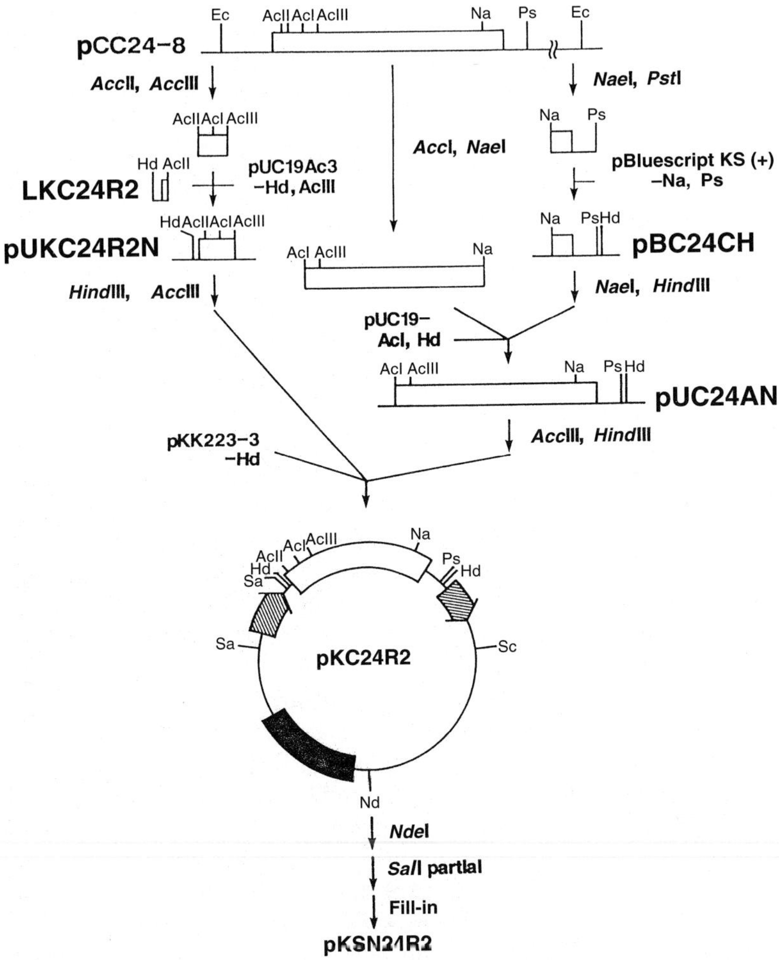

FIG. 2. Construction of a bacterial expression plasmid pKSN24R2 for mature 24-hydroxylase. The open, shaded, and closed boxes indicate the protein-coding regions for mature 24-hydroxylase, *tac* promoter and *rrnB* terminator, and the *rop* region, respectively. Restriction sites indicated are AcI, *Acc*I; AcII, *Acc*II; AcIII, *Acc*III; Ec, *Eco*RI; Hd, *Hind*III; Na, *Nae*I; Nd, *Nde*I; Ps, *Pst*I; Sa, *Sal*I. The synthesized linkers, AC3 and LKC24R2, are:

```
AC3,      5'-CTCCGGAG-3'
          3'-GAGGCCTC-5'
LKC24R2,      5'-AGCTTTTTTTTAATAAAATCAGGAGGAAAAAACATATGCG-3'
                  3'-AAAAAAATTATTTTAGTCCTCCTTTTTTGTATACGC-5'
```

(Reprinted from Ref. 30, with permission.)

TABLE II
KINETIC PARAMETERS OF RECOMBINANT 24-HYDROXYLASE FOR 25-HYDROXYVITAMIN D_3 AND 1,25-DIHYDROXYVITAMIN D_3

Substrate	K_m (μM)	V_{max} (mol/min/mol P450)
25-OH-D_3	3.1	10.4
1,25-$(OH)_2D_3$	0.23	4.3

hydroxylase are prepared from pBC24CH and pCC24-8, respectively, and doubly inserted between the *Acc*I and *Hind*III sites of pUC19 to yield pUC24ΔN. Finally, the *Hind*III–*Acc*III fragments of pUKC24R2N and pUC24ΔN are doubly inserted into the *Hind*III site of the expression vector pKK223-3 (Pharmacia, Sweden) to give a plasmid pKC24R2. To increase the plasmid copy number, the *rop* (repressor of primer) region, which is involved in the initiation of replication of a pBR-derived plasmid,[32] is deleted from pKC24R2. After *Nde*I digestion of the pKC24R2, the resulting fragment is partially digested by *Sal*I, filled in, and self-ligated to obtain the expression plasmid pKSN24R2. The structure of the constructed expression plasmid is confirmed by DNA sequencing. *Escherichia coli* JM109 is used as a host strain. Recombinant *E. coli* strains are grown in TB broth[33] containing 50 μg/ml ampicillin at 37° under good aeration. The induction of the *tac* promoter is initiated by addition of isopropylthio-β-D-galactopyranoside at a final concentration of 1 mM when the cell density (A_{660}) reached 0.5. After induction, cells are gently shaken for 24 hr at 28° to avoid over cultivation. They are then harvested and subcellular fractionation is carried out according to the method of Barnes *et al.*[31] The expressed soluble cellular protein can be used for spectral and immunochemical analysis as well as for kinetic studies. Kinetic parameters of this recombinant 24-hydroxylase for 25-hydroxyvitamin D_3 and 1,25-dihydroxyvitamin D_3 are shown in Table II.

Comments. Overproduction of the recombinant 24-hydroxylase protein may be toxic to the *E. coli* and will therefore be confined to insoluble inclusion bodies. The recombinant protein in these inclusion bodies is enzymatically inactive. Care should therefore be taken not to overculture the cells.

[32] A. J. Twigg and D. Sherratt, *Nature* (*London*) **283,** 216 (1980).
[33] K. D. Tartof and C. A. Hobbs, *Focus* (*Tokyo*) **9,** 12 (1987).

Conclusion

Regulation of vitamin D_3 hydroxylases plays a central role in calcium homeostasis. Both the 25- and 24-hydroxylases have been purified to homogeneity and their cDNA clones isolated. Subsequent studies carried out in this and other laboratories have shown that expression of the 24-hydroxylase is controlled at the promoter region of the gene and is dependent on the calcium status of the animal.[19,29,34] However, the most important hydroxylase in the metabolism of the vitamin, 25-hydroxyvitamin D_3 1α-hydroxylase, has not yet been purified or cloned.

[34] C. Zierold, H. M. Darwish, and H. F. DeLuca, *Proc. Natl. Acad. Sci. U.S.A.* **91,** 900 (1994).

[20] Role of 1α,25-Dihydroxyvitamin D_3 in Osteoclast Differentiation and Function

By TATSUO SUDA, EIJIRO JIMI, ICHIRO NAKAMURA, and NAOYUKI TAKAHASHI

Introduction

Osteoclasts are multinucleated giant cells that play a critical role in bone resorption. It is well known that the active form of vitamin D_3 [1α,25-dihydroxyvitamin D_3, 1α,25-$(OH)_2D_3$] stimulates osteoclastic bone resorption *in vivo* and *in vitro.*[1,2] We have developed a mouse marrow culture system and a coculture system of mouse osteoblastic cells and hemopoietic cells, in which osteoclasts are formed in response to 1α,25-$(OH)_2D_3$. These culture systems have made it possible to investigate each step of osteoclast development and function separately, which includes proliferation and differentiation of osteoclast progenitors, and activation of mature osteoclasts. Using the coculture system, we have shown that 1α,25-$(OH)_2D_3$ stimulates the differentiation process of the precursor cells into mononuclear and multinucleated cells with osteoclast phenotypes. The multinucleated cells formed here satisfied major criteria of osteoclasts such as tartrate-resistant acid phosphatase (TRAP, a marker enzyme of osteoclasts) activity, calcitonin receptors, $p60^{c\text{-}src}$, vitronectin receptors ($\alpha_v\beta_3$), and pit-forming activity.

[1] T. Suda, N. Takahashi, and T. J. Martin, *Endoc. Rev.* **13,** 66 (1992).
[2] T. Suda, N. Takahashi, and T. J. Martin, *Endocr. Rev. Monographs* **4,** 266 (1995).

0076-6879/97 $25.00

We have also developed a method for obtaining a large number of functionally active osteoclasts from cocultures performed on collagen gel-coated dishes. Using osteoclasts recovered from the collagen gel culture, a reliable pit formation assay was established to investigate the control mechanism of osteoclast function. Osteoclasts placed on plastic dishes were easily purified by treating with pronase–EDTA to remove osteoblastic cells. Although osteoclasts in this preparation could not be detached from the dish surface since they adhered to it very tightly, the purity in this preparation attained higher than 90%. This purified osteoclast preparation is expected to be utilized for biochemical studies on osteoclasts. We have established a method for obtaining an enriched preparation of functionally active osteoclasts as suspension from cocultures treated with $1\alpha,25\text{-(OH)}_2D_3$. The purity of osteoclasts in this enriched preparation was 50–70%. However, the enriched osteoclasts failed to form resorption pits on dentine slices. The pit-forming activity of enriched osteoclasts was strikingly increased by adding osteoblastic cells. These results indicate that not only osteoclast development, but also that osteoclast function is regulated by osteoblastic cells. In this paper, we describe the methods for examining osteoclast development and function.

Mouse Culture Systems for Investigating Osteoclast Development

Mouse Marrow Culture

The mouse bone marrow culture system for investigating osteoclast formation has been developed as a modification of the human marrow culture system originally established by the San Antonio group of Roodman and Mundy.[3,4] Tibiae are moved aseptically from 7- to 9-week-old male mice. We usually use ddY mice for our experiments. Other strains of mice such as BALB/c, C57BL, and ICR can be used for marrow cultures. Bone marrow cells collected in a tube are washed with α-minimum essential medium (αMEM) and cultured in αMEM containing 10% (v/v) fetal calf serum (FCS, GIBCO, Grand Island, NY) at 1.0×10^6 cells/0.5 ml/well in 24-well plates (Corning, Corning, NY) in a humidified atmosphere of 5% CO_2.[5] Cultures are fed every 3 days by replacing 0.4 ml old medium with fresh medium. To induce osteoclast formation, 10^{-8} *M* $1\alpha,25\text{-(OH)}_2D_3$

[3] B. R. MacDonald, N. Takahashi, L. M. McManus, J. Holahan, G. R. Mundy, and G. D. Roodman, *Endocrinology* **120,** 2326 (1987).

[4] G. R. Mundy and G. D. Roodman, *J. Bone Miner. Res.* **5,** 209 (1987).

[5] N. Takahashi, H. Yamana, S. Yoshiki, G. D. Roodman, G. R. Mundy, S. J. Jones, A. Boyde, and T. Suda, *Endocrinology* **122,** 1373 (1988).

TABLE I
CHARACTERISTICS OF CULTURE SYSTEMS FOR OSTEOCLAST FORMATION

Culture systems[a]	Culture period needed for osteoclast formation (days)	Number of osteoclasts formed (no./well of 24-well plate)	References
Bone marrow culture	6–8	50–200	Takahashi *et al.*[5]
Primary osteoblastic cells + spleen cells	6–8	100–500	Takahashi *et al.*[9]
Primary osteoblastic cells + bone marrow cells	6–8	400–2000	Akatsu *et al.*[13]
ST2 cells + spleen cells[b]	10–14	50–200	Udagawa *et al.*[10]
MC3T3-G2/PA6 cells + spleen cells[b]	10–14	50–200	Udagawa *et al.*[10]
KS-4 cells + spleen cells	6–8	100–500	Yamashita *et al.*[11]

[a] Cultures were performed as described in the text in the presence of 10^{-8} *M* 1α,25-$(OH)_2D_3$.
[b] Dexamethasone at 10^{-7} *M* and 10^{-8} *M* 1α,25-$(OH)_2D_3$ were added to cocultures of spleen cells and ST2 or MC3T3-G2/PA6 cells.

(Wako Pure Chemical Industry Co., Osaka, Japan) is added at the beginning of culture and each time the medium is changed. More frequent medium change increases the number of osteoclasts formed in the culture.[6] TRAP-positive mononuclear and multinucleated cells appear on days 3–4 and days 4–5, respectively.[5] The number of TRAP-positive multinucleated cells reaches maximum on days 6–8 (Table I). In mouse marrow cultures, alkaline phosphatase-positive cells (possibly osteoblastic cells) appear as colonies adjacent to TRAP-positive cells, suggesting that osteoblastic cells are involved in osteoclast development.[5] TRAP-positive multinucleated cells induced by 1α,25-$(OH)_2D_3$ possess many calcitonin receptors, and form resorption pits on dentin slices, satisfying major criteria of osteoclasts. Autoradiography using ^{125}I-labeled calcitonin as well as TRAP staining is recommended to identify multinucleated cells formed *in vitro* as osteoclasts.[7] Because FCS is one of the most important factors that affects osteoclast formation, a careful check of each FCS lot is recommended.

Identification of Osteoclasts Formed in Vitro

TRAP Staining. Cytochemical staining for TRAP is widely used for identifying osteoclasts *in vivo* and *in vitro.* Naphthol AS-MX phosphate

[6] T. Akatsu, N. Takahashi, N. Udagawa, K. Sato, N. Nagata, J. M. Moseley, T. J. Martin, and T. Suda, *Endocrinology* **125,** 20 (1989).
[7] N. Takahashi, T. Akatsu, T. Sasaki, G. C. Nicholson, J. M. Moseley, T. J. Martin, and T. Suda, *Endocrinology* **123,** 1504 (1988).

(5 mg, Sigma, St. Louis, MO) is resolved in 0.5 ml of *N,N*-dimethylformamide (Wako). Thirty milligrams of fast red violet LB salt (Sigma) and 50 ml of 0.1 *M* sodium acetate buffer (pH 5.0) containing 50 m*M* sodium tartrate are added to the mixture (the TRAP-staining solution). Cells are fixed with 3.7% (v/v) formaldehyde in Ca^{2+}- and Mg^{2+}-free phosphate-buffered saline [PBS(−)] for 10 min, fixed again with ethanol–acetone (50 : 50, v : v) for 1 min, and incubated with the TRAP-staining solution for 10 min at room temperature. TRAP-positive osteoclasts appear as red cells. The incubation period longer than 10 min should be avoided, since cells other than osteoclasts become weakly positive with time. After staining, cells are washed with distilled water, and TRAP-positive multinucleated cells having three or more nuclei are counted as osteoclasts under a microscope.[5]

Autoradiography for Calcitonin Receptors. Osteoclasts have been shown to possess abundant calcitonin receptors.[8] Expression of calcitonin receptors is one of the most reliable markers for identifying osteoclasts. Salmon calcitonin labeled with ^{125}I is prepared according to the method of Nicholson *et al.*[8] Specific activity of the labeled product is ~125 μCi/μg. For autoradiography of ^{125}I-labeled calcitonin, cultures are performed on plastic coverslips (13.5-mm, Sumitomo Bakelite, Tokyo, Japan) placed in 24-well culture plates. Cells grown on the coverslips are washed with αMEM, and incubated with salmon ^{125}I-labeled calcitonin (0.2 n*M*) in the presence or absence of 200 n*M* unlabeled salmon calcitonin in αMEM containing 0.1% bovine serum albumin (BSA) for 1 hr at 20°.[7] Cells are washed with PBS(−) and fixed for 5 min with 0.1 *M* cacodylate buffer (pH 7.4) containing 1% formaldehyde and 1% glutaraldehyde. The coverslips are then mounted on a glass slide, and dipped in NR-M2 emulsion (Sakura Photo Industry Co., Tokyo, Japan). They are stored in a shading box at 4° for 14 days and developed in Rendol (Fuji Photo Film Co., Tokyo, Japan).[7] Calcitonin receptors are identified by accumulation of dense grains due to ^{125}I-labeled calcitonin binding, which disappear from the specimen when incubated with excess unlabeled calcitonin.

Cocultures of Osteoblastic Cells and Hemopoietic Cells

Cocultures of primary osteoblastic cells with bone marrow cells or spleen cells produce more osteoclasts than bone marrow cultures do (Table I).[9] Primary osteoblastic cells are isolated from calvariae of 1- to 3-day-old

[8] G. C. Nicholson, J. M. Moseley, P. M. Sexton, F. A. O. Mendelssohn, and T. J. Martin, *J. Clin. Invest.* **78,** 355 (1986).

[9] N. Takahashi, T. Akatsu, N. Udagawa, T. Sasaki, A. Yamaguchi, J. M. Moseley, T. J. Martin, and T. Suda, *Endocrinology* **123,** 2600 (1988).

newborn ddY mice. Calvariae removed from 50 mice are put into a 50-ml centrifuge tube and incubated in 10 ml of αMEM containing 0.1% bacterial collagenase (Wako) and 0.2% dispase (Godo Shusei Co., Tokyo, Japan) for 10 min at 37° in a shaking waterbath (120 cycles/min). The incubation of calvariae with collagenase–dispase is repeated five times. Primary osteoblastic cells isolated in fractions 2–5 are collected and cultured for 3 days in αMEM with 10% FCS in 10-cm culture dishes (Corning) (cells from 10 calvariae/dish). Primary osteoblastic cells are then detached from the dish by treating with trypsin-EDTA (Gibco), and suspended in αMEM containing 20% FCS and 15% dimethyl sulfoxide (DMSO), and stored at −80° (1×10^6 cells/ml/freezing vial). The osteoclast-supporting activity of primary osteoblastic cells is not destroyed in the freezer for at least 3 months. Primary ostoblastic cells (2×10^4 cells/well) are cocultured with bone marrow cells (2×10^5 cells/well) or spleen cells (1×10^6 cells/well) in 24-well plates (0.5 ml/well) in αMEM containing 10% FCS and 10^{-8} *M* 1α,25-$(OH)_2D_3$. Cultures are fed every 3 days by replacing 0.4-ml old medium with fresh medium. The time course of changes in the appearance of TRAP-positive mononuclear and multinucleated cells is similar to that in bone marrow cultures.[5] When spleen cells and osteoblastic cells are cultured separately from each other by a membrane filter, no osteoclasts are formed even in the presence of 1α,25-$(OH)_2D_3$.[9] Therefore, cell-to-cell interaction between osteoclast progenitors and osteoblastic cells is required for their differentiation into osteoclasts. Like 1α,25-$(OH)_2D_3$, several hormones and cytokines such as parathyroid hormone (PTH), prostaglandin E_2 (PGE_2), interleukin-1 (IL-1), and IL-6 plus soluble IL-6 receptors stimulate osteoclast formation in marrow cultures and cocultures of primary osteoblastic cells with hemopoietic cells.[1,2]

Stromal Cells That Support Osteoclast Formation

We have reported that not only primary osteoblastic cells but also three established cell lines (ST2, MC3T3-G2/PA6, and KS-4) support osteoclast formation in cocultures with mouse spleen cells.[10,11] These established cell lines are very useful for investigating the origin of osteoclasts,[12] because the primary osteoblastic cell preparation contains a small number of osteoclast progenitors even if we prepare it carefully. However, several disadvantages

[10] N. Udaga, N. Takahashi, T. Akatsu, T. Sasaki, A. Yamaguchi, H. Kodama, T. J. Martin, and T. Suda, *Endocrinology* **125,** 1805 (1989).

[11] T. Yamashita, K. Asano, N. Takahashi, T. Akatsu, N. Udagawa, T. Sasaki, T. J. Martin, and T. Suda, *J. Cell. Physiol.* **145,** 587 (1990).

[12] N. Udagawa, N. Takahashi, T. Akatsu, H. Tanaka, T. Sasaki, T. Nishihara, T. Koga, T. J. Martin, and T. Suda, *Proc. Natl. Acad. Sci. U.S.A.* **87,** 7260 (1990).

arise from the use of these cell lines in experiments on osteoclast formation. The activity of MC3T3-G2/PA6 and ST2 to support osteoclast formation is much weaker than that of primary osteoblastic cells, and the culture period required to induce osteoclasts in cocultures with these cell lines is much longer compared with that in cocultures with primary osteoblastic cells (Table I).[10] Although KS-4 cells have a similar property in inducing osteoclasts in cocultures with spleen cells,[11] the supporting activity for osteoclast formation is unstable and tends to be lost with passage. Primary osteoblastic cells appear to be the best stromal cells to use for supporting osteoclast formation in cocultures with hemopoietic cells (Table I).

Assay Systems for Examining Osteoclast Function

Collagen Gel Culture

Osteoclasts formed on plastic culture dishes cannot be detached by treatment with either trypsin–EDTA or bacterial collagenase. To obtain functionally active osteoclasts formed in cocultures, a collagen gel culture has been developed.[13] Type I collagen solution is obtained from Nitta Gelatin Co. (Osaka, Japan). The collagen solution, 5× concentrated αMEM and 200 m*M* HEPES buffer (pH 7.4) containing 2.2% $NaHCO_3$ (7:2:1, v:v:v), is quickly mixed at 4°. A 10-cm culture dish (Corning) is coated with 4 ml of the collagen mixture at 4°. The dish is put in a CO_2 incubator for 30 min to make the aqueous type I collagen gelatinous at 37°. Primary osteoblastic cells (2×10^6 cells) and bone marrow cells (2×10^7 cells) are cocultured on a collagen gel-coated dish in 15 ml of αMEM containing 10% (v/v) FCS and 10^{-8} *M* 1α,25-$(OH)_2D_3$. The medium is changed every 2 days. After culture for 6 days, the dish is treated with 4 ml of 0.2% bacterial collagenase (Wako) for 20 min at 37° in a shaking waterbath (60 cycles/min). The cells released from the dish are collected by centrifugation at 250*g* for 5 min and suspended in 10 ml of αMEM containing 10% FCS (the crude osteoclast preparation). Usually, $4–10 \times 10^4$ osteoclasts are recovered from a 10-cm collagen gel-coated dish, and the purity is 2–3% in the crude osteoclast preparation (see Table II).[13] The crude osteoclast preparation is used for examining biological and biochemical studies on osteoclasts.

Actin Ring Formation Assay

Osteoclasts adhere to the bone surface through specialized discrete structures called *podosomes* in the clear zone, which consist mainly of dots

[13] T. Akatsu, T. Tamura, N. Takahashi, N. Udagawa, S. Tanaka, T. Sasaki, A. Yamaguchi, N. Nagata, and T. Suda, *J. Bone Miner. Res.* **7,** 1297 (1992).

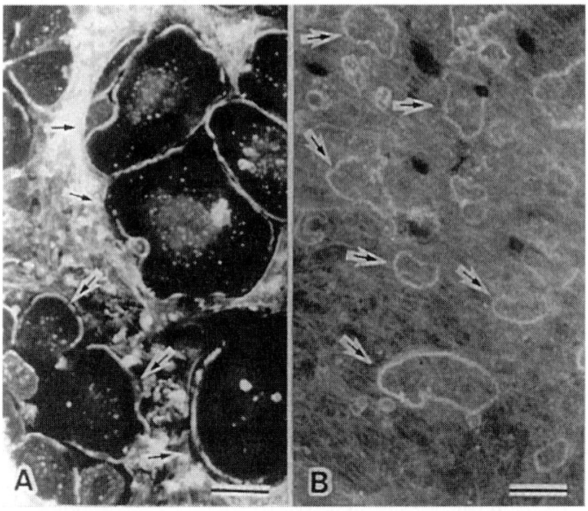

FIG. 1. Actin ring formation by osteoclasts. The crude osteoclast preparation obtained from a collagen gel culture was placed on (A) a plastic culture dish or (B) a dentine slice for 4 and 10 hr, respectively. The cells were then fixed, and incubated with rhodamine-conjugated phalloidin to visualize the distribution of F-actin. Arrows indicate osteoclasts having actin rings. Bars: 100 μm.

containing F-actin.[14] Therefore, the ringed structure of podosomes (actin ring) formed in osteoclasts is a characteristic of polarized osteoclasts. The actin rings are visualized by staining F-actin with rhodamine-conjugated phalloidin.[14] Rhodamine-conjugated phalloidin (Sigma) is dissolved in a small volume of methanol, diluted with PBS(−) to the final concentration of 0.3 μM, and stored at 4° in the dark. Cells are fixed with 3.7% (v/v) formaldehyde in PBS(−) for 10 min and permeated by treatment with 0.1% Triton X-100 in PBS(−) for 1 min. Cells are first stained for TRAP to identify osteoclasts, then incubated for 3 hr with the rhodamine-conjugated phalloidin solution.[15–18] The cells are washed with water, and actin rings formed by osteoclasts are detected with a fluorescence microscope (Olympus BX-FLA, Osaka, Japan) (Fig. 1). The rhodamine-conjugated phalloidin

[14] J. Kanehisa, T. Yamanaka, S. Doi, K. Turksen, J. N. M. Heersche, J. E. Aubin, and H. Takeuchi, *Bone* **11,** 287 (1990).

[15] H. Murakami, N. Takahashi, T. Sasaki, N. Udagawa, S. Tanaka, I. Nakamura, D. Zhang, A. Barbier, and T. Suda, *Bone* **17,** 137 (1995).

[16] S. Tanaka, N. Takahashi, N. Udagawa, H. Murakami, I. Nakamura, I. Kurokawa, and T. Suda, *J. Cell. Biochem.* **58,** 424 (1995).

[17] I. Nakamura, N. Takahashi, T. Sasaki, N. Udagawa, H. Murakami, S. Tanaka, K. Kimura, Y. Kabuyama, T. Kurokawa, T. Suda, and Y. Fukui, *FEBS Lett.* **361,** 79 (1995).

[18] D. Zhang, N. Udagawa, I. Nakamura, H. Murakami, S. Saito, K. Yamasaki, Y. Shibasaki, N. Morii, S. Narumiya, N. Takahashi, and T. Suda, *J. Cell. Sci.* **108,** 2285 (1995).

solution can be repeatedly used unless F-actin staining becomes weak. We have shown that actin rings formed in osteoclasts are disrupted by adding several inhibitors of bone resorption such as calcitonin, bisphosphonates,[15] herbimycin A (an inhibitor of tyrosine kinase),[16] wortmannin (an inhibitor of phosphatidylinositol 3-kinase),[17] and C3 exoenzyme (an inhibitor of the small GTP-binding protein, *rho* p21).[18]

Pit Formation Assay

When osteoclasts are placed on dentine slices, they form resorption pits within a few days.[19,20] We have developed a reliable pit formation assay system using the crude osteoclast preparation and dentine slices.[21] Dentine slices (ϕ 4 mm, 200 μm thick) are prepared from ivory blocks using a band saw (BS-3000, Exakt, Germany) and a cutting punch. They are cleaned by ultrasonication in distilled water, sterilized using 70% (v/v) ethanol, and dried under ultraviolet light. Dentine slices are placed in 96-well plates containing 0.1 ml/well of αMEM with 10% FCS (a slice/well). An aliquot of the crude osteoclast preparation (0.1 ml) is transferred onto the slices. After a setting period of 90 min at 37°, slices are removed, and placed into 24-well plates containing αMEM with 10% FCS (0.5 ml/slice/well). After incubation for 24–48 hr, the medium is removed and 1 *M* NH_4OH (1 ml/well) is added to the wells for 30 min. Dentine slices are then cleaned by ultrasonication, stained with Mayer's hematoxylin (hematoxylin, 1.0 g; $NaIO_3$, 0.2 g; $AlNH_4(SO_4)_2 \cdot 12H_2O$, 50 g; acetic acid, 7.5 ml/liter, pH 2.8) for 35–45 sec, and washed with distilled water. The resorption pits are clearly visualized with Mayer's hematoxylin under transmitted light (Fig. 2).[21] The resorbed area is measured using an image analysis system linked to the light microscope (LA-525; PIAS Co., Tokyo, Japan).

Purification of Osteoclasts

Purification of Osteoclasts for Biochemical Study

Because the purity of osteoclasts in the crude osteoclast preparation is only 2–3%, further purification is essential for biochemical studies on osteoclasts (Table II). Osteoclasts are easily purified from the crude osteoclast preparation placed on plastic dishes by treating with pronase–EDTA.[16–18] Pronase (Calbiochem, La Jolla, CA) is dissolved in PBS(−)

[19] A. Boyde, N. N. Ali, and S. J. Jones, *Br. Dent. J.* **152,** 216 (1984).

[20] T. J. Chambers, P. A. Revell, and N. A. Athanasou, *J. Cell. Sci.* **66,** 383 (1984).

[21] T. Tamura, N. Takahashi, T. Akatsu, T. Sasaki, N. Udagawa, S. Tanaka, and T. Suda, *J. Bone Miner. Res.* **8,** 953 (1993).

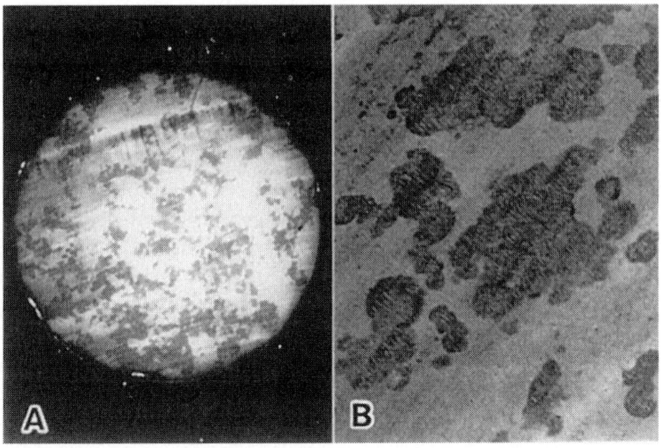

FIG. 2. Resorption pits formed by osteoclasts on dentine slices. The osteoclast preparation obtained from a collagen gel culture was placed on a dentine slice (ϕ 4 mm). After culture for 48 hr, cells were removed from the dentine slice. The slice was then stained with Mayer's hematoxylin to visualize resorption pits. (A) Whole surface of the dentine slice. Many resorption pits are observed on the slice. (B) Enlarged portion of the slice.

containing 0.02% EDTA to be the final concentration of 0.001% just before use. The crude osteoclast preparation placed on a 10-cm culture dish (Corning) for 6–10 hr in the presence of 10% FCS is washed with αMEM, and treated with 8 ml of PBS(−) containing 0.001% pronase and 0.02% EDTA for 10 min to remove osteoblastic cells. More than 90% of the adherent cells on the dish are TRAP-positive mononuclear and multinucleated cells (Fig. 3B, Table II). More than 95% of the total protein is estimated to be derived from osteoclasts. Using the purified osteoclast preparation, we have shown that osteoclasts possess focal adhesion kinase ($p125^{FAK}$),[16] phosphatidylinositol 3-kinase,[17] *rho* p21,[18] and $p60^{c\text{-}src}$.[22] We have also reported that an NF-κB-like transcription factor in osteoclasts is activated by adding IL-1 to the purified preparation.[23]

Purification of Osteoclasts for Biological Study

We have developed a method for obtaining a highly enriched preparation of functionally active osteoclasts[24] with a modification of the purifica-

[22] S. Tanaka, N. Takahashi, N. Udagawa, T. Sasaki, Y. Fukui, T. Kurokawa, and T. Suda, *FEBS Lett.* **313,** 85 (1992).

[23] E. Jimi, T. Ikebe, N. Takahashi, N. Hirata, T. Suda, and T. Koga, *J. Biol. Chem.* **271,** 4605 (1996).

[24] E. Jimi, I. Nakamura, H. Amano, Y. Taguchi, T. Turukai, N. Tamura, N. Takahashi, and T. Suda, *Endocrinology* **137,** 2187 (1996).

TABLE II
CHARACTERISTICS OF OSTEOCLAST PREPARATIONS OBTAINED FROM COLLAGEN GEL CULTURES

Osteoclast preparations	Purity of osteoclasts (%)	Recovery of osteoclasts (%)	Characteristics of osteoclast preparations	Refs.
Crude osteoclast preparations	2–3	~90	Applicable to pit formation assay and further purification of osteoclasts	Akatsu *et al.*[13] Tamura *et al.*[21]
Purified osteoclasts attached to plastic dishes	~90	~80	Applicable to the biochemical study but not to the study on osteoclast function	Tanaka *et al.*[16] Nakamura *et al.*[17] Zhang *et al.*[18] Tanaka *et al.*[22] Jimi *et al.*[23]
Enriched osteoclasts on Percoll density gradient	50–70	10–20	Applicable to the study on osteoclast function	Jimi *et al.*[24]

[a] The crude osteoclast preparation was obtained from the collagen gel culture as described in the text. An aliquot of the preparation was plated on a plastic culture dish for 4 hr, and stained for TRAP to determine the purity and recovery of osteoclasts in the preparation.
[b] The crude osteoclast preparation was plated on a plastic culture dish for 10 hr and treated with pronase-EDTA to remove osteoblastic cells. The cells remaining on the dish surface were stained for TRAP to determine the purity and recovery of osteoclasts in the preparation.
[c] The enriched osteoclast preparation was obtained from the collagen gel culture as described in the text. An aliquot of the preparation was placed on a plastic culture dish for 4 hr and stained for TRAP to determine the purity and recovery of osteoclasts in the preparation.

tion method for avian osteoclasts.[25] Primary osteoblastic cells and bone marrow cells are cocultured on collagen gel-coated dishes in αMEM with 10% FCS in the presence of $1\alpha,25\text{-(OH)}_2D_3$ at 10^{-8} *M* as described earlier. After culture for 6 days, collagen gel cultures are washed with PBS(−) containing 0.02% EDTA, and treated with PBS(−) containing 0.001% pronase and 0.02% EDTA for 10 min (5 ml/dish). Dishes are gently pipetted to remove some of the osteoblastic cells. The treatment with pronase–EDTA is repeated three times. Cocultures are then incubated with αMEM containing 10% FCS for 5 min to inactivate pronase followed by 0.2% bacterial collagenase to recover all the cells from the dish. Cells obtained from a collagen-coated dish (ϕ 10 cm) are centrifuged at 250*g* for 5 min and resuspended in 3 ml of αMEM. The osteoclast preparation (3 ml) is

[25] P. Osdoby, M. C. Martini, and A. I. Caplan, *J. Exp. Zool.* **224,** 331 (1982).

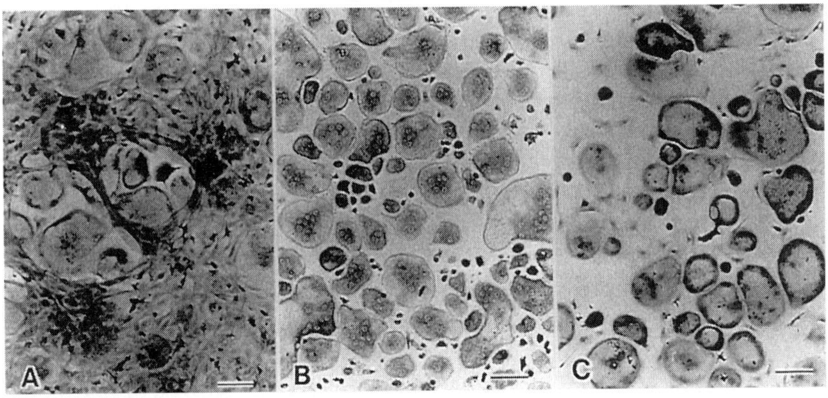

FIG. 3. Microscopic views of the different purified osteoclast preparations. (A,B) Purification of osteoclasts on a plastic culture dish. The crude osteoclast preparation obtained from a collagen gel culture was placed on two plastic dishes. After incubation for 10 hr at 37°, cells on one dish were fixed and stained for TRAP (A). The cells on the other dish were treated with pronase–EDTA to remove osteoblastic cells, and the remaining adherent cells were stained for TRAP (B). Although osteoclasts in this preparation could not be detached from the dish surface since they adhered to it very tightly, the purity of osteoclasts in preparation (B) attained about 90%. The purified osteoclasts are used for biochemical studies. Bars: 100 μm. (C) Enrichment of suspended osteoclasts on Percoll density gradient centrifugation. Osteoblastic cells and bone marrow cells were cocultured in the presence of 10^{-8} M 1α,25-$(OH)_2D_3$ on a collagen gel-coated dish for 6 days. The culture was then treated with pronase-EDTA to remove some osteoblastic cells, followed by collagenase digestion. The cell suspension was placed on 30% (v/v) Percoll solution, and centrifuged at 250g for 30 min at 20°. An aliquot of cells accumulating in the interface layer was placed on the plastic dish for 4 hr, and stained for TRAP. The purity of osteoclasts in preparation (C) was about 70%. The enriched osteoclasts formed very few resorption pits on dentine slices in the absence of osteoblastic cells. Bar: 100 μm.

overlaid on 8 ml of 30% Percoll (Pharmacia, Uppsala, Sweden) in PBS(−) and centrifuged at 250g for 30 min. The interface layer was collected, washed in αMEM, and resuspended in αMEM containing 10% FCS. The purity of osteoclasts in this preparation is 50–70% (Fig. 3C, Table II). The recovery of osteoclasts is 10–20% of the total osteoclasts formed in the collagen gel-coated dish. The treatment of cocultures on collagen gel-coated dishes with pronase-EDTA is an important step to purify osteoclasts, since this treatment markedly reduces cell aggregates, which are often observed in the crude osteoclast preparation after collagenase digestion (Fig. 3A). Using the enriched osteoclast preparation, we have shown that osteoblastic cells activate osteoclast function through a mechanism involving cell-to-cell contact.[24]

Role of 1α,25-Dihydroxyvitamin D_3 in Osteoclast Formation and Function

Using the culture systems described earlier, we have examined the role of vitamin D derivatives in osteoclast development and function. Neither 25-hydroxyvitamin D_3 nor 24,25-dihydroxyvitamin D_3 at 10^{-8} M showed a stimulatory effect on osteoclast formation in our culture systems. This indicates that osteoclast development induced by $1\alpha,25\text{-}(OH)_2D_3$ is mediated by the vitamin D receptor (VDR). The evidence that VDR is localized in osteoblastic cells but not in osteoclasts suggests that the major bone target cells for $1\alpha,25\text{-}(OH)_2D_3$ are osteoblastic cells. We have shown that $1\alpha,25\text{-}(OH)_2D_3$ specifically stimulates production of the third component of complement (C3) by mouse osteoblastic cells *in vitro*.[26] Treatment of mouse marrow cultures with various antibodies against C3 and its receptors inhibited osteoclast formation induced by $1\alpha,25\text{-}(OH)_2D_3$.[27] These findings suggest that the C3 produced by osteoblastic cells in response to $1\alpha,25\text{-}(OH)_2D_3$ has an important role in osteoclast development. Production of three noncollagenous proteins by osteoblastic cells, bone Gla protein (BGP, osteocalcin), matrix Gla protein (MGP), and osteopontin, is known to be regulated by $1\alpha,25\text{-}(OH)_2D_3$. However, there is no convincing evidence that these proteins are involved in osteoclast development. ST2 cells produced neither BGP nor MGP in response to $1\alpha,25\text{-}(OH)_2D_3$, but supported osteoclast formation in cocultures with spleen cells in the presence of $1\alpha,25\text{-}(OH)_2D_3$. Osteopontin appears to be involved in the attachment of osteoclasts to the bone surface. However, there is no evidence that the attachment of osteoclasts to the bone surface is controlled by $1\alpha,25\text{-}(OH)_2D_3$ through stimulating osteopontin production by osteoblastic cells. Although the physiologic role of BGP, MGP, and osteopontin in bone metabolism remains to be elucidated, the dependence of $1\alpha,25\text{-}(OH)_2D_3$ on the synthesis of those matrix proteins suggests that they are somehow involved in the regulation of osteoclast development and function.

The treatment of cocultures of osteoblastic cells and spleen cells or bone marrow cells with $1\alpha,25\text{-}(OH)_2D_3$ only for the last 2 days of the 6-day coculture period could induce significant numbers of osteoclasts.[28,29] When

[26] T. Sato, M. H. Hong, C. H. Jin, Y. Ishimi, N. Udagawa, T. Shinki, E. Abe, and T. Suda, *FEBS Lett.* **285,** 21 (1991).

[27] T. Sato, E. Abe, C. H. Jin, M. H. Hong, T. Katagiri, T. Kinoshita, N. Amizuka, H. Ozawa, and T. Suda, *Endocrinology* **133,** 397 (1993).

[28] S. Tanaka, N. Takahashi, N. Udagawa, T. Tamura, T. Akatsu, R. E. Stanley, T. Kurokawa, and T. Suda, *J. Clin. Invest.* **91,** 257 (1993).

[29] N. Takahashi, N. Udagawa, S. Tanaka, H. Murakami, I. Owan, T. Tamura, and T. Suda, *Dev. Biol.* **163,** 212 (1994).

hydroxyurea was added to suppress proliferation together with $1\alpha,25\text{-}(OH)_2D_3$ for the last 2 days, osteoclast formation similarly occurred. This indicates that $1\alpha,25\text{-}(OH)_2D_3$ stimulates the differentiation process of osteoclast precursors. We have shown that osteoblastic cells obtained from transgenic mice constitutively expressing human IL-6 receptors could support osteoclast development in the presence of human IL-6 alone in cocultures with normal spleen cells.[30] In contrast, osteoclast progenitors from splenic tissues of the transgenic mice could not differentiate into osteoclasts in response to human IL-6 in cocultures with normal osteoblastic cells.[30] Thus, the ability of IL-6 to induce osteoclast differentiation depends on the signal transduction mediated by IL-6 receptors expressed on osteoblastic cells but not on osteoclast progenitors. Recently Kato and his colleagues produced VDR deficient mice by targeted disruption of the gene encoding VDR.[31] Osteoblastic cells obtained from VDR (−/−) mice failed to support osteoclast development in cocultures with spleen cells obtained from the wild-type mice in response to $1\alpha,25\text{-}(OH)_2D_3$. In contrast, when spleen cells derived from VDR (−/−) mice were cocultured with osteoblastic cells obtained from the wild-type mice, TRAP-positive osteoclasts were formed in response to $1\alpha,25\text{-}(OH)_2D_3$. This strongly supports the hypothesis that the signals induced by $1\alpha,25\text{-}(OH)_2D_3$, are also transduced into osteoblastic cells to recruit osteoclasts, though the possibility that bone-resorbing factors directly act on hemopoietic cells cannot be completely disregarded. We have hypothesized that osteoblastic cells produce a factor(s) responsible for osteoclast differentiation in response to osteotropic factors including $1\alpha,25\text{-}(OH)_2D_3$.[1,2] Cloning of such a factor appears very important in osteoclast biology, which is currently being explored in our laboratory.

The resorbing activity of osteoclasts has been believed to be enhanced by osteotropic hormones such as PTH, PGE_2, and $1\alpha,25\text{-}(OH)_2D_3$ in the presence of osteoblastic cells. However, the resorbing activity of osteoclasts formed *in vitro* was not stimulated further by adding $1\alpha,25\text{-}(OH)_2D_3$ to the pit formation assay.[21] Osteoclasts may have already been stimulated during cocultures with osteoblastic cells, which are treated with a pharmacologic concentration (10^{-8} *M*) of $1\alpha,25\text{-}(OH)_2D_3$. Further studies are needed to elucidate the mechanism of action of $1\alpha,25\text{-}(OH)_2D_3$ in regulating osteoclast development and function.

[30] N. Udagawa, N. Takahashi, T. Katagiri, T. Tamura, S. Wada, D. M. Findlay, T. J. Martin, H. Hirota, T. Taga, T. Kishimoto, and T. Suda, *J. Exp. Med.* **182,** 1461 (1995).

[31] T. Yoshizawa, H. Handa, Y. Uematsu, K. Sekine, S. Takeda, Y. Yoshihara, T. Kawakami, H. Sato, K. Alioka, K. Tanimoto, A. Fukamizu, S. Masushige, T. Matsumoto, and S. Kato, *J. Bone Miner. Res.* **11,** S124 (1996).

[21] Assay of Direct Effect of 1,25-Dihydroxyvitamin D_3 on Calcium Ion Influx into Cultured Osteoblasts

By J. GARY MESZAROS and MARY C. FARACH-CARSON

Introduction

Calcium ion (Ca^{2+}) influx assays provide a useful means to assess the contribution of Ca^{2+} entry through the plasma membrane during a variety of membrane-initiated signaling processes including cellular responses to various hormones and cytokines. At the cellular level, Ca^{2+} signals involve a number of Ca^{2+}-mobilizing systems including ion channels in the plasma membrane and intracellular organelles, energy-dependent pumps, and ion exchangers.[1] These transmembrane proteins, which exist in a number of tissue-specific isoforms, cooperate functionally to elevate and lower intracellular concentrations from a typical resting concentration of 50–100 n*M* to an activating concentration that is significantly higher (>200 n*M*). Such signals contribute directly or indirectly to activation of a host of intracellular signal transducers including Ca^{2+}-dependent enzymes (kinases, phosphatases, phospholipases, and sphingomyelinases).[2,3] In the longer term, these signals determine cellular function and fate including patterns of gene expression.[4]

Historically, cells have been characterized throughout the literature as either "excitable" or "nonexcitable."[5] The former include cell types originating in tissues capable of generating action potentials such as the nervous system (central and peripheral) and muscle (skeletal, smooth, and cardiac). The latter include stromal cells from most glandular organs, epithelia, and fibroblasts. Skeletal tissues, neurosecretory cells, immune cells, and endothelial cells provide examples of a "hybrid" phenotype, possessing features of both excitable and nonexcitable cells. The emerging view represents cells of various origins as a continuum of phenotypes ranging from those most closely resembling "excitable" cells to those most closely resembling "nonexcitable" cells. A major distinguishing feature becomes the

[1] R. W. Tsien and R. Y. Tsien, *Annu. Rev. Cell Biol.* **6,** 715 (1990).
[2] N. Divecha and R. F. Irvine, *Cell* **80,** 269 (1995).
[3] J. H. Exton, *J. Biol. Chem.* **265,** 1 (1990).
[4] M. A. Thompson, D. D. Ginty, A. Bonni, and M. E. Greenberg, *J. Biol. Chem.* **270,** 4224 (1995).
[5] D. E. Clapham, *Cell* **80,** 259 (1995).

0076-6879/97 $25.00

number rather than the complement of cellular proteins involved in generation of Ca^{2+} signals.

Osteoblasts are the cells essential for the deposition of the mineralizing matrix unique to bone. The formation of this matrix, called *osteoid,* is under the influence of numerous calcitropic hormones most notably 1,25-dihydroxyvitamin D_3 [1,25-$(OH)_2D_3$], parathyroid hormone and calcitonin.[6] In the context of bone biology, it is critical to distinguish between *mineral* Ca^{2+}, that which will ultimately be deposited into mineralizing osteoid, and *signal* Ca^{2+}, that which is involved in activation of intracellular Ca^{2+}-dependent signaling pathways. This latter process is the subject of this chapter, which focuses on methods for measurement of Ca^{2+} entry during treatment of osteoblasts with calcitropic hormones such as 1,25-$(OH)_2D_3$.

Preparation of Cells for Calcium Influx Measurements

In addition to hormonal responsiveness, osteoblasts respond to mechanical forces including both stretch and strain.[7] Furthermore, attachment to an extracellular matrix is a requirement for appropriate biological response.[8] For these reasons, the preparation of osteoblastic cell cultures for influx measurements is a critical parameter in determining the type of data that will be obtained. For many cell types, cell suspensions provide a convenient means of measuring Ca^{2+} influx using large numbers of cells. The advantage of this type of preparation is that individual variability in responsiveness is eliminated as a variable, and results are indicative of the mean response of the cell population. In the case of enzymatically released osteoblastic cell suspensions, Ca^{2+} influx following treatment with calcitropic agents typically demonstrates diminished, variable, or nonmeasurable change.[9,10] For this reason, population influx studies are best performed using monolayer cultures in which cells remain attached to a substratum.[11] These studies can be performed either with resting cultures or cultures that have been subjected to forces of mechanical strain. These studies can also be performed using osteoblastic cells cultured on a variety of biological matrices. The procedure for conducting these influx assays is described later.

[6] P. J. Nijweide, E. H. Burger, and J. H. Feyen, *Physiol. Rev.* **66,** 855 (1986).
[7] L. V. Harter, K. A. Hruska, and R. L. Duncan, *Endocrinology* **136,** 528 (1995).
[8] A. L. Boskey, *Bone Miner.* **6,** 111 (1989).
[9] R. Civitelli, Y. S. Kim, S. L. Gunsten, A. Fujimori, M. Huskey, L. V. Avioli, and K. A. Hruska, *Endocrinology* **127,** 2253 (1990).
[10] M. Lieberherr, *J. Biol. Chem.* **262,** 13168 (1987).
[11] M. C. Farach-Carson, I. Sergeev, and A. W. Norman, *Endocrinology* **129,** 1876 (1991).

A second important decision concerns the cell line and the degree of confluency of the cells under study. It is clear that low-passage, growth phase cultures at low density (30–50% confluency) provide the most reliable and highest degree of measurable Ca^{2+} influx following treatment with 1,25-$(OH)_2D_3$.[12] This observation is most likely due to the coordinate expression of cell surface Ca^{2+} channels with the degree of cell differentiation. Supporting this idea, a number of osteoblastic-like cell lines have been compared with regard to expression of voltage-sensitive Ca^{2+} channels (VSCCs), the primary channels associated with Ca^{2+} entry into resting osteoblastic cells. Rat osteosarcoma lines including ROS 17/2.8 and UMR-106 express functional VSCCs at levels of 1500–2500 channels/cell, values at least an order of magnitude lower than excitable muscle cells.[13] MC3T3-E1 cells, an established line originating from newborn mouse calvaria, express slightly lower levels of mRNA encoding VSCCs than the osteosarcoma lines, but increased levels following treatment with the differentiating agent ascorbate.[14] Unfortunately, ascorbate treatment also stimulates the production of an elaborate collagenous extracellular matrix by these cells; this matrix serves as a Ca^{2+} *sink* (characteristic of its mineralization potential), which obscures any measurements of Ca^{2+} influx using radioactive tracers. For these reasons, multiple variables must be weighed when making a decision about how best to perform measurements of Ca^{2+} influx using osteoblastic cells.

Handling of Vitamin D-Related Compounds

Care must be exercised when using solutions of vitamin D-related compounds to prevent both degradation of the secosteroid and the loss of the compounds to the vessels in which they are stored or transferred. Optimal storage conditions for this family of compounds are provided by dark glass sealable containers, where they are best kept at concentrations of 10^{-4} or higher in absolute ethanol at $-20°$. Structural integrity and concentration should be routinely monitored from the absorption spectra and by calculation of the absorbance ratio at 264 and 228 nm, which should not be lower than 1.4. Concentrations can be calculated from the absorptivity at 264 nm using a molar absorptivity value of 18,200.

Vitamin D-related compounds in aqueous solution rapidly adhere to the walls of both glass and plastic containers, resulting in a drastic reduction in the effective concentration of compound from that originally present.

[12] M. C. Farach-Carson, J. Abe, Y. Nishii, R. Khoury, G. C. Wright, and A. W. Norman, *Am. J. Physiol.* (*Renal*) **265,** F705 (1993).

[13] J. M. Caffrey and M. C. Farach-Carson, *J. Biol. Chem.* **264,** 20265 (1989).

[14] R. T. Franceschi and B. Iyer, *J. Bone Miner. Res.* **7,** 235 (1992).

Caution must be taken to prevent vessel loss from occurring, because it will produce both misleading conclusions and erroneous dose responses. A suitable experimental protocol is to transfer the solution of the vitamin D-related compound immediately from the ethanol stock to the cell plate, making any necessary dilutions only into absolute ethanol. The concentration of vehicle in the final assay solution should always remain below 0.5% (v/v), with a vehicle control performed to guard against interpretation of vehicle-related effects on membrane permeability as being related to the test vitamin D compound.

Ion Tracer Influx Assays Using Monolayer Cultures of Osteoblastic Cells

This method is particularly useful for assaying populations of cells growing in monolayers attached to standard tissue culture dishes. Advantages are that the influx data obtained are representative of the entire cell population and that applied concentrations of test reagents are easily controlled. Disadvantages are the large number of replicate plates required to obtain meaningful data and the absolute requirement for standardization of plate cell number. Practically speaking, this means that these assays are most amenable to studies utilizing osteoblastic cell lines rather than primary cultures.

The first objective must be to identify the linear range of ion tracer uptake, in this case $^{45}Ca^{2+}$. For most cells, this range is limited to the first 2 min following initiation of the experiment at room temperature. Cells can be conveniently grown in replicate on 35-mm plastic culture dishes following controlled seeding at low density. At the start of the experiment, culture medium is removed and the cell monolayers are washed three times with room temperature Hanks' buffered saline solution to remove serum components including the serum protein capable of binding 1,25-$(OH)_2D_3$. Cells are then incubated for the desired period of time (usually 1 min) with either *resting* buffer (containing [in m*M*]: 132 NaCl, 5 KCl, 1.3 $MgCl_2$, 1.2 $CaCl_2$, 10 glucose, in 25 Tris-HCl, pH 7.4) or *stimulating* buffer (containing [in m*M*]: 5 NaCl, 132 KCl, 1.3 $MgCl_2$, 1.2 $CaCl_2$, 10 glucose, in 25 Tris-HCl, pH 7.4). Both solutions include 12.5 μCi/ml $^{45}Ca^{2+}$ and the desired concentration of test reagent, in this case 1,25-$(OH)_2D_3$. Uptake is terminated by the aspiration of the influx solution followed by three washes with ice-cold resting buffer. In some cases, it is desirable to include lanthanides in the wash buffer, but the usefulness of this addition should be determined empirically. Cell-associated Ca^{2+} is extracted by a 2-hr incubation of the culture dish with 0.5 *N* NaOH and measured by liquid scintillation counting. In the initial stages of refining the technique, protein assays should be performed on randomly selected plates to ensure that total protein/dish does not vary by more than 5%.

The depolarization-mediated stimulation (by the high K^+ buffer) of $^{45}Ca^{2+}$ influx obtained by this method when using osteoblastic monolayers should typically be 2.5- to 4-fold over that seen in resting buffer. This calculation is made after subtraction of both values by the background value obtained by adding and immediately aspirating the influx medium ("zero-time" control). The contribution of VSCC activity to the depolarization-mediated influx can be easily assessed by addition of 5–10 μM nitrendipine, an inhibitory dihydropyridine, to the solution containing stimulating buffer. Conversely, stimulation of VSCCs can be triggered by inclusion of 3 μM Bay K 8644 (Calbiochem, La Jolla, CA), a dihydropyridine agonist. It is precisely these experiments that have led to the conclusion that Ca^{2+} entry mediated by VSCCs represents a major influx pathway for osteoblastic cells.

Ion Tracer Assays Using Membrane Preparations

In some cases, it is desirable to measure Ca^{2+} flux in membrane preparations rather than in whole-cell assays. Membranes can be prepared according to standard procedures; a scaled-down version of published procedures has proven successful for several osteoblastic and muscle cell lines.[15] A procedure for measuring membrane Ca^{2+} flux was developed by adaptation of a protocol described for measurement of $^{22}Na^+$ flux.[16] Membranes are equilibrated overnight at 4° with 75 μCi/ml $^{45}Ca^{2+}$. Seven Whatman (Clifton, NJ) 934-AH glass fiber filter disks (2.4-cm diameter) are placed in a row on a strip of parafilm and a double layer of Whatman GF/C glass fiber filter disks 5 mm in diameter is centered onto the first 934-AH disk. The GF/C disks are presoaked with 10 μl of the resting buffer described earlier, after which 10 μl of the $^{45}Ca^{2+}$ equilibrated membrane preparation is added to the center of the GF/C disks. The loaded membranes are washed dropwise with 100 μl of resting buffer such that each drop covers the membranes on the GF/C disk prior to being absorbed into the underlying 934-AH disk. The GF/C disks along with the bound membranes are then transferred onto the next 934-AH membrane, and the washing process repeated through six filter stacks. At this point, essentially all of the extravesicular $^{45}Ca^{2+}$ is washed into the 934-AH membranes, leaving only the loaded $^{45}Ca^{2+}$ on the GF/C disk. The stimulated $^{45}Ca^{2+}$ efflux by 1,25-

[15] M. G. McNamee, *Biotechniques* **7,** 466 (1989).

[16] A. Paraschos, J. M. Gonzalez-Ros, and M. Martinez-Carrion, *Biochim. Biophys. Acta* **691,** 249 (1982).

$(OH)_2D_3$ or related compound is next readily measured by its addition to the resting buffer used for washing on the seventh membrane. An eighth and ninth membrane can be used if it is desired to measure the return to baseline. The number of wash steps prior to measurement of efflux can easily be varied; successful experiments have been obtained with four to six washes. Another useful measure for comparison is provided by the use of the stimulating buffer described earlier to replace resting buffer in the final wash. This depolarizing solution provides a measure of the contribution of depolarization-sensitive ion flux, a characteristic of VSCCs. Again, inclusion of dihydropyridine agonists and antagonists can be used to measure the contribution of VSCCs to the flux patterns observed in isolated membranes.

Measurements with Fluorescent Intracellular Calcium-Sensitive Dyes

The methodology for measurement of intracellular Ca^{2+} levels using calcium-sensitive dyes, such as Fura-2, has been well reviewed.[17] This section therefore focuses specifically on the strategy that may be adopted for use of this technology to assess the contribution of influx of extracellular Ca^{2+} to increases in cytoplasmic Ca^{2+} levels following treatment with calcitropic agents such as 1,25-$(OH)_2D_3$.

During development of a Ca^{2+} signal, an initial influx of Ca^{2+} typically occurs that is mediated by activation of VSCCs alone or in cooperation with receptor-operated channels.[1] Subsequent to this, there often occurs a rise in intracellular Ca^{2+}, often referred to as the Ca^{2+} transient, which includes the second messenger-mediated release of Ca^{2+} from intracellular stores. This Ca^{2+} transient can be followed by the development of a Ca^{2+} *wave* or Ca^{2+} *oscillations.*[1] Emptying of intracellular stores also serves as a signal for further influx of extracellular Ca^{2+} to refill the stores; this entry may be through VSCCs but more frequently involves so-called "capacitative" Ca^{2+} channels that have been identified in numerous cell types.[18]

To identify the contribution of Ca^{2+} influx to development of a Ca^{2+} signal detected using an intracellular dye such as Fura-2, several complementary strategies are recommended. First, the ability of depolarizing solutions of high K^+ or the VSCC agonist Bay K 8644 to trigger a Ca^{2+} transient should be assessed. A negative result does not necessarily mean that Ca^{2+} entry is not stimulated by depolarization, however, because it may only be that the level of Ca^{2+} entering the cell by this route alone is insufficient to be detected by the trapped cytosolic dye. For this, the $^{45}Ca^{2+}$ influx assays described earlier provide the needed information. The second strategy

[17] G. Grynkiewicz, M. Poenie, and R. Tsien, *J. Biol. Chem.* **260,** 3440, 1985.
[18] J. W. Putney, *Cell Calcium* **11,** 611 (1990).

should measure the ability of the test calcitropic agent, such as 1,25-$(OH)_2D_3$, to elevate intracellular Ca^{2+}. A positive control for osteoblastic cells is provided by addition of extracellular ATP, or by inclusion of thapsigargin, both of which efficiently trigger release of Ca^{2+} from stores in osteoblasts.[18] For an agent such as 1,25-$(OH)_2D_3$, which can produce a Ca^{2+} transient in some but not all cells, the requirement for activation of VSCCs in development of the Ca^{2+} signal can be measured by the addition of blocking concentrations of nitrendipine.

In the case of osteoblastic cells including ROS 17/2.8, UMR-106, and MC3T3-E1 cells, development of a Ca^{2+} signal following treatment with calcitropic hormones is strictly dependent on the cells being attached to a substratum. Cell suspensions are useful for measurements of activity of second messengers such as sphingolipids and phospholipids,[19] but are generally unreliable for measures of membrane-initiated hormonal responses. For these measurements, access to a computer-driven single-cell Ca^{2+} imaging system is far superior. Fortunately, this specialized equipment has become more economical.[19] The same general strategy described earlier can be used for measurements of Ca^{2+} influx using either suspended populations or attached single cells treated with calcitropic hormones.

Electrophysiologic Measurements

The last option to be discussed for measurements of Ca^{2+} influx is clearly the most specialized. Electrophysiologic patch clamp recording techniques (both whole-cell attached and single-channel configurations) provide a means of discerning the exact mechanism by which Ca^{2+} entry through VSCCs is stimulated by calcitropic hormones. In the case of osteoblastic cell lines treated with 1,25-$(OH)_2D_3$, it was found that (1) whole-cell inward Ca^{2+} currents were stimulated significantly by treatment with 1 n*M* 1,25-$(OH)_2D_3$; and (2) the increase in Ca^{2+} influx involved an increase in mean channel open time along with a shift in the current–voltage relation toward the resting potential.[13] The exact techniques and solutions used in performing these electrophysiologic measurements with 1,25-$(OH)_2D_3$-related compounds have been detailed in two publications.[13,20]

Structural Features of Vitamin D Analogs Stimulating Calcium Influx

A comprehensive review[21] describes hundreds of compounds structurally related to 1,25-$(OH)_2D_3$, some of which have been used for the elucida-

[19] R. Liu, M. C. Farach-Carson, and N. J. Karin, *Biochem. Biophys. Res. Commun.* **214,** 676 (1995).

[20] S. Yukihiro, G. H. Posner, and S. E. Guggino, *J. Biol. Chem.* **269,** 23889 (1994).

[21] R. Bouillon, W. H. Okamura, and A. W. Norman, *Endocr. Rev.* **16,** 200 (1995).

tion of the structural features of $1,25\text{-}(OH)_2D_3$ that are critical for stimulation of Ca^{2+} influx through the plasma membrane, presumably through a separate receptor system located in the plasma membrane. Although not complete, a picture is emerging of a subset of structural features that is associated with preferential activation of the membrane-initiated response. For example, in contrast to the nuclear receptor for $1,25\text{-}(OH)_2D_3$, the membrane system is fully responsive to compounds in the 6-s-*cis* (steroid-like) conformation. Furthermore, the 1α-OH is not strictly required. Analogs possessing these features provide an excellent experimental tool for dissection of the intracellular pathways that are activated in response to $1,25\text{-}(OH)_2D_3$. The reader is referred to a thorough review of this topic.[22]

Acknowledgments

The authors thank Mr. Ramzi Khoury for critical review of the manuscript. Studies were supported by USPHS grant DE 10318 (to M. C. F.-C.).

[22] A. W. Norman, J. E. Bishop, E. D. Collins, E.-G. Seo, D. P. Satchell, M. C. Dormanen, S. B. Zanello, M. C. Farach-Carson, R. Bouillon, and W. H. Okamura, *J. Steroid Biochem. Molec. Biol.* **56,** 13 (1996).

Section III

Vitamin E

[22] Vitamin E Status and Immune Function

By ALISON BEHARKA, SUSAN REDICAN, LYNETTE LEKA, and SIMIN NIKBIN MEYDANI

Introduction

Vitamin E is the generic name used to describe eight naturally occurring, fat-soluble compounds known as *tocopherols* and *tocotrienols.* Vitamin E has many functions in the body, including protecting cell membranes from oxidative damage.[1] Although vitamin E is a constituent of all cellular membranes, it is found in high concentrations in the membranes of immune cells because they are at especially high risk for oxidative damage.[2–4] In fact, vitamin E is essential for normal function of the immune system.[5,6] A deficiency in vitamin E diminishes the ability of the immune system to respond to infectious microorganisms, to produce a delayed-type hypersensitivity (DTH) reaction, or to mount an antibody response to antigen.[7–9] Under some conditions, a pharmacological level of vitamin E is needed to achieve an optimal immune response.[8,10] In several studies, data have shown that elderly humans, as well as laboratory and farm animals, consuming diets that contain more than five times the recommended dietary allowance (RDA) of vitamin E for their species had significantly increased humoral and cell-mediated immune responses and increased resistance to infectious diseases compared to nonsupplemented controls.[8,10–15] These results suggest

[1] L. Packer and V. E. Kagan, *in* "Vitamin E in Health and Disease" (L. Packer and J. Fuchs, eds.), p. 179. Marcel Dekker, New York, 1993.

[2] A. Coquette, B. Vray, and J. Vanderpas, *Arch. Int. Physiol. Biochem.* **94,** 529 (1986).

[3] L. J. Hatman and H. J. Kayden, *J. Lipid Res.* **20,** 639 (1979).

[4] L. J. Machlin, *in* "Handbook of Vitamins: Nutritional, Biochemical, and Clinical Aspects" (2nd ed.), p. 99. Marcel Dekker, New York, 1991.

[5] S. N. Meydani and R. P. Tengerdy, *in* "Vitamin E: Biochemical and Clinical Applications" (L. Packer and J. Fuchs, eds.), p. 549. Marcel Dekker. New York, 1991.

[6] S. N. Meydani and M. Hayek, *in* "Proceedings of the International Congress on Nutrition and Immunity," p. 105. ARTS Biomedical Publishers, St. Johns, Newfoundland, 1992.

[7] M. L. Scott, *Fed. Proc.* **39,** 2726 (1980).

[8] R. P. Tengerdy, *Avian Dis. J.* **34,** 848 (1990).

[9] K. N. Jeejeebhoy, *in* "Modern Nutrition in Health and Disease" (M. E. Shils, J. A. Olson, and M. Shike, eds.), p. 805. Lea and Febiger, Philadelphia, 1994.

[10] A. Bendich, G. E. Gabriel, and L. J. Machlin, *J. Nutr.* **116,** 675 (1986).

[11] S. Moriguchi, N. Kobayashi, and Y. Kishino, *J. Nutr.* **120,** 1096 (1990).

[12] R. P. Tengerdy, *in* "Antioxidant Nutrients and the Immune Response" (A. Bendich, M. Phillips, and R. Tengerdy, eds.), p. 103. Plenum Press, New York, 1989.

0076-6879/97 $25.00

that under some conditions the RDA for vitamin E might not be adequate for immunological vigor and health.[16]

The vitamin E requirement for ideal immune function depends on its interactions with other antioxidant and pro-oxidant nutrients, especially with polyunsaturated fatty acids, and on other factors that modulate the immune response, such as age and stress.[5,17] Elderly individuals respond to supplementation levels of vitamin E above the RDA with increased immunological vigor including enhanced DTH responsiveness.[18] Epidemiological studies have shown correlations between high plasma vitamin E levels and decreased incidence of infections in the elderly.[19,20] Recent studies also suggest that vitamin E supplementation above the RDA may improve the immune responsiveness of mice infected with the murine acquired immunodeficiency syndrome (AIDS) virus.[21]

Therefore, conventional methods for determining the RDA, while adequate for arriving at the level of vitamin E required to prevent clinical deficiency symptoms, may not adequately predict the optimal level of vitamin E needed to maintain immunological health. Because vitamin E is involved in the maintenance of immune function, which appears to be especially sensitive to changes in vitamin E status, it may be possible to utilize selected immune parameters as indicators of vitamin E status. This chapter provides protocols to study selected cell-mediated immune responses that are influenced by vitamin E status. In addition, quality control measures for human studies involving immune parameters are discussed.

General Quality Control Measures for Human Studies

Many measurements of immune function are subject to interpretation. Therefore, reliable test results can only be achieved by assay standardization. This includes, but is not limited to, the following criteria:

---

[13] A. Bendich, *Ann. N.Y. Acad. Sci.* **587,** 168 (1990).

[14] S. N. Meydani and J. B. Blumberg, *in* "Micronutrients in Health and in Disease Prevention" (A. Bendich and C. E. Butterworth, Jr., eds.), p. 289. Marcel Dekker, New York, 1991.

[15] M. Meydani, S. N. Meydani, L. Leka, J. Gong, and J. B. Blumberg, *FASEB J.* **7,** A415 (1993).

[16] L. M. Corwin and R. K. Gordon, *Ann. N.Y. Acad. Sci.* **393,** 437 (1982).

[17] M. L. Eskew, W. J. Scheuchenzuker, R. W. Scholz, C. C. Reddy, and A. Zarkower, *Environ. Res.* **40,** 274 (1986).

[18] S. N. Meydani and J. B. Blumberg, *in* "Nutritional Modulation of the Immune Response" (S. Cunningham-Rundles, ed.), p. 223. Marcel Dekker, New York, 1992.

[19] M. Chavance, G. Brubacher, B. Herbeth, *in* "Lymphoid Cell Functions in Aging" (A. L. Dewick, ed.), p. 231. Eurage, Interlaken, 1984.

[20] M. Chavance, G. Brubacher, B. Herbeth, G. Vernhers, T. Mistacki, F. Dete, C. Fournier, and C. Janot, *in* "Nutritional Immunity and Illness in the Elderly" (R. K. Chandra, ed.), p. 137. Pergamon Press, New York, 1985.

[21] Y. Wang, D. S. Hwang, B. Licing, and R. R. Watson, *J. Nutr.* **124,** 2024 (1994).

1. Potential subjects should be carefully screened before acceptance into a study. Exclusion criteria include most chronic medical conditions such as diabetes, cardiovascular or neoplastic diseases, calcium channel blocker therapy, steroidal or nonsteroidal antiinflammatory drug use, smoking or excessive alcohol use, nutrient supplementation.
2. At each visit, subjects should be asked a standardized set of questions regarding recent illness, medication, vaccinations, and stress. Protocols should be developed before the study begins specifying the conditions under which a subject may be excluded from the study. For example, a subject must not receive a flu vaccine less than 1 month before an immune test. In addition, the recorded answers to standardized questions may allow for interpretation of unusual results when the data are analyzed.
3. Theoretically, performance and evaluation of all tests should be done by the same technician; however, this may not be possible for large studies. During a longitudinal study, the subjects may need to be divided among more than one technician. Within a cross-sectional study, alternate technicians may be utilized as long as they have been identically trained and demonstrate low variability in replicating each other's observations.
4. Any test kits, reagents (especially mitogens), and media should be purchased from the same lot from the same company.
5. Appropriate controls, both positive and negative, must be included every time a test is performed.
6. A reliable baseline value for the parameter being tested must be determined for each subject. This will require multiple sampling. For example, three baseline blood draws within 2 weeks of each other would be utilized for tests involving blood mononuclear cells.
7. Common sense should be used at all times. For example, DTH skin test antigens should not be applied at sites that are difficult to assess due to pigmentation, inflammation, or extreme hairiness.

Cell-Mediated Immune Responses

General

Cell-mediated immunity is an immune response specific to antigens and mediated by lymphocytes and macrophages with minor participation by other cell types. Cellular immunity is responsible for DTH reactions, foreign graft rejection, resistance to many pathogenic microorganisms, and tumor immunosurveillance.[22]

[22] M. S. Meltzer and C. A. Nacy, *in* "Fundamental Immunology" (W. E. Paul, ed.), 2nd ed., p. 765. Raven Press, New York, 1989.

Delayed-Type Hypersensitivity Reaction

The DTH reaction is used as an *in vivo* assay to determine cell-mediated immune function. DTH is an antigen-specific, T-cell-dependent, recall response manifested as an inflammatory reaction that reaches peak intensity approximately 24–48 hr after antigenic challenge in primed animals. It is the kinetics of the inflammatory process following deposition of antigen in the skin that gives this response its name. Whereas allergic reactions occur within minutes and immune complex reactions occur within a day, DTH reactions peak at 2 days.[22,23]

The DTH test involves introducing a relatively small amount of soluble antigen into the epidermis and superficial dermal tissue by needle puncture. Circulating T cells sensitized to the antigen from prior contact react with the antigen in the skin and induce a specific immune response, which includes mitosis (blastogenesis) and the release of soluble mediators (cytokines). Concurrently, antigen is encountered and processed by macrophages. The macrophages are activated to produce and release cytokines such as interleukin 1 (IL-1) and tumor necrosis factor (TNF). The processed antigen is presented by macrophages to T lymphocytes, which then produce cytokines such as IL-2 and interferon γ (IFN). The cytokines produced by the activated macrophages and T cells are involved in the inflammation associated with a DTH reaction. The intensity of the overall dermal inflammation reaches its peak 24–48 hr after antigen application and is resolved within days or weeks.[22]

Measuring the intensity of DTH involves quantitating some aspect of the local inflammatory response. In humans and guinea pigs, this is readily done by measuring the redness and induration of an area of shaven skin. This approach has failed in mice; however, alternative assays that involve measuring changes in tissue thickness of the footpad with a caliper or determining cellular infiltration have proven reliable.[22–24]

DTH as a tool for assessing immunocompetence *in vivo* has been widely used for many years.[25] In humans, DTH was traditionally evaluated by the intradermal reactions observed following Mantoux-like techniques. However, this method has the potentially serious side effect of boosting DTH reactivity after repeated testing.[25,26] An alternative method that results in

[23] Y. Luo and M. E. Dorf, *in* "Current Protocols in Immunology" (J. E. Coligan, A. M. Kruisbeek, D. H. Margulies, E. M. Shevach, and W. Strober, eds.), p. 4.5.1. John Wiley & Sons, New York, 1995.

[24] M. A. Vadas, J. F. A. P. Miller, J. Gamble, and A. Whitelaw, *Int. Arch. Allergy Appl. Immunol.* **49,** 670 (1975).

[25] B. M. Lesourd, A. Wang, and R. Moulias, *Ann. Allergy* **55,** 729 (1985).

[26] E. C. Keystone, P. Demerieux, and D. Gladman, *Clin. Exp. Immunol.* **40,** 202 (1980).

minimal booster effect on DTH response is Multi-test CMI (Merieux Institute, Inc., Miami, FL), a multipuncture system that permits simultaneous application of seven antigens and a negative control.[27]

Multi-Test CMI for the Determination of DTH in Human Subjects

1. Select a test site. Preferred sites are the volar surfaces of the arms and the back. The site should be easily accessible for reading. Cleanse the site with ethanol and allow the area to dry. Identify the area by drawing lines above and below the test site.

2. A single-use disposable applicator is used to deliver several recall antigens and a control antigen into the epidermis and superficial dermal tissue by puncture. Commonly used recall antigens and respective concentrations include tetanus toxoid antigen (biologically equivalent to 5.5×10^5 Merieux tetanus units), diphtheria toxoid antigen (1.1×10^6 Merieux diphtheria units), *Streptococcus* antigen (2000 Merieux *Streptococcus* units), tuberculin antigen (3×10^5 U.S. tuberculin units), *Candida* antigen (2000 Merieux *Candida* units), *Trichophyton* antigen (150 Merieux *Trichophyton* units), and *Proteus* antigen (150 Merieux *Proteus* units). A negative control must also be applied and should contain the same solution that serves as a vehicle for the skin test antigens. In most cases, the vehicle will be a 70% (w/v) sterile glycerin solution.

3. Test sites are evaluated at both 24 and 48 hr. The larger reaction recorded from the two readings is used. The size of an indurated area is determined by inspection and palpation followed by measurements across two diameters at right angles. These measurements are averaged to determine mean diameter. An induration of 2 mm or greater at a test site is considered positive, providing there is no induration at the negative control site.

4. An antigen score is calculated as the total number of positive antigens, and the cumulative score is calculated as the total diameter of induration of all the positive reactions.

Interpretation of Delayed-Type Hypersensitivity Results. Delayed cellular hypersensitivity is a valuable measure of immune response because it involves a complex series of immunologic, cellular, mediator-associated, and vascular effects. Although numerous *in vitro* tests have been developed to assess T-cell activity, these *in vitro* assays do not necessarily monitor the same cells that mediate DTH nor do they consider the influences of regulatory events that may influence *in vivo* T-cell function. Therefore, local DTH

[27] W. T. Kniker, C. T. Anderson, J. L. McBryde, M. Roumiantzeff, and B. Lesourd, *Ann. Allergy* **52,** 75 (1984).

responses represent an important source of information concerning *in vivo* T-cell function.[22,23]

Clinically, DTH skin tests are of value in the overall assessment of immunocompetence. The inability of human adults to react to a battery of common skin antigens suggests hyporeactivity or anergy (nonresponsiveness to antigens). Cutaneous anergy may indicate functional impairment of or abnormalities in the cellular immune system. However, DTH is not an absolute determinant of immune system dysfunction, and any such interpretation must be avoided. DTH may be diminished or absent when there is *in vitro* evidence that T-cell function remains intact and when antibody-associated immunity and phagocytic function appear normal. Reactivity to DTH test antigens may decrease or disappear temporarily as a result of febrile illness; measles and other viral infections; or live virus vaccination including measles, mumps, and rubella. It is also possible to observe loss of reactivity in patients undergoing treatment with drugs such as corticosteroids or procedures that suppress immunity. Moreover, recent infections and vaccinations influence the magnitude of the response. A lack of previous exposure to the administered antigen will result in a negative DTH response to that antigen.[22,23,25]

Quality Control Measures for Multi-Test CMI[27]

1. All applicator kits should be from the same lot.
2. The same technician should administer and read the test results (see the General Guidelines section).
3. Sterility of the test antigens is essential: opened or damaged applicators should be discarded; and the selected test site should be cleaned with ethanol prior to application; the applicator should never be reused.
4. Systemic reactions can occur in those persons sensitive to allergenic media components. Subjects who react severely to any of the test antigens should only be tested after the test heads containing the problem antigens have been removed.
5. Individuals may acquire skin testing sensitivity resulting from either immunization or infection.

Effect of Vitamin E Status on Delayed-Type Hypersensitivity Response. The most widely accepted index of *in vivo* immune function is DTH.[28] DTH can be used as a predictor of morbidity and mortality in the elderly.[29]

[28] Report of an ILS/WHO Working Group, *Clin. Exp. Immunol.* **46,** 662 (1981).
[29] S. J. Wayne, R. L. Rhyne, P. J. Garry, and J. S. Goodwin, *J. Gerontol. Med. Sci.* **45,** M45 (1990).

Christou *et al.*[30] showed that in preoperative patients a significant correlation exists between DTH and mortality from sepsis. Improved DTH response in hospitalized patients has been associated with a decrease in sepsis and mortality.[31] Wayne *et al.*[29] showed that healthy elderly subjects who were anergic when tested for DTH had significantly higher mortality from all causes in subsequent years than those subjects with higher DTH responses.

Supplementation with vitamin E above the RDA can increase DTH response in some situations. This is especially true for elderly persons. DTH responses are significantly diminished in the elderly, but supplementation with vitamin E improves DTH responsiveness in both older laboratory animals and humans.[5,32] Old mice supplemented with 500 ppm vitamin E for 30 days had significantly higher DTH in response to 2,4-dinitrofluorobenzene than mice fed the RDA.[32] Short-term supplementation of healthy elderly persons with 800 mg vitamin E for 30 days increased DTH scores.[33] In addition, long-term supplementation with 60, 200, or 800 IU of vitamin E a day increased DTH scores over time.[34]

In the elderly, plasma vitamin E levels have been positively correlated with positive DTH responses to a variety of antigens, and epidemiologic studies indicate a lower incidence of infectious disease in elderly subjects with high plasma tocopherol levels.[20,35] Population groups maintaining high plasma tocopherol levels have also been noted to possess a lower incidence of cancer.[36–38]

Lymphocyte Responsiveness: Proliferation and Interleukin 2 Production

The cell-mediated immune responses involving lymphocyte functions are very sensitive to changes in vitamin E level. This section describes

[30] N. V. Christou, J. Rodriguez-Tellado, L. Chartrand, B. Giannas, B. Kapadia, J. Meakins, and H. Rode, *Ann. Surg.* **210,** 69 (1989).

[31] J. D. Fletcher, G. A. Koch, and E. Endres, *Aust. N.Z. J. Surg.* **56,** 17 (1986).

[32] S. N. Meydani, M. Meydani, C. P. Verdon, A. C. Shapiro, J. B. Blumberg, and K. C. Hayes, *Mech. Ageing Dev.* **34,** 191 (1986).

[33] S. N. Meydani, M. P. Barklund, S. Liu, M. Meydani, R. A. Miller, J. G. Cannon, F. D. Morrow, R. Rocklin, and J. B. Blumberg, *Am. J. Clin. Nutr.* **52,** 557 (1990).

[34] S. N. Meydani, M. Meydani, J. B. Blumberg, L. S. Leka, G. Siber, R. Loszewski, C. Thompson, M. C. Pedrosa, R. D. Diamond, and B. D. Stollar, *JAMA* **277,** 1380 (1997).

[35] M. Chavance, G. Brubacher, and B. Herbeth, *in* "Lymphoid Cell Functions in Aging" (A. L. DeWick, ed.), p. 231. Eurage, Paris, 1984.

[36] D. Trickler and G. Shklar, *JNC* **78,** 1615 (1987).

[37] M. S. Menkes, G. W. Comstock, J. P. Vuilleumier, K. J. Helsing, A. A. Rider, and R. Brookmezer, *N. Engl. J. Med.* **315,** 1250 (1986).

[38] R. M. Bostick, J. D. Potter, and D. R. McKenzie, *Cancer Res.* **53,** 4230 (1993).

procedures for assessing the basic immunologic function of lymphocytes: measurement of proliferative responses and cytokine production to a variety of stimuli.

Proliferation Assay

General. Measurement of proliferative responses by lymphocytes is a fundamental technique for the assessment of their biological responses to various stimuli. The assessment of cellular proliferation is perhaps the most often used technique in cellular immunology. Lymphocytes proliferate in response to mitogens, antigens, allogeneic or autologous cells, and many soluble factors. Quantitating the proliferative response involves measuring the number of cells present in a culture before and after the addition of a stimulating agent. Although cellular proliferation can be assessed by determining the increase in viable cell numbers through their actual enumeration using direct microscopy or automated cell counting, this approach can be time consuming. Therefore, the two most commonly used techniques are reduction of tetrazolium dyes (MTT) by active mitochondria or incorporation of tritiated thymidine into newly synthesized DNA.[39] Cell proliferation determined by estimating incorporation of [^{3}H]thymidine into DNA is a process that is related to underlying changes in cell number. As cells enter the S phase of the cell cycle, chromosome replication takes place through the incorporation of soluble nucleotide precursors into newly synthesized DNA. In this assay, dividing cells are incubated (pulsed) with radioactive thymidine for several hours after which the amount of radioactivity incorporated into their DNA is determined.[39]

Proliferative assays are applied in clinical studies as an assessment of overall immunologic competence of lymphocytes, as manifested by their ability to respond to proliferation signals. Defects in proliferation may be indicative of a fundamental cellular immunologic defect. Low proliferation is often found as a nonspecific secondary effect of chronic disease.[39]

Procedure for Mitogen-Induced Proliferation of Peripheral Blood Mononuclear Cells Utilizing Incorporation of [^{3}H]Thymidine into DNA

1. Peripheral blood mononuclear cells (PBMCs) are separated by density gradient from heparinized blood. Several procedures can be used to accomplish this.[40]

2. PBMCs are washed twice in RPMI 1640 medium supplemented with 100 mg/liter penicillin, 100 mg/liter streptomycin, 100 ml/liter heat-

[39] R. Fernandez and V. Vetvivka, "Methods in Cellular Immunology." CRC Press, Boca Raton, Florida, 1995.

[40] A. I. Boyum, *Scand. J. Clin. Invest.* **97** (Suppl. 21), 77 (1968).

inactivated autologous plasma, L-glutamine (final concentration 2 mmol/liter), and 25 mmol HEPES (final concentration 25 mmol/liter) (complete RPMI). Cells are resuspended in complete RPMI and counted under a light microscope. Cell viability is assessed by using the trypan blue exclusion method. Cell concentration is adjusted to 1×10^6 cell/ml with complete RPMI.

3. One hundred microliters of the cell suspension is pipetted into each well of a 96-well plate (1×10^5 cells/well). Triplicate wells are prepared for each experimental condition. Wells with no mitogen are included to measure background response.

4. Dilutions of mitogens between 1 and 100 μg/ml for PHA and concanavalin A (Con A) and 0.015% to 0.15% (w/v) for *Staphylococcus aureus* are prepared in complete RPMI. Optimal mitogen concentrations should be predetermined for each individual experimental condition. One hundred microliters of the diluted mitogen solution is added to each well.

5. Plates are incubated for 72 hr in a humidified 37°, 5% (v/v) CO_2 incubator.

6. Four hours before termination of incubation, wells are pulsed with 20 μl of 25 μCi/ml [^{3}H]thymidine.

7. Cells are harvested using an automated harvester, which aspirates the wells, lyses the cells, and retains DNA on glass filter paper, while allowing unincorporated [^{3}H]thymidine to wash through.

8. Filters are air dried and placed into a liquid scintillation cocktail.

9. Counting is done using a liquid scintillation counter.

10. Results are reported as corrected counts per minute (cpm): the average cpm of mitogen-stimulated cultures minus the average cpm of cultures without mitogens. Alternatively, proliferation data can be expressed as the stimulation index, calculated by dividing the proliferation of the stimulated cells in cpm by the background proliferation.

Interpretation of Proliferation Assay Results. The relative ease of acquiring cells and measuring lymphocyte proliferation has made this assay a popular measure of immune function. In many cases, the proliferation response appears to be correlated with the vitamin E content of the diet. However, results from this assay must be critically analyzed before a recommendation about the vitamin E status of an individual is made. Cell proliferation in response to external stimuli is a very complex process often involving delivery of a signal or set of signals to cell membranes, activation of intracellular enzymatic pathways which are not well understood, activation and transcription of multiple genes, DNA and protein synthesis, and, finally, cell division. The mechanism of how vitamin E status affects the proliferative response is not clear.

Experimental conditions must be thoroughly worked out for the proliferation assay to be of value. Cell populations used in this protocol must have high viability and contain an appropriate cell population for the response being tested. The composition of the culture medium, and especially the serum source, can be critical.[41] The selection of mitogen is important. There can be considerable variability in individual responses to particular activating agents; for example, a small portion of the normal human population has a low or absent response to pokeweed mitogen. In addition, the level of mitogen utilized becomes very important in the interpretation of proliferative responses.

Effect of Vitamin E on Mitogen-Induced Lymphocyte Proliferation. The level of *in vitro* proliferation by lymphocytes in response to mitogens is influenced by vitamin E level. Lymphocytes derived from animals maintained on vitamin E-deficient diets show depressed mitogenic response to T-cell mitogens, a defect that is reversible following vitamin E supplementation.[42] Eskew *et al.*[43] found that either vitamin E or selenium deficiency suppressed rat lymphocyte proliferation. Low levels of dietary vitamin E, which were sufficient to protect against many of the adverse effects of vitamin E deficiency, were insufficient to enhance lymphocyte proliferative responses even though the diet contained all other nutrients at recommended levels.

Addition of vitamin E above the recommended dietary levels enhances mitogenesis in mixed populations of lymphocytes from humans and a variety of animal species compared to controls.[5] Cells from healthy elderly humans consuming a diet supplemented with 800 IU/day of vitamin E demonstrated increased mitogenic responses to optimal doses of Con A compared to cells from elderly subjects consuming the RDA.[33] Moriguchi *et al.*[11] reported that as dietary intakes of vitamin E increased from 100 to 2500 mg/kg of diet, there was a corresponding increase in rat splenic lymphocyte responses to Con A. Lymphocytes from mice supplemented with varying levels of vitamin E only showed enhanced Con A proliferation when dietary levels were in excess of the vitamin content of normal chow diet.[44]

Interleukin-2 Production

General. The T-cell growth factor, IL-2, is a lymphokine produced by T-helper type 1 cells following induction by mitogens such as lectin or by

[41] S. P. James, *in* "Current Protocols in Immunology" (R. Coico, ed.), p. 7.10.1. John Wiley & Sons, New York, 1995.

[42] A. Bendich, *in* "Antioxidant Nutrients and Immune Functions" (A. Bendich, M. Phillips, and R. P. Tengerdy, eds.), p. 36. Plenum Press, New York, 1990.

[43] M. L. Eskew, R. W. Scholz, C. C. Reddy, D. A. Todhunter, and A. Zarkower, *Immunology* **54,** 173 (1985).

[44] M. Corwin and J. Shloss, *J. Nutr.* **110,** 916 (1980).

a specific antigen.[45] IL-2 induces proliferation and differentiation of both T and B cells and thus, to a major degree, it regulates the magnitude and duration of the immune response. In addition to promoting the proliferation of cells, IL-2 induces the release of cytokines such as interferon gamma from activated T lymphocytes.[46] Vitamin E level in the diet has been shown to influence IL-2 production. It has been hypothesized that vitamin E may exert some of its immunostimulatory effects by indirectly increasing IL-2 production.[32]

Expression, synthesis, and secretion of IL-2 can be investigated by a variety of techniques. The following assay can be used to determine biologically active, secreted IL-2 in culture supernatants or biological fluids. This assay utilizes clones of the murine CTLL cell line first described by Gillis *et al.*[47] Other cell lines that can be used include HT-2, FDC-2, or MT-1. CTLL cells proliferate in the presence of IL-2, and to a lesser extent, in the presence of murine IL-4[48] and IL-15.[49] Proliferation can be measured by using [^{3}H]thymidine or MTT.

Procedure for Production of Interleukin 2 in Vitro. Cells are cultured in the presence or absence of phorbol myristic acetate (PMA), PHA, or Con A in 12-well plates. Plates are incubated in a 5% CO_2 incubator at 37°. Following 48 hr of incubation, supernatants are collected and stored at −20° until analyzed.

Procedure for Detection of Biologically Active Human Interleukin 2[39,50]

1. The murine IL-2-dependent T-cell line, CTLL-2, is used to monitor IL-2 production. Prior to conducting the assay, logarithmically growing cells are washed three times with media to remove any residual IL-2.

2. Washed CTLL-2 cells are cultured at a concentration of 5×10^3 cells/well (100-μl well) in 96-well microtiter plates in the presence of serially diluted test or control supernatants (100-μl well).

3. The recombinant IL-2 standard and samples are diluted in complete RPMI. Culture supernatants are diluted 1:1–1:4. The IL-2 standard is titrated such that the least dilution allows the cells to achieve maximum DNA synthesis (positive control), usually at 20–40 U/ml (this is dependent

[45] K. A. Smith, *Science* **240,** 1169 (1988).

[46] E. S. Kimball and P. M. Grob, *in* "The Lymphocyte: Structure and Function" (J. J. Marchalonis, ed.), p. 71. Marcel Dekker, New York, 1988.

[47] S. Gillis, M. M. Ferm, W. Ou, and K. A. Smith, *J. Immunol.* **120,** 2027 (1978).

[48] R. Fernandez-Botran, P. H. Krammer, T. Diamanstein, J. W. Uhr, and E. S. Vitetta, *J. Exp. Med.* **164,** 580 (1986).

[49] K. H. Grabstein, J. Eisenman, K. Shanebeck, C. Rauch, S. Srinivasan, V. Fung, C. Beers, J. Richardson, M. Schoenborn, M. Ahdieh, L. Johnson, M. R. Alderson, J. D. Watson, D. M. Anderson, and J. G. Giri, *Science* **264,** 965 (1994).

[50] L. S. Davis, P. E. Lipsky, and K. Bottomly, *in* "Current Protocols in Immunology" (R. Coico, ed.), p. 6.3.1. John Wiley & Sons, New York, 1995.

on the brand of recombinant IL-2 used). A typical standard curve ranges from 20 U/ml down to 0.3 U/ml recombinant IL-2.

4. CTLL-2 cells cultured in the absence of IL-2 serve as background proliferation (negative control).

5. After an 18-hr incubation at 37° in 5% (v/v) CO_2 incubator, wells are pulsed with 0.5 μCi [^{3}H]thymidine. After 6 hr, the cells are harvested onto glass fiber filters with an automatic cell harvester and are then dried.

6. Isotope uptake (cpm) is determined using a liquid scintillation counter.

Interpretation of Interleukin-2 Bioassay Results. We have described a biological assay for IL-2. This bioassay has the advantage of being more sensitive than many of the other available techniques, and it can differentiate biologically active forms from inert forms of the cytokine. However, care must be taken to rule out the involvement of contaminating cytokines. Almost all the cloned murine cell lines used to measure IL-2 activity are also capable of responding to murine IL-4 and IL-15.[48,49,51]

Other compounds present in serum can potentially influence the outcome of an IL-2 bioassay.[39,51] CTLL cells are susceptible to inhibitors (e.g., cytokine antagonists, soluble cytokine receptors) present in biological fluids. In addition, stimulants that may be in the samples (e.g., mitogens) may have unforeseen effects on the CTLL responses. Therefore, appropriate controls must always be included. Controls should include negative samples such as medium alone, and medium plus any stimulants. Antibodies against IL-4 and IL-15 should be included in samples containing these cytokines in addition to IL-2. Appropriate controls must be included in every assay because CTLL cells are subject to assay-to-assay variability in their responsiveness to IL-2.

Effect of Vitamin E Status on Interleukin 2 Production. IL-2 production can be influenced by the level of available vitamin E. Elderly individuals produce less IL-2 than their young counterparts. This age-induced reduction in IL-2 production can be partially reversed by vitamin E supplementation above the RDA. For example, Meydani *et al.*[32] reported that older mice fed diets containing 500 ppm vitamin E had enhanced IL-2 production compared to mice fed a diet containing 30 ppm.[32] In addition, elderly healthy humans supplemented with 800 IU/day dl-α-tocopherol for 30 days had increased Con A-induced IL-2 production versus nonsupplemented subjects.[33]

[51] J. L. Rossio and A. J. H. Gearing, *in* "Clinical Applications of Cytokines: Role in Pathogenesis, Diagnosis and Therapy" (J. J. Oppenheim, J. Rossio, and A. Gearing, eds.), p. 16. Oxford University Press, New York, 1993.

Microbial disease can reduce IL-2 production. Human immunodeficiency virus (HIV)-positive persons and mice infected with murine acquired immunodeficiency (MAIDS) produce less IL-2 than uninfected controls. Vitamin E supplementation over the murine requirement increased IL-2 production by splenocytes from mice infected with the retrovirus that causes MAIDS.[52]

Mononuclear Phagocytic Cell Functions Affected by Vitamin E Status

General. Mononuclear phagocytic cells are often called *accessory cells* and include monocytes and macrophages. An accessory cell helps lymphocytes, enabling them to carry out their function of recognition and elimination of antigens. Accessory cells that can present antigens to lymphocytes are called antigen-presenting cells and include mononuclear phagocytes. These cells play a crucial role in the immune system through their functions as antigen-presenting cells and as regulatory cells, which synthesize and release soluble immune-enhancing and -suppressing cytokines. Because of the diversity of their functions, mononuclear phagocytes are key cellular elements in the immunological control of a number of diseases.

The mononuclear phagocyte is a highly secretory cell. The variety of secreted molecules is large and includes cytokines that are involved in the acute inflammatory response as well as lymphocyte response. Macrophage secretion of several cytokines appears to be affected by vitamin E level. However, the literature has been inconsistent in characterizing this effect. Two cytokines that consistently change concentration in response to vitamin E are TNF-α and IL-6.

Cytokine Production of Interleukin-6 and Tumor Necrosis Factor-α. IL-6 is produced by many cell types in response to a variety of immunological or pathological stimuli. The most potent sources include endotoxin-stimulated monocytes and macrophages and IL-1-stimulated fibroblasts. There are many assays available to detect IL-6, including commercially available enzyme-linked immunosorbent assay (ELISA) kits. Biologically active IL-6 can be detected using IL-6-dependent murine hybridomas. The most commonly used hybridoma lines are B-9 and 7TD1.[53,54] The general bioassay procedure was described previously in more detail for IL-2. Briefly, the hybridomas are maintained in recombinant IL-6, washed prior to assay,

[52] Y. Wang, D. S. Huang, S. Wood, and R. R. Watson, *Immunopharmacology* **29,** 225 (1995).

[53] J. Van Snick, S. Cayphas, A. Vink, C. Uyttenhove, P. Coulie, and R. Simpson, *Proc. Natl. Acad. Sci. U.S.A.* **83,** 9679 (1986).

[54] L. A. Aarden, E. R. De Groot, O. L. Schaap, and P. M. Lansdorp, *Eur. J. Immunol.* **17,** 1411 (1987).

and the cells are combined with the samples to be analyzed. Proliferation is measured 3 days later by [^{3}H]thymidine incorporation.[39]

TNF-α is produced by activated macrophages and other cells and has a broad spectrum of biological actions on many different target cells, both immune and nonimmune. TNF is considered a major inflammatory mediator. TNF can be measured by both ELISA and biological assay. There are many commercially available ELISA kits for detecting TNF. The most commonly used bioassay for TNF is a cytotoxicity assay based on murine fibroblasts such as L929 or L-M cells. These cells deteriorate and die over a 3-day period in the presence of TNF. Sensitivity levels vary with cell isolates, passage number, and temperature.[55]

EFFECT OF VITAMIN E STATUS ON TUMOR NECROSIS FACTOR-α AND INTERLEUKIN-6 SECRETION BY MONONUCLEAR PHAGOCYTES. Infection with the retrovirus that causes AIDS in humans and MAIDS in mice results in elevated levels of IL-6 and TNF-α as the disease progresses.[52,56,57] Supplementation over the murine requirement with vitamin E appears to lower the retrovirus-induced production of IL-6 and TNF in mice.[21,57] Wang *et al.*[21,52] reported that mice infected with MAIDS and supplemented with vitamin E at 15–450 times the NRC recommendation showed normalization of IL-6 and TNF-α produced by splenocytes during progression to MAIDS at all levels of vitamin E supplementation. Moreover, aging affects IL-6 and TNF-α production. Depending on the experimental conditions, aging decreased or increased production of IL-6 and TNF.[58,59] Vitamin E supplementation of elderly human subjects inhibited IL-6 production and prevented the exercise-induced increase in TNF-α production by peripheral blood mononuclear cells.[60,61]

Dietary alcohol (ethanol) consumption decreases production of IL-6 and TNF-α. When a 15-fold increase over the mouse NRC-recommended level of vitamin E was supplied to ethanol-fed mice for 10 weeks, production of IL-6 by Con A-stimulated splenocytes, and IL-6 and TNF-α by LPS-stimulated splenocytes was restored.[62]

[55] M. M. Hogan and S. N. Vogel, *in* "Current Protocols in Immunology" (R. Coico, ed.), p. 6.10.1. John Wiley & Sons, New York, 1993.

[56] F. Boue, C. Wallon, C. Goujard, F. Barresinouss, P. Galand, and J. F. Defraissy, *J. Immunol.* **148,** 3761 (1992).

[57] L. Y. Wang, D. S. Huang, P. T. Giger, and R. R. Watson, *Adv. Biol. Sci.* **86,** 335 (1993).

[58] R. B. Effros, R. L. Walford, R. Weindruch, and C. Mitcheltree, *J. Gerontol.* **46,** B142 (1991).

[59] J. G. Cannon, S. F. Orencole, R. A. Fielding, M. Meydani, S. N. Meydani, M. A. Fiatarone, J. B. Blumberg, and W. J. Evans, *Am. J. Physiol.* **259,** R1214 (1990).

[60] J. G. Cannon, S. N. Meydani, R. A. Fielding, M. A. Fitarone, M. Meydani, M. Farhangmehr, S. F. Orencole, J. B. Blumberg, and W. J. Evans, *Am. J. Physiol.* **260,** R1235 (1991).

[61] Y. Wang, D. S. Huang, P. T. Giger, and R. R. Watson, *Alcohol. Clin. Exp. Res.* **18,** 64 (1994).

[62] S. M. Wahl, J. B. Allen, S. Gartner, J. M. Orenstein, M. Popovic, D. E. Chenoweth, L. O. Arthur, W. L. Farrar, and L. M. Wahl, *J. Immunol.* **142,** 3553 (1989).

Phagocytosis and Chemotaxis

Phagocytosis is the engulfment of cells, microorganisms, and particulate materials. Phagocytosis is very important in a variety of circumstances including the clearance of microbial infections and in the daily turnover of senescent tissues and cells. Chemotaxis is important because the phagocytes must be able to move to where they are needed. Factors that inhibit macrophage migration can have a deleterious effect on host protective function.[63] Phagocytosis can be measured a variety of ways. The following protocol is used for the measurement of Fcγ receptor-mediated binding and phagocytosis.[64]

PHAGOCYTOSIS ASSAY.

1. Sheep red blood cells (SRBC) are maintained in Alsever's solution (1:1, v/v) at 4°.
2. Five milliliters SRBC are placed into a 15-ml conical centrifuge tube, diluted to 15 ml with physiological saline, and centrifuged at 300*g* for 10 min at room temperature. Discard supernatant and repeat wash two times. Count using a hemacytometer. Resuspend SRBC at 1×10^8 cells/ml in saline.
3. Opsonization is accomplished by incubating 10 ml of SRBC suspension with 200 μl of rat anti-SRBC antiserum (heat inactivated: optimum concentration previously determined) for 60 min at 37°.
4. Radiolabeling of the opsonized SRBC is accomplished by incubation with 200 μCi of sodium [^{51}Cr]chromate for 1 hr at 37°. The opsonized ^{51}Cr-labeled SRBC cultures are then washed three times with RPMI to remove excess ^{51}Cr, and the final volume is adjusted to give a 0.6% suspension of SRBC.

Prepare macrophage/monocyte samples.

1. Purified macrophages/monocytes are washed two times by adding 10 ml PBS, centrifuging at 1200 rpm for 10 min at 4°, and discarding supernatant. Cells are resuspended in complete RPMI to a final concentration of 2×10^6 cells/ml.
2. One hundred microliters of cell suspension is added to each well of a 96-well flat bottom tissue culture plate. Prepare at least three replicate wells for each treatment and include positive and negative controls. Allow cells to adhere (4 hr in a 5% CO_2, 37° incubator), then add treatment (interferon) if needed.
3. The opsonized [^{51}Cr]SRBC cultures are added. After 2 hr at 37°, the

[63] A. D. Politis and S. N. Vogel, *in* "Current Protocols in Immunology" (R. Coico, ed.), p. 14.8.1. John Wiley & Sons, New York, 1995.

[64] R. E. Harris, L. A. Boxer, and R. L. Bahner, *Blood* **55,** 338 (1980).

cultures are rinsed once with distilled water to lyse nonphagocytosed SRBC and then washed twice with 0.1 mol/liter phosphate buffer, pH 7.2. All remaining cells are lysed with 0.1 mol/liter NaOH.

4. The radioactivity of the lysate is measured in a gamma counter. The stimulation index is calculated by assigning a value of 1 to the phagocytic activity of control macrophages incubated with medium only, then by comparing this value to the phagocytic activity of macrophages from each treatment.

EFFECT OF VITAMIN E ON PHAGOCYTIC CELL RESPONSES. The effects of dietary vitamin E on phagocytic cell function have been well documented. Rats fed vitamin E-deficient diets exhibit impaired macrophage chemotaxis, reduced ability to ingest complement-coated beads, and decreased protection from autooxidative damage.[65] Dietary vitamin E supplementation has been shown to ameliorate endotoxin-induced inhibition of monocyte migration and phagocytosis in rats,[66] to reduce immunosuppressive cytokine production by macrophages in mice, and to normalize monocyte chemotaxis in humans with diabetes.[67] The ability of alveolar macrophages to phagocytose opsonized sheep red blood cells has been reported to increase with increasing concentrations of vitamin E in diets and showed a fivefold increase in rats fed the diet with the highest vitamin E content compared to rats in the control group.[11] Yano and Ichikawa[67] reported that alveolar macrophages from vitamin E-supplemented rats demonstrated enhanced phagocytosis of opsonized particles.

Mechanism

The mechanism underlying the immunostimulatory effect of vitamin E has not been completely delineated.[6] Although some of the effects of vitamin E on immune cell functions can be attributed to its antioxidant activities, other mechanisms appear to be involved.[68,69] It has been hypothesized that one of these alternative mechanisms is the reduction of immunoinhibitory molecules secreted by immune cells themselves, particularly by macrophages. Activated macrophages secrete molecules such as H_2O_2[70,71] and prostaglandin E_2 (PGE_2), which have been shown to depress lymphocyte

[65] N. P. Rocha, *Brazilian J. Med. Biol. Res.* **22,** 1401 (1989).
[66] H. R. Hill, N. H. Augustine, M. L. Rallison, and J. I. Santos, *J. Clin. Immunol.* **3,** 70 (1983).
[67] T. Yano and T. Ichikawa, *Nutr. Res.* **14,** 1387 (1994).
[68] L. M. Corwin and J. Shloss, *J. Nutr.* **110,** 2497 (1980).
[69] H. Sies, *Am. J. Med.* **91,** 31 (1991).
[70] R. I. Fisher and F. Bostick-Bruton, *J. Immunol.* **129,** 1770 (1982).
[71] Z. Metzger, J. T. Hoffeld, and J. J. Oppenheim, *J. Immunol.* **124,** 938 (1980).

proliferation,[70,72,73] lymphokine production,[74] and the generation of cytotoxic T cells.[75] Vitamin E modulates the formation of these potentially immunosuppressive molecules, particularly PGE_2. Peritoneal macrophages and splenocytes from mice supplemented with vitamin E produced less PGE_2 after stimulation than unsupplemented control mice.[32,76] Additional research is needed to understand fully the mechanisms of the effect of vitamin E on immune responsiveness.

Summary

Evidence from animal and human studies indicates that vitamin E plays an important role in the maintenance of the immune system. Even a marginal vitamin E deficiency impairs the immune response, while supplementation with higher than recommended dietary levels of vitamin E enhances humoral and cell-mediated immunity. The current RDA level of vitamin E prevents clinical deficiency syndrome but in some situations, especially in older subjects or in a disease state, fails to maintain optimal host defense. The immunological parameters reviewed are all sensitive to changes in the availability of vitamin E and, therefore, may reflect the vitamin E status of a given individual more accurately than conventional methods.

Acknowledgments

This project has been funded at least in part with federal funds from the U.S. Department of Agriculture, Agricultural Research Service, under contract 53-K06-01. The contents of this publication do not necessarily reflect the views or policies of the U.S. Department of Agriculture, nor does mention of trade names, commercial products, or organizations imply endorsement by the U.S. government. The authors thank Timothy S. McElreavy for preparation of this manuscript.

[72] M. Rola-Plezczunski, *Immunol. Today* **6,** 302 (1985).
[73] D. R. Webb, T. J. Rogers, and E. Nowoiejski, *Proc. Natl. Acad. Sci. U.S.A.* **332,** 260 (1980).
[74] R. P. Phipps, S. H. Stein, and R. L. Roper, *Immunol. Today* **12,** 349 (1991).
[75] M. Plaut, *J. Immunol.* **123,** 692 (1979).
[76] E. H. Romach, S. Kidao, B. G. Sanders, and K. Kline, *Nutr. Cancer* **20,** 205 (1993).

[23] Inhibition of Platelet Adhesion as Functional Test for Vitamin E Status

By MANFRED STEINER

Introduction

The development of thrombotic events in the arterial circulation, for example, the coronary arteries or the cerebral vasculature, is mediated by platelets. Their adherence to sites of vascular endothelial cell discontinuity is usually the initial event. Circulating platelets become activated by contact with an adhesive surface, for example, the collagenous subendothelium of blood vessels, thereby changing their discoid shape into a spherical form with protrusion of long slender filopodia. The latter have the ability to anchor the platelet to a surface, a process that eventually leads to the spreading of the platelet body on the adhesive surface.[1] Agents affecting platelet function have been found to be far better suited to disrupt the development of arterial thromboses than anticoagulants and heparin. Thus, aspirin, a potent inhibitor of cyclooxygenase, strongly inhibits the oxidative conversion of arachidonic acid, released from phospholipids in the course of platelet activation, into thromboxane, thereby effectively inhibiting the propagation of platelet activation. Aspirin, however, does not inhibit activation of platelets produced by interaction with adhesive surfaces.[2] On the other hand, vitamin E is a very potent inhibitor of this process.[3] Studies have shown that supplemental administration of this vitamin alters the activation-induced shape change, specifically the development of normal pseudopodia, and strongly inhibits adhesion of platelets to adhesive surfaces. There was a good correlation between the inhibition of platelet adhesion and the serum level of vitamin E. Therefore, measurement of this platelet function *ex vivo* becomes a feasible test for evaluating the status of vitamin E in humans.

[1] S. F. Mohammad, *in* "Principles of Cell Adhesion" (P. D. Richardson and M. Steiner, eds.), p. 349. CRC Press, Boca Raton, Florida, 1995.

[2] D. D. Dawicki, *in* "Principles of Cell Adhesion" (P. D. Richardson and M. Steiner, eds.), p. 337. CRC Press, Boca Raton, Florida, 1995.

[3] J. Jandak, M. Steiner, and P. D. Richardson, *Blood* **73,** 141 (1989).

METHODS IN ENZYMOLOGY, VOL. 282
0076-6879/97 $25.00

Measurement of Platelet Adhesion

Methods to Evaluate Adhesion

Measurement of adhesion at its simplest is performed by allowing a suspension of cells to settle and attach themselves to the wall of the vessel. After a predetermined time the adherent cells are enumerated. Adhesion is usually documented by showing that the shape of the attached cell has been altered and by demonstrating that it takes a specific force applied to the system to dislodge the cells.[4] Note that adhesion and aggregation are distinct processes, the former denoting attachment of a cell to a surface other than that of a like cell, which can be maintained in the presence of a finite disturbing force. Aggregation, however, refers to the attachment of cells to like cells, usually in the fluid phase of blood or of cell suspensions.

In the past it has been assumed that adhesion was a one-time phenomenon for a specific site and involved four distinct phases: adsorption, contact, attachment, and spreading.[5] The minimum force required for detachment can at least theoretically be assessed by applying a shearing flow over the attached cells and quantifying the necessary shear stress to remove them. Directly pulling the cell from its attachment site is yet another method to evaluate adhesive forces.[6] Cell-to-cell encounters induced by Brownian motion, shear enhanced laminar motion, or centrifugation have been recognized as important factors for promoting efficiency of aggregation or adhesion. Centrifugation of platelet suspensions as used in the George test[7] and exposure of platelets to rotating fibrinogen-coated tubes[8] are older tests; more recently the plate–cone viscometer has been used to expose platelets to varying shear stress.[9]

Evaluation of Adhesion in Presence of Flow over Surface

Implantation of substances in the form of rings of material into blood vessels provided a method of measuring the adhesion of platelets and subsequent thrombus formation in a test commonly known as the Gott ring test.[10] The glass bead column is yet another test vehicle measuring aggregation and adhesion phenomena in a system providing nonuniform

[4] P. D. Richardson, *in* "Principles of Cell Adhesion" (P. D. Richardson and M. Steiner, eds.), p. 3. CRC Press, Boca Raton, Florida, 1995.

[5] F. Grinell, *Int. Rev. Cytol.* **53,** 65 (1978).

[6] E. Evans, D. Berk, and A. Leung, *Biophys. J.* **59,** 838 (1991).

[7] J. N. George, *Blood* **40,** 862 (1972).

[8] J. P. Cazenave, M. A. Packham, and J. F. Mustard, *J. Lab. Clin. Med.* **82,** 978 (1973).

[9] P. Y. Huang and J. D. Hellums, *Biophys. J.* **65,** 344 (1993).

[10] V. L. Gott, M. D. Ramos, J. L. Allen, and K. E. Becker, *J. Surg. Res.* **6,** 274 (1966).

flow distribution around the packed beads.[11] To define flow profiles more clearly Petschek *et al.*[12] prepared everted arteries that could be continually observed for evidence of adhesion and aggregation of platelets. This system is now favored by some investigators because of the well-known flow characteristics near a solid surface called the axisymmetric forward stagnation point flow.

Various flow chambers have been developed over the years: some tubular, others flat. Tubular channels do not require as much flow as flat chambers to achieve a certain velocity gradient near the wall. Use of this type of chamber is exemplified by the Baumgartner technique in which an everted blood vessel is mounted around the inner cylinder.[13] Flat flow chambers make it easier to observe larger surface areas in the focal plane of a microscope. Richardson *et al.*[14] have described a flow chamber that, from the standpoint of fluid dynamics, belongs to the Hele–Shaw class of flow chambers. The insertion of a central disk creates variations of local shear rates along the streamlines that are followed by platelets or other cells in the perfusate, which are readily computable on the basis of potential flow theory.[15]

A schematic of this flow chamber and the actual model are shown in Fig. 1. The actual dimensions of the chamber are approximately those of a wide glass microscope slide with recessed margins that leave room for a gasket that provides a vacuum-maintained seal for the upper deck of the chamber. The ability to detach and to coat the surface of the upper deck of the flow chamber with adhesive proteins represents a distinct advantage in the use of this device. A defined area of the surface can be observed under the microscope over extended periods of time and photographic recordings of the events can be made. Not only can platelet-rich plasma (PRP) be perfused through the chamber, but whole blood is also suitable for this system. Use of the former is more advantageous when trying to define the characteristics of platelet adhesion from a kinetic standpoint and for tracing the life history of individual platelets adhering to specific sites.

It has been observed that platelets have a tendency to depart from sites of adhesion after variable lengths of occupancy and that they are

[11] C. R. Robertson and H. N. Chang, *Ann. Biomed. Eng.* **2,** 361 (1974).

[12] H. Petschek, D. Adams, and A. R. Kantrowitz, *Trans. Am. Soc. Artif. Intern. Organs* **14,** 256 (1968).

[13] H. R. Baumgartner, *Microvasc. Res.* **5,** 167 (1973).

[14] P. D. Richardson, S. F. Mohammad, and R. G. Mason, *Proc. Eur. Soc. Artif. Organs* **4,** 175 (1977).

[15] G. K. Batchelor, "An Introduction to Fluid Dynamics." Cambridge University Press, Cambridge, UK, 1967.

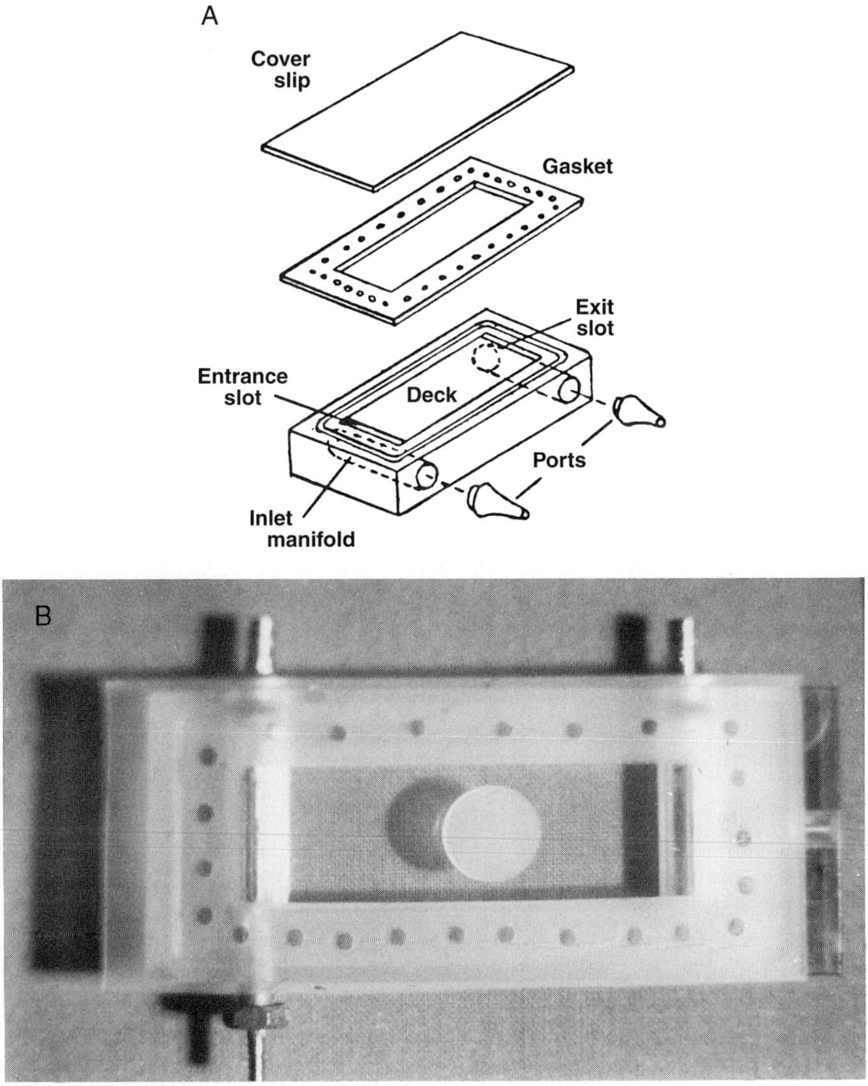

FIG. 1. (A) Schematic drawing of the flat, parallel wall Hele–Shaw type flow chamber. The chamber, which consists of a 70- × 30-mm machined piece of methyl methacrylate has a 43- × 14-mm flow channel with a 0.4-mm gap between the upper deck and the base. A silicone rubber disk, 10 mm in diameter and of the same depth as the channel gap, is placed in the middle of the chamber (if variable shear rate measurements are to be performed in the assay). The glass slide coated with an (adhesive) protein of choice, which constitutes the upper deck of the chamber, is held in place by a perimeter vacuum pump. A silicone rubber gasket with perforations seals the chamber. Inflow and outflow ports are at either end of the chamber. An additional port opposite the inlet may be used for measuring temperature or for recording flow rate. (B) The actual flow chamber.

subsequently replaced by newly arriving cells.[16] A distinct predilection to attach themselves to sites that have been previously occupied by platelets was recognized. By choosing different pumps for propelling the perfusate through the flow chamber one can change the flow from continuous to pulsatile, the latter mimicking perfusion through blood vessels under *in vivo* conditions. Such flow chambers thus are extremely versatile and adaptable to the particular need of an investigator.

Methodology for Evaluating Platelet Adhesion Using Laminar Flow Chamber

Platelet-Rich Plasma as Perfusate

Platelet-rich plasma is prepared from whole blood that is usually collected in sodium citrate. Other anticoagulants such as heparin or specific inhibitors of thrombin (e.g., hirudin or D-phenylalanyl-L-prolyl-L-arginine chloromethyl ketone) can also be used. A number of different centrifugation methods for the separation of PRP have been described. Although plasma-suspended platelets are usually the preferred perfusate for obtaining time-resolved observation on platelet adhesion, one can also use platelets suspended in buffer media that may be prepared by passage of PRP over a Sepharose column. For optimal results, the platelet suspension should be allowed to "rest" for a period of time (e.g., 30 min) before adhesion measurements are begun.

Whole Blood as Perfusate

Whole blood is more suitable for evaluating solely the quantitative aspect of platelet adhesion. The rate of transport of platelets to the flow chamber wall is greatly enhanced by the presence of red cells, roughly a hundred times the transport rates expected on the basis of platelet diffusivity assessed from their molecular weight. Whole blood is anticoagulated as described earlier. An obvious difference between whole blood versus PRP as perfusate of the laminar flow chamber is the problem of being unable to make direct observations of the adherent platelets in the presence of red cells. Introduction of a washing step with an isosmotic buffer medium or platelet-poor plasma readily clears the contaminating red cells from the field of observation.

[16] P. D. Richardson, R. Kane, and K. Agarwal, *Trans. Am. Soc. Artif. Intern. Organs* **27,** 203 (1981).

Performance of Flow Study

The flow chamber shown in Fig. 1B is primed by filling the chamber with isosmotic solutions, for example, lactated Ringer's solution, pH 7.2, containing 4 U heparin/ml. This procedure ensures that there are no air bubbles present in either the inflow line or the chamber and also eliminates the creation of an air–blood interface. If PRP is the perfusate of choice, it is usually loaded into a plastic syringe of appropriate volume (e.g., 35 ml) that is connected by silicone rubber tubing with a 1/8-inch internal diameter to the inlet port of the flow chamber. For optimal performance, both syringe and inflow tubing leading to the chamber should be surrounded by jackets that can be perfused with water at 37°. The prefilled flow chamber with the upper deck glass plate coated with the protein of choice and held in place by suction (provided by an aspirator filter pump or a regular laboratory vacuum pump) is placed on the stage of an inverted microscope. For analyses of the temporal profile of platelet adhesion and the determination of the individual life history of platelets as well as kinetic measurements of the adhesion process a Hoffman modulation contrast system (Modulation Optics Inc., Greenvale, NY) can be used to obtain optimal photomicrographic documentation. This optical system provides an almost three-dimensional image of cells and thus facilitates recognition of platelets.

PRP or whole blood is pumped through the chamber at the desired flow rate using either a syringe infusion pump or a roller pump. If the rate of perfusion is relatively slow, the flow chamber may be drained by gravity through the outflow line into a waste container. If a high flow rate is desired, that is, to generate high shear rates, it may be more convenient and economical to recirculate the perfusate emerging from the outflow port of the flow chamber through the water-jacketed pump reservoir. This does not influence the adhesion measurements in a recognizable manner because the overall deposition of platelets on the adhesive surface is small compared to the total number of platelets in the perfusate. When whole blood is the intended perfusate the chamber is set up in a manner similar to that used for the preparation with PRP described above. Practice is required to achieve a tight seal on the flow chamber. Careful observation of the gasket holding the upper deck of the flow chamber in place is advisable throughout the perfusion run to detect leaks early.

To record the adhesion process, one can use a camera mounted on the microscope. We routinely make observations in an area that covers 70,000 μm^2, which is approximately 1/8000 of the total chamber area available for adhesion. Shear rates can be calculated according to a variety of methods but we have found those described by Batchelor for Hele–Shaw cells most convenient.[15] If a central disk is placed in the flow chamber, it is possible to make observations on the effect of varying shear rates on adhesion on the same slide. To correlate the adhesion at a particular location, the wall shear rate has to be calculated at that position using the above-mentioned method. If time-resolved measurements of adhesion are to be performed, a 30-sec interval between photographic frames will capture ≥80% of the adhesion events that actually take place. Thus ≤20% of the potentially adherent platelets may attach and depart within the 30-sec interval and therefore escape our observation. When documenting the adhesion process photographically, it is important to focus on the plane in which the platelets are stationary and adhere to the surface. Moving platelets will be blurred.

Quantification of Adherent Cells

Various methods are available to quantify adherent platelets. We have used computer-aided analyses of the individual photomicrograph frames resulting from each experimental run.[3] Projection of the frames (approximately 12 × 18 cm) onto a digitizing tablet and use of a transparent crosshair cursor allows one to click on individual platelets. With this system the average diameter of a projected platelet is about 3 mm. Because the resolution limit of the digitizing tablet is less than that, it is possible to recognize abutting platelets as separate entities. This system makes it possible to perform a site-by-site spatial and temporal analysis of adherent platelets. We have used a computer program written in BASIC by Kane that gives the following information for each discrete time period: (1) occupied sites, that is, the total number of sites currently occupied by platelets; (2) fresh adhesion sites, that is, the number of currently occupied sites that were not occupied during the immediately preceding time period; (3) new sites, that is, the number of currently occupied sites that had never been occupied until the present time; (4) cumulative sites, that is, a running total of fresh adhesion sites.[17] From these data, it is possible to determine sites of adhesion that were occupied once, twice, three times, or more and thus provide an indication of the relative reuse of sites by adherent platelets.

When whole blood is the perfusate, qualitative investigations of the type mentioned previously are generally not possible; only quantitative measurements at given points in time can be performed. A wide variety of methods is available to analyze quantitative aspects of platelet adhesion. Although digitization methods can be used to obtain purely quantitative evaluations of platelet adhesion, the increased deposition of platelets on adhesive surfaces when whole blood is perfused makes this method very

[17] R. L. Kane, Master's thesis, Spatio-temporal dynamics of human platelet adhesion in a Hele-Shaw flow chamber. Brown University (1981).

tedious and extremely time consuming. Under these circumstances labeling the platelets with a fluorophore-conjugated antibody to one of the prevalent surface proteins, for example, $gpII_b/III_a$ or gpI_b or gpIX, and subsequent analysis of the fluorescence with an imaging system can readily quantify the adherent platelets. If such systems are not available, it is possible to estimate the number of platelets by eluting the cells (and the protein coating) from the glass slide that constitutes the upper deck of the flow chamber using a detergent solution. From the latter, the proteins can be precipitated, for example, with an equal volume of acetone, to prepare a concentrated solution for resolution by sodium dodecyl sulfate (SDS)–polyacrylamide gel electrophoresis (PAGE). After electrophoretic transfer onto an appropriate membrane, an abundant platelet protein (e.g., actin) can be identified with specific antibody, which can be visualized and densitometrically evaluated by use of chemiluminescence detection methods. Many samples can be analyzed at the same time together with a series of standard platelet concentrates for calibration.

[24] Inhibition of Plasma Cholesterol Ester Hydroperoxide and Phosphatidylcholine Hydroperoxide Formation as Measures of Antioxidant Status

By NORIKO NOGUCHI and ETSUO NIKI

Introduction

With increasing evidence demonstrating the involvement of the oxidation of lipids, proteins, and deoxyribonucleic acid (DNA) induced by free radicals and active oxygen species in a variety of diseases and aging, the role of antioxidants has received renewed attention. The results of *in vitro* experimental studies, clinical therapeutic results, and epidemiological data support the beneficial effects of antioxidant vitamins and related biofactors.[1] It is also known that cigarette smoking, one of the typical examples of oxidative stress, lowers the plasma level of antioxidants. Accordingly, it is important and desirable to measure the extent of oxidative stress and defense potency by antioxidants. Various methods have been devised and applied for such purposes.[2]

[1] N. I. Krinsky and H. Sies, eds. Antioxidant vitamins and β-carotene in disease prevention. *Am. J. Clin. Nutr.* **62,** Suppl. 12995 (1995).
[2] W. A. Pryor and S. S. Godber, *Free Rad. Biol. Med.* **10,** 177 (1991).

0076-6879/97 $25.00

The oxidation of lipids produces lipid hydroperoxides and cyclic peroxides as primary products, as well as secondary products such as alcohols, aldehydes, epoxides, ketones, and carboxylic acids.[3] The products of protein oxidation are less well understood than those of lipids and are only poorly characterized.[4] The formation of hydroperoxides, carbonyl compounds, and disulfide has been observed following protein oxidation.[4,5] The oxidation of DNA yields both single- and double-strand breakage and numerous oxygenated products of DNA bases.[6] Of these, 8-hydroxyguanine and thymine glycol are the most well characterized.

Analyses of plasma lipid oxidation products have been extensive, and many methods for their quantitation have been devised and applied. The iodometric titration method has been used for quantitative measurement of total peroxide.[7] This method is specific and quantitative,[8] but may not be sensitive enough for biological fluids. The so-called thiobarbituric acid (TBA) test has been employed to a great degree since it was introduced by Kohn and Liversedge[9] to quantify lipid peroxidation products.[10] Although this method may be useful in certain cases, the TBA test is either specific nor quantitative and TBA values may vary with changes in analytical conditions.[11]

Other methods for measuring hydroperoxides have utilized cyclooxygenase,[12] methylene blue derivatives,[13] Fe(III)-xylenol orange,[14] and specific oxidation products such as 4-hydroxynonenal,[15,16] F_2-isoprostanes,[17] or oxysterols.[18] Hydrocarbon gases such as ethane and pentane[19] and dityro-

[3] N. A. Porter, S. E. Caldwell, and K. A. Mills, *Lipids* **30,** 277 (1995).
[4] E. R. Stadtman, *Annu. Rev. Biochem.* **62,** 797 (1993).
[5] S. Gebicki and J. M. Gebicki, *Biochem. J.* **289,** 743 (1993).
[6] A. P. Breen and J. M. Murphy, *Free Rad. Biol. Med.* **18,** 1033 (1995).
[7] S. M. Thomas, W. Jessup, J. M. Gebicki, and R. T. Dean, *Anal. Biochem.* **176,** 353 (1990).
[8] W. Jessup, R. T. Dean, and J. M. Gebicki, *Methods Enzymol.* **233,** 289 (1994).
[9] H. I. Kohn and M. Liversedge, *J. Pharm. Exp. Ther.* **83,** 292 (1944).
[10] D. R. Janero, *Free Rad. Biol. Med.* **9,** 515 (1990).
[11] H. Kosugi, T. Kojima, and K. Kikugawa, *Lipids* **24,** 873 (1989).
[12] R. J. Kulmacz, J. F. Miller, R. B. Pendleton, and W. E. M. Lands, *Methods Enzymol.* **186,** 371 (1990).
[13] K. Yagi, K. Kiuchi, Y. Saito, A. Miike, N. Kayahara, T. Tatano, and N. Ohishi, *Biochem. Int.* **12,** 367 (1986).
[14] B. L. Gupta, *Microchem. J.* **18,** 363 (1973).
[15] H. Esterbauer and K. H. Cheeseman, *Methods Enzymol.* **186,** 407 (1990).
[16] K. Uchida and E. R. Stadtman, *Methods Enzymol.* **233,** 371 (1994).
[17] J. D. Morrow, K. E. Hill, R. F. Burk, T. M. Nammour, K. F. Badr, and L. J. Roberto, *Proc. Natl. Acad. Sci. U.S.A.* **87,** 9383 (1990).
[18] L. L. Smith, *Lipids* **31,** 453 (1996).
[19] C. M. F. Kneepkens, G. Lepage, and C. C. Roy, *Free Rad. Biol. Med.* **17,** 127 (1994).

sine[20] have also been utilized, as have immunohistochemical techniques.[21,22]

The identification and quantification of different classes of lipid hydroperoxides have also been performed by high-performance liquid chromatography (HPLC). Detection has been accomplished by conjugated diene,[23] chemiluminescence,[24,25] electrochemistry,[26] fluorescence,[27] and iron thiocyanite assays.[28] These methods, employing separation with HPLC followed by postcolumn detection, enable investigators to determine the origin of lipid oxidation products with high sensitivity and have been successfully applied. The identification of regio- and stereoisomers of cholesterol ester hydroperoxides has also been performed using a chiral HPLC column[29]; however, only those products that elute under the HPLC analytical conditions employed have been analyzed. Furthermore, the peroxides measured, due to their rapid clearance and secondary reactions, may only partially reflect the peroxidation taking place. Tissue antioxidant status has also been accepted as a marker of oxidative stress or as a measure of defense potency against oxidative damage. The antioxidant enzymes, vitamins, and related compounds and their redox state have been analyzed as a measure of antioxidant status. Some examples of the techniques used are described next.

Assay of Cholesteryl Ester Hydroperoxide and Hydroxide

Cholesteryl ester hydroperoxide (CEOOH) and hydroxide (CEOH) are separated by an appropriate HPLC column and elution program, and detected by their 234-nm absorbance.[30] The lower limit of detection with ultraviolet absorption is less than 5–10 pmol. This analytical procedure is therefore useful for the measurement of oxidized plasma or lipoprotein.

Samples (100 μl) are extracted by the addition of 2.5 ml of ice-cold methanol and 10 ml hexane in sequence, vortexed vigorously (10 and 30 sec, respectively), and then centrifuged at 1500 rpm at 4° for 3 min

[20] C. Giulivi and K. J. A. Davies, *Methods Enzymol.* **233,** 363 (1994).

[21] G. Jurgens, Q. Chen, H. Esterbauer, S. Mair, G. Ledinski, and H. P. Dinges, *Arterioscler. Thromb.* **13,** 1689 (1993).

[22] H. Itabe, E. Takeshima, H. Iwasaki, Y. Yoshida, T. Imanaka, and T. Takano, *J. Biol. Chem.* **269,** 15274 (1994).

[23] C. G. Crawford, R. D. Plattner, D. J. Sessa, and J. J. Rackis, *Lipids* **15,** 91 (1980).

[24] Y. Yamamoto, M. H. Brodsky, J. C. Baker, and B. N. Ames, *Anal. Biochem.* **160,** 6 (1987).

[25] T. Miyazawa, K. Yasuda, and K. Fujimoto, *Anal. Lett.* **20,** 915 (1987).

[26] J. H. Song, C. O. Chang, J. Terao, and D. K. Park, *Biosci. Biotech. Biochem.* **57,** 479 (1993).

[27] K. Akasaka, T. Suzuki, H. Ohrui, and H. Meguro, *Anal. Lett.* **20,** 731 (1987).

[28] A. Mullertz, A. Schmeds, and G. Holmer, *Lipids,* **25,** 415 (1990).

[29] J. A. Folcik, R. A. Nivar-Aristy, J. P. Krazeuski, and M. K. Cathcart, *J. Clin. Invest.* **96,** 504 (1995).

[30] L. Kritharides, W. Jessup, J. Gifford, and R. T. Dean, *Anal. Biochem.* **213,** 79 (1993).

using a Beckman GS-6R centrifuge (Palo Alto, CA). Samples can also be extracted with two volumes of chloroform/methanol (2/1, v/v).

Eight milliliters of the hexane layer or an aliquot of the chloroform layer is removed, evaporated under vacuum, and then redissolved in 100 μl of HPLC eluent or methanol. The conditions of the HPLC analytical system are as follows:

Column:	Octadecylsilica; LC-18 (Supelco, Tokyo, Japan; 5-μm particle, 0.46 $\times$ 25 cm with 5-cm guard column)
Eluent:	Acetonitrile/2-propanol/water (44/54/2, v/v/v)
Flow rate:	1.0 ml/min
Wavelength:	234 nm

The elution patterns in Fig. 1 show the chromatogram of an extract of an low-density lipoprotein (LDL) suspension that was oxidized by a radical initiator. A subclass of CEOOH and a small amount of CEOH are detected (Fig. 1A). In this system, it is difficult to distinguish between cholesteryl docosahexaenoate hydroxide (Ch22:6-OH) and cholesteryl arachidonate hydroperoxide (Ch20:4-OOH), and between cholesteryl arachidonate hydroxide (Ch20:4-OH) and cholesteryl linoleate hydroperoxide (Ch18:2-OOH). After reduction by triphenylphosphine, only the CEOH species are detected (Fig. 1B).

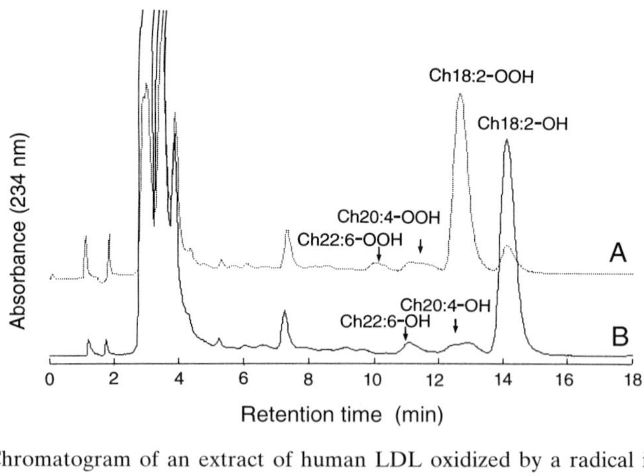

FIG. 1. Chromatogram of an extract of human LDL oxidized by a radical initiator with an octadecylsilyl column using acetonitrile/2-propanol/water (44/54/2, by volume) as an eluent, detected by absorption at 234 nm. LDL suspension was extracted with chloroform/methanol (2/1, by volume) in the absence and presence of 200 μM triphenylphosphine (A and B, respectively). After reduction by triphenylphosphine, Ch22:6-OOH and a large peak of Ch18:2-OOH disappears and that of Ch18:2-OH appears. It is difficult to separate Ch22:6-OH from Ch20:4-OOH and Ch20:4-OH from Ch18:2-OOH, respectively.

When higher sensitivity is needed to detect smaller amounts of hydroperoxide, chemiluminescence is useful.[24,25] Chemiluminescence, which is approximately 50–100 times more sensitive than UV detection, is detected using postcolumn on-line detection with a chemiluminescence detector. The schematic diagram of the chemiluminescence–HPLC system is shown in Fig. 2.

Column:	Octylsilica; LC-8 (Supelco; 5-μm particle, 0.46 × 25 cm with 5-cm guard column)
Eluent:	A: Methanol/*tert*-butanol (19/1, v/v)
	B: 50 m*M* sodium borate, pH 10, 1 m*M* isoluminol, and 5 mg/liter microperoxidase (heme catalyst, MP-11, Sigma, St. Louis, MO) in 50% aqueous methanol solution (v/v)
Flow rate:	A: 1.0 ml/min
	B: 1.5 ml/min

Assay of Phosphatidylcholine Hydroperoxide

Samples are extracted by vortexing vigorously with four volumes of methanol followed by centrifugation at 12,000 rpm (Beckman GS-15R

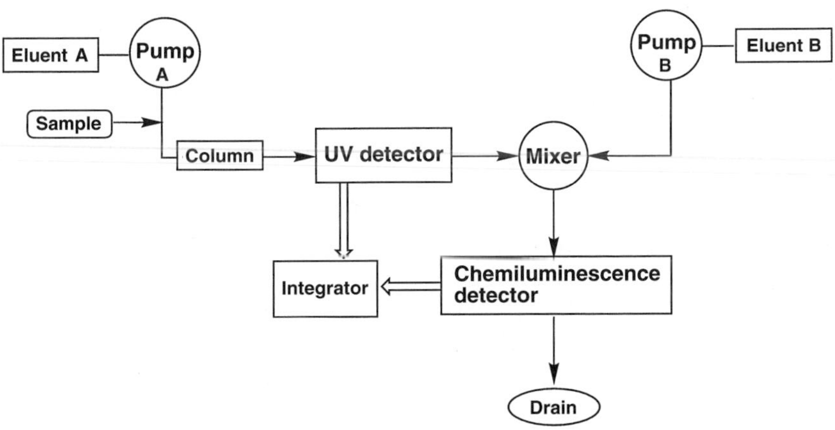

FIG. 2. Schematic diagram of chemiluminescence–HPLC system. Eluent A, containing mobile phase suitable for lipid hydroperoxide, is carried at a flow rate of 1.0 ml/min by pump A. The injected sample is separated with an HPLC column, and conjugated dienes (hydroperoxide + hydroxide) are detected by their 234-nm absorbance with a UV detector. The hydroperoxide is mixed at the T-joint mixer with chemiluminescence reagent containing 50 m*M* sodium borate buffer, pH 10, 1 m*M* isoluminol, and 5 mg/liter microperoxidase in 50% aqueous methanol solution (v/v) pumped at a flow rate of 1.5 ml/min by pump B. The chemiluminescence detector consists of a spiral flow cell and a photomultiplier.

centrifuge) at 4° for 3 min. Phosphatidylcholine hydroperoxide (PCOOH) is detected by a chemiluminescence–HPLC system as follows:

Column: Silica; LC-SI (Supelco; 5-μm particle, 0.46 × 25 cm with 5-cm guard column)

Eluent: A: methanol/*tert*-butanol/40 m*M* monobasic sodium phosphate (6/3/1, v/v/v)

B: 50 m*M* sodium borate, pH 10, 1 m*M* isoluminol, and 5 mg/liter microperoxidase (heme catalyst, MP-11, Sigma, St. Louis, MO) in 50% aqueous methanol solution (v/v)

Flow rate: A: 1.0 ml/min

B: 1.5 ml/min

PCOOH and phosphatidylcholine hydroxide (PCOH) have the same retention time, and the sum of the two is detected by the UV detector. The amount of PCOH is calculated by subtracting the amount of PCOOH from the total coeluting amount of PCOOH and PCOH measured by the detection at 234 nm. PCOOH has been measured with a chemiluminescence–HPLC system by several different analytical conditions.[24,25] Note that the sensitivity of the assay depends on the analytical reagents and also apparatus alterations such as the distance between mixer and cell.[31]

Assay of Vitamin E

α-Tocopherol is the most abundant lipophilic antioxidant in plasma, and many analytical methods have been reported for its determination. The measurement using HPLC separation with electrochemical detection (ECD), which is sensitive and simple, is described here. The extraction methods can be classified as those utilizing direct extraction with appropriate organic solvent or employing a saponification step. Saponification is not necessary for the biological fluids such as plasma or lipoprotein suspensions. The samples are extracted by using methanol and hexane as described for CEOOH and CEOH or two volumes of chloroform/methanol (2/1, v/v). The hexane or chloroform extract is evaporated under nitrogen or under vacuum and redissolved in methanol. The analytical condition of the HPLC-ECD system is as follows:

Column: Octadecylsilica; LC-18 (Supelco; 5-μm particle, 0.46 × 25 cm with 5-cm guard column)

Eluent: 50 m*M* sodium perchlorate monohydrate in methanol/*tert*-butanol (9/1, v/v)

[31] N. Noguchi and E. Niki, *Free Rad. Res.* **23,** 329 (1995).

Flow rate: 0.8–1.0 ml/min
ECD: 800 mV

γ-Tocopherol can be simultaneously detected under these conditions, and the analytical conditions for detection of stereoisomers of α-tocopherol using a chiral column have been reported.[32]

Assay of Ascorbic Acid

Ascorbic acid is one of the hydrophilic antioxidants contained in high concentration in plasma. Ascorbic acid has been known to reduce α-tocopheroxyl radical to regenerate α-tocopherol. Ascorbic acid can be detected with HPLC–UV or HPLC–ECD[33] after separation of protein by precipitation with four volumes of methanol as described for the extraction of PCOOH. Note that ascorbic acid is quite unstable, especially in solution containing metal ions. The analytical condition of the HPLC system is as follows:

Column: Aminopropylsilica; LC-NH_2 (Supelco; 5-μm particle, 0.46 × 25 cm with 5-cm guard column)
Eluent: Methanol/*tert*-butanol (9/1, v/v)
Flow rate: 1.2 ml/min
Wavelength: 265 nm

Assay of Allantoin: Oxidation Products of Uric Acid

Uric acid is another important hydrophilic antioxidant and is present at a concentration of 160–450 μ*M* in human plasma. Analytical procedures for uric acid determination have been reported in the literature,[34] and the procedure for measurement of allantoin, the oxidation product of uric acid, is described.

Plasma is ultrafiltered through a Millipore Ultrafree C-3LTK filter (Nihon Millipore) at 5000 *g* for 30 min, and a 20-μl aliquot of the filtrate is injected directly into an ion-exchange column (SAX-1010, 5-μm particle, 0.45 × 25 cm IRIKA, Kyoto, Japan). The mobile phase is 125 m*M* NaCl containing 1 m*M* K_2HPO_4 adjusted to pH 6.75 with HCl at a flow rate of 0.7 ml/min. A fraction covering the retention time range of 5.5–10 min, where allantoin is known to be eluted, is collected and then evaporated to dryness at 40° under vacuum. The residue is reconstituted in 80 μl of 0.5 *M* NaOH, heated in a boiling water bath for 20 min, then treated with 10 μl

[32] H. Yamaguchi, Y. Itakura, and K. Kunihiro, *Iyakuhin Kenkyu* **15,** 536 (1984).
[33] K. P. Mitton and J. R. Trevithick, *Methods Enzymol.* **233,** 523 (1994).
[34] H. Ogihara, T. Ogihara, M. Miki, H. Yasuda, and M. Mino, *Pediatr. Res.* **39,** 117 (1996).

of 5 *M* HCl followed by 10 μl of a 3 m*M* solution of 2,4-dinitrophenylhydrazine in 1 *M* HCl with heating for 10 min. After the mixture had been allowed to cool, an aliquot is injected onto the HPLC. Glyoxylate 2,4-dinitrophenylhydrazone is detected as the allantoin derivative. The analytical condition of the HPLC system is as follows:

Column:	Octadecylsilica; LC-18 (Supelco; 5-μm particle, 0.46 × 25 cm with 5-cm guard column)
Eluent:	30 m*M* sodium citrate in 27.7 m*M* sodium acetate buffer (pH 4.75)/methanol (7/3, v/v)
Flow rate:	1.0 ml/min
Wavelength:	360 nm

[25] α-Tocopherol-Binding Proteins: Purification and Characterization

By ASIM K. DUTTA-ROY

I. Introduction

α-Tocopherol[1] and γ-tocopherol are the most common of the eight naturally occurring vitamin E homologs (α-, β-, γ-, and δ-tocopherol and α-, β-, γ-, and δ-tocotrienol) in the human diet. The most biologically active of these compounds is α-tocopherol.[2] Although γ-tocopherol is a more effective free radical scavenger than α-tocopherol *in vitro*,[3] the reverse is true *in vivo*.[4] Experimental studies[5–7] have shown limited uptake of γ-tocopherol into peripheral tissues and its rapid clearance from plasma within 2–3 days of eating large amounts of γ-tocopherol. Normally therefore γ-tocopherol, although efficiently absorbed, accounts for only 10–15% of plasma tocopherol.[8] This suggests that there is a selection process discrimi-

[1] This work was supported by the Scottish Office of Agriculture, Environment and Fisheries Department, and the Henkel Corporation, USA.

[2] M. L. Scott, *in* "The Fat Soluble Vitamins" (H. F. Delucia, ed.), p. 133. Plenum Press, New York, 1978.

[3] G. G. Duthie, B. M. Gonzalez, P. C. Morrice, and J. R. Arthur, *Free Rad. Res. Commun.* **15,** 35 (1991).

[4] G. G. Duthie, *Chem. Industr.* **16,** 598 (1992).

[5] G. J. Handelman, L. J. Machlin, K. Fitch, J. J. Weiter, and E. A. Dratz, *J. Nutr.* **115,** 807 (1985).

[6] H. G. Baker, G. J. Handelman, L. J. Machlin, E. A. Bhagavan, E. A. Dratz, and O. Frank, *Am. J. Clin. Nutr.* **43,** 382 (1986).

[7] M. G. Traber and H. G. Kayden, *Ann. N.Y. Acad. Sci.* **570,** 95 (1989).

[8] W. A. Behrens and R. Madere, *J. Am. Coll. Nutr.* **5,** 91 (1986).

0076-6879/97 $25.00

nating against the uptake or accumulation of γ-tocopherol and reducing its *in vivo* effectiveness. α-Tocopherol is the major lipid-soluble antioxidant present in the membranes of cells and cellular organelles, where it plays an important role in the suppression of lipid peroxidation.[9–11] Protection against the peroxidation of membrane lipids by α-tocopherol is dependent on its incorporation into membranes. The extent of this protection is related to the quantity of α-tocopherol present in the membranes.[10] Thus the regulation of α-tocopherol concentrations in the plasma as well as in cell membranes may be important in moderating the free radical-induced events involved in the pathogenesis of diseases in man and animals.[12]

Unlike other fat-soluble vitamins, vitamin E does not have a special plasma carrier protein and is transported in plasma by lipoproteins such as high-density lipoprotein (HDL), low-density lipoprotein (LDL), and very low density lipoprotein (VLDL).[13] Dietary studies indicate that there is little or no discrimination between the vitamin E homologs in the intestine, because tocopherols are absorbed from the gut in micelles whose formation depends on bile salts and pancreatic lipase.[14,15] Micelles containing vitamin E may passively diffuse through the brush border but the mechanism by which vitamin E is then transported across the intestinal epithelial cells is poorly understood. The vitamin is released from the enterocyte into lymph within chylomicrons, which subsequently appear in the circulation where they are catabolized by lipoprotein lipase.[16] Once transported into the liver, there is a clear differentiation between the homologs with α-tocopherol being the exclusive isomer incorporated with VLDL for secretion into the plasma,[17,18] whereas the other isomers are excreted through the biliary canaliculi.[19,20]

Studies on the absorption and transport of various forms of vitamin E indicate that plasma α-tocopherol concentration is regulated specifically by the preferential incorporation into nascent VLDL of α-tocopherol rather

[9] H. Sies and M. E. Murphy, *J. Photochem. Photobiol. B: Biol.* **8,** 211 (1991).
[10] D. J. Kornburst and R. D. Mavis, *Lipids* **15,** 315 (1980).
[11] V. E. Kagan, "Lipid Peroxidation in Biomembranes." CRC Press, Boca Raton, Florida, 1988.
[12] T. Nakamura and F. Masugi, *Int. J. Vit. Nutr. Res.* **49,** 364 (1979).
[13] L. K. Bjornson, J. H. Kayden, E. Miller, and A. N. Moshell, *J. Lipid Res.* **17,** (1976).
[14] H. Gallo-Torres, *Lipids* **5,** 379 (1970).
[15] P. Mathias, J. Harris, T. Peters, and D. P. R. Muller, *J. Lipid Res.* **22,** 829 (1981).
[16] A. Bjorneboe, E. A. Bjornboe, E. Bodd, B. F. Hagen, HN. Kveseth, and C. A. Drevon, *Biochim. Biphys. Acta* **889,** 310 (1986).
[17] J. Davignon, M. Ray, R. Doufour, and G. Roederer, *in* "Primary Hyper-Lipoproteineamias" (G. Steiner and E. Shafrir, eds.), p. 201. Mc-Graw Hill, New York, 1991.
[18] M. G. Traber and H. G. Kayden, *Am. J. Clin. Nutr.* **40,** 747 (1984).
[19] W. Cohn and H. Kuhn, *Ann. N.Y. Acad. Sci.* **570,** 61 (1989).
[20] I. R. Peake, H. Winduller, and J. Bieri, *Biochim. Biophys. Acta* **260,** 679 (1972).

than the other homologs of vitamin E (tocopherols and tocotrienols).[21,22] The hepatic α-tocopherol-binding protein (TBP) or α-tocopherol transfer (α-TTP) of 30-kDa molecular mass is believed to be involved in this regulation.[21,22] The 30-kDa TBP is found only in the liver cytosol in various species.[23–29] Because α-tocopherol is widely distributed in organelles of various tissues but the 30-kDa TBP resides only in hepatocytes,[23–29] its role as a general intracellular carrier of α-tocopherol in different tissues seems unlikely. However, translocation of lipophilic compounds through aqueous compartments is often mediated by specific carrier proteins. The recently identified low molecular mass (~15-kDa) TBP[30–33] in the cytosol of various tissues including liver is suggested to be involved in the intracellular distribution and metabolism of α-tocopherol in tissues.[33]

II. Purification and Characterization of Hepatic 30-kDa α-Tocopherol-Binding Protein

Identification of 30-kDa TBP in Liver Cytosol

The 30-kDa TBP is isolated and purified from human and rat liver.[23–29] Rat liver cytosol is obtained from 33% (w/v) homogenate in a buffer consisting of 0.01 *M* sodium phosphate buffer, pH 7.4, and 0.25 *M* sucrose. Cytosol is prepared from this homogenate by ultracentrifugation at 100,000*g*

[21] H. G. Kayden and M. G. Traber, *J. Lipid Res.* **34,** 343 (1993).

[22] M. G. Traber, *Free Rad. Biol. Med.* **16,** 229 (1994).

[23] N. Kaplowitz, H. Yoshida, J. Kuhlenkamp, B. Slitsky, I. Ren, and A. Stolz, *Ann. N.Y. Acad. Sci.* **570,** 85 (1989).

[24] G. L. Catignani and J. G. Bieri, *Biochim. Biophys. Acta* **497,** 349 (1977).

[25] Y. Sato, K. Hagiwara, H. Arai, and K. Inoue, *FEBS Lett.* **288,** 41 (1991).

[26] H. Yoshida, M. Yusin, I. Ren, J. Kuhlenkamp, T. Hirano, A. Stolz, and N. Kaplowitz, *J. Lipid Res.* **33,** 343 (1992).

[27] G. L. Catignani, *Biochem. Biophys. Res. Commun.* **67,** 66 (1975).

[28] J. Kuhlenkamp, M. Ronk, M. Yusin, A. Stolz, and N. Kaplowitz, *Protein Expr. Purif.* **4,** 382 (1993).

[29] Y. Sato, H. Arai, A. Miyata, S. Tokita, K. Yamamoto, T. Tanabe, and K. Inoue, *J. Biol. Chem.* **268,** 17705 (1993).

[30] A. K. Dutta-Roy, D. J. Leishman, M. J. Gordon, and G. G. Duthie, *Biochem. Biophys. Res. Commun.* **196,** 1108 (1993).

[31] A. K. Dutta-Roy, M. J. Gordon, D. J. Leishman, B. J. Paterson, G. G. Duthie, and W. P. T. James, *Mol. Cell. Biochem.* **123,** 139 (1993).

[32] M. J. Gordon, F. M. Campbell, G. G. Duthie, and A. K. Dutta-Roy, *Arch. Biochem. Biophys.* **318,** 140 (1995).

[33] A. K. Dutta-Roy, M. J. Gordon, F. M. Campbell, G. G. Duthie, and W. P. T. James, *J. Nutr. Biochem.* **5,** 562 (1994).

for 60 min at 4°. The cytosol (~100 ml) is then mixed with 5 μCi α-[^{3}H]tocopherol (~8 Ci/mmol), incubated while stirring for 30 min at 4°, and eluted on a 5- × 100-cm Sephacryl S-200 gel column with 0.01 *M* sodium phosphate, pH 7.4, at 4°. Ten-milliliter fractions are collected and the absorbance at 280 nm and radioactivity of each fraction measured. The second radioactive elution peak corresponding to proteins with molecular mass of 30–40 kDa is used for the purification of the TBP.[23–29] The first peak, corresponding to the void volume, did not bind tocopherol specifically and was discarded.

Affi-Gel Blue Affinity Chromatography

The pooled fraction (180 ml) is then mixed with 100 μl of α-[^{3}H]tocopherol, incubated for 1 hr, and eluted on an Affi-Gel blue column (3 × 40 cm) with 0.01 *M* sodium phosphate buffer, pH 7.4, at 4°.[26] The column is eluted with a linear NaCl gradient by mixing 400 ml of the initial buffer and the same volume of 0.01 sodium phosphate containing 0.55 *M* NaCl and 2 m*M* EDTA. After completion of the gradient, the column is eluted with 5 m*M* NaCl in 0.01 *M* phosphate buffer. Four-milliliter fractions are collected and monitored for radioactivity, absorbance at 280 nm, and conductivity (Fig. 1). The fractions corresponding to each peak of the radioactivity are pooled. Each pool is separately dialyzed in Spectrophor membrane (Spectrum, Houston) (molecular weight cutoff of 12,000) against 0.05 m*M* Tris-HCl (pH 7.5) and concentrated by ultrafiltration using Diaflo YM10 membranes. The pooled fractions from Affi-Gel chromatography is further incubated with α-[^{3}H]tocopherol and subjected to chromatofocusing in a FPCL Mono P HR5/20 column equilibrated with 0.025 *M* Tris-HCl containing 0.02% Triton X-100, pH 6.4, at 4°. The column is eluted with 1:10 diluted Polybuffer 74 containing 0.02% Triton X-100, pH 4.0. One-milliliter fractions are collected and monitored for pH, protein content, and radioactivity. The fractions are immediately neutralized with 100 μl of 0.5 *M* sodium phosphate, pH 7.4. α-Tocopherol-binding activity is resolved into two different peaks, whose isoelectric points are 5.0 and 5.1, respective.[25]

The presence of two isoforms of TBP has been reported in rat liver by different investigators.[25,26] The Western blot, 2D gel electrophoresis, and other data indicate that these are naturally occurring isoforms of TBP and are not the artefacts of protein purification. With regard to human liver TBP, chromatofocusing of the Affi-Gel blue pooled peak revealed a major radioactive peak eluting at pH 5.9, in contrast with the rat liver TBP, which was eluted at pH 5.3.

Microbore HPLC is utilized to further purify the pH 5.9 radioactive peak.[28] Two major protein peaks (TBP-1 and TBP-2) emerge from HPLC

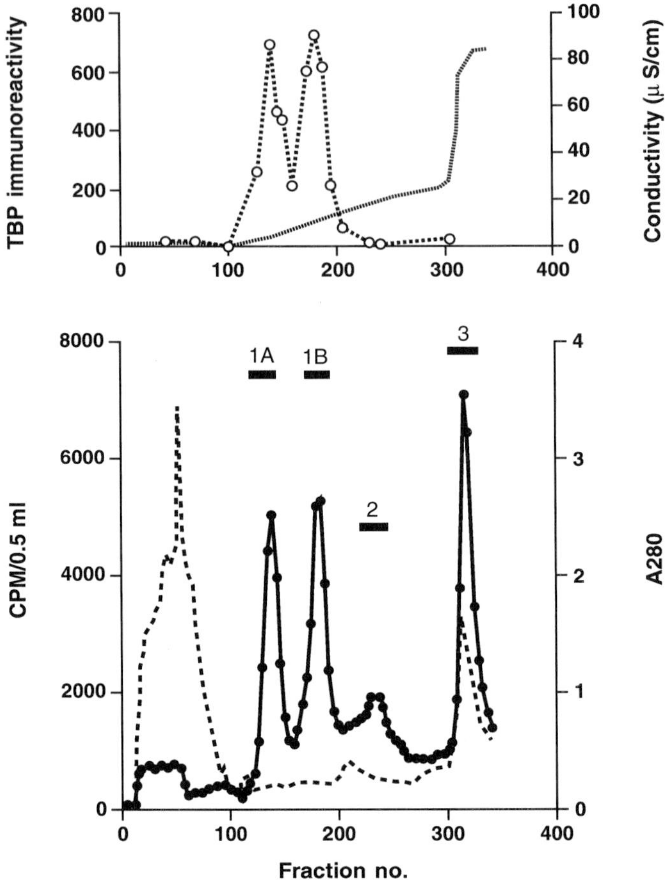

FIG. 1. Affi-Gel blue affinity chromatography of the Y′ fraction of rat liver cytosol. The lower panel shows elution of α-[^{3}H]tocopherol (closed circles) and absorbance at 280 nm (dotted line). The upper panel shows elution of immunoreactivity with anti-TBP antisera (open circles, arbitrary unit) and conductivity (dotted line). The fractions pooled for further purification are indicated by horizontal bars and are designated AFB-1A, 1B, 2, and 3. Using steeper conductivity gradients AFB-1A and AFB-1B appeared as a single peak (AFB-1B), accounting for the nomenclature that was used. [Reprinted by permission of the publisher from H. Yoshida *et al., J. Lipid Res.* **33,** 343 (1992).]

and both exhibit α-tocopherol-binding activity. Rat TBP has striking similarities in composition but differs in content of histidine, methionine, and proline compared to human TBP.[28] TBP represents only 0.2% of total cytosol proteins of liver.[26]

Amino Acid and DNA Sequence and Cloning of TBP Gene of Rat and Human Liver

A cDNA clone for TBP from rat liver is isolated.[29] Polyclonal antibody against rat liver TBP (α-TTP) is used for immunological screening of λgt11 rat liver cDNA library.[29] Plaque hybridization screening of the rat liver λgt11 cDNA library with the positive clone is performed.[29] The inserts are subcloned in pBluescript II SK(−). DNA sequence is determined using a Sequenase DNA sequencing kit with [α^{32}P]dCTP or a *Taq* dye primer cycle sequencing kit on a DNA sequencer.[29] Nucleotide and deduced amino acid sequences are then analyzed using a computer program.[29] The amino acid sequence from the clone contains all sequences determined by direct analysis of the purified protein. The isoform contains 278 amino acid residues having a calculated molecular weight of 31,845. Comparison of the primary structure of rat TBP with those of other known proteins reveals a notable homology with cellular retinaldehyde-binding proteins (CRALBP) from bovine and human (Fig. 2).

Both the isoforms of rat liver TBP are structurally related with respect to amino acid sequence, amino acid composition, and molecular weight and substrate specificity.[29] All of the clones isolated share the same nucleotide sequences in their coding region.[29] It is suggested that both isoforms can be translated from a single mRNA followed by differential posttranslational modification in the liver cells. The isolated cDNA clone has a relatively large 3′-untranslated region. The 3′-untranslated region is believed to have a role in RNA translation, transport, or stability. The region of the isolated cDNA has four AUUUA motifs.[29] AUUUA sequence motifs in mRNA are generally involved in the selective degradation of the messenger, causing marked reduction in mRNA stability.

A fragment of the rat TBP (α-TPP) cDNA insert (λrTPP4-2) is used to screen a human liver λgt11 cDNA library.[34] Screening of ~100,000 clones resulted in the identification of one positive clone, λhTPP 1-2 (Fig. 3). Further screening of the human liver cDNA library using this clone resulted in the isolation of several nearly full-length cDNA encoding TPP (TBP). A total of eight independent clones was isolated, most exceeded 2.5 kb in length; one of them contained sequences corresponding to the N-terminal end of human TBP. The human TBP cDNA contains an open reading frame of 832 bases, which predicts a 31,749-Da protein composed of 278 amino acids.[34] The human TBP cDNA is cloned into the bacterial expression vector, pUCP$_L$-cI under the control of the P$_L$ promoter. The resulting

[34] M. Arita, Y. Sato, A. Miyata, T. Tanabe, E. Takahashi, H. J. Kayden, H. Arai, and K. Inoue, *Biochem. J.* **306,** 437 (1995).

```
α-TTP(38-246)     GVPETPQPLTDAFLLRFLRARDFDLDLAWRLMKNYYKWPAECPELSADLH   87
b-CRALBP(82-290)  AVAERVQGKDSAFFLRFIRARKFHVGRAYELLRGYVNFPLQYPELFDSLS  131
h-CRALBP(82-290)  AVAERVQEKDSGFFLRFIRARKFNVGRAYELLRGYVNFPLQYPELFDSLS  131

                  PRSILGLLKAGYHGVLRSRDPTGSRVLIYRISYWDPKVFTAYDVFRVSLI  137
                  PEAVRCTVEAGYPGVLSTRDKYGRVVMLFIIENWDSEEITFDEILQAYCV  181
                  PEAVRCTIEAGYPGVLSSRDKYGRVVMLFNIENWQSQEITFDEILQAYCF  181

                  TSELIVQEVETQRNGVKAIFDLEGWQTSHAFQITPSVAKKIAAVVTDSFP  187
                  ILEKLLENEETQINGFCIIENFKGFTMQQAAGLRPSDLRKMVDMLQDSFP  231
                  ILEKLLENEETQINGFCIIENFKGFTMQQAASLRTSDLRKMVDMLQDSFP  231

                  LKVRGIHLINEPVIFHAVFSMIKPFLTEKIKGRIHLHGNNYKSSLLQHF-  236
                  ARFKAIHFIYQPWYFTTTYNVVKPFLKSKLLQRVFVHGEEL-SSFYQEFD  280
                  ARFKAIHFIHQPWYFTTTYNVVKPFLKSKLLERVFVHGDDL-SGFYQEID  280

                  PDILPLEYGG                                          246
                  EDILPSDFGG                                          290
                  ENILPSDFGG                                          290
```

FIG. 2. Alignment of the amino acid sequences of the α-TTP (TBP) and bovine and human CRALBP. The sequences of bovine and human CRALBPs have been taken from Ref. 27. One-letter amino acid notation is used. The position number of the residues at the right end of each line corresponds to that in Fig. 3. of this chapter or Ref. 27, respectively, Gaps (−) have been inserted to achieve maximum homology. Residues identical with those of α-TTP are enclosed with a solid line, and residues showing conservative substitution are indicated by shaded. Conservative amino acid substitutions are defined as pairs of residues belonging to one of the following groups: S, T, P, A and G, N, D, E and Q, H, R and K, M, I, L and V, E, Y and W (41). [Reprinted by permission of the publisher from Y. Sato *et al., J. Biol. Chem.* **268,** 17706 (1993).]

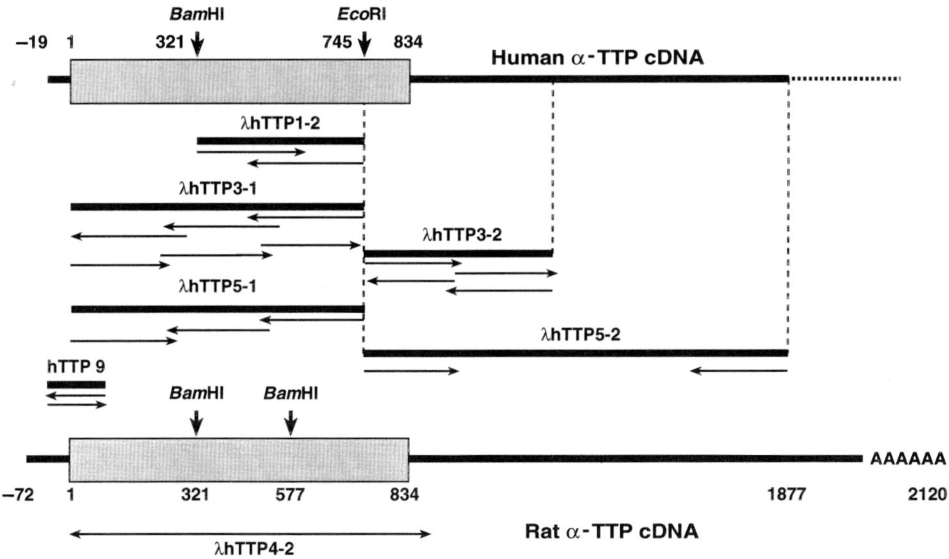

FIG. 3. Sequencing strategy of human α-TTP (TBP) cDNA. The structure of human α-TTP (TBP) is shown above and that of rat TBP cDNA below. The shaded boxes indicate the open reading frame, and the thin bars indicate the 5′- and 3′-untranslated regions. Rat TBP cDNA fragment λrTTP4-2, which was used for screening, is indicated. hTPP9 was obtained by the method of 5′ RACE. Arrows indicate the direction and extent of sequencing. [Reprinted by permission of the publisher from M. Arita *et al., Biochem. J.* **306,** 437 (1995).]

plasmid is introduced into *Escherichia coli* W3110 competent cells. Human recombinant TBP exhibited substantial α-tocopherol transfer activity from liposome to the heavy membrane fraction.[34] Northern blot analysis of TBP is performed using a fragment of TBP cDNA (coding region), hybridized to poly(A)$^+$ RNAs of several human tissues (heart, brain, placenta, lung, skeletal muscle, and kidney). The results showed that TBP is expressed in the liver but not in any other human tissues. In fact both Western and Northern blot analyses revealed that the 30-kDa TBP is expressed exclusively in the hepatocytes and is absent from tissues such as heart, spleen, and lung and from other cells (endothelial, Kupffer cells) in human and rat.[23–29,34]

Like rat liver TBP, human liver TBP also shows remarkable similarity with CRALBP, and also with yeast SEC14p (the SEC14 gene product). SEC14p has phosphatidylinositol/phosphatidylcholine transfer activity and is required for protein secretion through the Golgi complex in yeast.[35]

[35] V. A. Bankitis, J. R. Aitken, A. E. Cleves, and W. Dowhan, *Nature* **347,** 561 (1990).

The chromosomal locus of human liver TBP is determined using a panel of a blot containing *Eco*RI-digested DNA samples from human–human hybrid cells lines as well as human and hamster controls.[34] A fragment of human TBP cDNA (coding Asp-185–Glu-249) is used as a probe for hybridization, and a 5.0-kb genomic fragment, which specifically hybridized to human TBP gene but not hamster counterpart, was detected. The human TBP gene was detected only in hybrid cell lines containing human chromosome 8 indicating that the human TBP gene is localized in chromosome 8. Fluorescence *in situ* hybridization also revealed a single TBP gene corresponding to the 8q13.1–13.3 region of chromsome 8,[34] which is identical to the locus of a clinical disorder, ataxia, with selective vitamin E deficiency.[36,37]

III. Purification, Structure, and Function of Low Molecular Weight (~15-kDa) α-Tocopherol-Binding Protein from the Cytosol of Heart and Liver

Identification of Low Molecular Weight TBP

Heart or liver cytosol is prepared by centrifuging tissue homogenate at 110,000*g* at 4° for 80 min. For identification of α-[^{3}H]tocopherol-binding proteins, 1 ml of the supernatant (~25 mg protein) is incubated with 100 n*M* of α-[^{3}H]tocopherol (specific activity > 55 Ci/mmol) for 30 min at 37° and subjected to gel-permeation chromatography using an FPLC Sephacryl S-300 column (2.6 × 60 cm) preequilibrated and then eluted at 4° with 10 m*M* Tris-HCl buffer, pH 7.4, containing 5 m*M* 2-mercaptoethanol and 100 m*M* KCl and 5% (v/v) glycerol. The flow rate is 0.75 ml/min. Five-milliliter fractions are collected for measurement of protein absorbance at 280 nm and radioactivity.[30–33] Gel filtration of heart cytosol revealed two radioactive peaks of radioactivity, one corresponding to a molecular mass of 12–16 kDa and the other eluted in the void volume (Fig. 4). Similar fractionation of liver cytosol revealed the presence of the same peaks with an additional 30–40 kDa α-[^{3}H]tocopherol-binding fraction (Fig. 4). The 12–15 kDa fraction of both heart and liver, and 30–40 kDa fraction of liver cytosol bound α-[^{3}H]tocopherol whereas there was no α-tocopherol specific binding by the hepatic or heart proteins eluted at the void volumes.[30–33] In

[36] B. M. Hamida, S. Belal, G. Sirugo, C. B. Hamida, K. Panayides, P. Ionannou, J. Beckmann, J. L. Mandel, F. Hentati, M. Koenig, and L. Middleton, *Neurology* **43,** 2179 (1993).

[37] C. B. Hamida, N. Doerflinger, S. Belal, C. Linder, I. Teutenauer, C. Dib, G. Gyapay, A. Vignal, D. Lepaslier, D. Cohen, M. Pandolfo, V. Mokini, G. Novelli, F. Hentati, B. M. Hamida, J. L. Mendel, and M. Koenig, *Nature Genet.* **5,** 195 (1993).

subsequent studies, the peak corresponding to low molecular size fractions (240–310 ml) was used for purification of low molecular weight TBP.

Purification of Low Molecular Weight (~15-kDa) TBP

The supernatant obtained after centrifugation of cytosol at 100,000*g* is treated with 70% saturated ammonium sulfate solution. After centrifugation at 30,000*g* for 30 min, the supernatant is dialyzed against 5 m*M* Tris-HCl buffer, pH 7.4, containing 5 m*M* 2-mercaptoethanol at 4° for 24 hr. The dialyzed fraction is then freeze-dried. The concentrated fraction (~30 mg of protein) is applied to an FPLC Sephacryl S-300 column.[30–33] The fractions that emerged in the elution volume of 240–310 ml are pooled and dialyzed against 5 m*M* imidazole, pH 7.0, containing 5 m*M* 2-mercaptoethanol. The dialyzed protein sample is then freeze-dried and used for further purification of TBP either by using anion-exchange column method (FPLC Mono Q HR column)[30] or by chromatofocusing (FPLC Mono P column).[32] TBP is eluted from a Mono Q HR column using a 20 m*M* 0–2 *M* NaCl linear gradient in the above buffer.[30]

For chromatofocusing, proteins are delipidated using Lipidex 1000 at 37°,[38] before being subjected to chromatofocusing on a FPLC Mono P column (HR 5/20) that had been equilibrated with the starting buffer, pH 7.0 (15 m*M* imidazole, 0.02% sodium azide, 5 m*M* 2-mercaptoethanol). The column is then washed with 8 ml of starting buffer to remove any protein with a p*I* value higher than 7.0. The protein was then eluted from the column with 10% Polybuffer 74 containing 5 m*M* 2-mercaptoethanol, pH 4. One-milliliter fractions are collected, and A_{280} and pH determined.[32]

Chromatofocusing of the 12–18-kDa fractions eluted from the Sephacryl S-300 column resolved into several peaks eluting at different pH. The peak that emerged around pH 4.5 showed α-[^{3}H]tocopherol-binding activity (Fig. 5). The protein peaks that emerged around pH 5.0 to 6.0 were fatty acid-binding protein (FABP) as indicated by Western blot using anti-FABP antibody. The fraction that emerged at around pH 4.5 also showed a single band on sodium dodecyl sulfate–polyacrilamide gel electrophoresis (SDS–PAGE) with silver staining corresponding to a molecular mass of 15 kDa (Fig. 6). Anti-heart FABP antibody did not recognize the heart TBP in Western blots, indicating that it is not a heart FABP (~15 kDa), which may be exclusively involved in the intracellular transport of α-tocopherol. The Western blot analysis showed that antiheart FABP antibody did not recognize the TBP of bovine heart cytosol.

[38] J. F. C. Glatz, A. M. Janssen, C. C. F. Baerwaldt, and J. H. Veerkamp, *Biochim. Biophys. Acta* **837,** 57 (1985).

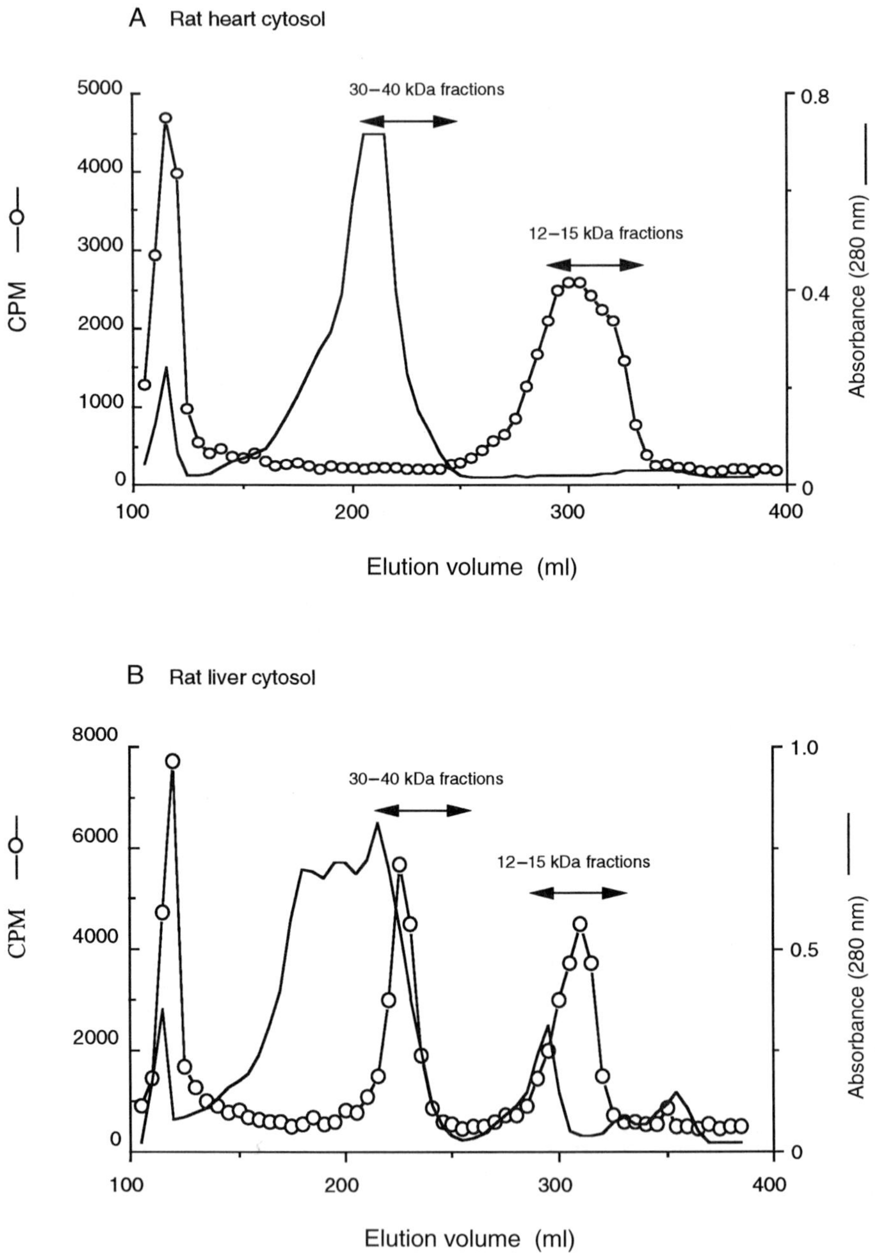
A Rat heart cytosol
30–40 kDa fractions
12–15 kDa fractions
CPM
Absorbance (280 nm)
5000
4000
3000
2000
1000
0
0.8
0.4
0
100
200
300
400
Elution volume (ml)
B Rat liver cytosol
30–40 kDa fractions
12–15 kDa fractions
CPM
Absorbance (280 nm)
8000
6000
4000
2000
0
1.0
0.5
0
100
200
300
400
Elution volume (ml)

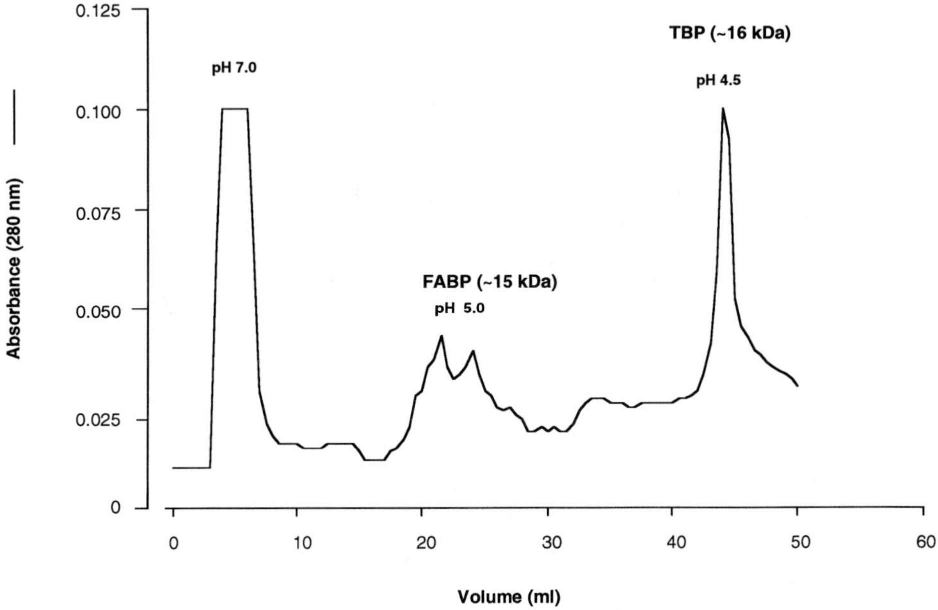

FIG. 5. Isochromatofocusing FPLC of Sephacryl S-300 purified protein fraction. The low molecular mass fraction from the Sephacryl S-300 column is delipidated and then applied to a Mono P column (HR 5/20). The column is equilibrated start buffer (15 m*M* imidazole, 0.02% sodium azide, 5 m*M* 2-mercaptoethanol, pH 7.0). The protein sample is then applied to the column, which was washed with 8 ml of starting buffer to remove any protein with a p*I* > 7.0. The proteins are eluted with Pharmacia (Piscataway, NJ) Polybuffers (pH 7–4). The protein absorbance is measured at 280 nm. [Reprinted by permission of the publisher from M. J. Gordon *et al., Arch. Biochem. Biophys.* **318,** 140 (1995).]

Preparative Isoelectric Focusing on Rotofor System

Because the yield of TBP in the preceding procedure was poor (Table I), a preparative isoelectric focusing step was included using a Rotofor cell system. Approximately 55 mg (2.5 ml) delipidated 70% ammonium sulfate

---

FIG. 4. Gel filtration of rat liver and heart cytosol on Sephacryl S-300 column with α-[^{3}H]tocopherol. α-[^{3}H]Tocopherol (100 n*M*) is incubated with 1 ml of cytosol of liver or heart (~25 mg protein) for 30 min at 37°. After the incubation, the mixture is then applied to an FPLC Sephacryl S-300 column (2.6 × 60 cm) for gel permeation chromatography. The protein is eluted from the column with 10 m*M* Tris-HCl buffer, pH 7.4, containing 5 m*M* 2-mercaptoethanol, 100 m*M* KCl, and 5% glycerol at 4°. Five-milliliter fractions are collected, and A_{280} and radioactivity were determined. (A) Heart cytosol; (B) liver cytosol. [Reprinted by permission of the publisher from A. K. Dutta-Roy *et al., J. Nutr. Biochem.* **5,** 562 . Copyright 1994 by Elsevier Science Inc.]

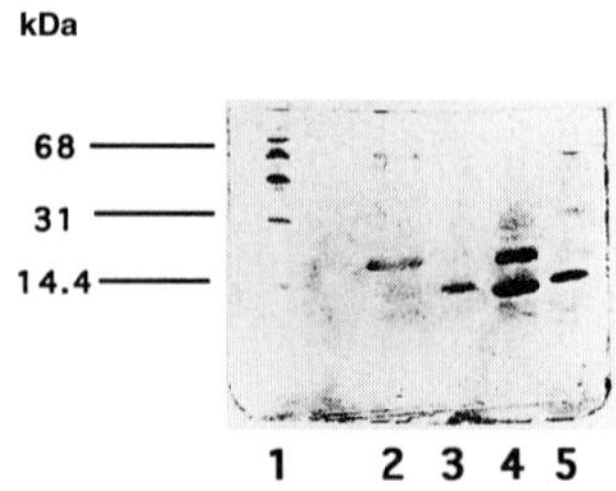

FIG. 6. SDS–PAGE of proteins. Lane 1, molecular mass markers, Phosphorylase *b* (97.4 kDa), bovine albumin (68 kDa), ovalbumin (45 kDa), glyceraldehyde-3-dehydrogenase (35 kDa) (not visible), carbonate dehydratase (31 kDa), trypsin inhibitor (20.1 kDa) (not visible), lysozyme (14.2 kDa); lane 2, purified TBP; lane 3, bovine heart FABP; lane 4, proteins emerged in the void volume; lane 5, recombinant bovine heart FABP. [Reprinted by permission of the publisher from M. J. Gordon *et al., Arch. Biochem. Biophys.* **318,** 140 (1995).]

protein fraction is used for Rotofor fractions (Bio-Rad, Richmond CA) on the linear portion of the pH gradient. The Rotofor cell is prefocused with a 40-ml ampholyte solution containing 1.5% Bio-Lyte (pH 3–10) and 5% glycerol for 1 hr at 15 W constant power. Then 2.5 ml of protein solution is added to the Rotofor cell. Isoelectric focusing in the Rotofor cell required almost 3 hr at 15 W constant power at 4°. Twenty fractions are collected, their pH values determined, and aliquots examined for protein concentration. The yield is significantly increased by using the Rotofor cell fractionation system. TBP appeared in the fraction that emerged at pH 5.0, whereas FABPs emerged at pH > 6.0. The improvement of yield was around 85-fold when compared with that of the chromatofocusing step.

TABLE I
PURIFICATION OF TBP FROM BOVINE HEART[a]

Purification steps	Protein (mg)	cpm/mg Protein (10^{-3})	Purification (-fold)	Yield (%)
110,000*g* Supernatant	2614	0.46	1	100
70% $(NH_4)_2SO_4$ fraction	907.5	1.49	3.2	112
Sephacryl S-300 column	43.35	173	376	623
Mono P column	0.061	404	878	2.10

[a] Reprinted by permission of the publisher from M. J. Gordon *et al., Arch. Biochem. Biophys.* **318,** 140 (1995).

α-[³H]Tocopherol-Binding Assay

Binding of α-[^{3}H]tocopherol to the protein was carried out using a Lipidex-1000 column method[39] with modifications.[30–32] In brief, 10–25 μg of protein is incubated in 50 m*M* Tris-HCl buffer, pH 7.4, 100 m*M* KCl in the presence of 4 n*M* α-[^{3}H]tocopherol for 30 min at 37°, with occasional shaking. Triton X-100 (100 μ*M*) is also included in the incubation mixture to avoid the binding of α-[^{3}H]tocopherol to the vessel wall. After incubation, the assay mixture is applied to a Lipidex-1000 column (2.0 ml) at 4°. The protein is then eluted with the same assay buffer. The fractions (0.5 ml) are analyzed for protein content and radioactivity. The binding of α-[^{3}H]tocopherol to the protein is measured by determining the radioactivity bound to the protein, which appeared in the void volume (Fig. 7). Specificity of α-[^{3}H]tocopherol binding to the purified protein is determined by incubating the protein with 4 n*M* α-[^{3}H]tocopherol in the presence of a 500-fold excess of unlabeled α-, γ-, or δ-tocopherol in the assay mixture. The residual binding of α-[^{3}H]tocopherol to the protein is estimated and the percentage displacement of the bound α-[^{3}H]tocopherol to the protein by individual tocopherol is calculated assuming total binding of α-[^{3}H] tocopherol to the protein in the absence of unlabeled tocopherols as 100%.

The binding of α-[^{3}H]tocopherol to the purified protein was found to be rapid, saturable and reversible. The binding generally approached saturation within 5 n*M* α-[^{3}H]tocopherol. The α-[^{3}H]tocopherol binding attained equilibrium within 30 min of incubation at 37°. Direct binding studies showed that TBP did not bind [^{14}C]oleate. Unlabeled α-tocopherol at 500- or 1000-fold excess could only displace around 55% of the radiolabeled ligand bound to the protein, whereas the γ- and δ-tocopherol did not displace the bound α-[^{3}H]tocopherol suggesting a high specificity for α-tocopherol. The inability to displace the bound α-[^{3}H]tocopherol by any more than 55% may be due to the microaggregation of α-tocopherol molecules at such a high concentration (500- or 1000-fold excess) in the assay mixture, which may interfere with the reversibility of binding. The binding of α-[^{3}H]tocopherol to the purified protein was analyzed by Scatchard plot. Scatchard analysis showed one class of binding sites for α-tocopherol in the protein. The dissociation constant (K_d) of α-tocopherol binding was estimated to be 2.56 ± 0.45 n*M* with a maximum binding capacity (B_{max}) of 0.89 ± 0.01 mole per mole of the purified protein.[32]

[39] A. K. Dutta-Roy, N. Gopalswamy, and D. V. Trulzsch, *Eur. J. Biochem.* **162,** 615, (187).

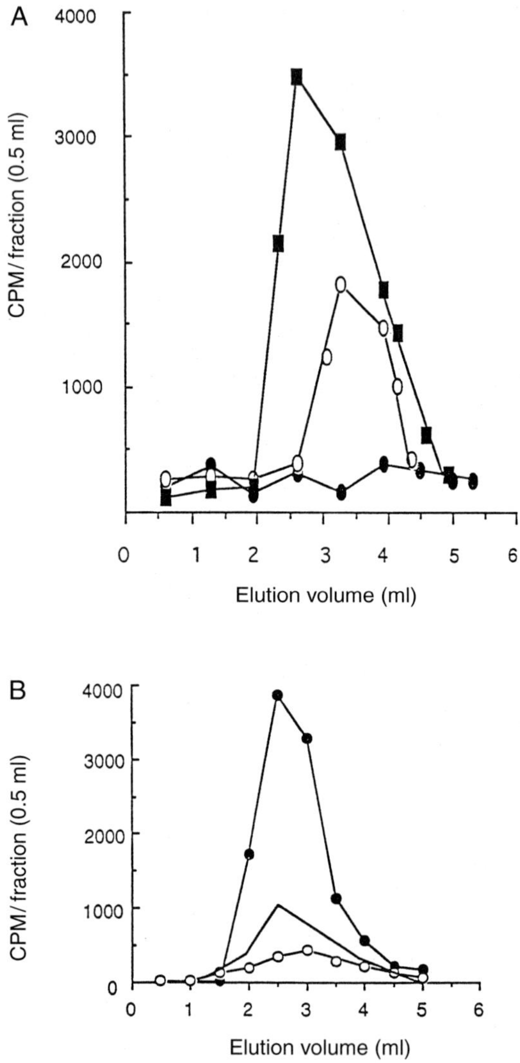

FIG. 7. (A) Binding of α-[^{3}H]tocopherol to the 12–18 kDa (40 μg protein) and 30–40 kDa (25 μg protein) molecular weight fractions of the rat liver cytosol in the Lipidex-1000 column: 30–40 kDa protein fractions (–○–), 12–15 kDa protein fraction (–■–); D-α-[^{3}H]tocopherol only (–●–). (B) Binding of D-α-[^{3}H]tocopherol to the rat heart (6 μg) in the absence and presence of excess unlabeled α-tocopherol. Low molecular weight TBP with radiolabeled α-tocopherol (–●–). The 14.2-kDa TBP with radiolabeled α-tocopherol plus 500-fold excess unlabeled α-tocopherol (———), D-α-[^{3}H]tocopherol only (–○–). [Reprinted by permission of the publisher from A. K. Dutta-Roy *et al., Biochem. Biophys. Res. Commun.* **196,** 1108 (1993).]

Extraction and Analysis of α-[^{3}H]Tocopherol from the Incubation Mixture

The purity of D-α-[^{3}H]tocopherol is checked during the course of the binding studies by HPLC and/or by TLC and stored at −80° under nitrogen before use. The possibility that α-[^{3}H]tocopherol may be degraded during the incubation period is investigated by incubating the α-[^{3}H]tocopherol with the protein in the incubation mixture. After the incubation, radioactivity is extracted by making the sample 50% (v/v) in ethanol, adding approximately 45 μg of α-tocopherol, then extracting twice with an equal volume of hexane. The sample is dried under a stream of N_2, and redissolved in a small volume of hexane. The sample is then analyzed by thin-layer chromatography (TLC) using a solvent system of hexane/dichloromethane/acetic acid (70/30/0.5). The tocopherol is visualized under an ultraviolet lamp and the silica gel marked. Each segment is scraped from the plate directly into a scintillation vial and the radioactivity measured. More than 94% of radioactivity was recovered unchanged as α-tocopherol.

Amino Acid Analysis

Amino acid analysis of the 15-kDa TBP is carried out on an Applied Biosystem 420H amino acid analyzer (Foster City, CA) with automatic hydrolysis and derivitization. The PTC amino acids generated are identified on-line, using a 130A separation system employing a C_{18} reversed-phase narrow bore cartridge. The system is calibrated using norleucine as an internal standard. The amino acid composition of the TBP is given in Table II. TBP is enriched with Ser, Gly, Try, and Ala residues, whereas it contains only one Met and two Cys residues. The number of Ala, Asx, Glx, Gly, His, Ser, Thr, Tyr, and Val residues in these proteins differs considerably. Direct amino-terminal sequencing of TBP was not possible because its amino terminal was found to be blocked.

IV. Comparative Biochemistry of α-Tocopherol-Binding Proteins with Other Intracellular Lipid-Binding Proteins

The 30-kDa TBP is a monomeric protein[23–29] and the 15-kDa binding protein cannot be its subunit. The 15-kDa TBP present in various tissues including liver is suggested to be responsible for intracellular transport and metabolism of α-tocopherol. This protein specifically binds α-tocopherol but not the γ or δ homolog and, moreover, it stimulats by 10-fold its transfer from liposomes to mitochondria.[30–33] However, the size of the low molecular weight TBP is closer to the size of the FABP (~15 kDa), a highly abundant

TABLE II
AMINO ACID COMPOSITION OF TBP AND FABPs FROM BOVINE HEART[a]

Amino acid residues	TBP (p*I* 4.5)	FABP	
		p*I* 4.9[b]	p*I* 5.1[b]
Ala	16	6	6
Arg	4	4	4
Asx	9	16	15
Cys	2	3	4
Glx	5	14	13
Gly	28	12	11
His	7	2	2
Ile	4	10	10
Lys	16	13	13
Met	1	2	2
Phe	2	6	6
Pro	11	2	2
Ser	26	7	7
Thr	10	19	20
Tyr	4	2	2
Val	7	14	14

[a] Values (mol/mol protein) refer to residues determined by amino acid analysis as described in the "Methods" section. Values are given to the nearest integer. Reprinted by permission of the publisher from M. J. Gordon *et al., Arch. Biochem. Biophys.* **318,** 140 (1995).
[b] Data from Ref. 24.

lipid-binding protein, present in almost all tissues.[33,40,41] Therefore, it was important to compare various structural and functional properties of TBP with the similar size (~15 kDa) cytosolic FABP in a tissue. The amino acid composition of the bovine heart TBP (15 kDa) was found to be significantly different than that of both FABPs (15 kDa, p*I* 4.9 and 5.1) of the same tissue.[32,42] In particular, TBP has a large number of Ala, Gly, Pro, Tyr, and Ser residues but a lower number of Phe, Thr, Val, Asx, Glx, and Ile residues compared with the FABPs of this tissue. The differences between the TBP and the FABPs are also observed in the immunological cross-reactivity. The polyclonal antisera against bovine heart FABP could not recognize

[40] J. H. Veerkamp, R. A. Peeters, and R. G. H. J. Maatman, *Biochim. Biophys. Acta* **1081,** 1 (1991).
[41] A. K. Dutta-Roy, Y. Huang, B. Dunbar, and P. Trayhurn, *Biochim. Biophys. Acta* **1169,** 73 (1993).
[42] G. Jagschies, M. Reers, C. Unterberg, and F. Spener, *Eur. J. Biochem.* **152,** 537 (1985).

the TBP in the Western blot. All these comparative data suggest that TBP (~15 kDa, p*I* 4.5) is different from the bovine heart FABP (~15 kDa, p*I* ~4.9 and 5.1) despite their similar size and p*I* values. Heart FABP appears to bind only fatty acids and not any other heterogeneous ligands such as prostaglandins, cholesterol, heme, and retinoids as is the case with liver FABP.[33,39,40] In addition, FABP does not bind tocopherols.[33,40] The purified TBP binds only α-tocopherol but not the fatty acids, again confirming that this protein is distinct from the FABP. In addition to FABP, heart is reported to express very low levels (0.06%) of sterol carrier protein-2 (SCP-2)[43] but does not express cellular retinol-binding and retinoic acid-binding proteins (CRBP and CRABP).[44] However, the size (13.2 kDa) and p*I* value (8.2) of SCP-2[43] are quite different from that of low molecular weight TBP (~15 kDa, p*I* 4.5). In addition, the amino acid compositions of these lipid-binding proteins (CRBP, CRABP, and SCP-2)[45,46] are different from that of TBP. All these data indicate that TBP is quite distinct from these lipid-binding proteins, however, further analysis on amino acid and cDNA sequence of the 15-kDa TBP is required for definitive conclusions.

As mentioned earlier, the 30-kDa TBP of rat and human liver is closely related to CRALBP from bovine and human[29] and SEC14p.[34] Because CRALBP has only been found in visual organs and selectively binds 11-*cis*-retinaldehyde,[47] it is considered a substrate-specific carrier protein in the visual cycle. CRALBP, however, does not belong to the superfamily of lipid-binding proteins that includes CRBP, CRABP, FABP, and peripheral nerve myelin P2 protein.[40,46] The 30-kDa TBP in the liver and CRALBP in the visual tissue, however, may form a novel family of proteins that binds a certain class of hydrophobic ligands and transports the ligand within specific cell type. In addition, both human and rat TBPs exhibit high-sequence homology with SEC14 gene product, SEC14p.[34] SEC14P is found in both a cytoplasmic pool and in a stable, apparently specific peripheral association with the yeast Golgi complex.[35] SEC14P is suggested to play an essential role in the Golgi secretory function by regulating Golgi-membrane phospholipid composition. In this way it maintains the essential secretory

[43] S. Kesav, J. McLaughlin, and T. J. Scallen, *Biochem. Soc. Trans.* **20,** 818 (1992).

[44] G. Wolf, *Nutr. Rev.* **49,** 1 (1991).

[45] M. Hiemberg, E. H. Goh, H. A. Klausner, C. Soler-Argilaga, I. Weinstein, and H. G. Wilcox, *In* "Disturbances in Lipid and Lipoprotein Metabolism" (J. M. Dieschy, A. M. Gotto, Jr., and J. A. Onthro, eds.), p. 251. American Physiological Society, Williams & Wilkins, Baltimore, MD, 1978.

[46] J. Sundelin, S. R. Das, U. Eriksson, L. Rask, and P. A. Peterson, *J. Biol. Chem.* **260,** 6494 (1985).

[47] J. C. Saari, and D. L. Bredberg, *J. Biol. Chem.* **262,** 7618 (1986).

function of the membrane. The N-terminal 129 residues of the SEC14P are sufficient to direct this protein to the Golgi complex.[48] The most conserved domains in the 30-kDa TBP and SEC14P are located in the N-terminal (positions 47–66) and C-terminal (positions 211–214) regions. However, further work is necessary on the mechanism of association of the 30-kDa TBP with the Golgi complex in hepatocytes and the α-tocopherol transfer activity of the protein.

V. Roles of 15- and 30-kDa α-Tocopherol-Binding Proteins in Transport and Metabolism of α-Tocopherol

The liver discriminates among the tocopherols and tocotrienols by secreting only α-tocopherol in nascent VLDL, despite the presence of other absorbed forms of vitamin E circulating in chylomicrons. The liver contains both the 30- and 15-kDa TBP, whereas heart contains only 15-kDa TBP. Both proteins bind α-tocopherol specifically and do not bind other homologs. Since the 30-kDa TBP is only selectively found in the liver, it is tempting to speculate that its presence relates to a specific hepatic role, for example, the secretion of α-tocopherol into nascent VLDL, and thereby maintains plasma levels of α-tocopherol. The distribution of these proteins in tissues and their roles in hepatic α-tocopherol transport and metabolism are now known. Patients with familial isolated vitamin E deficiency have similar neurological abnormalities and no abnormality in gastrointestinal function or in lipoprotein metabolism, but have remarkably low levels of plasma α-tocopherol.[33,36,37] In these patients, both α- and γ-tocopherol were equally absorbed by the intestine into chyolmicrons, but both forms of tocopherol disappeared rapidly from the plasma, in contrast to the normal situation. This faster decrease was attributed either to a defective liver TBP or to the absence of liver TBP, leading to a lack of incorporation of α-tocopherol to nascent VLDL in the liver. Recently, patients in Tunisia who had neurological symptoms resembling those of Freidreich ataxia but had normal structure of chromosome 9 were found to have very low levels of α-tocopherol in the plasma.[36,37] The locus of abnormality was found on chromosome 8q.[36,37] Since TBP gene is located also on chromosome 8, they may have a genetic defect resulting in a complete lack of functionality of TBP.

The 15-kDa TBP, however, exists in all the tissues where it has so far been measured so it seems reasonable that it is involved in intracellular distribution and metabolism of α-tocopherol in all tissues including the

[48] H.B. Skinner, J. G. Alb, Jr., E. A. Whitters, G. M. Helmkemp, Jr., and V. A. Bankitis, *EMBO J.* **12,** 4775 (1993).

liver. Because the 15-kDa TBP is thought to be responsible for intracellular transport as well as the retention of α-tocopherol in the tissue, expression and function of this TBP may be crucial for the regulation of α-tocopherol levels in the tissues. The activity of both the higher and low molecular weight TBPs may therefore be crucial for effective regulation of α-tocopherol levels in plasma, membranes, and cellular organelles. The presence of the 15-kDa TBP in the cytosol of hepatocytes and cardiomyocytes supports the conventional view that it plays a key role in the delivery of α-tocopherol to microsomal membranes, over and above its ability to enhance the retention of α-tocopherol within these cells after its uptake across the plasma membrane. If reduced activity of the 30-kDa TBP in the liver lead to signs of vitamin E deficiency,[36,37] because of impaired incorporation of α-tocopherol into nascent VLDL, abnormalities in the function of the 15-kDa TBP may have a similar outcome, despite the normal plasma levels of vitamin E. They may have individual roles or act together in facilitating maintenance of plasma levels of α-tocopherol, targeting α-tocopherol to organelle membranes, regenerating α-tocopherol, and altering the membrane structure and function.

[26] α-Carboxyethyl-6-Hydroxychroman as Urinary Metabolite of Vitamin E

By MANFRED SCHULTZ, MARCEL LEIST,* ANGELIKA ELSNER, and REGINA BRIGELIUS-FLOHÉ

Introduction

The National Research Council of the United States and the DGE (German Nutritional Society) have recommended an intake of 10–12 mg vitamin E per day to prevent deficiencies.[1,2] Because vitamin E has been reported to have beneficial effects in diseases related to oxidative stress,

* Present address: Faculty of Biology, University of Konstanz, Konstanz, Germany.

[1] National Research Council, "Recommended Daily Allowances. Vitamin E," 10th ed., pp. 99–114. National Academy Press, Washington, DC, 1989.

[2] Deutsche Gesellschaft für Ernährung, "Empfehlungen für die Nährstoffzufuhr," Umschau, Frankfurt, 1991.

0076-6879/97 $25.00

such as cardiovascular disease[3,4] or cancer,[5] some authors have recommended doses of up to 100 mg/day, that is, doses significantly beyond those sufficient to prevent deficiencies.[6] In contrast, some oxidative events in the organism such as the immune response[7,8] or signal transduction pathways[9] have a physiological function that must not be suppressed. Thus, there is a need for markers indicating when the vitamin E level of an organism is adequate and when supplementation is required.

In search of such an indicator of vitamin E status, we have studied the excretion of vitamin E metabolites (for a review, see Ref. 10) into human urine after increased supplementation with α-tocopherol to volunteers.[11] The nature of urinary metabolites of α-tocopherol was first described by Simon *et al.* in 1956.[12] In their scheme, the first step was the hydrolytic opening of the chroman ring, presumably during oxidation of α-tocopherol to α-tocopherylquinone. This was followed by the oxidation of the terminal methyl group in the side chain and the degradation of the resulting carboxylate by β-oxidation. α-Tocopherylquinone was later shown to be the product of the reaction of α-tocopherol with superoxide anion[13] or peroxynitrite.[14] The products thus formed, α-tocopheronic acid and the lactone derived therefrom (Scheme 1, left-hand pathway), were then excreted into the urine. On the basis of those data there was no reason to doubt the original assumption that Simon metabolites are produced from α-tocopherol after its action as an antioxidant, that is, that β-oxidation of the side chain followed ring opening. In our study, however, we found that the primary human urinary metabolite of α-tocopherol at high supplementation doses was 2,5,7,8-tetramethyl-2-(2′-carboxyethyl)-6-hydroxychroman (α-CEHC) (Scheme 1, right-hand pathway). The identification of α-CEHC as a human

[3] M. Stampfer, C. Hennekins, J. Manson, G. Colditz, B. Rosner, and W. Willett, *New Engl. J. Med.* **328,** 1444 (1993).

[4] E. Rimm, M. Stampfer, A. Ascherio, E. Giovannucci, G. Colditz, and W. Willet, *New Engl. J. Med.* **328,** 1450 (1993).

[5] T. Byers and N. Guerrero, *Am. J. Clin. Nutr.* **62,** 1385S (1995).

[6] A. T. Diplock, *VitaMinSpur* **8,** 11 (1993).

[7] B. M. Babior, R. S. Kipnes, and J. T. Curnutte, *J. Clin. Invest.* **52,** 741 (1973).

[8] A. Bendich, *in* "Nutrition and Immunology" (D. M. Klurfeld, ed.), Vol. 8, p. 217. Plenum Press, New York, 1993.

[9] R. Schreck, P. Rieber, P. A. Baeuerle, *EMBO J.* **10,** 2247 (1991).

[10] B. Gassmann, M. Schultz, M. Leist, and R. Brigelius-Flohé, *Ernährungs-Umschau* **42,** 80 (1995).

[11] M. Schultz, M. Leist, M. Petrzika, B. Gassmann, and R. Brigelius-Flohé, *Am. J. Clin. Nutr.* **62,** 1527S (1995).

[12] E. J. Simon, A. Eisengart, L. Sundheim, and A. T. Milhorat, *J. Biol. Chem.* **227,** 807 (1956).

[13] D. C. Liebler, *Crit. Rev. Toxicol.* **23,** 147 (1993).

[14] N. Hogg, V. M. Darley-Usmar, M. T. Wilson, and S. Moncada, *FEBS Lett.* **326,** 199 (1993).

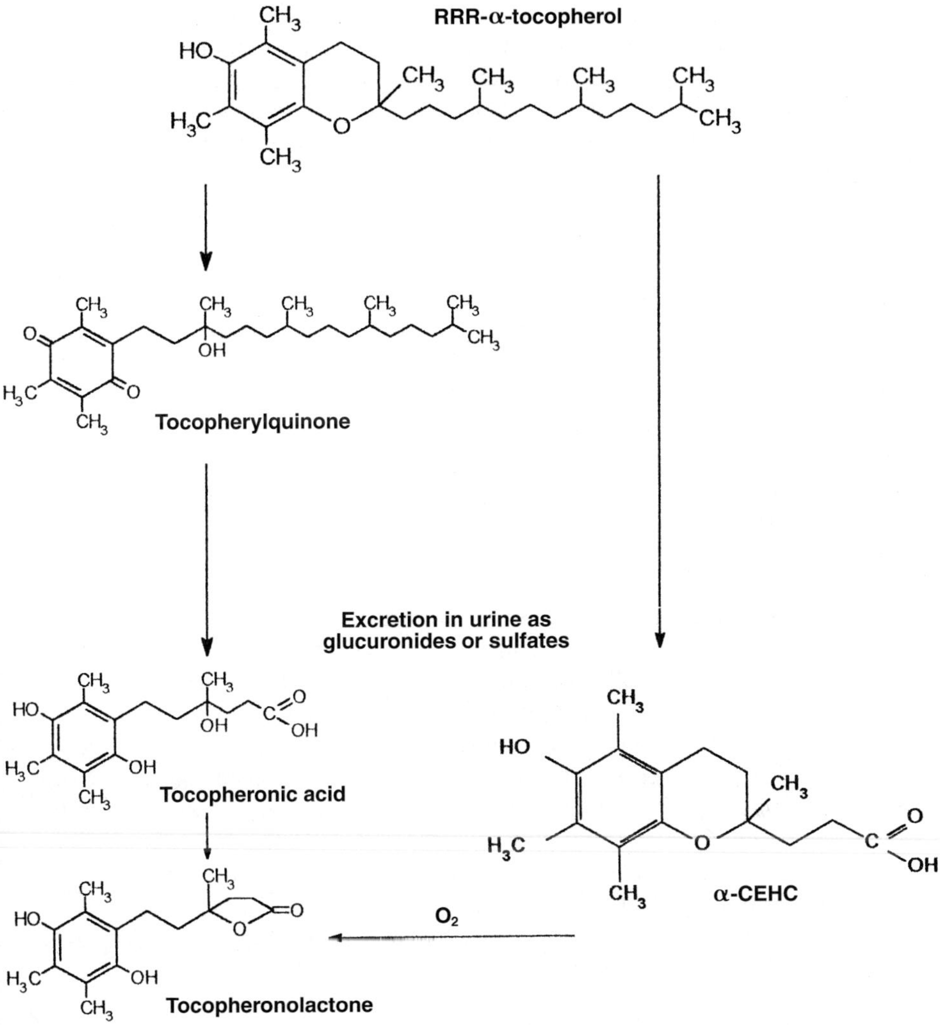

SCHEME 1. Possible urinary metabolites of α-CEHC. The left-hand pathway requires the cleavage of the chroman structure and leads to the so-called Simon metabolites, α-tocopheronic acid and α-tocopheronolactone. The right-hand pathway leads to α-CEHC without opening of the chroman ring. The α-CEHC can be converted to α-tocopheronolactone by oxygenation. Thus, α-CEHC may be the primary urinary excretion product of α-tocopherol. For further details see text.

α-tocopherol metabolite with an intact chroman structure, but a degraded side chain, clearly shows that α-tocopherol can be metabolized without previous oxidative opening of the chroman ring. The finding that oxygenation of α-CEHC led to the transformation of α-tocopheronolactone led us to speculate that the Simon metabolites could have been produced during sample processing when oxygenation is not avoided. That Simon metabolites might indeed be methodological artefacts was first suggested by Schmandke *et al.*[15] and later corroborated by Chiku *et al.* for δ-tocopherol.[16] From the findings that α-CEHC was excreted after a certain threshold of plasma α-tocopherol was exceeded, we concluded that α-CEHC excretion might be a useful marker for a vitamin E supply exceeding saturation levels in individual human subjects.[11] We describe here methods for the identification and estimation of urinary α-CEHC suitable to further investigate the conditions for α-CEHC excretion in humans and animals.

Sample Preparation

Urine Sampling

Urine is collected over a 24-hr period into sealable vessels containing sodium azide to provide a final concentration of about 0.02% (w/v). About 1.5 liter urine can be collected during 1 day. At the end of collection, the samples are immediately lyophilized under nitrogen and then pulverized in liquid nitrogen to guarantee homogeneity. Urine powder is stored at $-20°$ under N_2 until further treatment. The dry substance recovered from human urine is in the range of 55–70 g per 24 hr and a total α-CEHC content of about 10 mg (160 μg/g powder) may be expected at a supplementation dose of 800 mg α-tocopherol/day. An aliquot of urine powder corresponding to about 20 ml 24-hr urine is sufficient for the estimation of α-CEHC. Lyophilized urine powder may be stored under nitrogen over months without significant loss of α-CEHC.

Extraction

Basic Procedure. Urine powder (4 g) is suspended in 40 ml methanol and if α-CEHC is to be quantified via the internal standard mode, the standard is added. The suspension is shaken for 30 min at room temperature, centrifuged at 500 *g* for 10 min at room temperature, and the supernatant is collected. The extraction is repeated three times and the pooled supernatants are concentrated to 10 ml. The whole extraction procedure is carried

[15] H. Schmandke, *Int. Z. Vitaminforsch.* **35,** 346 (1965).

[16] S. Chiku, K. Hamamura, and T. Nakamura, *J. Lipid Res.* **25,** 40 (1984).

out under nitrogen. The sample size can be reduced to 0.5–0.8 g urine powder together with a corresponding reduction in all other chemicals without changing results.

Alternative Extraction Procedures. Method O: No further extraction before hydrolysis. *Method A:* Hexane extraction can be used to remove hydrophobic substances that might prevent efficient enzymatic hydrolysis. One milliliter of ethanol and 0.2–0.5 ml water are added to the dried methanolic extract and vortexed. The ethanolic phase is then extracted three times with 2 ml hexane each and the phases separated by centrifugation. The ethanolic phase is then dried. All extraction procedures should be carried out under nitrogen or argon. *Method B:* An alkaline extraction can be used to separate acidic from alkaline compounds. The dried urine is extracted three times with 6 ml ethanol containing 0.1% of 1 *M* NaOH by vigorously shaking and subsequent centrifugation to remove insoluble compounds. The supernatants are then combined in a glass tube and dried. All extraction procedures should be carried out under nitrogen or argon. *Method C:* Method A is followed by method B. In this case it is not necessary to dry the sample after treatment according to method A.

Testing Efficiency of Extraction Procedures. The efficiency of extraction of α-CEHC conjugates from the pulverized urine was tested. Urine powder from seven volunteers supplemented with 800 mg α-tocopherol for 7 days was pooled and used as a test substance. All samples were previously shown to contain a substantial amount of α-CEHC (40–220 μg/g urine powder). Pooling was performed to obtain a large stock of a homogeneous sample. Methods A, B, and C were applied and compared to method O. In all methods, α-tocopherol acetate was added as an internal standard. The extracts obtained by the different methods were enzymatically hydrolyzed, analyzed by GC/MS, and quantified (see below). The results (Table I) obtained with the different extraction procedures did not show great variation. However, amounts of α-CEHC measured after extraction with method C were slightly higher than those obtained with method O. The maximal improvement was 25%.

Hydrolysis

α-CEHC is excreted as a sulfate or glucuronic acid conjugate. These conjugates must be hydrolyzed before determination to yield either the methyl esters or the free metabolites.

Procedure Leading to α-CEHC Methyl Ester. Hydrolysis with HCl in the presence of methanol results in the formation of the methyl ester of α-CEHC (α-CEHC-M). For instance, 4.2 ml 10 *M* HCl (final concentration, 3 *M*) is added to the extracts (10 ml) prepared according to the basic

TABLE I
EFFICIENCY OF DIFFERENT EXTRACTION PROCEDURES ON AMOUNT OF α-CEHC DETECTED IN HUMAN URINE BY GC/MS[a]

Method	Extraction	α-CEHC (μg/g urine powder)
O	—	161.1 ± 20.2
A	Hexane	175.8 ± 9.3
B	Ethanol/NaOH	192.6 ± 10.1
C	Hexane + ethanol/NaOH	200.7 ± 11.4

[a] A pool of dried urine from seven volunteers supplemented for 7 days with 800 mg α-tocopherol per day was taken to test the efficiency of extraction procedures. A 0.5-g sample was extracted by the methods indicated and enzymatically hydrolyzed. Identification and quantification of α-CEHC was performed by GC/MS with α-tocopherol acetate as internal standard. Values are means ± SD from three individual procedures. For further details see text.

procedure (see previous section), the mixture is shaken for 20 hr at room temperature, then 14 ml water is added and the suspension is extracted three times with 50 ml ether each. The ether phases are combined and dried. The dry residue is dissolved in 2 ml methanol and stored for subsequent analysis. The whole hydrolysis procedure and the storage are carried out under nitrogen.

Procedure Leading to Free α-CEHC. Hydrolysis with a glucuronidase/sulfatase mixture leads to free α-CEHC. The dry extracts are taken up in 3 ml sodium acetate (80 m*M*, pH 4.5) containing 8 mg enzyme (133 U sulfatase; 2700 U glucuronidase, Sigma, St. Louis, MO), incubated in a shaking water bath for 5 hr at 37°, and extracted five times with 5 ml ether. The pooled ether phases are evaporated in a vacuum concentrator. The residue is taken up in 1 ml methanol and stored at −20° for subsequent analysis. All procedures are carried out under nitrogen.

Testing Efficiency of Enzymatic Hydrolysis. Due to the lack of a standard α-CEHC conjugate, it was not possible to test the recovery of α-CEHC released from conjugates. The conditions of enzymatic hydrolysis were therefore varied to determine if they were optimal. Increasing amounts of enzyme were used to hydrolyze a fixed amount of urine powder (0.8 g) for increasing times, and the resulting α-CEHC was measured by means of gas chromatography/mass spectrometry (GC/MS) (Fig. 1). From Fig. 1, we see that the time for enzymatic hydrolysis may be reduced to 2 hr without significant changes in recovery for most samples. Hydrolysis for only 1 hr with the usual amount of enzyme (8 mg) gave a recovery of 80%, and a

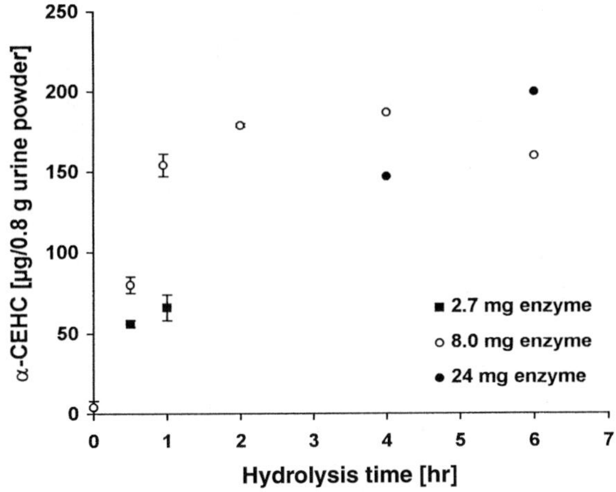

FIG. 1. Efficiency of the enzymatic hydrolysis. Urine (24 hr) was collected from a volunteer after 7 days of supplementation with 800 mg α-tocopherol per day, and an aliquot of 0.8 g urine powder was taken for analysis. The powder was extracted according to method O, hydrolyzed with 2.7 mg (■), 8 mg (○), and 24 mg (●) enzyme for the times indicated. Released α-CEHC was quantified by GC/MS. Amounts are expressed as μg α-CEHC per 0.8 g urine powder; values are means ± SD from triplicate measurements. For further details see text.

30-min incubation with 8 mg enzyme or a 1-hr incubation with 2.7 mg enzyme liberated 40 or 30%, respectively. Based on these data and the data shown in Table I we performed a routine analysis with 0.5–0.8 g urine powder, 8 mg enzyme, 5-hr hydrolysis time, and extraction according to method C. The absolute recovery following hydrolysis of the α-CEHC conjugates has not yet been precisely determined because standard conjugates of glucuronic or sulfuric acid are not available. However, removal of putative inhibitors by preextracting the samples (method C), increasing the amount of enzyme, or increasing hydrolysis time, did not cause a substantial augmentation of the yield of α-CEHC. Therefore, the protocol suggested appears to be adequate to liberate the majority of α-CEHC conjugates present in human urine.

Analysis

Reversed-Phase High-Performance Liquid Chromatography

A high-performance liquid chromatography (HPLC) system 440 (Kontron Instruments, Eching, Germany) equipped with a diode array detector (DAD), a fluorescence (SFM 25) detector, and a solvent degasser was

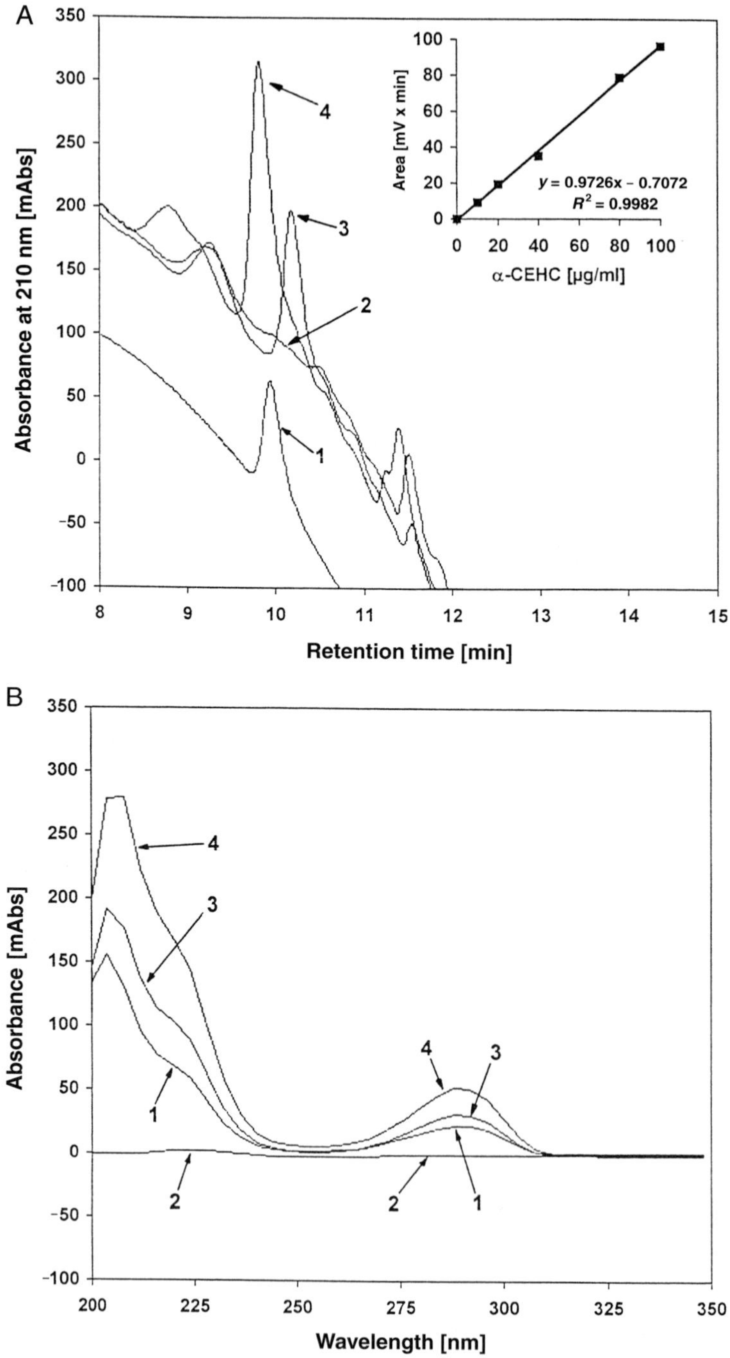
A
Absorbance at 210 nm [mAbs]
Retention time [min]
Area [mV x min]
α-CEHC [μg/ml]
y = 0.9726x − 0.7072
R² = 0.9982
B
Absorbance [mAbs]
Wavelength [nm]

utilized with a 4-μm, 125- × 4-mm LiChrospher RP 8 column (Merck, Darmstadt, Germany). The α-CEHC standard was synthesized by LABORAT GmbH (Berlin, Germany), and the α-CEHC methyl ester standard was synthesized by Synthese Labor A. Weiss (Kleinmachnow, Germany). Acetonitrile McIlvain buffer, (0.02 *M* sodium phosphate/0.01 *M* citric acid, pH 4.2) and all other solvents were of HPLC grade.

The α-CEHC-M is determined using the fluorescence detector (ex. 230, em. 330 nm). The urine hydrolyzate is dried under nitrogen and the pellet resolved in acetonitrile. An aliquot (10 μl) is injected onto the column and eluted by isocratic elution with acetonitrile/water (40/60, v/v) using a flow rate of 0.5 ml/min. The retention time of α-CEHC-M in our system is usually around 12 min. Pure standards of α-CEHC-M (100–800 ng) are used to verify a linear relationship between concentration and peak area. At the beginning and end of each series of analyses, standards (400 ng) are run to standardize the system.

The hydrolysis of α-CEHC conjugates with methanolic HCl has been used by several authors to esterify the carboxylic acid derivatives of α-tocopherol and prevent lactonization.[16,17] The procedure requires long reaction times during which oxygenation cannot be entirely excluded. Oxygenation would facilitate opening of the chroman structure and therefore destroy α-CEHC. Despite these methodological limitations, we prepared α-CEHC methyl esters as described and analyzed them by different methods. However, the recovery was usually about 50% of that obtained with the enzymatic procedure. Thus, it seems likely that during the drastic acidic hydrolysis the samples were destroyed or the reaction was incom-

[17] M. Watanabe, M. Toyoda, I. Imada, and H. Morimoto, *Chem. Pharm. Bull.* **22,** 176 (1974).

FIG. 2. HPLC of urinary samples for the identification and quantification of α-CEHC. (A) α-CEHC eluting from HPLC. Urinary samples were treated according to the basic procedure and enzymatically hydrolyzed. HPLC was loaded with sample corresponding to 8 μg urine powder and eluted with a gradient of acetonitrile–McIlvain buffer. Detection was at 210 nm with the diode array detector. Sample 1, standard α-CEHC (400 ng); sample 2, urinary hydrolyzate of a volunteer not supplemented with α-tocopherol; sample 3, urinary hydrolyzate of the unsupplemented volunteer plus 500 ng standard α-CEHC/ml (the recovery was 103%); sample 4, urinary hydrolyzate obtained from a volunteer after a 7-day period of supplementation with 800 mg α-tocopherol per day. The α-CEHC content in this sample was 88 μg/g dried urine. *Inset:* Calibration curve of α-CEHC eluting around 10 min and detected at 210 nm. Concentration–peak area relationship is taken for quantification. (B) Absorption spectra of α-CEHC. Spectra (diode array detection) are taken in the peak maxima from compounds eluting at a retention time of around 10 min shown in Fig. 1A. Samples 1–4 correspond to the samples 1–4 of part A. For further details see text.

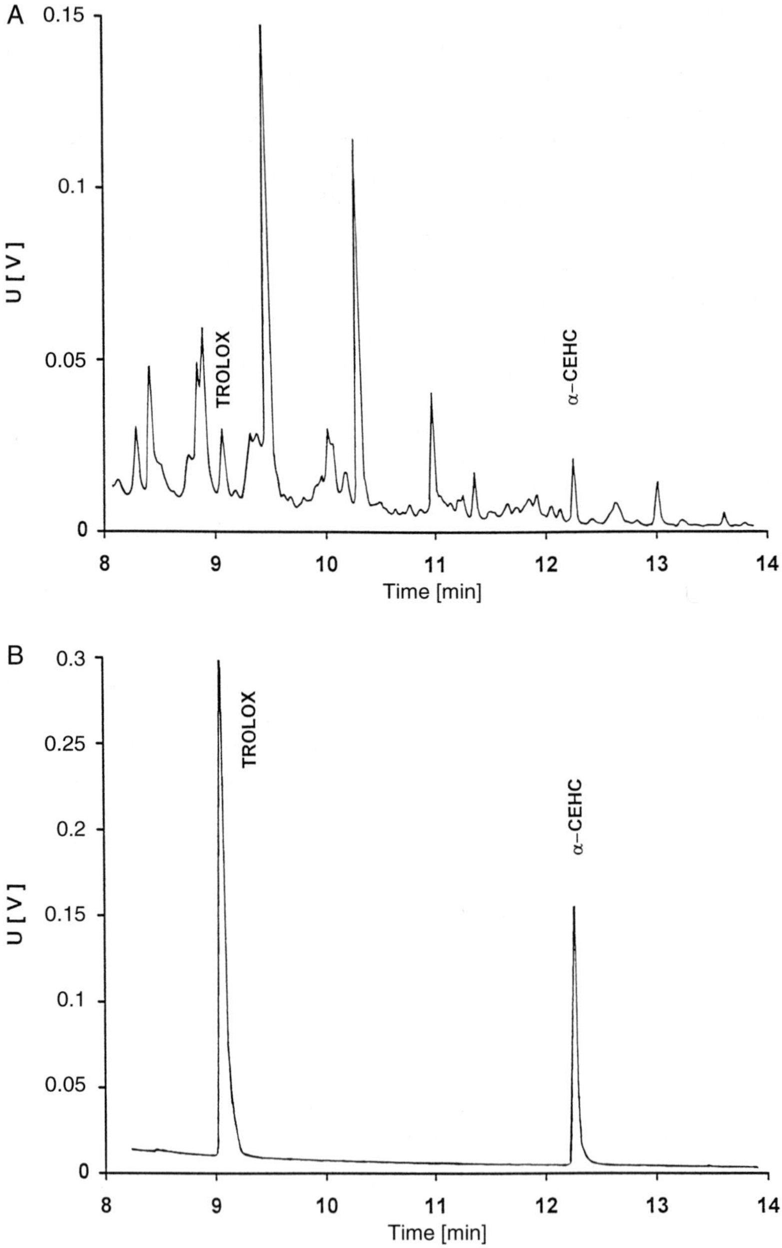
A
0.15
0.1
0.05
0
U [V]
TROLOX
α–CEHC
8
9
10
11
12
13
14
Time [min]
B
0.3
0.25
0.2
0.15
0.1
0.05
0
U [V]
TROLOX
α–CEHC
8
9
10
11
12
13
14
Time [min]

plete. We therefore recommend the use of enzymes for cleavage of the conjugates.

Free α-CEHC is determined by UV absorbance (210 or 290 nm, see spectra in Fig. 2B). The urine hydrolyzate is dried under nitrogen and the pellet resolved in acetonitrile. An aliquot is injected onto the column and eluted by a gradient to acetonitrile–McIlvain buffer as follows. Acetonitrile (30%) for 5 min, then linear increased to 10% within 8 min; 100% acetonitrile is kept constant for 3 min, then decreased to 30% within 3 min. To regenerate the column, it is flushed with 30% acetonitrile for 4 min further. The flow rate is 0.5 ml/min. The retention time of α-CEHC in our system is usually around 10 min. Standards of α-CEHC from 100–1000 ng are injected to provide a linear concentration versus peak area graph (see Fig. 2A, inset) and to set up the system.

Testing HPLC for Measuring α-CEHC. HPLC chromatograms of a variety of samples utilizing a diode array detector at 210 nm are shown in Fig. 2. Figure 2A shows an α-CEHC standard eluting at the typical retention time of around 10 min (curve 1). Urinary samples of a volunteer not supplemented with α-tocopherol did not contain any α-CEHC (curve 2), whereas in a urine sample obtained from a volunteer with a daily intake of 800 mg α-tocopherol for 7 days we found 87.5 μg α-CEHC/g urine powder (curve 4). Fifty milligrams of standard α-CEHC was added per milliliter of hydrolyzed urine sample from the unsupplemented person (500 ng injected) and the recovery was 103% (curve 3). An α-CEHC calibration curve is shown in the insert of Fig. 2A. Figure 2B shows the absorption spectra of compounds eluting at the retention time of around 10 min of Fig. 1A. Curves 1, 3, and 4, which contained α-CEHC, showed identical spectra. This demonstrates that standard α-CEHC and compounds prepared from urine samples are identical. It further demonstrates that HPLC can be used for quantitation of α-CEHC.

FIG. 3. Gas chromatograms of urinary samples for the identification and quantification of α-CEHC. (A) Elution profile of a urinary sample. The 24-hr urine of a volunteer who had taken 150 mg α-tocopherol/day for 7 days was lyophilyzed. An aliquot of 0.5 g was used for extraction according to method C, enzymatic hydrolysis, and processing for detection by GC. A 1-μl sample was injected to GC and eluted as described. The amount of α-CEHC was calculated to be 71.2 ng/μl test volume corresponding to 71.2 μg/g urine powder. The amount of Trolox corresponded to 30 μg added to 0.5 g urine powder before starting the sample processing. (B) Elution profile of standard compounds. One microliter of a mixture containing 100 ng α-CEHC and 200 ng Trolox, both derivatized with TMS, was injected, eluted as described, and identified by typical retention times. For further details see text.

TABLE II
COMPARISON OF GC/MS AND GC ANALYSIS[a]

Volunteer	α-Tocopherol supplementation (mg/day)	α-CEHC (μg/g urine powder)	
		GC/MS	GC
A	0	ND[b]	ND
A	50	ND	ND
A	150	61.4/73.0	71.2/73.7
A	350	86.0/100.6	90.3/100.0
Pool of 7 persons	800	145.5/150.8	164.7/153.3

[a] Seven volunteers were supplemented with the indicated doses of α-tocopherol for 7 days each. Aliquots of lyophilyzed 24-hr urine were extracted according to method C and enzymatically hydrolyzed. Analysis was performed with GC/MS and GC in duplicate. For further details see text.

[b] ND, not detectable.

Gas Chromatography/Mass Spectrometry

Derivatization. An internal standard is added to the pulverized urine aliquots if samples are to be analyzed by GC/MS. After hydrolysis the samples are mixed with 200 μl hexane, 250 μl BSTFA [*N,O-*

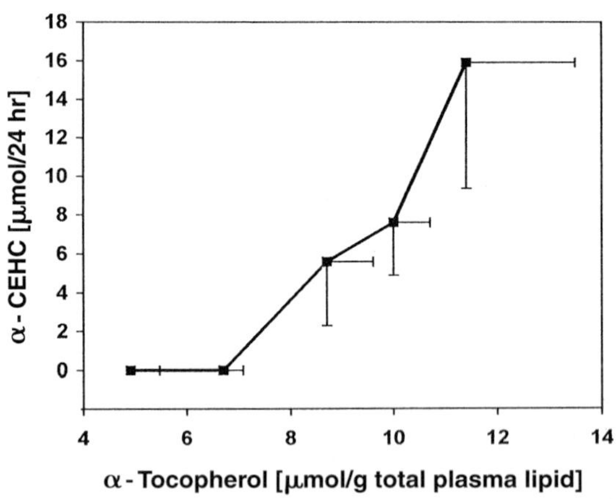

FIG. 4. α-CEHC excretion depends on plasma α-tocopherol content. Seven volunteers were supplemented with different doses of α-tocopherol for 7 days each. Aliquots of lyophilized 24-hr urine were extracted according to the basic procedure and enzymatically hydrolyzed. Samples were analyzed by HPLC. Blood was taken on the same day as urine was collected and the α-tocopherol content in plasma was estimated[11] and standardized for total lipid. Values are means of seven volunteers with the biological variation shown as deviation bars.

bis(trimethylsilyl)trifluoroacetamide (Fluka, Neu Ulm, Germany)], 40 ml BSA [*N,O*-bis(trimethylsilyl)acetamide (Aldrich, Steinheim, Germany)], and 10 μl TMCS [trimethylchlorosilane (Merck, Darmstadt, Germany)]. The mixtures are vortexed and incubated at 50° for 30 min. The derivatization reagent is removed with an argon stream and 500 μl hexane is added to the sample. For internal standards, Trolox [2,5,7,8-tetramethyl-2-carboxy-6-hydroxychroman (Hoffmann-LaRoche, Grenzach-Wyhlen, Germany) 30 μg/0.5 g sample] or dl-α-tocopherol acetate [(Merck, Darmstadt, Germany) 102 μg/0.5 g sample] can be used.

Running the GC/MS. The derivatized samples are analyzed in GC/MS system (SSQ 710 MAT; Finnigan MAT GmbH, Bremen, Germany) by electron ionization at 70 eV under the following operating conditions: a fused silica DB-5MS capillary column (30 m, 0.25 mm, 0.25 μm; J&W Scientific, Folsom, CA), a temperature program from 180° (2 min) to 280° (10°/min, 20 min isothermic), an injector temperature of 260°, a transfer line temperature of 300°, an emission current of 200 μA, a scan range of 50–700 atomic mass unit [mass per charge ratio, (m/z)], and a scan time of 0.5 sec. Derivatives of α-CEHC (422 m/z and 237 m/z) or Trolox (394 m/z and 277 m/z) are identified by their specific ionized fragments. The main ion of α-tocopherol acetate (not derivatized) is 430 m/z, which corresponds to α-tocopherol following cleavage of the acetate residue. Quantification can be performed from the mass chromatograms via the ratio of sample peak area to internal standard peak area. The α-CEHC amount present in the samples is calculated from the response factor and the peak areas. A calibration curve with α-tocopherol acetate was linear between 8 and 400 ng α-CEHC per injection; a calibration curve obtained with Trolox was linear between 10 and 100 ng α-CEHC.

Gas Chromatography

Derivatized (see above) samples are analyzed by GC (Varian STAR 3400 CX, Varian Chromatography Systems, Walnut Creek, CA) with a flame ionization detector under the following conditions: a fused silica DB-5MS capillary column (30 m, 0.25 mm, 0.25 μm; J&W Scientific), a temperature program from 180° (2 min) to 280° (10°/min, 20 min isothermic), an injector temperature of 260°, and a detector temperature of 300°. α-CEHC and internal standards are identified via the retention times of the respective standards. In our system the retention time for α-CEHC was 12.4 min and for Trolox 9.2 min. Quantification is carried out as described for GC/MS with trolox as an internal standard. A linear calibration curve was obtained using 100–300 ng α-CEHC per injection. A typical chromatogram is shown in Fig. 3, where α-CEHC can be clearly identified and quantified by using standard substances.

Testing GC/MS and GC for Measuring α-CEHC. Both GC/MS and GC yielded identical results (Table II). To verify this, urine samples of volunteers supplemented with 50, 150, 350, or 800 mg α-tocopherol per day for 7 days were extracted by method C, enzymatically hydrolyzed, and measured with both methods. A comparison of the values obtained with GC/MS and GC is shown in Table II. Identical values were obtained. Therefore, if a GC/MS station is not available, GC is suitable for the estimation of α-CEHC, however, a standard is needed.

The methods described can be used to study biologically relevant processes. The urinary α-CEHC contents of the seven volunteers supplemented with 0, 50, 150, 350, or 800 mg α-tocopherol per day [11] were plotted against the plasma α-tocopherol content standardized for plasma lipids. As shown in Fig. 4, α-CEHC excretion started only after a certain threshold of plasma α-tocopherol was exceeded. Estimation of α-CEHC in human urine may therefore be a reasonable method to characterize the vitamin E status of individuals. It may also be useful to determine if the requirement for vitamin E is modified under conditions of enhanced oxidative stress.

Section IV

Vitamin K

[27] Purification of Vitamin K-Dependent Carboxylase from Cultured Cells

By KATHLEEN L. BERKNER and BETH A. MCNALLY

Introduction

The isolation of the carboxylase from tissue culture cell lines offers a number of advantages over its isolation from tissue. Analysis of the recombinant human carboxylase (r-carboxylase), and of mutated forms, is now possible. Cell lines lacking endogenous vitamin K-dependent (VKD) proteins have been identified[1] and can be used to coexpress a single VKD protein with the carboxylase. These cell lines make it possible to analyze the interaction of the carboxylase with individual VKD proteins. In contrast, in the isolation of carboxylase from tissues such as liver, the carboxylase is distributed among several different VKD proteins. In addition to being able to isolate pure carboxylase from tissue culture cell lines, it is now possible to isolate the carboxylase–VKD protein complexes from vitamin K-depleted cells. This ability to isolate and characterize the enzyme–substrate complex should be valuable in defining the mechanism of carboxylation. Finally, the ability to isolate free carboxylase or the carboxylase in complex allows their direct comparison, for example, in studies on the regulation of the carboxylase by VKD propeptides. The main advantage in isolating the carboxylase from tissue is the amount of protein that can be obtained.[2,3] However, with high-level r-carboxylase expressing cell lines, it is now possible to isolate milligram amounts of the carboxylase and, as described later, the purification scheme is considerably easier.

All VKD proteins have in common an approximately 18 amino acid sequence, which in most cases is a propeptide removed during secretion.[4-6] The observation that this sequence is observed even in VKD proteins which otherwise share little homology led to the proposal that this propeptide is a recognition signal for the carboxylase,[6] and this hypothesis was supported

[1] S. E. Lingenfelter and K. L. Berkner, *Biochemistry* **35,** 8234 (1996).
[2] S. M. Wu, D. P. Morris, and D. W. Stafford, *Proc. Natl. Acad. Sci. U.S.A.* **88,** 2236 (1991).
[3] K. L. Berkner, M. Harbeck, S. Lingenfelter, C. Bailey, C. M. Sanders-Hinck, and J. W. Suttie, *Proc. Natl. Acad. Sci. U.S.A.* **89,** 6242 (1992).
[4] B. Furie and B. C. Furie, *Cell* **53,** 505 (1988).
[5] G. Manfioletti, C. Brancolini, G. Avanzi, and C. Schneider, *Mol. Cell Biol.* **13,** 4976 (1993).
[6] L. C. Pan and P. A. Price, *Proc. Natl. Acad. Sci. U.S.A.* **82,** 6109 (1985).

0076-6879/97 $25.00

by subsequent mutational analyses.[7,8] Identification of the propeptide as a carboxylase recognition signal has been important for the purification of the carboxylase. Two approaches have been developed, one based on the initial isolation of a carboxylase–VKD protein complex followed by carboxylase displacement using a propeptide,[1,3,9–11] and the other approach based on carboxylase adsorption to an immobilized propeptide ligand.[2,12] We have adapted both methods for the isolation of carboxylase from tissue culture cell lines, and describe both approaches in this chapter. We have also purified the carboxylase using an immobilized anticarboxylase antipeptide antibody column, and this procedure is also described.

Two different expression systems for obtaining r-carboxylase are presented. We have overexpressed the r-carboxylase in mammalian cell lines, which also contain endogenous carboxylase, either alone or coexpressed with a r-VKD protein. The second system is the transient expression of carboxylase and factor IX in baculovirus-infected insect cells, which do not otherwise contain endogenous carboxylase or VKD proteins.[13] The relative merits of these two expression systems are discussed at the end of the chapter.

Purification of Carboxylase from Mammalian Cell Lines

The carboxylase purification is based on the initial isolation of a carboxylase–VKD protein complex, from cell lines stably transfected with r-VKD proteins, on immobilized anti-VKD protein antibody columns. We have performed this isolation using BHK (baby hamster kidney) cell lines or 293 cell lines that express r-factor IX, r-factor VII, r-protein C, or r-prothrombin.[1,14] The purification scheme we have developed is described for a 293 cell line expressing r-factor IX, however, the results obtained from the other cell lines are very similar.

[7] D. C. Foster, M. S. Rudinski, B. G. Schach, K. L. Berkner, A. A. Kumar, F. S. Hagen, C. A. Sprecher, M. Y. Insley, and E. W. Davie, *Biochemistry* **26,** 7003 (1987).

[8] M. J. Jorgensen, A. B. Cantor, B. C. Furie, C. L. Brown, C. B. Shoemaker, and B. Furie, *Cell* **48,** 185 (1987).

[9] M. DeMetz, C. Vermeer, B. A. M. Soute, G. J. M. Van Scharrenburg, A. J. Slotboom, and H. C. Hemker, *FEBS Lett.* **123,** 215 (1981).

[10] J. C. Swanson and J. W. Suttie, *Biochemistry* **21,** 6011 (1982).

[11] M. C. Harbeck, A. Y. Cheung, and J. W. Suttie, *Thromb. Res.* **56,** 317 (1989).

[12] B. R. Hubbard, M. M. W. Ulrich, M. Jacobs, C. Vermeer, C. Walsh, B. Furie, and B. C. Furie, *Proc. Natl. Acad. Sci. U.S.A.* **86,** 6893 (1989).

[13] D. A. Roth, A. Rehemtulla, R. J. Kaufman, C. T. Walsh, B. Furie, and B. C. Furie, *Proc. Natl. Acad. Sci. U.S.A.* **90,** 8372 (1993).

[14] K. L. Berkner, unpublished data (1993).

The generation and characterization of r-VKD protein expressing cell lines have been described in detail elsewhere.[15] A critical requirement for a suitable line is that it does not express endogenous VKD proteins, so that virtually all of the carboxylase will be in complex with the r-VKD protein introduced exogenously. Cell lines expressing sufficient levels of VKD protein for the carboxylase isolation are easily obtained; even with low levels of production the intracellular r-VKD proteins are usually still in excess of the carboxylase. For example, in a BHK cell line expressing low levels of r-factor IX (0.3 μg/ml/day), the intracellular factor IX levels were approximately 20-fold higher than that of the carboxylase. Almost all (>95%) of the carboxylase was bound to factor IX, as determined by adsorption of carboxylase peptide activity to anti-factor IX antibody.[14]

Microsomes are prepared from confluent r-factor IX-expressing 293 cells (1–4 $\times$ 10^9, which corresponds to ~40 to 160–150-mm plates) by dislodging the cells with EDTA (Versene, GIBCO, Grand Island, NY). Cells are pelleted at 2000*g* for 5 min, washed twice in phosphate-buffered saline (PBS) (50–200 ml, Gibco-BRL, Gaithersburg, MD) and recentrifuged at 2000*g* for 5 min. All steps are performed at 4° unless otherwise specified. The cells are resuspended in 0.25 *M* sucrose, 0.025 *M* imidazole, pH 7.3, and 1 m*M* phenylmethylsulfonyl fluoride (PMSF) and then sonicated (Heat Systems Cell Disrupter, model W-220F, Plainview, NY), while on ice, using 4–15-sec bursts with 30-sec intervals between each sonication. Cell breakage is monitored by vital dye exclusion, using trypan blue (0.4%, Gibco-BRL). The sonicate is Dounce homogenized (Bellco, model 0212, Vineland, NJ) and pelleted at 4000*g* for 15 min. The postnuclear fraction is then spun at 100,000*g* for 1 hr, the supernatant is discarded, and the pellets are quick frozen in liquid nitrogen and stored at −80°. The carboxylase activity in the microsomes is remarkably stable: we have used replicate tubes up to 3 years old with no observed difference in total carboxylase activity or in percent carboxylase in complex with factor IX.

To solubilize the carboxylase, microsomes are resuspended in 50 m*M* Tris, pH 7.4, 100 m*M* NaCl, and 1 m*M* PMSF (10–25 ml for 1–4 $\times$ 10^9 cells). Optimal solubilization conditions using (3-[(3-cholamidopropyl)dimethylammonio]-1-propane-sulfonate) (CHAPS) (Pierce, Rockford, IL) and various NaCl concentrations are determined,[16] as is described in detail for the r-carboxylase in the following two sections. This step is one of the most critical parts of the isolation, since optimal carboxylase solubilization is a balance between maximizing total activity recovery while avoiding carboxylase inactivation that occurs when the detergent concentration is

[15] K. L. Berkner, *Methods Enzymol.* **222,** 450 (1993).
[16] J.-M. Girardot and B. C. Johnson, *Anal. Biochem.* **121,** 315 (1982).

too high. Even with optimal conditions, it is not yet possible to avoid some carboxylase inactivation (described in detail below). Once the optimal CHAPS and NaCl concentrations have been determined at a given protein concentration, it is important to perform subsequent carboxylase isolations under identical conditions to obtain reproducible solubilization. The protein concentration in the microsomal preparations can be reproducibly generated if care is taken to monitor the initial cell number.

For 293 cells expressing r-factor IX and endogenous carboxylase (4×10^9 cells, which yields an initial microsomal suspension of ~10 mg/ml protein for 25 ml), we use two sequential solubilizations: resuspended microsomes are adjusted to 0.1% CHAPS, rocked (on a nutator, Thermolyne Vari-Mix, Dubuque, IA) for 1 hr, then centrifuged at 100,000*g* for 1 hr. The pellet is resuspended in 50 m*M* Tris, pH 7.4, 200 m*M* NaCl, and 1 m*M* PMSF, and then adjusted to 0.5% CHAPS. After 1 hr of rocking, the sample is recentrifuged at 100,000*g* for 1 hr. To determine the carboxylase recoveries, the following peptide assay is performed on the 0.1% (w/v) suspension, the 0.1% (w/v) supernatant, the 0.5% (w/v) suspension, and the 0.5% (w/v) supernatant: aliquots (0.2–1 mg protein) are assayed for up to 1 hr in a 150-μl cocktail of 1.2 *M* ammonium sulfate, 0.06% (w/v) phosphatidylcholine IIIE (Sigma, St. Louis, MO), 0.06% CHAPS (Pierce), 0.06% sodium cholate (Calbiochem, La Jolla, CA), 2.5 m*M* $NaHCO_3$ (50 mCi/mmol, Amersham, Arlington Heights, IL), 5 m*M* dithiothreitol (DTT), 2.5 mM Boc-glu-glu-leu-OMe (EEL, Sigma), 50 m*M* (*N,N*-bis[2-hydroxyethyl]-2-aminoethane sulfonic acid) (BES), pH 7.2, with or without 20 μM factor X propeptide (SLFIRREQANNILARVTR), purified on a preparative C_{18} column). After incubation at room temperature, the samples are precipitated with trichloroacetic acid (TCA) on ice for 15 min (10%, 1 ml) and $^{14}CO_2$ is removed by boiling the TCA supernatant with a Teflon chip (Norton, Akron, OH), to near dryness. Factor IX recovery is monitored by enzyme-linked immunosorbent assay (ELISA).[1] Very little carboxylase is solubilized in 0.1% CHAPS (Table I), in contrast to factor IX, where 50–90% solubilization is observed (data not shown). Sequential solubilization is thus useful for removing excess factor IX (i.e., factor IX not complexed with the carboxylase), thereby reducing the amount of anti-factor IX antibody resin subsequently required for purifying the factor IX–carboxylase complex, and lowering the nonspecific protein background.

The 0.5% supernatant is applied to an affinity purified anti-factor IX polyclonal antibody column (5 ml, 5 mg antibody per milliliter), the column is rocked overnight, and the flow-through is collected. When the 0.5% supernatant and flow-through are analyzed for factor IX by ELISA, less than 0.01% factor IX is detected in the flow-through.[1] The column is washed, at room temperature, with 100 ml of 50 m*M* Tris, pH 7.4, 500 m*M* NaCl, 0.1% CHAPS, 0.1% phosphatidylcholine IIIE, 5 m*M* DTT (buffer A), then

TABLE I
PURIFICATION OF CARBOXYLASE FROM 293 CELLS[a]

Fraction	Carboxylase activity (cpm/hr)	Protein (mg)	Specific activity (cpm/hr/μg)	Purification (-fold)
A. Native factor IX-293 cells				
0.1% Suspension	2.2×10^7	260	85	1
0.1% Supernatant	1.2×10^6	120	—	—
0.5% Suspension	2.0×10^7	140	—	—
0.5% Supernatant	1.2×10^7	60	200	2.4
α-Factor IX-Sepharose	1.0×10^7	—	—	—
Flow-through	4×10^5	60	—	
Propeptide eluant	9.0×10^6	0.003	3×10^6	3.5×10^4
B. Untransfected 293 cells				
0.1% Suspension	1.1×10^7	220	50	
0.1% Supernatant	8×10^5	100	—	
0.5% Suspension	6.4×10^6	140	—	
0.5% Supernatant	4.0×10^6	60	67	
α-Factor IX-Sepharose	0	—	—	
Flow-through	4.0×10^6	60	67	
Propeptide eluant	0	—	—	

[a] The protein concentration determinations for precolumn samples were performed by BCA. The propeptide eluant protein concentration was determined by densitometry of gel-electrophoresed samples, using BSA as a standard, as previously described [K. L. Berkner, M. Harbeck, S. Lingenfelter, C. Bailey, C. M. Sanders-Hinck, and J. W. Suttie, *Proc. Natl. Acad. Sci. U.S.A.* **89,** 6242 (1992)]. The samples were assayed in the presence of 20 μM propeptide.

50 ml buffer A containing 1 m*M* ATP and 5 m*M* $MgCl_2$, then an additional 50 ml of buffer A. The ATP–$MgCl_2$ wash removes BiP,[17] a major contaminant in the preparation. We have also tried column washes using buffer A containing 0.1–0.5% (w/v) Triton X-100. Use of this detergent increased the purity of carboxylase isolated from bovine liver[2] (Fig. 1B); however, in our isolations from tissue culture cell lines it has not been required (Fig. 1A).

After the column washes, factor X propeptide (100 μM in 5 ml of buffer A) is added and the column is rocked overnight, at room temperature. After the propeptide eluant is collected, an additional equal amount of propeptide is added and the column is rocked for another 24 hr. Approximately 70 and 30% carboxylase activity are recovered in the first and second elutions, respectively. When the propeptide eluant is analyzed by sodium dodecyl sulfate–polyacrylamide gel electrophoresis (SDS–PAGE) an apparently homogenous preparation of protein is obtained (Fig. 1A). The

[17] S. Munro and H. R. B. Pelham, *Cell* **46,** 291 (1986).

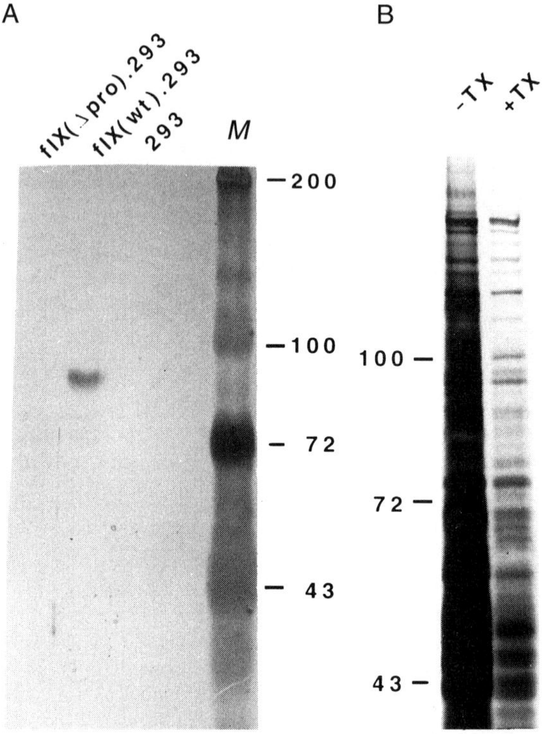

FIG. 1. Analysis of propeptide eluants by SDS–PAGE. Propeptide eluants isolated from (A) tissue culture microsomes absorbed to anti-human factor IX antibody resin or (B) bovine liver microsomes adsorbed to anti-bovine prothrombin antibody resin were silver stained after gel electrophoresis. In (A) the propeptide eluant is from three different microsome preparations: untransfected 293 cells or 293 cells stably transfected with either native factor IX (wt) or a factor IX variant lacking the propeptide (Δpro). In (B) the column washes were performed with or without a 0.1–0.5% Triton X-100 (TX) gradient.

molecular weight of this protein is identical to that of tissue-isolated carboxylase.[2,3] Factor IX is not detected in the propeptide eluant, either by gel analysis or by a sensitive ELISA. A 3.5×10^4-fold purification from the starting 0.1% suspension, or 1.5×10^4-fold purification from the 0.5% solubilized microsomes, is obtained (Table I). These numbers are consistent with the 7000-fold purification from solubilized bovine liver microsomes[2] and the observation that starting bovine liver microsomes contain two to three times as much carboxylase activity as 293 microsomes[1–3] (Table I). The final specific activity is based on a protein concentration determined using SDS–PAGE and referenced to coelectrophoresed bovine serum albumin (BSA) standards, because of the propeptide present in the final eluant.

The carboxylase activity in the propeptide eluant is extremely stable. We have observed 100% retention of activity with preparations stored at −80° for as long as 2 years, even after multiple freeze-thawing of the samples.

We have shown that the apparently single molecular weight form observed in the purified carboxylase preparation (Fig. 1A) actually comprises two proteins of near-identical molecular weights.[1] One protein is the carboxylase.[18] The other is a protein we have called GCAP, for gamma-carboxylase-associated protein, whose role in carboxylation is currently being investigated.[19] GCAP copurified with the carboxylase from both bovine liver and 293 cells.[1,3,19] The stoichiometric amounts of GCAP and carboxylase in the 95-kDa band are presently unknown. GCAP is not detected in BHK- or insect-derived microsomes, and propeptide eluants isolated from these sources, which appear to be homogenous for carboxylase, do not contain detectable GCAP (data not shown).

The anti-factor IX resin is regenerated by washing the column with 50 m*M* Tris, pH 7.4, 100 m*M* NaCl, then with 4 *M* guanidine hydrochloride (GuCl), and then with 50 m*M* Tris, pH 7.4, 100 m*M* NaCl. The polyclonal anti-factor IX antibody is stable to the GuCl treatment. We have reused the same anti-factor IX resin approximately 25 times, with no observed decrease in column capacity.

An unusual feature of the carboxylase purification is the lengthy incubations required both for adsorption and for elution of the carboxylase from the antibody column. We have determined that the minimum length of time necessary for adsorption of ~90% of the carboxylase is 4 hr. The carboxylase activity in the solubilized microsomes is fairly stable. For example, we assayed solubilized extract, stored at 4°, on a daily basis, and observed only a 50% decrease in activity after 1 week. Because of this stability, carboxylase adsorption to anti-factor IX resin is performed overnight, for convenience. Propeptide elution also occurs slowly, requiring 2 days for full recovery. The elution is specifically due to the propeptide; when a random peptide was used only a small amount (~5%) of the carboxylase "eluted" from the anti-factor IX antibody column. In addition to the factor X propeptide, we have also tested the factor IX-, protein C-, and prothrombin-propeptide sequences and we obtain similar recoveries of carboxylase using these peptides.[14] Both the temperature (20–22°) and a high NaCl concentration (0.5 *M*) are important for optimal recovery during propeptide elution; for example, at 4° or 0.1 *M* NaCl only 10–20% activity is recovered.

[18] S. M. Wu, W. F. Cheung, D. Frazier, and D. W. Stafford, *Science* **254,** 1634 (1991).

[19] S. L. Lingenfelter, B. A. McNally, S. Mathewes, J. Johnson, K. L. Walker, C. Bailey, P. O'Hara, and K. L. Berkner, (submitted).

We have performed the carboxylase isolation from matched sets of 293 cell lines that are either untransfected or stably transfected with r-factor IX (Table I). Carboxylase adsorption to anti-factor IX resin is only observed in the isolation from factor IX-containing cells. We have also performed an isolation comparison between 293 cell lines stably transfected with native r-factor IX or with a r-factor IX variant lacking the propeptide.[1] Again, carboxylase adsorption to anti-factor IX resin is only observed in the isolation from native factor IX-containing microsomes. This comparison shows that the carboxylase adsorption to anti-factor IX is specific, and is due to its association with factor IX containing the propeptide. This specificity is also clear from an analysis of the propeptide eluants obtained from untransfected 293 cells or 293 cells stably transfected with native- or propeptide-deleted factor IX, where carboxylase is only observed in the isolation from native factor IX-containing 293 cells (Fig. 1A).

Although almost all of the carboxylase in native factor IX-expressing 293 cells appears to be bound to factor IX, as indicated by >95% adsorption of carboxylase activity to anti-factor IX resin (Table IA), a substantial amount of inactive carboxylase that does not copurify with factor IX is detected by Western analysis. Figure 2 shows that when the carboxylase fractionation on the anti-factor IX column is monitored by a Western blot using an anticarboxylase antipeptide antibody, a substantial amount of carboxylase protein is observed in the flow-through. A mixed population of active and inactive enzyme is not surprising, because the carboxylase can be inactivated during its solubilization from microsomes (as described later). This mixture is important to consider with regard to the different approaches that can be used to isolate the carboxylase. For example, if an anticarboxylase antibody (as described later) or an antibody to an epitope tag is used, both inactive and active carboxylase will be purified. Thus, isolation of carboxylase via its association with factor IX, which apparently fractionates inactive from active carboxylase (Fig. 2, Table I), will yield a carboxylase preparation with a higher specific activity.

Purification of r-Carboxylase from Mammalian Cell Lines

Purification Using Initial Isolation of Factor IX–Carboxylase Complex

The main criterion of a suitable cell line for purifying r-carboxylase based on the isolation of a carboxylase–VKD protein complex is that the VKD protein is in sufficient intracellular excess over the carboxylase so that most of the carboxylase will be bound to it. The screen for cell lines coexpressing r-carboxylase and r-factor IX therefore includes evaluating intracellular r-factor IX levels as well as r-carboxylase expression.

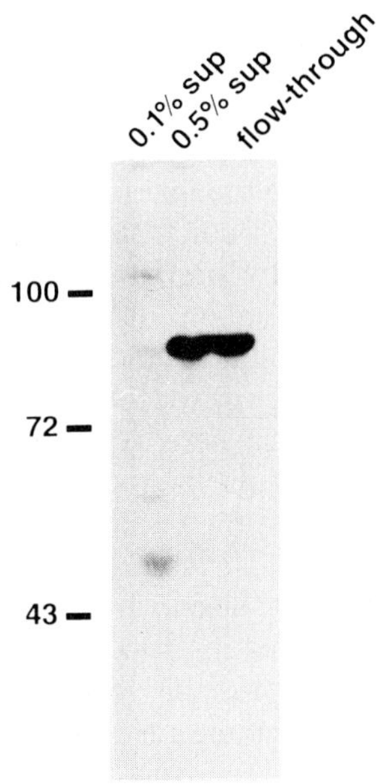

FIG. 2. Western blot analysis of solubilized microsomes from native factor IX-293 cells chromatographed on immobilized anti-factor IX polyclonal antibody. Microsomes prepared from 293 cells expressing r-factor IX were solubilized in 0.1% and then 0.5% CHAPS, as described in the text. The 0.5% supernatant was adsorbed to polyclonal anti-factor IX resin, and the starting material and flow-through were analyzed along with the 0.1% supernatant (0.5 mg each) in a Western blot using an anticarboxylase antipeptide antibody and detection with ^{125}I-labeled protein A.

The 293 cells are cotransfected with 10 μg each of r-factor IX/pD5[15] and the carboxylase cDNA subcloned behind the metallothionien promoter in ZEM228, a vector that also contains a cassette for expressing a selectable marker (neomycin phosphotransferase) from the same plasmid.[20] After selection using G418, stably transfected clones are isolated and expanded (to ~2 × 10^5 cells) and the expression of factor IX and carboxylase is then measured. Factor IX levels are determined using either an ELISA or Western blot, which are both sufficiently sensitive assays for detecting positive clones, even with the small number of cells analyzed. To measure carboxylase peptide activity from small numbers of cells, we have developed a quick, reliable screen using cell lysates. Cells are dislodged in Versene (1 ml), rinsed twice in 1 ml cold PBS, and then resuspended in 1 ml cold

[20] S. J. Busby, E. Mulvihill, D. Rao, A. A. Kumar, P. Lioubin, M. Heipel, C. Sprecher, L. Halfpap, D. Prunkard, J. Gambee, and D. C. Foster, *J. Biol. Chem.* **266,** 15286 (1991).

0.25 *M* sucrose, 0.025 *M* imidazole, pH 7.3, and 1 m*M* PMSF. Cells are freeze-thawed twice, alternating between dry ice–ethanol and 4° incubation, and the crude lysates are then assayed. Carboxylase activity is stable to freeze-thawing, even with multiple (e.g., four) cycles, and most of the carboxylase activity is released in the first freeze-thaw. The results obtained with this assay fairly accurately predict the carboxylase levels of cell lines that are subsequently scaled up for preparation of microsomes. The advantage of screening such small numbers of cells is that the amount of cell culture required to identify a suitable clone is considerably reduced.

Figure 3A shows an example of two 293 cell lines that express r-factor IX and that express carboxylase activity 5- or 20-fold higher than the endogenous levels. The intracellular factor IX levels for each cell line are approximately 2-fold higher than the carboxylase levels (data not shown), when a specific activity of 3×10^6 cpm/μg/hr is used to determine the amounts of carboxylase. We have also overexpressed the r-carboxylase and r-factor IX in BHK cells. In this instance, the r-carboxylase cDNA was subcloned into ZEM229, an expression vector similar to ZEM228, which contains a selection cassette encoding resistance to methotrexate. Because

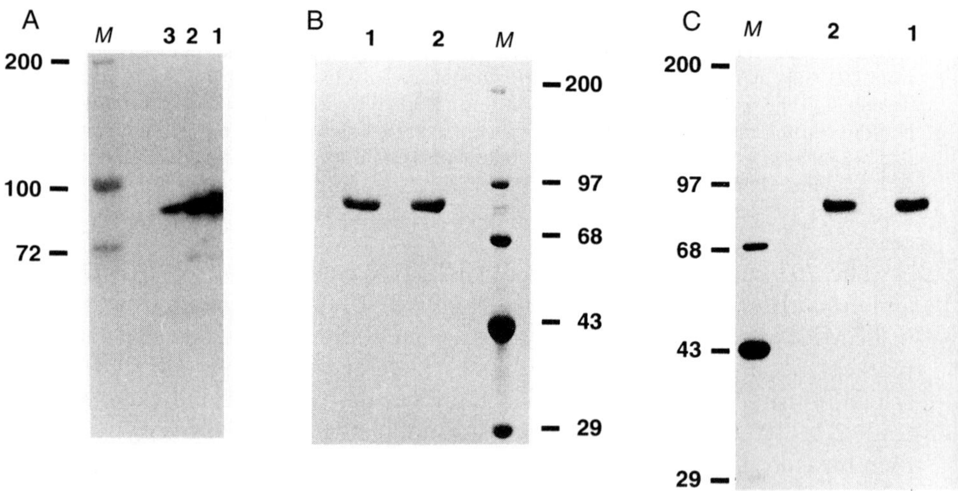

FIG. 3. Analysis of microsomes and purified carboxylase from 293 cells expressing r-carboxylase and r-factor IX. (A) Solubilized microsomes (0.5 mg) isolated from 293 cells stably transfected with r-factor IX (3) or r-factor IX and r-carboxylase at 5-fold (2) or 20-fold (1) higher levels than endogenous carboxylase were analyzed in a Western blot using an anticarboxylase antipeptide antibody and detection with ^{125}I-labeled protein A. (B, C) Propeptide eluants derived from the 293 cells stably transfected with r-factor IX and r-carboxylase (1,2) were gel-electrophoresed and then analyzed by staining with Coomassie (B) or by silver staining (C).

BHK cells amplify this plasmid in response to methotrexate selection, much higher levels of r-carboxylase expression can be achieved in BHK cells than in 293 cells. For example, we have obtained r-carboxylase expression that is 70-fold higher than the endogenous BHK carboxylase levels.

The purification of carboxylase from the r-carboxylase-, r-factor IX–stably transfected 293 cell lines is performed as described in the previous section, with one modification. Because the intracellular factor IX levels are no longer in vast excess over carboxylase, the first CHAPS extraction (at 0.1%) is omitted. A cell line is scaled up (to 2×10^9 cells) and microsomes are prepared as above. After resuspension of the microsomes in 50 m*M* Tris, pH 7.4, 100 m*M* NaCl, 1 m*M* PMSF (15 ml) the optimal solubilization conditions are determined by incubating 500-μl aliquots with varying concentrations of CHAPS and NaCl for 1 hr at 4° followed by centrifugation for 1 hr at 45,000 rpm at 4°, using an SW55 rotor and adapters (Beckman Instruments, Fullerton, CA, model number 456860) for centrifuging small volumes. To determine total activity and percent solubilization, both the suspension and supernatant are assayed for carboxylase activity. As shown in Table II, as much as 70% solubilization of carboxylase activity is obtained. At higher concentrations of CHAPS, a substantial decrease in both total (i.e., suspension) and solubilized carboxylase activity is observed (data not shown). These optimal solubilization conditions (i.e. 0.7% CHAPS,

TABLE II
SOLUBILIZATION OF MICROSOMES FROM A R-CARBOXYLASE-, R-FACTOR IX-EXPRESSING 293 CELL LINE

CHAPS(%)	NaCl (*M*)	Activity: In suspension ($cpm/hr \times 10^{-3}$)[a]	Activity: In supernatant ($cpm/hr \times 10^{-3}$)[a]	Solubilization (%)
0.3	0.1	179	14	8
	0.2	169	39	23
	0.3	183	57	31
	0.4	169	68	40
0.5	0.1	178	32	18
	0.2	179	90	50
	0.3	220	101	46
	0.4	174	115	66
0.7	0.1	209	55	26
	0.2	200	106	53
	0.3	178	117	66
	0.4	161	118	73

[a] The suspension and supernatant samples (20 μl) contained 110 and 34 μg protein, respectively, as determined by BCA (Pierce).

0.3–0.4 *M* NaCl) depend on the protein concentration of the microsomal suspension (~6 mg/ml in this case). Thus, once the solubilization conditions have been established, it is important to generate microsomes with the same protein concentration, which, as indicated above, is done by carefully monitoring and using the same number of cells each time.

The carboxylase–factor IX complex from solubilized microsomes (0.7% CHAPS, 0.4 *M* NaCl, Table II) is adsorbed to polyclonal anti-factor IX antibody resin, then washed and eluted exactly as described in the previous section. When analyzed by SDS–PAGE, an apparently homogenous preparation of protein is obtained (shown for two different preparations in Figs. 3B and 3C), representing an approximate 900-fold increase in specific activity. Pure preparations of r-carboxylase have also been obtained from BHK cell lines (data not shown). When compared with the purified endogenous carboxylase (previous section), a proportional increase in Coomassie staining material with respect to peptide activity is observed (data not shown). Thus, the specific activities of r-carboxylase and endogenous 293 carboxylase are indistinguishable.

Affinity Purification of Free Carboxylase Using Immobilized Propeptide Ligand

BHK cell lines expressing r-carboxylase are generated by transfecting the human carboxylase cDNA in ZEM229 (20 μg) into BHK cells, followed by selection using 1 μM methotrexate. The BHK cell lines do not contain detectable levels of VKD proteins, as determined by *in vitro* carboxylation of microsomal BHK preparations using $^{14}CO_2$, followed by SDS–PAGE (data not shown). Thus, the r-carboxylase BHK clones express carboxylase that is not bound to any VKD protein. Individual clones are screened for carboxylase activity using the lysate screen described above for assaying small numbers (e.g., 2×10^5) of cells. The highest expressing clones are subsequently analyzed in Western blots using an anti-C-terminal anticarboxylase antibody (described later) and also by peptide activity assays on sonicated cell lysates. For this assay, cells (2×10^7) are resuspended in 5 ml 0.25 *M* sucrose, 0.025 *M* imidazole, pH 7.3, 1 m*M* PMSF, sonicated for 4–15-sec bursts and then centrifuged at 4000*g* for 15 min at 4°. Aliquots are stored at −80°, and reproducible carboxylase peptide activity values can be obtained with samples stored for at least a year.

A BHK cell line expressing the highest amount of r-carboxylase is scaled up (to 2×10^9 cells), and microsomes are prepared as described in the previous section. Optimal solubilization conditions with NaCl and CHAPS are determined, as described in the first part of this section. Solubilized carboxylase (i.e., the 100,000*g* supernatant) is adsorbed to immobilized propeptide ligand by rocking the supernatant with the resin overnight at

TABLE III
PURIFICATION OF R-CARBOXYLASE ON PROPEPTIDE LIGAND OR ANTICARBOXYLASE ANTIPEPTIDE ANTIBODY COLUMNS

Fraction	Carboxylase activity (cpm/hr × 10^{-3})
A. 0.5% Supernatant	20
Immobilized propeptide ligand	3
Flow-through	16
DTT eluant	3
B. 0.5% Supernatant	10.8
Anticarboxylase antipeptide antibody	9.2
Flow-through	0.6
Peptide eluant	9.2

4°. We have used two different methods to immobilize the propeptide. A factor IX propeptide (TVFLDHENANKILNRPKRY) is coupled to CNBr-activated Sepharose via the amino groups.[21] Alternatively, a factor IX propeptide with a *cys* appended to the C terminus is coupled via the thiol,[12] to activated thiol-Sepharose 4B (Pharmacia, Piscataway, NJ). Although the thiol-coupled propeptide ligand would be predicted to be more accessible to the carboxylase, surprisingly both immobilized ligands work equally well. Both factor IX and factor X propeptides have been tested and yield essentially identical results. We have varied the concentration of propeptide coupled to resin (from 1–10 mg/ml factor IX propeptide attached via amino or thiol groups). Even at 1 mg/ml the propeptide is in vast excess (~100-fold) over the carboxylase, and similar adsorption efficiencies for the carboxylase and ultimate recoveries of activity are obtained over the range tested.

After carboxylase adsorption to the resin, the column containing propeptide coupled to CNBr-activated Sepharose is washed in buffer A (50 m*M* Tris, pH 7.4, 500 m*M* NaCl, 0.1% CHAPS, 0.1% phosphatidylcholine IIIE, 5 m*M* DTT, e.g., 100 ml for a 5-ml column) and eluted with propeptide as described in the previous section. The propeptide–thiol-Sepharose is washed in buffer A that lacks DTT, and the carboxylase is eluted with buffer A. Carboxylase adsorption is monitored by assaying the starting material, the flow-through, and the carboxylase-bound resin. Carboxylase elution is monitored by assaying the propeptide or DTT eluant and the resins pre- and postelution. Table III (A) shows the results obtained when 5 mg/ml solubilized microsomes (using 0.5% CHAPS, 0.3 *M* NaCl, followed by centrifugation at 100,000*g* for 1 hr at 4°) are applied to a factor IX

[21] E. Harlow and D. Lane, "Antibodies: A Laboratory Manual." Cold Spring Harbor Laboratory Press, Cold Spring Harbor, New York, 1988.

propeptide–thiol-Sepharose column (5 mg propeptide/ml, 5 ml resin), and the resin is washed and eluted as described earlier. Only ~15% of the carboxylase activity is adsorbed to the resin, and virtually all of this activity is recovered in the eluant. No increase in carboxylase binding is observed if the adsorption step is performed for an additional 24 hr (data not shown). Presumably the nonquantitative binding is due to the carboxylase orientation in the micelle, which limits carboxylase accessibility to the immobilized propeptide ligand. We have tried to increase the binding by using higher concentrations of CHAPS in the initial solubilization and shorter times for incubating the resin and solubilized microsomes. However, these efforts have not been successful, and result only in the irreversible loss of carboxylase activity.

Purification of Carboxylase on Immobilized Anticarboxylase Antipeptide Antibody Column

Rabbit polyclonal antibody to a KLH-coupled human carboxylase derived C-terminal peptide (SNPPESNPDPVHSEF) is generated and affinity purified on an immobilized peptide column (10 mg peptide per milliliter, the peptide is coupled to CNBr-activated Sepharose).[21] Affinity purified antibody is then coupled to Sepharose (to 5 mg/ml). Microsomes from a BHK cell line expressing r-carboxylase are solubilized as described in the first part of this section and, after centrifugation (100,000*g*, 1 hr, 4°), the supernatant is rocked overnight with the anticarboxylase antipeptide antibody resin at 4°. The resin is washed, as described in the previous section, and then eluted with a mixture of 100 μM carboxylase–C-terminal peptide and 100 μM factor X propeptide in 50 m*M* Tris, pH 7.4, 500 m*M* NaCl, 0.1% phosphatidylcholine IIIE, 0.1% CHAPS, and 5 m*M* DTT. Propeptide is included to stabilize the carboxylase. Elution is at 22° for 24 hr, with the sample rocking. As seen in Table III (B) the affinity of the anti-C-terminal carboxylase antipeptide antibody is sufficient for immunopurification, with virtually all of the carboxylase bound to the antibody resin. Moreover, the elution yields a near quantitative recovery. We have also used this antibody resin to isolate carboxylase from r-carboxylase-, r-factor IX-expressing 293 cell lines, where the carboxylase is in nearly complete complex with factor IX (data not shown). Quantitative adsorption and recovery of carboxylase from this source is also obtained, showing that the carboxylase is recognized by this antibody even when bound to factor IX.

This affinity purification thus provides a reliable alternative to the isolation of carboxylase from carboxylase–VKD protein complexes, presented above. This purification will be particularly useful for isolating carboxylase variants with mutations that affect binding to VKD proteins. As discussed, the anticarboxylase antipeptide antibody will recognize both active and inactive carboxylase, while the purification based on the initial isolation of a carboxylase–factor IX complex fractionates the two different populations.

Purification of r-Carboxylase from Baculovirus-Infected Insect Cells

Generation of Baculovirus Containing Factor IX or Carboxylase

A *Bam*HI fragment encoding full-length factor IX (1.4-kb cDNA) or carboxylase (2.5-kb cDNA) is subcloned into the polylinker of pBacPAK8 (Clontech, Palo Alto, CA), placing these cDNAs downstream of the polyhedron promoter. Plasmids (0.5 μg) are incubated with *Bsu*36I-digested viral DNA (BacPAK6, Clontech, 100 ng) in 100 μl 0.05% lipofectin for 15 min at room temperature. The transfection mix is then added dropwise to the medium of freshly plated spodoptera frugiperda (SF) 21 cells (1×10^6) and after 5 hr at 26° the medium is changed.

*Bsu*36I cleaves the viral DNA three times, essentially lowering the wild-type virus background to zero in subsequent progeny. Digested DNA can be purchased from Clontech, or alternatively a large stock (e.g., 1 mg) of viral DNA can be prepared and digested.[22,23] We have observed similar transfection efficiencies and wild-type virus backgrounds using either source.

Transfected cells are incubated at 26° for 3 days, and the media are harvested and stored at 4°. Clonal recombinant viruses are obtained by serially diluting the media (10^{-1} to 10^{-3}), infecting 1.5×10^6 SF21 cells with 100 μl of the inoculum for 1 hr, followed by overlaying the cells with 1.5 ml 1% SeaPlaque agarose (FMC, Rockland, ME) in insect cell medium. After 5 days, plaques are visualized by staining with a 0.03% neutral red overlay in PBS (1 ml). After 4 hr the stain is aspirated and the following day approximately 20 well-separated plaques are picked into 500 μl media. Although the wild-type virus background is low and most plaques are the r-viruses, a large number of plaques are still isolated because we have occasionally observed large variability in the subsequent expression of the virus (e.g., shown for a carboxylase-containing baculovirus in Table IV). The virus stocks are stable at 4° indefinitely. Thus, if low levels of expression are observed for an individual isolate of virus, additional plaques from these stocks can be evaluated.

[22] D. R. O'Reilly, L. K. Miller, and V. A. Luckow, "Baculovirus Expression Vectors: A Laboratory Manual" W. H. Freeman, New York, 1992.

[23] M. D. Summers and G. E. Smith, Texas Agricultural Experiment Station Bulletin No. 1555 (1988).

TABLE IV
CARBOXYLASE ACTIVITY IN BACULOVIRUS (CARBOXYLASE)-INFECTED SF21 CELLS[a]

Virus	Peptide activity (cpm/hr/μg)
Carboxylase 1-1	3800
Carboxylase 1-2	4200
Carboxylase 2-1	90
Carboxylase 2-9	3000
Wild type	0
Mock-infected	0

[a] SF21 cells (2×10^7) were infected with four clonal isolates of baculovirus-containing carboxylase or with wild-type virus (multiplicities of infection of 10 for each). Aliquots (50 μl) of sonicated cell lysates (5 ml total volume, prepared as described in the text) were assayed for carboxylase peptide activity.

To prepare and screen P1 stocks of carboxylase-containing baculovirus, aliquots of the plaque isolates (100 μl) are used to infect 5×10^5 SF21 cells for 1 hr. Media is added and after 4 days at 26° the media are collected and stored at 4°. To screen for recombinant virus, 10 μl of each medium is dotted onto nitrocellulose, and the blot is air dried overnight, followed by UV cross-linking (Stratalinker 2400, Stratagene, La Jolla, CA). The nitrocellulose is processed as in Southern blot analysis[24] using a randomly primed carboxylase cDNA (8×10^8 cpm/μg) as a probe. Controls include mock-infected cell supernatants and carboxylase cDNA (1 ng). Several positive P1 viruses are then used to generate P2 stocks, which are then screened for carboxylase expression by assaying carboxylase peptide activity in P2-infected SF21 cell lysates (Table IV). We have also screened for carboxylase expression by Western blot analysis. We have observed a significant background problem with infected insect cells, with some of the affinity purified antipeptide antibodies, using either ^{125}I-labeled protein A or commercially available detection systems (e.g., Immun-Lite chemiluminescent assay, Bio-Rad). Prominent bands around the carboxylase molecular mass form (95 kDa) are observed, which are also observed in the mock- or wild-type-infected SF21 cell lysates, and the molecular mass forms differ for the various antipeptide antibodies used. To improve the background,

[24] F. M. Ausubel, R. Brent, R. E. Kingston, D. D. Moore, J. G. Siedman, J. A. Smith, and K. Struhl, "Current Protocols in Molecular Biology." John Wiley & Sons, New York, 1995.

quantitative immunoprecipitation is performed with one anticarboxylase antipeptide antibody and a different antibody is used as the probe in Western blot analysis.

To screen for baculovirus-containing factor IX, P1 stocks are generated as above and the media are assayed by ELISA.[1] Several positive stocks are subsequently analyzed for expression levels after the generation of P2, and then P3, virus stocks.

Both the P2 and P3 stocks of baculovirus-containing factor IX or carboxylase are titered, using 10^{-4} to 10^{-6} serially diluted virus. This method of quantitation is accurate enough to obtain reproducible expression levels when SF21 cells are infected with each individual virus.

Infection Conditions

The optimal expression times for carboxylase and factor IX are determined by measuring carboxylase peptide activity or factor IX antigen levels at 12-hr intervals following infection. SF21 cells (2×10^7) are infected at multiplicities of infection (MOI) of 10 for each virus and at the appropriate times the supernatants (for factor IX) or cell lysates (for the carboxylase, 5-ml sonicates prepared as described for r-carboxylase in mammalian cells, in the previous section) are prepared and assayed. Expression onset is observed around 12 hr and continues to increase out to the last time measured, 72 hr. Cell lysis in infected cells becomes a problem at late infection times, and we have observed proteolytic degradation of factor IX after ~48 hr (Fig. 4A). Consequently, we use an earlier harvest time, 42 hr, where factor IX degradation has not been observed, thereby optimizing the quality of the preparation at the expense of decreasing the overall yield (by about two-fold).

In coinfection experiments, at MOIs of 10 each for factor IX- and carboxylase-containing viruses, SF21 cell lysates (from 2×10^7 cells in a final 5 ml volume) contain 600–700 ng/ml of each protein (Table V). However, these coinfection experiments are complicated by the fact that each virus (factor IX- or carboxylase-containing) suppresses the expression of the other during infection (Table V). Thus, for example, infection of just the factor IX virus results in a secretion level of 4.4 μg/ml factor IX, in contrast to 1.6 μg/ml factor IX when equal amounts of factor IX and carboxylase viruses are coinfected. When equal amounts of factor IX and carboxylase virus are used, carboxylase expression is reduced to 72% of the level of carboxylase when expressed alone (Table V). Moreover, with factor IX:carboxylase MOIs of 2:1 or 3:1, the carboxylase levels are further reduced, to 8–10% (data not shown). Because the MOIs are determined by an assay that is not particularly quantitative (e.g., in duplicate

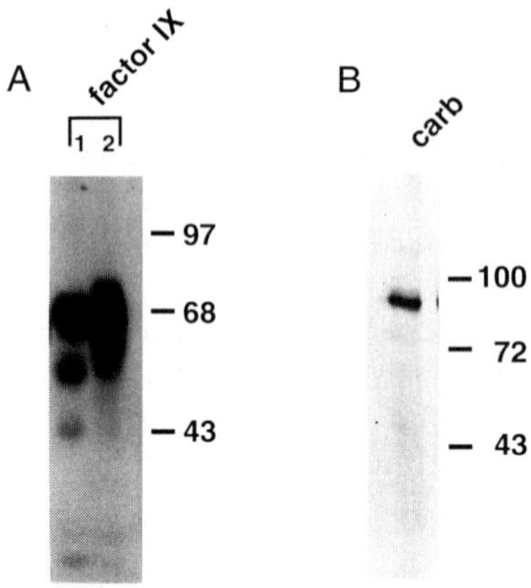

FIG. 4. Analysis of r-factor IX and r-carboxylase produced in infected SF21 cells. (A) Factor IX from SF21 cells infected with factor IX-containing baculovirus was purified on an anti-factor IX monoclonal antibody column (ESN1, American Diagnostica, Greenwich, CT, 2 mg antibody per milliliter, 2 ml resin) and then analyzed in a Western blot using affinity purified anti-factor IX polyclonal antibody and detection by chemiluminescence (Immun-Lite, Bio-Rad). Proteolytic digestion of factor IX was observed at late times (>48 hr) in infection, as shown in lane 1, but not at earlier times (42 hr, lane 2). (B) The propeptide eluant derived from SF21 cells coinfected with baculoviruses-containing factor IX and carboxylase, isolated as described in the text, was analyzed by SDS–PAGE and silver staining.

TABLE V
EXPRESSION OF FACTOR IX AND CARBOXYLASE IN COINFECTED SF21 CELLS[a]

Multiplicities of infection (factor IX : carboxylase)	Secreted factor IX (μg/ml)	Intracellular factor IX (μg/ml)	Carboxylase activity (μg/ml)[b]	Carboxylase specific activity (cpm/hr/μg)
10:0	4.4	2.1	0.0	0
0:10	0.0	0.0	0.8	2500
10:10	1.6	0.7	0.6	1800
7:10	1.2	0.5	0.6	1800
4:10	0.6	0.3	0.6	1700
2:10	0.2	0.1	0.7	2200
1:10	0.1	0.0	0.7	2100

[a] SF21 cells (2×10^7) infected with factor IX- and carboxylase-containing baculoviruses were assayed for factor IX expression by an ELISA in both media (10 final volume) and in sonicated cell lysates (5 ml, prepared as described in the text).

[b] The carboxylase peptide activity, measured on cell lysates, was determined using a specific activity of 3×10^6 cpm/hr/μg.

titer determinations the numbers can differ by two- to threefold), the suppression of expression of one virus by another introduces unacceptable variability in coinfection experiments. To overcome this potential problem, the factor IX and carboxylase virus stocks are titered and then always optimized for coexpression by performing a series of infections at different MOIs and quantitating the consequent carboxylase and factor IX levels, as shown in Table V. Clearly, preparing a large stock of each virus is desirable, since so much characterization is required.

Solubilization of Carboxylase from Coinfected SF21 Cells

SF21 cells (2×10^8) are coinfected with baculoviruses containing factor IX and carboxylase (MOIs of 10, pretested as described earlier in small-scale experiments) and after 42 hr microsomes are prepared. The cells are dislodged by scraping, centrifuged at 2000*g* for 5 min at 4° and then rinsed twice with cold PBS (100 ml). After resuspension in 10 ml of 0.25 *M* sucrose, 0.025 *M* imidazole, pH 7.3, and 1 m*M* PMSF, the cells are sonicated, centrifuged, and the microsomal pellet frozen, as described earlier.

The optimal solubilization conditions for infected SF21 microsomes are determined after resuspending the microsomal pellet in 50 m*M* Tris, pH 7.4, 100 m*M* NaCl, and 1 m*M* PMSF. Because the intracellular factor IX and carboxylase levels are similar and sequential solubilization steps are not required to remove the excess factor IX, only one solubilization is performed. Aliquots (500 μl) with different concentrations of CHAPS and NaCl are rocked at 4° for 1 hr and then centrifuged for 1 hr at 100,000*g* at 4°. The suspension and supernatants are assayed for peptide activity. As seen in Table VI, the optimal solubilization conditions for a 2 mg/ml microsomal suspension are approximately 0.3% (w/v) CHAPS, 0.6 *M* NaCl. With these conditions only ~20% carboxylase activity is solubilized, in contrast to the 70% solubilization we obtained from mammalian cells (Table II). When higher concentrations of CHAPS or NaCl are used to try to effect more solubilization, a large decrease in carboxylase activity is observed (Table VI). As emphasized for the mammalian cell experiments, reproducible solubilization depends on generating reproducible protein concentrations in microsomes.

Purification of the Carboxylase

To purify the carboxylase, a 2 mg/ml suspension of microsomes (10 ml) is solubilized in 0.4% CHAPS, 0.4 *M* NaCl for 1 hr and then centrifuged at 100,000*g* for 1 hr, both at 4°. The supernatant is adsorbed to polyclonal anti-factor IX antibody (5 mg/ml, 1 ml), then washed and eluted with propeptide, as described in the second section. More than 90% of the

TABLE VI
SOLUBILIZATION OF R-CARBOXYLASE-, R-FACTOR IX-EXPRESSING SF21 CELLS

CHAPS(%)	NaCl (M)	Activity: In suspension (cpm/hr × 10^{-3})[a]	Activity: In supernatant (cpm/hr × 10^{-3})[a]
0.3	0.2	118	10
	0.4	118	18
	0.6	118	20
0.5	0.2	111	20
	0.4	111	21
	0.6	111	12
0.7	0.2	63	5
	0.4	63	3
	0.6	63	2
0.9	0.2	30	2
	0.4	30	1
	0.6	30	1

[a] The suspension and supernatant samples (20 μl) contained 44 and 22 μg protein, respectively, as determined by BCA (Pierce).

carboxylase binds to the anti-factor IX resin (data not shown), indicating that most of the carboxylase is in complex with factor IX. When the propeptide eluant is analyzed by SDS–PAGE, a near homogenous preparation of protein is obtained (Fig. 4B), with a similar molecular weight to that of tissue- or mammalian cell-isolated carboxylase[1–3] (Figs. 1A, 3B and 3C).

Practical Considerations in Isolating Carboxylase from Mammalian Cells versus Insect Cells versus Tissue

The generation and scaleup of mammalian cell lines expressing r-carboxylase and/or r-factor IX requires considerably less effort than generating and titering baculoviral stocks. Moreover, the supply of virus needs to be replenished continually for scaled up carboxylase production. With the r-carboxylase stably transfected 293 cell lines, a comparable expression level is obtained to that of SF21 cells infected with carboxylase-containing baculovirus (Tables II and VI). With r-carboxylase-, r-factor IX-stably transfected BHK cell lines, where the r-carboxylase expression is amplified, levels four to five times higher than infected SF21 cells have been obtained (data not shown). In our hands, the solubilization of mammalian cell microsomes gives a better recovery of total carboxylase activity than using infected SF21 cells (Tables II and VI). Thus, for the amount of effort as well as for ultimate recoveries, the mammalian cell source is preferred for producing large amounts of protein. However, for analysis of mutant carboxylase forms, the insect cell expression system provides a distinct advantage in lacking endogenous carboxylase. The insect cell system also has the advantage of more readily allowing the combinatorial coexpression of different carboxylase variants and VKD proteins by coinfection of appropriate stocks, as opposed to having to generate a new mammalian cell line each time for a given set of VKD or carboxylase proteins.

The preparation of microsomes from tissue culture cells requires substantially less time and effort than from tissue. Moreover, the single step purification from cell line microsomes is much faster than from tissue microsomes,[1-3] and this purification is highly reproducible for obtaining a homogenous preparation of protein (Figs. 3B and 3C). Thus, the major cost from tissue isolation is for labor while the main cost for cell lines is for the cell culture. By manipulating the culture conditions, for example, adapting the cells to growth in 5% serum or eliminating the G418 selection agent during the cell scaleup (where the time frame is too short to affect expression stability), the cost for isolating the carboxylase from tissue versus cultured cells is not substantially different.

[28] Purification of Native Bovine Carboxylase and Expression and Purification of Recombinant Bovine Carboxylase

By B. C. Furie, A. Kuliopulos, D. A. Roth, I. Sugiura, C. T. Walsh, and B. Furie

Introduction

Posttranslational conversion of glutamic acid to γ-carboxyglutamic acid in specific proteins is the only biosynthetic reaction known to be dependent on vitamin K. In addition to CO_2 the enzyme that catalyzes this reaction, the vitamin K-dependent γ-glutamyl carboxylase, requires stoichiometric amounts of reduced vitamin K hydroquinone and O_2. In the presence of CO_2, O_2, reduced vitamin K, and a glutamate-containing peptide, the products of the enzyme reaction catalyzed by the carboxylase are γ-carboxyglutamate in the peptide substrate, vitamin K epoxide, and H_2O. *In vivo* the vitamin K epoxide is recycled to the hydroquinone by reduction, a reaction inhibited by the vitamin K antagonist warfarin.[1,2] The protein

[1] J. W. Suttie, *Annu. Rev. Biochem.* **54,** 459 (1985).
[2] C. Vermeer, *Biochem. J.* **266,** 625 (1990).

0076-6879/97 $25.00

substrates of the carboxylase are identified by a γ-carboxylation recognition site usually within the propeptide of the substrate proteins.[3,4] Although the vitamin K-dependent carboxylase is widely distributed in mammalian cells, substrates for the enzyme have been identified in only a limited number of tissues. Known protein substrates of the vitamin K-dependent carboxylase include precursor forms of the blood coagulation proteins factor VII, factor IX, factor X, and prothrombin; two protein regulators of blood coagulation protein C and protein S; two bone proteins osteocalcin and matrix Gla protein; and Gas6, a putative ligand for receptor tyrosine kinases.[5–9]

Although the carboxylase has been actively studied for more than 20 years, isolation of essentially homogeneous enzyme and determination of the primary sequence of the enzyme from the sequence of cDNA clones have only recently been achieved. The initial success in purification was based on the strategy of using a peptide containing a γ-carboxylation recognition site as an affinity ligand for the enzyme.[10,11] Although these procedures can provide essentially homogeneous enzyme they are tedious and yield is generally low. The availability of a cDNA clone for the enzyme has provided the opportunity to synthesize the carboxylase with an epitope tag to assist in isolation of recombinant enzyme.[12,13] This chapter describes the purification of the vitamin K-dependent carboxylase from bovine liver[11] and purification of recombinant epitope-tagged bovine carboxylase expressed in Chinese hamster ovary (CHO) cells.[14] We have also found it useful in our studies of the enzyme to have an expression system for recombinant enzyme in which the carboxylase is not endogeneously ex-

[3] B. Furie and B. C. Furie, *Cell* **53,** 505 (1988).
[4] B. Furie and B. C. Furie, *Blood* **75,** 1753 (1990).
[5] G. Manfioletti, C. Brancolini, G. Avanzi, and C. Schneider, *Mol. Cell. Biol.* **13,** 4976 (1993).
[6] T. N. Stitt, G. Conn, M. Gore, C. Lai, J. Bruno, C. Radziejewski, K. Mattsson, J. Fisher, D. R. Gies, P. F. Jones, P. Masiakowski, T. E. Ryan, N. J. Tobkes, D. H. Chen, P. S. Distefano, G. L. Long, C. Basilico, M. P. Goldfarb, G. Lemke, D. J. Glass, and G. Yacopoulos, *Cell* **80,** 661 (1995).
[7] B. C. Varnum, C. Young, G. Elliot, A. Garcia, T. D. Bartley, Y.-W. Fridell, R. W. Hunt, G. Trail, C. Clogston, R. J. Toso, D. Yanagihara, L. Bennett, M. Sylbar, L. A. Merewether, A. Tseng, E. Escobar, E. T. Liu, and H. D. Yanmano, *Nature* **373,** 623 (1995).
[8] K. Ohashi, K. Nagata, J. Toshima, et al., *J. Biol. Chem.* **270,** 22681 (1995).
[9] P. J. Godowski, M. R. Mark, J. Chen, M. D. Sadick, H. Raab, and R. G. Hammonds, *Cell* **82,** 355 (1995).
[10] S.-M. Wu, D. P. Morris, and D. W. Stafford, *Proc. Natl. Acad. Sci. U.S.A.* **88,** 2236 (1991).
[11] A. Kuliopulos, C. E. Cieurzo, B. Furie, B. C. Furie, and C. T. Walsh, *Biochemistry* **31,** 9436 (1992).
[12] S.-M. Wu, W.-F. Cheung, D. Frazier, and D. W. Stafford, *Science* **254,** 1634 (1991).
[13] A. Rehemtulla, D. A. Roth, L. C. Wasley, A. Kuliopulos, C. T. Walsh, B. Furie, B. C. Furie, and R. J. Kaufman, *Proc. Natl. Acad. Sci. U.S.A.* **90,** 4611 (1993).
[14] I. Sugiura, B. Furie, C. T. Walsh, and B. C. Furie, in preparation.

pressed. For this purpose, we have used baculovirus-infected insect cells.[15] The methods for bovine carboxylase expression in insect cells are also included in this chapter.

Assay of γ-Glutamyl Carboxylation

Assay of the formation of γ-carboxyglutamic acid by the vitamin K-dependent carboxylase is performed as previously described in this series with minor modifications.[16] Reduction of vitamin K is carried out as described earlier except that all of the sodium borohydride (8 mg) used to reduce 1 ml of vitamin K (phytonadione injection, 10 mg/ml, USP, Abbott Laboratories, Chicago, IL) is added initially with 5 μl of 2-mercaptoethanol. The reaction is allowed to proceed for 30 min under an N_2 atmosphere at ambient temperature in a foil-wrapped capped vial.

Preparation of reaction mixtures and assay of incorporated $^{14}CO_2$ are as described in Volume 222 of this series except that the assay is run at a final vitamin K concentration of 220 μ*M* rather than the 880 μ*M* concentration used previously. The higher concentrations of vitamin K have been found to be inhibitory.

General Methods

Several pieces of equipment and several reagents are used in more than one of the procedures described in this chapter. These are identified or defined here. When homogenization is described as being carried out using a Polytron the model used is PT 3000 from Brinkmann (Farmingdale, NY). Sonication is performed with an Ultrasonic Processor W-220 fitted with a microprobe. The instrument and probe are supplied by Heatsystems-Ultrasonics, Inc. (Littau, Switzerland). A 10-fold concentrate of protease inhibitors, 10× PIC, is prepared to contain 20 m*M* dithiothreitol (DTT), 20 m*M* EDTA, 1.25 μg/ml FFRCK (Phe-Phe-Arg-chloromethyl ketone, Bachem Bioscience, King of Prussia, PA), 1.25 μg/ml FPRCK (Phe-Pro-Arg-chloromethyl ketone, Bachem Bioscience), leupeptin (5 μg/ml, Sigma, St. Louis, MO), pepstatin (7 μg/ml, Boehringer Mannheim, Indianapolis, IN), aprotinin (20 μg/ml, Boehringer Mannheim), phenylmethylsulfonyl fluoride (340 μg/ml, Sigma) and stored at −20°. For purposes of comparison of the specific activities of the bovine vitamin K-dependent carboxylase

[15] D.A. Roth, A. Rehemtulla, R. J. Kaufman, C. T. Walsh, B. Furie, and B. C. Furie, *Proc. Natl. Acad. Sci. U.S.A.* **90,** 8372 (1993).

[16] K. J. Kotkow, D. A. Roth, T. J. Porter, B. C. Furie, and B. Furie, *Methods Enzymol.* **222,** 435 (1993).

obtained from the different procedures described in this chapter it is important to note that the specific activity of the $NaH^{14}CO_3$ (Amersham Life Science, Buckinghamshire, England) used for assay of the enzyme does not vary much from batch to batch as supplied (e.g., four different batches prepared between July 1995 and April 1996 had reported specific activities between 52 and 56 mCi/mmol). This minimal variation makes it possible to compare the specific activities of enzyme preparations assayed at different times.

Purification of Bovine Liver Carboxylase

The purification of the bovine carboxylase described in this section is based on the method of Wu *et al.*[10] as modified by Kuliopulos *et al.*[11] As indicated earlier, the yields of this purification procedure vary between 4 and 35% and the purity of the vitamin K-dependent carboxylase obtained from this procedure has been reported from 50–90%. A 5000- to 7000-fold purification from bovine liver microsomes has been reported by this procedure and the isolated enzyme has a specific activity of about 2×10^9 cpm/30 min/mg when γ-carboxylase activity is assayed as described previously.

Preparation of Microsomal Enzyme

The preparation of bovine liver microsomes has been previously described in this series.[16] For purification of carboxylase the procedure in Volume 222 (this series) is followed up to the point of isolation of the soft microsomal pellet by ultracentrifugation. The microsomal pellet from 1 liter of homogenized liver is resuspended in 50 m*M* Tris-HCl, 1 *M* NaCl, pH 7.4, using the Polytron. Enzyme activity in this microsomal suspension is stable for greater than 1 year when the suspension is stored at −80°.

Solubilization of Microsomal Carboxylase

The carboxylase is solubilized by a procedure that is modified slightly from that described in Volume 222 of this series. The microsomes are pelleted at 150,000 *g* for 60 min at 4° and resuspended in 25 m*M* Tris-HCl, 0.5 *M* NaCl, pH 7.4. Solid 3-[(3-cholamidopropyl)dimethylammonio]-1-propane sulfonate (CHAPS, Sigma, St. Louis, MO) is added to a final concentration of 1% (w/v) and allowed to stir on ice for 15 min. Powdered ammonium sulfate (grade III, Sigma) is added to 30% saturation to the stirring microsome preparation, and the mixture is stirred on ice for an additional 15 min. The material precipitated at 30% ammonium sulfate is removed by centrifugation at 150,000 *g* for 1 hr at 4°, and the supernatant is adjusted while stirring to 57.5% saturation with powdered ammonium sulfate. The pH is maintained at 7.4 by addition of 1 *M* Tris-HCl, pH 7.4. The mixture is stirred for 15 min on ice and the precipitated proteins isolated by centrifugation at 30,000 *g* for 30 min at 4°. The floating pastelike pellets are pooled and resuspended in 150 ml of 25 m*M* MOPS, 0.5 *M* NaCl, 20% (v/v) glycerol, 0.1% (w/v) phosphatidylcholine, and 0.1% (w/v) CHAPS containing 1× PIC. The protein concentration of the solubilized microsomal preparation is determined using a Bradford assay and immunoglobulin G (IgG) as a standard.

Preparation of Affinity Chromatography Matrix

The affinity chromatography matrix used for isolation of the bovine liver carboxylase employs a 59-residue peptide based on the sequence of human factor IX propeptide and the first 41 residues of the γ-carboxyglutamic acid-rich domain. The peptide has been either synthesized by direct peptide synthesis or generated as a fusion protein in *Escherichia coli.* FIXQS contains four mutations from the native sequence. Two of them, Arg to Gln at −4 and Arg to Ser at −1, are made in the peptide to prevent cleavage of the propeptide by contaminating proteases during affinity chromatography of the crude bovine carboxylase. The two additional mutations, Thr to Ala at −18 and Met to Ile at 19, are important for expression of the protein as a fusion protein, which will be cleaved by cyanogen bromide.

Direct Peptide Synthesis of FIXQS

FIXQS can be synthesized using *N*-(-9-fluorenyl)methoxycarbonyl/*N*-methylpyrrolidone (Fmoc/NMP)-based chemistry. The synthesis performed in our laboratory used an Applied Biosystems (Foster City, CA) model 430A peptide synthesizer.[15] Amino acids are coupled as 1-hydroxybenzotriazole esters onto 0.25 mmol of *p*-hydroxymethylphenoxymethyl polystyrene resin. Side-chain-protecting groups include 2,2,5,7,8-pentamethylchroman-6-sulfonyl (arginine), O*t*Bu (aspartic acid and glutamic acid), trityl (cysteine), *tert*-butoxycarbonyl (lysine), and *t*Bu (serine, threonine, and tyrosine). Following each coupling step all uncoupled a-NH_2 termini are acetylated. Activation is performed with 1 ml of *N*-hydroxybenzotriazole/*N*,*N*-dicyclohexylcarbodiimide/*N*-*N*-methylpyrrolidone dissolved in 0.4 ml dichloromethane. The coupling reaction is allowed to proceed for 36.5–61.5 min depending on the amino acid residue. Samples are obtained after coupling and prior to acetylation for ninhydrin assays to determine coupling efficiencies. Cleavage of the peptide from its solid support and simultaneous side-chain deprotection is performed in thioanisole, ethyl methyl sulfide, water, and trifluoroacetic acid (5 : 2.5 : 5 : 87.5) for

3 hr at 25°. The resin is removed by filtration in a fritted funnel, 50 ml dichloromethane is added to the cleavage reaction supernatant, and the volume is reduced to about 2 ml on a Buchi Rotavapor-R (Brinkmann, Westbury, NY). The crude peptide is precipitated using cold anhydrous ethyl ether and washed several times with ethyl ether, then dried under vacuum. The crude deprotected peptide is purified by high-performance liquid chromatography (HPLC) using a Hi-Pore 318 column (250 × 21.5 mm, Bio-Rad, Hercules, CA), A linear gradient of 20–40% solvent B (solvent A: 0.1% trifluoroacetic acid, water; solvent B: 0.1% trifluoroacetic acid, acetonitrile) over 60 min is employed. The absorption of the peptide is monitored at 214 nm. The peptide is collected, lyophilized, and dissolved in buffer appropriate for linkage to a chromatography matrix.

Construction and Expression of Recombinant Steroid Isomerase–Factor IX Fusion Protein in Escherichia coli and Purification of FIXQS

The polymerase chain reaction (PCR) is used to create a 186-nucleotide fragment which contains an *Nco*I site at the 5′ end of the fragment, an in-frame methionine residue at codon −19 of factor IX, a mutation of Thr to Ala at codon −18, and a TGA stop codon and a *Bam*HI site after the codon for Phe-41 in the factor IX cDNA. The forward deoxynucleotide primer for the PCR reaction is

5′-GCCAAGCTTCCATGGCAGTTTTCTTGATCATG-3′

the reverse primer is

5′-CAAAGCATGCGGATCCTCAAAATTCAGTTGTCTTA-3′

and the template is human factor IX cDNA. The PCR-amplified product is digested with *Nco*I and *Bam*HI and ligated into sites placed at the stop codon of the KSI gene within pAK1370Y14F[17] creating a fusion construct, which becomes the template for further mutation of Met-19 to Ile, Arg-4 to Gln, and Arg-1 to Ser by Kunkel mutagenesis as described by Wu *et al.*[18]

The fusion construct containing the four mutations from the factor IX sequence (FIXQS) is expressed under the control of the lac promoter in the pUC-derived vector pAK1370Y14F. The steroid isomerase–factor IX fusion protein is expressed in the protease-deficient *E. coli* strain BL21 (DE3) and processed into inclusion bodies. The isolation of inclusion bodies and cleavage at the junctional methionine between the N-terminal KSI

[17] A. Kuliopulos, A. S. Mildvan, D. Shortle, and P. Talalay, *Biochemistry* **28,** 141 (1989).
[18] S.-M. Wu, B. Soute, C. Vermeer, and D. W. Stafford, *J. Biol. Chem.* **265,** 13124 (1990).

fragment and the C-terminal FIXQS is performed as described by Wu *et al.*[18] except that the CNBr/protein ratio is 1/2 (w/w). The FIXQS is purified to homogeneity on a 1- × 10-cm HR10/10 Mono Q anion-exchange column (Pharmacia, Uppsala, Sweden).

Preparation of Affinity Chromatography Resin

The FIXQS is linked to an *N*-hydroxysuccinimide-modified, cross-linked agarose through covalent bonds formed with primary amines in the peptide. The affinity matrix is prepared by coupling 100 mg of FIXQS to 25 ml of resin (Affi-Gel 10, Bio-Rad). The peptide is dissolved in 33.5 ml of 10 m*M* NaOH, the pH adjusted to between pH 7.5 and 8.0 by the addition of 3.35 ml of 1 *M* HEPES, pH 8.0. The solution is made 80 m*M* in $CaCl_2$ and cooled to 4°. The resin is washed with 100 ml of cold HPLC-grade H_2O and combined with the peptide within 20 min of washing the resin. Peptide and resin are gently agitated for 4 hr at 4°. Unbound sites on the resin are blocked with a one-tenth volume of 1 *M* ethanolamine at pH 8.0. The efficiency of coupling is 85–90%.

Affinity Chromatographic Isolation of Bovine Microsomal Carboxylase

Solubilized microsomal protein (150 ml) containing 8 g of total protein in buffer A [0.1% CHAPS, 25 m*M* MOPS, pH 7.0, 0.5 *M* NaCl, 20% glycerol (v/v), 0.1% phosphatidylcholine, 1× PIC] is placed on ice and sonicated at a scale setting of 9 for 100 2-sec pulses. The sonicated solubilized microsomal proteins are loaded onto the FIXQS affinity resin, which is equilibrated with buffer A. The sample is loaded at a flow rate of 10 ml/hr at 4°. The loaded affinity resin is washed with 200 ml of equilibration buffer. The affinity resin is further washed with a linear gradient (50 ml total) of 0.05–0.5% Triton X-100 in 25 m*M* MOPS, pH 7.0, 50 m*M* NaCl, 20% glycerol, 0.2% phosphatidylcholine, 1× PIC followed by 50 ml of 0.5% Triton X-100 in the same buffer. Washing of the column is continued with a linear gradient (50 ml total) of Triton X-100 from 0.5–1.0% and then the Triton X-100 concentration is reduced with a linear gradient (50 ml total) of 1.0–0.5% Triton X-100 both in the same buffer as the initial Triton X-100 gradient. The carboxylase is eluted from the column with a linear gradient (200 ml total) of 0.1–1.0% CHAPS (200 ml) in 25 m*M* MOPS, pH 7.0, 50 m*M* NaCl, 20% glycerol (v/v), 0.1% phosphatidylcholine, 1× PIC, and 2.0 μ*M* proFIX18 followed by 50 ml of 1% CHAPS in the same buffer. Several hundred micrograms of carboxylase can be obtained from this procedure. The isolated protein is stable at −80° for at least 6 months.

Baculovirus Expression of Recombinant Bovine Carboxylase

An excellent description of general laboratory methods and guidelines for recombinant protein expression using baculovirus-infected insect cells has been provided by O'Reilly *et al.*[19] The construction of the appropriate recombinant baculovirus transfer vectors and the expression of bovine carboxylase in insect cells is carried out using plasmids and cells supplied by Invitrogen (San Diego, CA). In this system the nonessential polyhedrin gene of the baculovirus *Autographa californica* multiple nuclear polyhedrosis virus (AcMNPV) is replaced by the bovine carboxylase cDNA coding sequence by homologous recombination. Sf9 cells (*Spodoptera frugiperda,* fall armyworm cells) are cotransfected with linearized AcMNPV DNA and a recombinant baculovirus transfer vector pBLII/bCbx containing both the bovine carboxylase cDNA under the control of the polyhedrin promoter and the *E. coli lacZ* gene under the control of an early viral promoter, Petl.[15] Because the polyhedrin protein is responsible for the production of viral occlusions but not necessary for viral infection or replication in tissue culture cells, clones of virally infected Sf9 cells containing recombinant virus without the polyhedron gene have the occlusion-negative phenotype. In addition recombinant viral plaques can be identified by their blue color in the presence of 5-bromo-4-chloro-3-indolyl-β-D-galactoside (X-Gal). The bovine carboxylase can be expressed by infection of Sf9 cells with plaque purified recombinant virus.

Plasmid Construction

The cDNA encoding bovine liver carboxylase is subcloned into the *Nhe*I site of the baculovirus transfer vector pBLII (Invitrogen) to generate pBLII/bCbx by standard methods as described later.[20] The cDNA is digested with *Nco*I. A 3.5-kb fragment containing the entire coding region is blunt-ended using the large fragment of DNA polymerase I and ligated into pBLII, which has been linearized with *Nhe*I and blunt-ended using the large fragment of DNA polymerase I. Recombinant plasmid pBLII/bCbx is isolated by CsCl gradient ultracentrifugation and the proper orientation of the cDNA insert with respect to the polyhedrin promoter in pBLII can be confirmed by restriction enzyme analysis and DNA sequence analysis.

[19] D. R. O'Reilly, L. K. Miller, and V. A. Luckow, "Baculovirus Expression Vectors: A Laboratory Manual." W. H. Freeman, New York, 1992.

[20] J. Sambrook, E. F. Fritsch, and T. Maniatis, "Molecular Cloning: A Laboratory Manual." Cold Spring Harbor Laboratory Press, New York, 1989.

Cell Culture

Sf9 cells are grown in complete TNM-FH [supplemented Grace's insect medium (Invitrogen), 10% (v/v) fetal bovine serum (Hyclone, Logan, UT), gentamicin (50 μg/ml, Life Technologies, Gaithersburg, MD)] at 27° in a standard laboratory incubator. Carbon dioxide supplementation is not required, however, a pan with water should be kept in the incubator to maintain a humidified atmosphere. Cells can be propagated as adherent monolayers in loosely capped plastic tissue culture flasks. Alternatively, for large-scale expression of recombinant carboxylase, suspension cultures are preferred. Suspension cultures are grown in glass spinner flasks (Bellco, Vineland, NJ) and stirred at 60 rpm on a Bell Stir MultiStir Spinner plate (Bellco). The speed of the magnetic stir plate must be carefully calibrated to avoid high speeds, which result in hydrodynamic sheering of the cells. Slower speeds result in cell clumping, inadequate oxygenation, and low cell viability. If suspension cultures of greater than 100 ml are used, the medium should include 0.1% pluronic surfactant F-68 (JRH Biosciences, Lenexa, KS) to protect the cells from hydrodynamic stress observed in large suspension cultures.

Production of Recombinant Baculovirus

Cotransfection of 2.0×10^6 Sf9 cells, grown as an adherent monolayer, is performed with 1 μg of linearized wild-type AcMNPV baculovirus DNA (Invitrogen) and 3 μg of pBLII/bCbx using cationic liposomes (Invitrogen). The tissue culture medium harvested from cotransfected cells is used to isolate recombinant viruses from contaminating wild-type virus by the plaque assay technique. An adherent monolayer of Sf9 cells is infected with an aliquot of this medium, and overlayed with a thin layer of X-Gal supplemented agarose to inhibit viral diffusion across the plate during subsequent incubation. After several days, infected cells lyse, resulting in plaques underneath the agarose. Viral plaques of recombinant baculovirus resulting from a double-crossover homologous recombination event demonstrate blue color in the presence of X-Gal and an occlusion-negative phenotype, which is easily confirmed by microscopic analysis. Viral clones of recombinant baculovirus, vbCbx/AcMNPV, encoding carboxylase are considered plaque-pure after three or four rounds of plating, titered by plaque assay and amplified by passage in Sf9 cells infected in suspension cultures at low multiplicity of infection [0.1–0.5 plaque-forming unit (pfu) per cell] using standard techniques.[19] We have not routinely concentrated our amplified viral stocks prior to storage. Stocks of amplified virus are harvested 4–5 days following infection of cells after cell lysis is near complete, centrifuged at 1000 *g* for 10 min to remove cellular debris, and the tissue culture supernatant containing high titer budded virus is transferred to sterile plastic Falcon tubes and stored at 4° in the dark. Aliquots of virus in tissue culture medium are maintained at −80° for long-term storage.

Expression of Recombinant Protein

Large-scale protein expression is performed with Sf9 cells propagated and infected in suspension culture. Cells are infected at high multiplicity of infection (10 pfus per cell) using high-titer viral stocks titered within 2–3 months of use, to ensure a synchronous infection and reproducible kinetics of protein expression following inoculation of cultured cells. Cells are harvested 48–66 hours after infection, based on the results of a time course of carboxylase expression, which demonstrated maximal carboxylase enzyme activity correlating with the expected kinetics of expression of a gene under control of the polyhedrin promoter. Infected cells are harvested by centrifugation at 1000 *g* for 10 min at 4°. All further processing is done with ice-cold buffers or at 4°, unless otherwise stated. Cell pellets are washed once by resuspension in phosphate-buffered saline (137 m*M* NaCl, 2.7 m*M* KCl, 4.3 m*M* Na_2PO_4, 1.5 m*M* KH_2PO_4, pH 7.4), followed by a repeat centrifugation. Washed cell pellets are resuspended in a hypotonic lysis buffer (10 m*M* MOPS, 10 m*M* KCl, 1 m*M* $MgCl_2$, pH 7.0) supplemented with 1× PIC. After a 5-min incubation, cells are lysed by sonication. Nuclei and insoluble debris are sedimented at 600 *g* for 5 min. Microsomes are collected from the postnuclear supernatant by centrifugation at 100,000 *g* for 60 min. Microsomal pellets are resuspended in 2–3 volumes of 25 m*M* MOPS, 1 *M* NaCl, 10% glycerol, 1× PIC, pH 7.4, and sonicated at 4°.

The recombinant carboxylase can be solubilized by mixing the resuspended microsomes with an equal volume of 1.5% CHAPS, 1.5% phosphatidylcholine, 25 m*M* MOPS, 1 *M* NaCl, 1× PIC, pH 7.4, and sonicated again with 10 5-sec pulses. Insoluble material is pelleted by ultracentrifugation at 100,000 *g* for 60 min at 4°, and the solubilized enzyme preparation is aliquoted and stored at −80°. The soluble enzyme preparation is stable to multiple freeze-thaw cycles provided the glycerol concentration is at least 5% (v/v). In an independent study, recombinant bovine carboxylase was expressed in insect cells with a His_6-T7 tag, and the concentration of enzyme in the cell lysate established using Western blot analysis and a monoclonal anti-T7 antibody.[21] The specific activity of the His_6 carboxylase is 0.6×10^9 cpm/30 min/mg, very similar to that determined for native bovine carboxylase.

[21] D. A. Roth, M. L. Whirl, L. J. Velazquez-Estades, C. T. Walsh, B. Furie, and B. C. Furie, *J. Biol. Chem.* **270,** 5305 (1995).

Isolation of Recombinant, Epitope-Tagged Bovine Carboxylase

Preparation of Expression Vector for FLAG–Bovine Carboxylase

The cDNA of bovine γ-glutamyl carboxylase is inserted into the cloning vector pGEM-7Zf(+) (Promega, Madison, WI) to create pG7/CBX. The sequence for FLAG epitope (DYKDDDDK) is introduced between the codon for the initiator Met of the carboxylase and the second codon for Ala using PCR-based mutagenesis. A 60-bp forward primer (5′-GAC GTC GCA TGC GTC GAC ATG gac tac aag gac gat gac gat aag GCG GTC TCC GCT CGG-3′) encoding the FLAG sequence (shown in lowercase) located 5′ to a 15-nucleotide sequence complementary to nucleotides 18–33 of the carboxylase cDNA is used. An oligonucleotide, 5′-AGC ATG ACG TAG GGG-3′, which is complementary to nucleotides 911–926, is used as a reverse primer. In our laboratory the oligonucleotides are synthesized on an Applied Biosystems 381A DNA synthesizer using standard methods. The PCR reaction is performed using 1 unit of Vent DNA Polymerase (2 unit/μl) and the buffer provided by the manufacturer (New England Biolabs, Beverly, MA) in the presence of 200 μM dNTP, 50 pmol of each primer, 0.5 μg of plasmid DNA in a final volume of 100 μl for 30 cycles (94° for 1 min, 56° for 1 min, 72° for 2 min) and an additional extension step (72°, 7 min). The PCR product is inserted into pG7/CBX using *Sph*I and *Eco*RI to restrict both the PCR fragment and the vector creating pG7/FLAG-CBX. Using *Sal*I and *Sma*I sites, the FLAG–carboxylase cDNA from pG7/FLAG-CBX is excised and is inserted into a similarly restricted mammalian expression vector pED producing pED/FLAG-CBX. The mammalian expression vector pED was a kind gift from Genetics Institute, Cambridge, MA.

Cell Culture, Transfection, and Cell Line Selection

Dihydrofolate reductase-deficient Chinese hamster ovary cells (CHO-Dukx-B11, a kind gift from Genetics Institute[22]) are employed to express the FLAG–carboxylase. The cells are transfected using Lipofectin (Life Technologies, Gaithersburg, MD). CHO cells are plated in 100-mm dishes with MEM medium [αMEM, 10% fetal bovine serum (v/v), 2 mM L-glutamine, 10 mM HEPES, 1 unit/ml penicillin, 100 μg/ml streptamycin] at a density that will provide cells at about 60% confluence in 16–24 hr of growth. Lipofectin (75 μl) and 75 μl of pED/FLG-CBX DNA at 1 μg/5 μl in sterile TE (10 mM Tris-HCl, pH 7.6, 1 mM EDTA) are mixed in a 15-ml polystyrene tube and allowed to incubate at ambient temperature

[22] G. Urlaub and L. A. Chasin, *Proc. Natl. Acad. Sci U.S.A.* **77,** 4216 (1980).

for 15 min. The mixture of DNA and Lipofectin will be cloudy. The CHO cells are harvested and washed twice with 6 ml of serum-free MEM medium. Serum-free MEM (8 ml) is added to the DNA/Lipofectin, the mixture is added to the pelleted CHO cells, the cells suspended by several inversions of the tube and incubated for 24 hr at 37°. The cells are transfered to complete MEM medium and allowed to grow at 37° for 48 hr. The transfected cells are transfered to selective medium [MEM medium containing adenosine (10 μg/ml), deoxyadenosine (10 μg/ml), and thymidine (10 μg/ml)] and plated in 10 100-mm tissue culture dishes. It is advisable to split cells at dilutions of 1:40, 1:60, and 1:160 to ensure the growth of well-separated colonies that can be cleanly isolated. Colonies form in about 1 week.

Clones expressing FLAG–carboxylase can be identified by performing Western blots on cell lysates using the anti-FLAG M2 monoclonal antibody (Kodak, Rochester, NY). Cells from a stable colony producing the highest level of enzyme are selected stepwise with increasing concentrations of methotrexate, up to 0.15 μM, to augment carboxylase production.[23]

Preparation of Solubilized FLAG–Bovine Carboxylase

CHO cells expressing FLAG–carboxylase are grown in a humidified incubator [37°, 5% (v/v) CO_2] in 500-cm^2 tissue culture dishes with MEM medium containing 11.1 μM vitamin K and 0.15 μM methotrexate. Confluent cells (10 tissue culture dishes) are harvested with PBS (2.7 mM KCl, 1.5 mM KH_2PO_4, 137 mM NaCl, 6.5 mM Na_2HPO_4) containing 5 mM EDTA, and washed twice with 200 ml of PBS, separating the cells from the wash buffer by centrifugation at 280 *g*. To prepare microsomes approximately 3×10^8 cells resuspended at a cell density of 1×10^8 cells/ml in PBS, 20% glycerol (v/v), 1× PIC are homogenized with 30 strokes of a Potter tissue grinder (4 ml; Kontes, Vineland, NJ) with the Teflon pestle attached to a Con-Torque power unit (Eberbach, Ann Arbor, MI). The homogenate is subjected to centrifugation at 900 *g* for 5 min and the supernatant recovered. The homogenization and centrifugation procedure are repeated twice more using the postcentrifugation cell pellets. The combined supernatants from the three centrifugations are subjected to centrifugation at 150,000 *g* for 1 hr at 4°. The postcentrifugation pellet containing the microsomes is resuspended in 3 ml of PBS, 20% (v/v) glycerol, 1× PIC.

To solubilize the enzyme the resuspended microsomes are diluted with an equal volume of PBS, PBS, 20% (v/v) glycerol, 1% CHAPS, 0.2% phosphatidylcholine, the solution placed in an ice bath and sonicated at level 4

[23] F. M. Ausubel, R. Brent, R. E. Kingston, D. D. Moore, J. G. Seidman, J. A. Smith, and K. Struhl, "Current Protocols in Molecular Biology." John Wiley & Sons, New York, 1994.

with two 5-sec pulses. After incubation at 4° for 30 min the solution is subjected to centrifugation at 150,000 *g* for 1 hr at 4°. The postcentrifugation supernatant containing solubilized enzyme is recovered and the pellet containing unsolubilized enzyme resuspended in 1 ml of PBS, 20% glycerol (v/v), and solubilized as before by addition of an equal volume of PBS, 20% glycerol, 1% CHAPS, 0.2% PC, and sonication. The solubilized enzyme is recovered by centrifugation as before and the combined postcentrifugation supernatants stored at −80°.

Purification of FLAG–Bovine Carboxylase by Affinity Chromatography

Chromatography is carried out at 4°. Anti-FLAG M2 monoclonal antibody affinity resin (1.5-ml packed volume, Kodak, Rochester, NY) is washed twice with 20 ml of PBS, 20% glycerol (v/v), 0.5% CHAPS. The washed resin is incubated with 5 ml of solubilized enzyme for 2 hr with gentle shaking. The loaded resin is packed in a chromatography column (1.6 × 13 cm) and unbound material allowed to flow through the column. The resin is washed five times with 10 ml of PBS, 20% glycerol (v/v), 1 m*M* EDTA, 0.5% CHAPS, 0.1% phosphatidylcholine, and once with 10 ml of PBS, 20% glycerol (v/v), 0.5% CHAPS, and 0.2% phosphatidylcholine. The bound FLAG-carboxylase is eluted by sequential addition of 6 ml of PBS, 20% glycerol (v/v), 0.5% CHAPS, 0.2% phosphatidylcholine containing 5, 10, 20, 50, 75, 100, 150, 200, 300, and 400 μg/ml of FLAG–peptide (Kodak). At each step the column bed is incubated with the eluting buffer for 15 min. The isolated enzyme is stored at −80°. The column bed is regenerated with 0.1 *M* glycine-hydrochloride, pH 3.0, 0.5% CHAPS followed by 10 ml of PBS and stored in PBS, 0.2% NaN_3 at 4°.

About 150 μg of FLAG–carboxylase can be isolated by this procedure. The initial fractions of carboxylase eluted from the anti-FLAG M2 monoclonal antibody affinity column contain significant levels of an unknown protein with a molecular weight of 45,000. Fractions eluted later are free of this contaminant and are at least 90% pure. The highly purified carboxylase represents about three-quarters of the enzyme eluted from the column. The specific activity of the enzyme is 1.8×10^9 cpm/30 min/mg. The enzyme is stable for at least 6 months.

Final Remarks

The specific activities of the purified bovine liver vitamin K-dependent carboxylase and the recombinant FLAG–carboxylase are equivalent. The advantages and disadvantages of isolating native versus recombinant enzyme are thus dependent on the skills of the laboratory. Fresh bovine liver is a readily available source of the enzyme so long as an abbatoir is within convenient reach. However, if the required molecular biology and cell culture skills are available, preparation of epitope-tagged recombinant carboxylase provides a more reproducible enzyme preparation in both quality and yield and is far less tedious than the isolation of the enzyme from bovine liver.

[29] Purification of γ-Glutamyl Carboxylase from Bovine Liver

By SHEUE-MEI WU, VASANTHA P. MUTUCUMARANA, and DARREL W. STAFFORD

Introduction

The vitamin K-dependent γ-glutamyl carboxylase is a moderately rare integral membrane protein that is present in the endoplasmic reticulum of different tissues. Among these, liver has the highest γ-glutamyl carboxylase activity and is therefore the usual model. Since the cDNA cloning of this enzyme, recombinant γ-glutamyl carboxylase has been successfully expressed in human kidney 293 cells, Chinese hamster ovary (CHO) cells, and insect Sf9 (*Spodoptera frugiperda,* fall armyworm ovary) cells.[1–3] Furthermore, a built-in artificial tag allows the recombinant carboxylase to be identified and purified using commercially available antibodies and affinity matrices.[4] In spite of the great potential of using the tagged recombinant carboxylase, bovine liver is still the most economic resource for purifying γ-glutamyl carboxylase.

In this chapter, we describe in detail an affinity purification specifically designed for bulk isolation of γ-glutamyl carboxylase from bovine liver. The principle of this methodology is based on two factors: (1) the affinity of γ-glutamyl carboxylase to a recombinant substrate FIXQ/S and (2) the different biophysical properties of different detergents. Starting from 8 g of microsomal proteins, we routinely achieve a 7000-fold purification with

[1] S.-M. Wu, W.-F. Cheung, D. Frazier, and D. Stafford, *Science* **254,** 1634 (1991).

[2] A. Rehemtulla *et al., Proc. Natl. Acad. Sci. U.S.A.* **90,** 4611 (1993).

[3] D. A. Roth *et al., Proc. Natl. Acad. Sci. U.S.A.* **90,** 8372 (1993).

[4] B. C. Furie, A. Kuliopulos, D. A. Roth, C. T. Walsh, and B. Furie, *Methods in Enzymol.* **282** [28], 1997 (this volume).

about 30% recovery, which is equivalent to 300–400 μg of γ-glutamyl carboxylase with 80–90% purity.[5]

Preparation of Affinity Column

FIXQ/S is a recombinant peptide that contains the propeptide and Gla domain sequences of human factor IX.[6] Because of the affinity between the propeptide and γ-glutamyl carboxylase, FIXQ/S works well as an affinity ligand in the purification. Factor IXQ/S contains two mutations (R^{-4} to Q and R^{-1} to S) in the propeptide sequences, which were introduced to improve the proteolytic stability of this peptide without changing its affinity to γ-glutamyl carboxylase. This modification increases the lifetime of the Affi-FIXQ/S column.

FIXQ/S is expressed in *Escherichia coli* BL21(DE3) as an insoluble fusion protein with T7 gene 10. The expression of the fusion protein is under the control of T7 gene 10 promoter, a tightly regulated, strong promoter that is only recognized by T7 RNA polymerase. In this system, the expression of T7 RNA polymerase is initiated by isopropylthiogalactoside (IPTG) induction, which in turn results in overproduction of FIXQ/S fusion protein. The aggregates of fusion protein form inclusion bodies that can be separated from cellular proteins by centrifugation after cell lysis. Following CNBr cleavage, gene 10 peptides are separated from FIXQ/S by dialysis precipitation and FIXQ/S is further purified on DEAE-Sepharose.

Growth of pMcFixQ/S Transformed BL21(DE3)

Materials

LB medium:	0.5% NaCl, 1% Bacto-tryptone, 0.5% Bacto-yeast extract
Fermentation medium:	0.5% NaCl, 1% Bacto-tryptone, 2% Bacto-yeast extract supplemented with 0.5% (v/v) glycerol and 100 m*M* potassium phosphate, pH 7.4
Chloramphenicol:	34 mg/ml in ethanol
IPTG:	500 m*M*

Procedure

1. Inoculate a single colony of pMcFIXQ/S-transformed BL21(DE3) into 10 ml of LB containing 10 μg/ml chloramphenicol. Incubate at 37° for 8 hr with vigorous shaking.

[5] S. M. Wu, D. P. Morris, and D. W. Stafford, *Proc. Natl. Acad. Sci. U.S.A.* **88,** 2236 (1991).
[6] S. M. Wu, B. A. Soute, C. Vermeer, and D. W. Stafford, *J. Biol. Chem.* **265,** 13124 (1990).

2. Inoculate the 10-ml preculture into 500 ml of LB containing 10 μg/ml chloramphenicol. Incubate at 37° overnight with vigorous shaking.
3. Inoculate the 500-ml preculture into 10 liters of fermentation medium containing 10 mg/liter chloramphenicol. Grow the cells in a New Brunswick Microferm fermenter at 37°, 400 rpm, air flow 4.0 SLPM. Add a few drops of antifoam if needed.
4. Follow cell growth by OD_{600}.
5. Add IPTG to a final concentration of 0.5 m*M* when OD_{600} reaches 7.5; induce for 2 hr.
6. Harvest cells by centrifugation: Sorvall HG-4L rotor, 3500 rpm, 20 min, 4°.
7. Discard the supernatant and weigh cell pellets.
8. Freeze cell pellets in liquid nitrogen and store at −80° until use.

Comment. The most frequent problem when expressing the fusion protein in this system is to lose the plasmid that contains the complete expression cassette. Because minor promoter leakage exists in this system and it is to the advantage of the bacteria not to express foreign proteins, a bacterium that has lost the expression cassette during an early stage of the culture can outgrow and dominate the population. Therefore, healthy preculture and correct induction time are the two most important factors for a successful production. We routinely check fusion protein levels from preinduction and postinduction cultures by sodium dodecyl sulfate–polyacrylamide gel electrophoresis (SDS–PAGE) analysis. Average yield is around 20–30% of the total protein although the yield of the fusion protein can vary. One hundred and fifty grams of wet cell pellet is expected from a 10-liter culture.

Preparation of Inclusion Bodies

Materials

Lysis buffer:	50 m*M* Tris-HCl, pH 8.0, 25% sucrose, 1 m*M* EDTA
Lysozyme:	20 mg/ml in lysis buffer (freshly made or can be stored at −20°)
10× DNase I buffer:	100 m*M* $MgCl_2$, 10 m*M* $MnCl_2$
DNase I:	2 mg/ml in 150 m*M* NaCl and 50% (v/v) glycerol
Detergent buffer:	20 m*M* Tris-HCl, pH 7.5, 200 m*M* NaCl, 2 m*M* EDTA, 1% sodium deoxycholate, 1% Nonidet P-40
Wash solution:	1% Triton X-100, 1 m*M* EDTA

Procedure. All procedures are carried out at 0–4°.

1. Thaw the frozen cell pellets.
2. Resuspend cells in 100 ml of lysis buffer by homogenization using a Brinkmann Polytron PT 3000 with generator PT-DA 300712.
3. Dilute cell suspension with appropriate amount of lysis buffer to make a final volume of 360 ml.
4. Add 40 ml of lysozyme solution. Stir for 1 hr.
5. Add 44 ml of 10× DNase I buffer and 2.2 ml of DNase I solution. Stir for 1 hr.
6. Add 800 ml of detergent buffer. Stir for 1 hr.
7. Isolate inclusion bodies from cell lysate by centrifugation: Sorvall GSA rotor, 10,000 rpm, 30 min.
8. Discard the supernatant.
9. Resuspend inclusion bodies in 500 ml of wash solution by homogenization.
10. Repeat steps 7 and 8.
11. Weigh the inclusion bodies.
12. Freeze the inclusion bodies in liquid nitrogen and store at −80° until use.

Comment. The purity of inclusion bodies should be examined by SDS–PAGE analysis. We prefer to start with inclusion bodies that have a purity greater than 90% for CNBr cleavage. If necessary, thaw the inclusion bodies and repeat steps 9 and 10 several times to wash off trapped soluble proteins. This procedure yields 15 g of inclusion bodies.

CNBr Cleavage of FIXQ/S Fusion Protein

Materials

CNBr solution:	23 g in 50 ml of 70% formic acid (freshly made)
Guanidine hydrochloride:	8 *M*

Procedure. All procedures are carried out at room temperature in the hood.

1. Dissolve 15 g of inclusion bodies in 100 ml of 88% formic acid.
2. Transfer dissolved fusion protein into a round bottle.
3. Add 50 ml of CNBr solution; stir.
4. Seal and keep in the dark for 12 hr.
5. Repeat steps 3 and 4.
6. Dry down the cleaved peptide at 42° with a vacuum rotary evaporator.

7. Redissolve the peptides in 150 ml of 77% formic acid.
8. Repeat steps 6, 7, and 6 again.
9. Dissolve dried down peptides in 85 ml of 8 *M* guanidine hydrochloride.
10. Transfer the peptides into a dialysis tubing (width 45 mm, molecular weight cutoff 3500, Spectra/Por).

Comment. Complete CNBr cleavage is driven by using 100-fold excess molar ratio of CNBr to Met under acidic conditions. However, this harsh condition also leads to undesirable side reactions. Formylation and dimerization of FIXQ/S are two major observed side reactions. Although formylation of FIXQ/S can be reversed by dithiothreitol (DTT), dimerization of FIXQ/S appears to be irreversible.

Purification of FIXQ/S by DEAE-Sepharose

Materials

Dialysis buffer:	20 m*M* MOPS, pH 8.0, 50 m*M* NaCl
Dithiothreitol (DTT):	1 *M*
Buffer L:	25 m*M* MOPS, pH 8.0, 50 m*M* NaCl, 2 m*M* DTT, 1 m*M* EDTA
Buffer H:	25 m*M* MOPS, pH 8.0, 500 m*M* NaCl, 2 m*M* DTT, 1 m*M* EDTA

Procedure. All procedures are carried out in the cold room.

1. Dialyze the sample against 5 liters of dialysis buffer.
2. Change the buffer every 6–8 hr, four times.
3. Remove precipitated gene 10 peptides from FIXQ/S by centrifugation: Sorvall T647.5 rotor, 40,000 rpm, 30 min.
4. Save the supernatant.
5. Resuspend the pellet in 100 ml of dialysis buffer. Stir overnight.
6. Repeat step 3.
7. Combine the supernatant with the one from step 4.
8. Add DTT to a final concentration of 2 m*M*.
9. Adjust the pH to 8.0 and conductivity to below the equivalent of 50 m*M* NaCl.
10. Equilibrate a DEAE-Sepharose CL-6B column (2.9 × 20 cm) with buffer L.
11. Load the reduced supernatant on the column at a flow rate of 1 ml/min and record OD_{254}.
12. Wash the loaded column with 200 ml of buffer L.
13. Elute bound peptides with an 800-ml linear gradient of buffer L to buffer H (400 ml each).

14. Determine the peptide concentration in the eluate by Bradford assay using bovine serum albumin as the standard.

Comment. FIXQ/S was eluted approximately at 125 mM NaCl as the single major peak in the OD_{254} elution profile. It was often followed by a small peak of FIXQ/S dimer, which is an irreversible side product from CNBr cleavage. In spite of the difference in molecular weight, FIXQ/S dimer shares similar kinetic parameters with the monomer in *in vitro* γ-carboxylation assay. This procedure yields 300–400 mg of FIXQ/S monomer.

Coupling FIXQ/S to Affi-Gel 10

Materials

FIXQ/S:	100 mg
Affi-Gel 10 (Bio-Rad, Richmond, CA):	25 ml
Buffer L:	25 mM MOPS, pH 8.0, 50 mM NaCl, 2 mM DTT, 1 mM EDTA

Procedure. All procedures are carried out in the cold room.

1. Wash 25 ml of Affi-Gel with 300 ml of ice-cold water in a glass fritted funnel. Apply vacuum to accelerate washing.
2. Combine DEAE fractions equivalent to 100 mg of FIXQ/S. Adjust pH to 6.0.
3. Mix washed Affi-Gel 10 and FIXQ/S. Gently agitate on a rocker for 24 hr.
4. Add 1 M Tris-HCl, pH 8.8, to a final concentration of 50 mM. Agitate for 2 hr to complete blocking.
5. Transfer the gel into a 2.9- × 15-cm column and wash with 500 ml of buffer L.

Comment. Coupling efficiency can be determined by using trace amount of ^{125}I-labeled FIXQ/S or by Bradford assay. Greater than 90% coupling efficiency was routinely obtained after 24 hr. FIXQ/S coupled Affi-Gel 10 (abbreviated as Affi-FIXQ/S) is now ready for affinity purification of carboxylase or can be stored with 0.02% sodium azide in the cold room.

Purification of γ-Glutamyl Carboxylase

All buffers are precooled and all procedures are carried out at 0–4° unless stated otherwise.

Bulk Preparation of Ground Bovine Liver

Materials

Buffer A: 50 m*M* Tris-HCl, pH 7.5, 100 m*M* NaCl

Procedure

1. At the slaughter house, wrap the individual bovine liver (30-min postmortem) in a plastic bag and chill in ice–water for 5 min.
2. Remove the membranes surrounding the liver, connective tissues, and large vessels.
3. Cut the soft tissues into 1- $\times$ 2- $\times$ 6-inch strips.
4. Divide liver strips into 1.2 kg per bag. Store in ice.
5. Add 2400 ml of buffer A and 1.2 kg of liver strips into a 4-liter Waring blender (model CB-5); grind for 30 sec at the lowest speed.
6. Filter the liver slurry through a single layer of cheesecloth.
7. Pour the filtered liver slurry directly into a large volume of liquid nitrogen.
8. Sieve out "popcorn" (frozen liver slurry, small bubble-like nodules) from liquid nitrogen and store at $-80°$.

Comment. Step 1 chills the liver surface and helps to remove membranes from soft tissues.

Isolation of Microsome

Materials

"Popcorn":	1200 g for two loads of Sorvall rotor T647.5
Buffer B:	50 m*M* MOPS, pH 7.5, 1 *M* NaCl
Buffer C:	100 m*M* MOPS, pH 7.5

Procedure

1. Thaw 1200 g of "popcorn" in a 4-liter glass beaker in a 37° waterbath. Stir constantly to keep the entire sample below 4°.
2. Divide the liver slurry into 400-ml fractions and homogenize each for 1 min at speed 3, using a Brinkmann Polytron PT 3000 with generator PT-DA 300712.
3. Centrifuge at 13,000 rpm ($RCF_{avg} = 17{,}300g$) for 30 min in a Sorvall GSA rotor.
4. Collect the supernatant without taking any of the loose greenish-brown interface.
5. Ultracentrifuge postmitochondria at 45,000 rpm ($RCF_{avg} = 150{,}000g$) for 1.5 hr in two Sorvall T647.5 rotors.

6. Discard the supernatant and gently wash off the loose interface with 10 ml of buffer B.
7. Resuspend each pellet in 30 ml of buffer B by homogenization using the conditions described in step 2.
8. Combine the microsome suspension and ultracentrifuge at 45,000 rpm (RCF_{avg} = 150,000g) for 1.5 hr in a Sorvall T647.5 rotor.
9. Repeat step 6.
10. Combine the pellets and resuspend in 65 ml of buffer C by homogenization using the conditions described in step 2.
11. Refine the suspension with a 55-ml Potter–Elvehjem tissue grinder, two strokes.
12. Add appropriate amount of buffer C to make the protein concentration 60 mg/ml.
13. Divide microsomal homogenate into 45-ml fractions and store in 50-ml conical tubes, save 200 μl for assay.
14. Freeze in liquid nitrogen and store at −80°.

Extraction of γ-Glutamyl Carboxylase from Washed Microsomes

Materials

10× PIC (protease inhibitor cocktail, 150 ml):	12 ml of 250 mM EDTA, 3 ml of 1 M DTT, 75 μl of Phe-Phe-Arg-chloromethyl ketone (2.5 mg/ml in 1 mM HCl), 75 μl of Phe-Pro-Arg-chloromethyl ketone (2.5 mg/ml in 1 mM HCl), 1.5 ml of leupeptin (0.5 mg/ml), 1.5 ml of pepstatin (0.7 mg/ml in methanol), 3 ml of phenylmethyl sulfonyl fluoride (PMSF) (17 mg/ml in 2-propanol), 300 μl of aprotinin (10 mg/ml), and 128.55 ml of H_2O
NaCl:	4M
CHAPS:	10% (w/v)
Buffer D:	25 mM MOPS, pH 7.5, 500 mM NaCl

Procedure

1. Thaw 200 ml of microsome homogenate in ice–water.
2. Prepare solubilization cocktail by mixing 44 ml of 10× PIC, 44 ml of 10% CHAPS, 110 ml of 4 M NaCl, and 42 ml of H_2O.
3. Add 240 ml of ice-cold solubilization cocktail to 200 ml of microsome homogenate. Stir for 60 min.
4. Ultracentrifuge in a Sorvall T647.5 rotor at 45,000 rpm (RCF_{avg} = 150,000g) for 90 min.
5. Transfer the supernatant into a 1-liter flask.

6. Add 140 g of $(NH_4)_2SO_4$ slowly into the supernatant while stirring. Stir for 30 min longer.
7. Centrifuge in a Sorvall GSA rotor at 10,000 rpm ($RCF_{avg} = 10{,}000g$) for 30 min.
8. Carefully aspirate the supernatant without removing the "floating" pellet.
9. Dissolve the floating pellet in 60 ml of buffer D with 1× PIC.
10. Measure the protein concentration by Bradford assay.

Comment. "Floating" pellet is a pink sheet that loosely floats on the top of the solution after centrifugation. This extraction yields approximately 8 g of microsomal proteins at a protein concentration of around 95 mg/ml.

Affinity Purification of γ-Glutamyl Carboxylase

Materials. All phosphatidylcholine (PC) containing buffers are sonicated to maximal clarity before use.

proFIX19:	AVFLDHENANKILNRPKRY, synthetic peptide
5× Buffer D:	125 m*M* MOPS, pH 7.5, 2.5 *M* NaCl
5% CHAPS/PC	
Equilibration buffer (300 ml):	25 m*M* MOPS, pH 7.5, 500 m*M* NaCl, 0.1% CHAPS, 0.1% PC, 1× PIC, 20% (v/v) glycerol
WI buffer (60 ml):	25 m*M* MOPS, pH 7.5, 50 m*M* NaCl, 0.05% Triton X-100, 0.2% PC, 1× PIC, 20% (v/v) glycerol
WII buffer (60 ml):	25 m*M* MOPS, pH 7.5, 50 m*M* NaCl, 0.85% Triton X-100, 0.2% PC, 1× PIC, 20% (v/v) glycerol
WIII buffer (60 ml):	25 m*M* MOPS, pH 7.5, 50 m*M* NaCl, 0.1% CHAPS, 0.1% PC, 1× PIC, 20% glycerol
WIV buffer (60 ml):	25 m*M* MOPS, pH 7.5, 50 m*M* NaCl, 1% CHAPS, 0.1% PC, 1× PIC, 20% glycerol
EI buffer (60 ml):	25 m*M* MOPS, pH 7.5, 500 m*M* NaCl, 0.1% CHAPS, 0.1% PC, 1× PIC, 20% glycerol
EII buffer (260 ml):	25 m*M* MOPS, pH 7.5, 500 m*M* NaCl, 1% CHAPS, 0.1% PC, 1× PIC, 20% glycerol, 2 μ*M* proFIX19

Procedure

1. Equilibrate a 25-ml Affi-FIXQ/S column with 125 ml of equilibration buffer.

2. Prepare the loading sample by mixing 85 ml of solubilized microsome (95 mg/ml), 3 ml of 5% CHAPS/PC, 30 ml of glycerol, 6.5 ml of 10× PIC, 13 ml of 5× buffer D, and 12.5 ml of H_2O.
3. Sonicate the sample using a standard probe attached to a programmable ultrasonic processor (Heat Systems, model XL2020) at scale 6 for 6 min: programmed as 240 cycles of 1.5 sec on followed by 8.5 sec off.
4. Set the flow rate at 9.6 ml/hr and fraction size 10 ml throughout the affinity purification.
5. Load the sample onto Affi-FIXQ/S column and determine the binding efficiency by measuring the remaining carboxylase activity in the flowthrough.
6. Wash the loaded column with 150 ml of equilibration buffer.
7. Wash the column with 100 ml of Triton X-100 linear gradient from WI buffer to WII buffer.
8. Wash the column with 100 ml of CHAPS linear gradient from WIII buffer to WIV buffer.
9. Elute γ-glutamyl carboxylase with 100 ml of CHAPS/proFIX19 double gradients in high salt from EI buffer to EII buffer.
10. Complete the elution with additional 100 to 200 ml of EII buffer.
11. Freeze the carboxylase containing fractions in liquid nitrogen and store at −80°.

Comment. Many factors affect the purity and yield of a carboxylase preparation. In our hands, protein to detergent ratio, sufficient sonication, correct flow rate, and salt concentration are the most important factors. It is also important not to let the eluted fractions stand in the cold room but to freeze the eluted carboxylase every 4–6 hr, because prolonged incubation of carboxylase with the high concentration of CHAPS results in irreversible denaturation. To follow the recovery and purity, we remove 50 μl of sample from each step for *in vitro* carboxylation assay[7] and SDS–PAGE analysis. Because many reagents used in the purification affect carboxylase activity in the assay, we include 0.8 *M* $(NH_4)_2SO_4$, 16 μM proFIX19, 0.32% CHAPS, and 0.16% PC to standardize the reaction. A reducing SDS–PAGE analysis of the microsomal fraction, the wash fractions, and the purified carboxylase is shown in Fig. 1; Table I gives the relative purification achieved in our preparation. These figures were taken from our original PNAS publication.[5] Figure 2 shows a PAGE analysis of a carboxylase preparation.

[7] R. J. T. J. Houben, B. A. M. Soute, and C. Vermeer, *Methods in Enzymol.* **282** [30], 1997 (this volume).

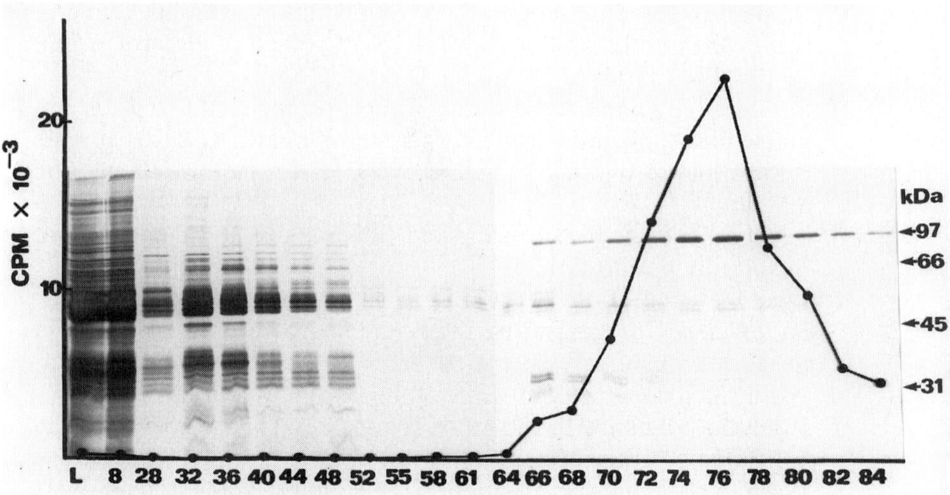

FIG. 1. Activity profile and reducing SDS–PAGE analysis (10% silver-stained gel) of fractions from Affi-FIXQ/S chromatography elution II. A 5.0 sample of each fraction was used for SDS–PAGE analysis except that the loading material and fraction 8 were diluted 1:100 before analysis. The fraction number is shown on the *x* axis. The *y* axis represents the carboxylase activity in each lane. The carboxylase activity was determined by the $^{14}CO_2$ incorporation into FLEEL in the standard assay. Lane L is microsomal loading material; fractions 1–20 are flow-through (because each fraction is equivalent, only one is shown); fractions 21–30 are wash; fractions 31–49 are Triton X-100 gradient; fractions 56–63 are CHAPS gradient; fractions 64–75 are CHAPS/proFIX19 double gradient; fractions 76–90 are 1% CHAPS/2 μM proFIX19 elution. Molecular mass ($\times 10^{-3}$) are shown on the right (Data from Wu *et al.*[5]).

TABLE I
PURIFICATION OF CARBOXYLASE

Sample	Total protein (mg)	Total carboxylase activity (cpm/30 min)	Recovery of activity (%)	Specific activity (cpm/mg/hr)	Purification (-fold)
Solubilized microsomes (load)	8100	1.14×10^9	100	2.81×10^5	1
Flow-through of Affi-FIXQ/S	8090	8.08×10^8	70	2×10^5	0.7
Bound to Affi-FIXQ/S	4.7	3.3×10^8	30	1.4×10^8	502
Affinity-purified carboxylase	0.402	3.88×10^8	34	1.93×10^9	7000

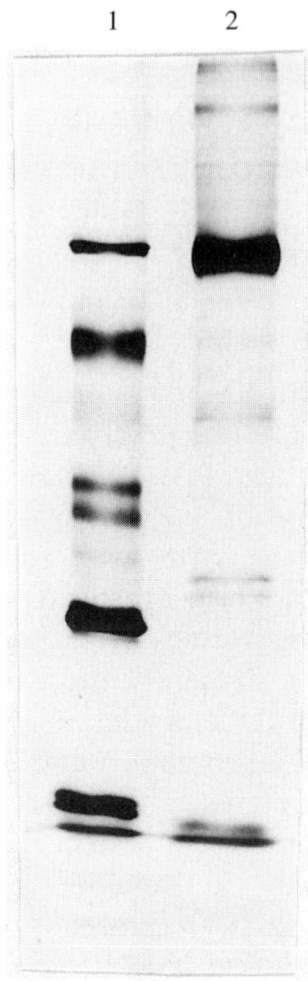

FIG. 2. Reduced SDS–PAGE analysis of a concentrated carboxylase preparation. Lane 1, molecular weight standards; lane 2, γ-glutamyl carboxylase.

Regeneration of Affi-FIXQ/S Column

Procedure

1. Wash with 4–10 column volumes of 3 *M* NaSCN.
2. Wash with 4–10 column volumes of 1 *M* NaCl and 0.1 *M* sodium acetate, pH 3.2.
3. Wash with 4–10 column volumes of 1 *M* NaCl and 0.1 *M* Tris-HCl, pH 8.8.
4. Reequilibrate with starting buffer or store in the presence of 0.02% NaN_3.

Comment. Affi-FIXQ/S can be regenerated for more than 10 times without showing any detectable loss in binding capacity or quality of purification.

[30] Assay of Vitamin K-Dependent Carboxylase Activity in Hepatic and Extrahepatic Tissues

By ROGER J. T. J. HOUBEN, BERRY A. M. SOUTE, and CEES VERMEER

Introduction

Vitamin K-dependent carboxylase is present in a wide variety of mammalian cells and tissues, where it is involved in the production of different kinds of γ-carboxyglutamic acid (Gla) proteins. In some cases these Gla proteins are unique products of one cell type; examples of this category are prothrombin and osteocalcin, which are exclusively synthesized by liver (hepatocytes) and bone (osteoblasts), respectively. Other Gla proteins, such as protein S and matrix Gla protein, are expressed at low levels in many different extrahepatic tissues. Presently, there is a discrepancy between the tissue carboxylase content and the production of known Gla proteins, such that it is to be expected that new Gla proteins will be discovered. One way of identifying these new proteins is to induce the accumulation of their intracellular precursors by treatment of experimental animals with vitamin K antagonists. These precursor proteins may then be specifically labeled by *in vitro* carboxylation using $^{14}CO_2$, and subsequently purified and characterized using denaturing techniques. Reliable procedures for isolating and testing the vitamin K-dependent systems from various tissues is a first requirement for these investigations.

The vitamin K-dependent carboxylase is a typical integral membrane protein, localized in the endoplasmic reticulum. After tissue homogenization, its activity can only be preserved by leaving the carboxylase within the microsomal membrane remnants, or by incorporating the enzyme into well-defined detergent micelles. Although there are no indications that carboxylases from different tissues are different gene products, accompanying proteins may affect the substrate specificity or affinity for vitamin K of carboxylase. Besides carboxylases in various states of purification, research of this kind also requires the availability of potential substrates and substrate analogs, differing from each other in their primary structure and their affinity for carboxylase. Substrates for carboxylase may vary from simple, synthetic peptides (e.g., Phe-Leu-Glu-Glu-Leu, FLEEL) to polypeptides of more than 60 amino acid residues, which are generally produced by molecular biology techniques. The cofactors required are either vitamin K hydroquinone (for assaying the carboxylase exclusively) or one of the oxidized forms (vitamin K quinone or epoxide). In the latter case, the

0076-6879/97 $25.00

combined action of carboxylase and vitamin K reductase would be measured. In this chapter, the preparation procedures for enzymes, substrates, and cofactors are detailed.

Preparation Procedure for Carboxylase from Different Tissues

The best sources of vitamin K-dependent carboxylase are soft organ tissues, and the preparation procedures are very similar. Here we describe the preparation of carboxylase from bovine liver, lung, kidney, and testis, but the same procedure can be used for other tissues as well. Tissue homogenates may be frozen in liquid nitrogen and stored at −80° for several years without loss of carboxylase activity. For experiments in which a pretreatment of the animals precedes their use as organ donors, it may be practical to work with small animals such as rats. In those cases in which the availability of the same (frozen) material over a long period is preferred, the cow is the experimental animal of choice. Also if large amounts of tissue are required (e.g., for purifications) the least expensive and most practical tissue donor is the cow.

Washed Microsomes

Bovine liver, testis, kidney, and lung are collected at the abbatoir, immediately after slaughter. The different tissues are cooled on ice and sliced in small pieces, following removal of the tougher parts like membranes, veins, and connective tissue. All subsequent steps are performed at 4° unless stated otherwise. In a typical procedure, the sliced tissue is added to two volumes of buffer A (100 m*M* NaCl, 50 m*M* Tris-HCl, pH 7.5) and minced in a blender. The mixed slurry is then homogenized in a Potter–Elvehjem tube (equipped with a Teflon pestle) at 300 rpm with two strokes (up and down) of the pestle. At this stage the solution thus obtained can be either frozen or used immediately for the preparation of microsomes (see later discussion). In the latter case, 500 ml of tissue homogenate is centrifuged for 15 min at 10,000*g*, after which the red supernatant fluid (postmitochondrial fraction) is collected and centrifuged for 1 hr at 105,000*g*. The supernatant fluid and a loose interface are discarded, and the pellets are transferred to the Potter tube, homogenized in 450 ml of buffer A, and centrifuged again for 1 hr at 105,000*g* (first washing step). Washing with buffer A is repeated until a colorless supernatant fluid is obtained (at least three washing cycles are required), and is followed by a washing step with buffer B (1 *M* NaCl, 50 m*M* Tris-HCl, pH 7.5). Finally the pellet is resuspended in buffer C (0.5 *M* NaCl, 25 m*M* Tris-HCl, pH 7.5) at a final protein concentration of approximately 40 mg/ml. Washed microsomes thus obtained can be either subsampled and stored at −80° or used for further purification.

Partly Purified Carboxylase

During purification, carboxylase is increasingly sensitive to proteases. We therefore recommend the use of a protease inhibitor cocktail in all further steps described in this chapter. Unless mentioned otherwise, all buffers used in our purification procedure contain the following protease inhibitors: benzamidine (1 m*M*), aprotinin (0.3 μM), phenylmethylsulfonyl fluoride (PMSF, 12 μM), dithiothreitol (DTT, 2 m*M*), and EDTA (2 m*M*). Washed microsomes are supplemented with an equal amount of solubilization buffer [2% (w/v) 3-[(3-cholamidopropyl)dimethylammonio]-1-propane sulfonate (CHAPS), 2 *M* NaCl], stirred for 30 min, and centrifuged for 2 hr at 105,000*g*. The supernatant fluid (containing most of the carboxylase activity) is transferred into a precooled flask to which ammonium sulfate is added with constant stirring to a final concentration of 55% saturation. The solution is kept at 4° for at least 30 min and subsequently centrifuged for 30 min at 20,000*g*. The resulting "floating pellet" contains the major part of the carboxylase activity and is dissolved in buffer C to a final protein concentration of 18 mg/ml. Aliquots of 1 ml may be frozen in liquid nitrogen and stored at −80° for several years.

The preparation thus obtained is called "partly purified carboxylase" and was used in the experiments described later. Carboxylase has been identified in most soft tissues, except in brain and muscle, and the procedure described here can be used for nearly all tissues except those which are hard or tough (bone, cartillage, tendon, and vessel wall). Partly purified carboxylase can also serve as the starting material for further purification of the enzyme. A protocol for the affinity purification of hepatic carboxylase to homogeneity is described elsewhere in this volume. It should be noted, however, that for the affinity purification of *extra*hepatic carboxylases this protocol requires substantial adaptations; for instance, the concentrations of salt and detergent concentrations must be optimimzed for each type of tissue.

Substrates for Carboxylase

In vivo substrates for carboxylase invariably contain the so-called "pro-sequence," which serves as a recognition site for carboxylase. In most substrates the pro-sequence is located in the leader peptide, immediately preceding the NH_2 terminus in the mature protein.[1] Shortly before cellular

[1] B. Furie and B. C. Furie, *Blood* **75,** 1753 (1990).

excretion the pro-sequence is removed from the precursor protein by limited proteolytic degradation. Only in the case of matrix Gla protein (MGP) does the pro-sequence form an integral part of the mature protein.[2] For *in vitro* carboxylation one can rely on endogenous precursor proteins that accumulate in the tissues during vitamin K deficiency or warfarin treatment of the donor animal.[3] In general this leads to a rapid carboxylation reaction, which will come to an abrupt stop when the substrate is depleted. The carboxylation reaction continues for a prolonged period of time when reaction mixtures are supplemented with exogenous substrates. These substrates can be prepared by thermal decarboxylation of isolated Gla proteins, by peptide synthesis, or by the production of recombinant Gla proteins in carboxylase-deficient systems.

Decarboxylation of Gla Proteins

In general decarboxylated Gla proteins will not contain the pro-sequence, which is the main reason why they are poor substrates for carboxylase. Unfortunately the only exception (MGP) is highly insoluble and difficult to isolate.[2] Therefore, this technique can only be recommended for one other protein, osteocalcin (also known as bone Gla protein, BGP). As can be seen in many reviews, the pro-sequences of all known Gla proteins have a number of strictly conserved amino acid residues, for example, Ala-10 and Phe-16, which implies that these residues are probably critical for substrate recognition by carboxylase. Osteocalcin forms an exception to this rule, because the Ala-10 has been replaced by Gly. Substitution of either Ala-10 or Phe-16 by other amino acid residues strongly impaired substrate carboxylation of blood coagulation factors.[4] Hence it might be expected that the pro-sequence of osteocalcin has a relatively low affinity for carboxylase. This could be why the osteocalcin molecule itself has developed in such a way that the mature sequence contributes significantly to the recognition by carboxylase. Whatever the reason may be, decarboxylated osteocalcin is a fairly good and easily prepared substrate for *in vitro* carboxylation.[5] Bovine tibia may be obtained from the abbatoir, cleaned, and defatted in acetone before being ground in a bone mill. Further removal of traces of fat can be accomplished in a second acetone wash, after which the powder is extracted in three subsequent steps of 24 hr each, with a

[2] P. A. Price, J. D. Fraser, and G. Metz-Virca, *Proc. Natl. Acad. Sci. U.S.A.* **84,** 8335 (1988).

[3] B. A. M. Soute, M. M. W. Ulrich, and C. Vermeer, *Thromb. Haemostas.* **57,** 77 (1987).

[4] M. J. Jorgensen, A. B. Cantor, B. C. Furie, C. L. Brown, C. B. Shoemaker, and B. Furie, *Cell* **48,** 185 (1987).

[5] C. Vermeer, B. A. M. Soute, H. Hendrix, and M. A. G. de Boer-van den Berg, *FEBS Lett.* **165,** 16 (1984).

solution containing 1 *M* EDTA, 0.2 *M* KCl, 10 m*M* benzamidine, 5 m*M* ε-aminocaproic acid, 1 m*M* *p*-hydroxymercuribenzoate, and 2 mg/liter aprotinin, pH 8.0.

In a typical preparation procedure, 500 g of bone powder and 3 × 2 liters of extraction buffer are used. After each extraction step, the residue is removed by centrifugation (5,000*g*), and the supernatant is diluted 20 times with distilled water. Then, preswollen QAE-Sephadex slurry (in 20 m*M* Tris-HCl, 0.1 *M* NaCl, pH 8.0; 10 ml per liter of diluted extract) is added followed by 2 hr of stirring; the Sephadex is isolated by filtration over nylon cloth, and eluted with 1 *M* NaCl. Osteocalcin is purified from the solution obtained by size-exclusion chromatography on Sephadex G-75 (Pharmacia Biotech AB, Uppsala, Sweden), and by high-performance liquid chromatography (HPLC) using a Mono Q column. Fractions are tested using a commercial test kit with cross-reactivity for bovine osteocalcin (e.g., Incstar, Stillwater, OK). For decarboxylation the peak fractions are pooled, dialyzed against phosphate-buffered saline (PBS: 0.14 *M* NaCl, 2.7 m*M* KCl, 8.1 m*M* Na_2HPO_4, pH 7.4), and brought to pH 1 with 0.1 *M* HCl. This preparation is lyophilized to dryness, and subsequently heated to 105° overnight, under vacuum. Optimal results are obtained if during the entire heating period there is constant suction using a powerful vacuum pump. The preparation thus obtained is reconstituted with water to its original volume, and the pH is adjusted to 7.4. The precise concentration of osteocalcin must be determined by at least two independent methods (e.g., protein measurement and a specific osteocalcin assay).

Peptide Synthesis

It is surprising how few structural elements are required to permit the carboxylation of a peptide substrate. Of course, the predominant requirement is the presence of a glutamate residue. Glu derivatives such as Boc-Glu methyl ester (Boc-Glu-Me) and short peptides such as Phe-Leu-Glu-Glu-Leu (FLEEL) are commercially available (Bachem, Bubendorf, Switzerland), and others (Boc-Glu-Glu-Val[6]) can be synthesized. These small substrates do not contain the pro-sequence, but since they can be added in high concentration, they allow for relatively high carboxylase activities. Different chain lengths of polyglutamates may also serve as a substrate for carboxylase.[7] A common characteristic of all these substrates is that their apparent K_m (K_m^{app}) is rather high (in the millimolar range). With present technology it is possible also to synthesize substrates containing the pro-sequence of one of the coagulation factors, for instance, the substrate known

[6] F. Acher and R. Azerad, *Intl. J. Pept. Prot. Res.* **37,** 210 (1991).

[7] B. A. M. Soute, R. Budé, and C. Vermeer, *Biochim. Biophys. Acta* **1073,** 434 (1991).

as proPT-28 containing the amino acid residues −18 through +10 of the human prothrombin precursor.[8]

Molecular Biology

Because prokaryotes do not contain the complicated machinery for posttranslational processing of proteins, expression of cDNAs from all Gla proteins in prokaryotic systems will result in noncarboxylated proteins still containing the pro-sequence. This has been reported for osteocalcin[9] as well as for coagulation factors and their Gla domains. In particular, we draw attention to the peptide consisting of the pro-sequence and the first 41 residues of human coagulation factor IX. The preparation of this polypeptide (designated as proIX-59) and its crucial importance for the purification of carboxylase have first been described by Wu *et al.*[10,11] These authors reported the construction of a cDNA coding for a chimeric protein composed of proIX-59, linked to a phage T7 capsid protein.[10] The recombinant protein product accumulated in inclusion bodies, which were harvested and subjected to CNBr-mediated cleavage to liberate the recombinant proIX-59. After an anion-exchange purification step, proIX-59 formed an excellent carboxylase substrate, all 12 glutamate residues of which may be carboxylated after prolonged incubation.[10] Even more important was the observation that proIX-59—in contrast to proPT-28—could be used as an affinity ligand for purifying carboxylase.[11]

Coenzymes for Carboxylase

Vitamin K hydroquinone (KH_2) is the active coenzyme for carboxylase, and it can be prepared by incubating 2.5 m*M* of a detergent-solubilized vitamin K (e.g., Konakion from Hoffmann-La Roche, Basel, Switzerland) in 150 m*M* DTT, pH 8.5, at 37° overnight in a light-protected tube. The resulting KH_2 is colorless and gives the best results in CO_2 incorporation studies. In nonpurified systems, vitamin K quinone (K) and vitamin K 2,3-epoxide (KO) can be used as coenzymes for carboxylase as well. In these cases, however, the coenzymes have to be reduced by the enzyme KO reductase, which is present in washed microsomes from all tissues, and also in partly purified carboxylase from some sites (e.g., liver, testis). So if the carboxylase reaction is performed with either K or KO instead of KH_2 the

[8] M. M. W. Ulrich, B. Furie, M. R. Jacobs, C. Vermeer, and B. C. Furie, *J. Biol. Chem.* **263,** 9697 (1988).
[9] M. E. Benton, P. A. Price, and J. W. Suttie, *Biochemistry* **34,** 9541 (1995).
[10] S.-M. Wu, B. A. M. Soute, C. Vermeer, and D. W. Stafford, *J. Biol. Chem.* **265,** 13124 (1990).
[11] S.-M. Wu, D. P. Morris, and D. W. Stafford, *Proc. Natl. Acad. Sci. U.S.A.* **88,** 2236 (1991).

sequential activities of carboxylase and KO reductase will be tested. By comparing these data with those obtained for the KH_2-initiated reaction, an impression can be obtained concerning the KO reductase content of the preparation. KO is prepared by dissolving 20 mg of phylloquinone in 5 ml of 2-propanol/hexane in a ratio of 2:1 (v/v), to which 0.1 ml of 0.5 *M* NaOH in 0.2 *M* Na_2CO_3 and 0.3 ml of 30% H_2O_2 are added.[12] After an overnight incubation in a light-protected tube at 37°, the mixture is supplemented with 3 ml of water, vortex-mixed for 1 min, and the hexane phase is collected. The latter is washed twice with 5 ml of water and evaporated to dryness with gentle heating under a constant stream of nitrogen. The residue is dissolved in ethanol to a concentration of 10 mg/ml.

Optimal Conditions for *in Vitro* Carboxylation of Endogenous and Exogenous Substrates

Endogenous substrates are protein precursors that have accumulated during *in vivo* treatment of the donor animals with vitamin K antagonists (warfarin, brodifacoum). At least part of these precursor proteins remain complexed to carboxylase during the purification procedure and may be carboxylated *in vitro* under the conditions described later. In general, the supply of endogenous substrate is rapidly exhausted, and the addition of an exogenous substrate (e.g., the pentapeptide FLEEL) substantially increases the total amount of CO_2 fixed. After the carboxylation reaction is completed, the (pro-containing) endogenous substrate can be separated from the short peptide substrate by trichloroacetic acid (TCA) precipitation (see later discussion). An important difference between substrates that do and those that do not contain the pro-sequence is that the carboxylation of the latter ones is greatly enhanced (10- to 20-fold) by the presence of 1 *M* $(NH_4)_2SO_4$. The mechanism behind this stimulatory effect is not quite clear, but kinetic analysis has shown that the high salt concentration affected the V_{max}, and not the K_m^{app}.[13] No such effect was observed for the carboxylation of endogenous precursor proteins or for pro-containing substrates such as proPT-28 and proIX-59. Another point to realize is that if radiolabeled bicarbonate is the only source of $^{14}CO_2$, the concentration of bicarbonate is the rate-limiting step in the carboxylation. The carboxylation rate will substantially increase by adding 5 m*M* of nonlabeled NH_4HCO_3, but obviously the amount of incorporated label will decrease. Carboxylase is not very stable at 37°, which is why most investigators work at 20° or lower.

[12] L. F. Fieser, M. Tishler, and W. L. Sampson, *J. Biol. Chem.* **137,** 659 (1941).

[13] B. A. M. Soute, F. Acher, R. Azerad, and C. Vermeer, *Biochim. Biophys. Acta* **1034,** 11 (1990).

Maximal substrate carboxylation may even be obtained in overnight incubations at 10°.[14]

To allow the carboxylation reaction to proceed, the following ingredients should be pipetted in the order indicated to give 0.125-ml reaction volumes of the following composition: 0.45 mg of microsomal proteins, a suitable substrate (either 4 m*M* of a substrate devoid of propeptide, or 20 μM of a pro-containing one), 4 m*M* DTT, 1 *M* $(NH_4)_2SO_4$, 0.5 *M* NaCl, 25 m*M* Tris-HCl, and 5 μCi (185 kBq) of $NaH^{14}CO_3$. Reactions are initiated by the addition of 5 μl (220 μM) of KH_2, K, or KO. Incubation is performed for 30 min at 20°, and the reaction is stopped by adding 0.1 ml of the reaction mixture to 0.8 ml of 5% (w/v) TCA in a glass vial, containing antibumping granules. The vials are boiled for a short time (2 min) on a hot plate to remove the last traces of unbound $^{14}CO_2$, after which scintillation liquid is added for radiolabel counting.

Optimal Conditions for Measuring KO Reductase in Microsomal Fractions

During the carboxylation reaction, KH_2 is oxidized into KO, which can be recycled via the action of the enzyme KO reductase. Several procedures for the assessment of KO reductase have been described; the data shown later were produced using the method described by Thijssen.[15] Reaction mixtures (0.25 ml) contained 0.9 mg of microsomal proteins, 1 *M* $(NH_4)_2SO_4$, 0.5 *M* NaCl, 4 m*M* DTT, and 25 m*M* Tris-HCl, pH 7.5. After a preincubation of 2 min at 20°, reactions were started by adding 10 μl of 220 μM KO in ethanol. After incubation periods of 0, 5, 10, and 20 min, 50-μl aliquots were taken and extracted with 1 ml 2-propanol/hexane (2:1, v/v) containing 5 μg (+)-α-tocopherol as an internal standard. After the addition of 1 ml of water, the mixtures were vortexed and 0.2 ml of the hexane phase was taken and evaporated to dryness under a gentle stream of N_2 at room temperature. The residue was dissolved in 50 μl 2-propanol of which 20 μl was used for HPLC analysis.[15] The enzyme activity was deduced from the initial reaction rate during the first 10 min of incubation.

Comparison of Enzymatic Activities of Hepatic and Extrahepatic Carboxylases

The enzymes of the vitamin K cycle obtained from different tissues can be compared (1) by measuring the carboxylase and KO reductase per

[14] B. A. M. Soute, R. Budé, H. Buitenhuis, and C. Vermeer, *Analyt. Biochem.* **182,** 207 (1989).
[15] H. H. W. Thijssen, *Biochem. Pharmacol.* **35,** 3277 (1986).

TABLE I
PARTLY PURIFIED CARBOXYLASES FROM DIFFERENT ORIGINS: COMPARISON OF TISSUE CONTENT[a]

Origin of carboxylase	KH_2 carboxylase (pmol CO_2 fixed)	KO carboxylase (pmol CO_2 fixed)	KO reductase (pmol K formed)
Liver	0.93	0.24	69
Testis	0.78	0.19	496
Kidney	0.36	0	0
Lung	0.22	0	0

[a] Carboxylase activity is expressed as picomoles CO_2 incorporated per milligram of microsomal protein and per minute of incubation at 20°, using the pentapeptide FLEEL as a substrate. KH_2 carboxylase means that the reaction was initiated with vitamin K hydroquinone, KO carboxylase stands for the carboxylation reaction initiated with vitamin K epoxide. KO reductase is expressed as picomoles vitamin K quinone formed per milligram of microsomal protein and per minute of incubation at 20°, using vitamin K epoxide as a substrate.

milligram of protein, and (2) by assessing their respective kinetic constants. In these measurements, it is important that substrates and cofactors be present in excess and that the rate of product formation be assessed under linear reaction rate conditions. Carboxylase activity is assessed by initiating the carboxylation reaction with KH_2. A test for the simultaneous activities of carboxylase and KO reductase is performed by initiating the carboxylase reaction with KO. If KO-stimulated carboxylase is less than 25% of the KH_2-driven reaction, this is indicative of a relative KO reductase deficiency of the preparation. This can be assessed in a more direct way by measuring the conversion of vitamin K epoxide into the corresponding quinone. Table I summarizes these three enzymatic activities for the four partly purified carboxylase preparations presented in this chapter. It is clear that partly purified carboxylase was obtained from all four tissues and that the activities of the extrahepatic preparations ranged from 23 to 84% of that obtained from the liver. KO reductase was present in washed microsomes from all tissues but, on fractionation, the enzyme activity was lost in the preparations from kidney and lung. This loss was due to inactivation (probably because of the high detergent concentration) rather than to separation in a different fraction. Partly purified carboxylase from liver and testis contained high levels of KO reductase, and it appears that the testis especially is the tissue of choice for the purification of KO reductase. Another striking point is that there is no apparent stoichiometry between carboxylase and reductase: in terms of picomoles of product formed, the enzymatic activity of reductase exceeds that of carboxylase by 75- (liver) to 635- (testis) fold. The reason

TABLE II
PARTLY PURIFIED CARBOXYLASES FROM DIFFERENT ORIGINS: COMPARISON OF SUBSTRATE SPECIFICITY[a]

	Origin of carboxylase							
	Liver		Testis		Kidney		Lung	
Substrate used	K_m^{app}	V_{max}	K_m^{app}	V_{max}	K_m^{app}	V_{max}	K_m^{app}	V_{max}
FLEEL	2.2 m*M*	1.4	2.7 m*M*	1.0	2.1 m*M*	0.6	2.2 m*M*	0.4
Boc-EEV	1.6 m*M*	1.1	1.3 m*M*	0.9	1.2 m*M*	0.5	1.3 m*M*	0.2
Boc-Glu-Me	2.4 m*M*	1.0	2.5 m*M*	0.8	2.7 m*M*	0.4	2.7 m*M*	0.2
D-Osteocalcin	8.2 μM	0.19	6.8 μM	0.11	7.2 μM	0.08	6.3 μM	0.04
proGlu-10	4.1 μM	0.21	4.9 μM	0.12	3.5 μM	0.06	2.7 μM	0.05
proPT-28	2.1 μM	0.24	3.5 μM	0.14	2.7 μM	0.07	3.0 μM	0.03
proIX-59	0.8 μM	0.25	1.1 μM	0.13	1.2 μM	0.08	1.5 μM	0.03

[a] K_m^{app} for the first three substrates is expressed in *milli*molar, for the last three substrates in *micro*molar; V_{max} is expressed in pmol CO_2 incorporated per minute and per milligram of microsomal protein at 20°; proPT-28 stands for the prothrombin precursor sequence −18 to +10; proIX-59 stands for the factor IX precursor sequence −18 to +41; proGlu-10 stands for a polypeptide consisting of the prothrombin propeptide with 10 glutamate residues at its carboxy-terminal site.

for this large excess of reductase is not known, but obviously it will result in a rapid reduction of any KO formed.

Different carboxylase preparations can also be compared by measuring their kinetic constants using various substrates. In Table II, we show the results for six different substrates tested in the four partly purified carboxylase preparations. Peptide substrates lacking the pro-sequence have K_m^{app} values in the millimolar range, except for decarboxylated osteocalcin the mature sequence of which seems to contribute substantially to the enzyme–substrate recognition. The reason for this unique property is presently unknown. The K_m^{app} values for pro-containing substrates are at least three orders of magnitude lower than those for the short peptides. The fact that there is very little difference between a pro-sequence connected to a sequence derived from a clotting factor and that linked to poly(L-glutamate) demonstrated that the pro-sequences in the clotting factor precursors are the major (if not the only) structural requirement for enzyme–substrate interaction. Strikingly, there is a large difference between the carboxylation of peptide substrates in the presence of noncovalently bound propeptide, and the carboxylation of polypeptides containing both the pro-sequence as well as the carboxylatable glutamate residues. As was reported by Ulrich *et al.*,[8] the K_m^{app} for small substrates such as FLEEL and PT/1-10 (decarboxyprothrombin residues 1-10) decreased by a factor of 3 with the addition of nonbound propeptide, but it was at least 1000 times lower if the propeptide was covalently attached. Similarly, pro-containing substrates decreased the K_m^{app} of carboxylase for its cofactor KH_2 about 20-fold, whereas the combination of propeptide + peptide substrate did not.[16] Thus, the covalent attachment of the propeptide and the substrate glutamate residues is a critical requirement for efficient carboxylation of the substrate.

As has been detailed elsewhere in this volume, hepatic carboxylase has now been purified to homogeneity, and may be subject to physicochemical characterization and further investigations for its reaction mechanism. The fact that the kinetic characteristics of carboxylases form various tissues are very similar is consistent with the idea that all are products of the same gene. It is to be expected, therefore, that the various extrahepatic Gla proteins rather than the extrahepatic carboxylases may be a subject of interest in forthcoming years. Warfarin treatment of animals, followed by endogenous substrate labeling in the *in vitro* carboxylase reaction is a promising technique to recognize these proteins during their purification and characterization, even if they would have lost all functional activity. Washed microsomes and partly purified carboxylase may remain useful to test multienzyme systems in their mode of action and mutual interaction. One example is the combined activity of carboxylase and KO reductase, but this may well be extended to other posttranslational steps such as prolyl hydroxylation (in osteoblasts), aspartate hydroxylation (in hepatocytes), or disulfide bond formation.

Acknowledgment

The work in the author's laboratory was supported by grant 93.003 from the Netherlands Thrombosis Foundation.

[16] B. A. M. Soute, A. D. J. Watson, J. E. Maddison, M. M. W. Ulrich, R. Ebbering, and C. Vermeer, *Thromb. Haemostas.* **68,** 521 (1992).

[31] Expression of Human Anticoagulation Protein C and γ-Carboxyglutamic Acid Mutants in Mammalian Cell Cultures

By FRANCIS J. CASTELLINO and JIE-PING GENG

Introduction

Protein C (PC), the zymogen of the potent anticoagulant enzyme, activated protein C (APC), contains 419 amino acids in the mature protein.[1] Its amino acid sequence is provided in Fig. 1. Considerable posttranslational and cotranslational modifications occur in this protein, which include signal polypeptide and propeptide release, endoproteolytic-catalyzed liberation of the dipeptide, K156R157,[1] β-hydroxylation of D71,[2] and oligosaccharide assembly on N97, N313, and partially on N329.[3] In addition, an array of nine γ-carboxyglutamic acid (Gla) residues is present, all of which exist within the amino-terminal 29 residues of the light (nonprotease) chain of the protein. These residues are located at sequence positions 6, 7, 14, 16, 19, 20, 25, 26, and 29.[4] The disposition of Gla residues in human PC represents their minimal arrangement in other vitamin K-dependent coagulation proteins, such as prothrombin and factors VII, X, and X, which contain these same nine Gla residues, as well as others downstream of Gla29 of PC. At least in the case of PC, the Gla residues are primarily responsible for binding of Ca^{2+} to PC and APC, the result of which is to induce a Ca^{2+}-dependent conformation in the protein, which is essential for the phospholipid (PL)-dependent anticoagulant activity of APC. Approximately seven g-atoms of Ca^{2+} per mole of protein interact through a variety of types of binding modalities with these Gla residues,[5,6] and it is a formidable task to identify the specific functions of each of the Gla residues, as well as those of each Ca^{2+} atom. One approach to solution of this problem is through site-directed mutagenesis strategies, wherein Gla residues are

[1] D. C. Foster, C. A. Sprecher, R. D. Holly, J. E. Gambee, K. M. Walker, and A. A. Kumar, *Biochemistry* **29,** 347 (1990).

[2] T. Drakenberg, P. Fernlund, P. Roepstorff, and J. Stenflo, *Proc. Natl. Acad. Sci. U.S.A.* **80,** 1802 (1983).

[3] W. Kisiel, *J. Clin. Invest.* **64,** 761 (1979).

[4] R. J. Beckmann, R. J. Schmidt, R. F. Santerre, J. Plutzky, G. R. Crabtree, and G. L. Long, *Nucleic Acids Res.* **13,** 5233 (1985).

[5] M. Soriano-Garcia, K. Padmanabhan, A. M. deVos, and A. Tulinsky, *Biochemistry* **31,** 2554 (1992).

[6] W. T. Christiansen, A. Tulinsky, and F. J. Castellino, *Biochemistry* **33,** 14993 (1994).

METHODS IN ENZYMOLOGY, VOL. 282
0076-6879/97 $25.00

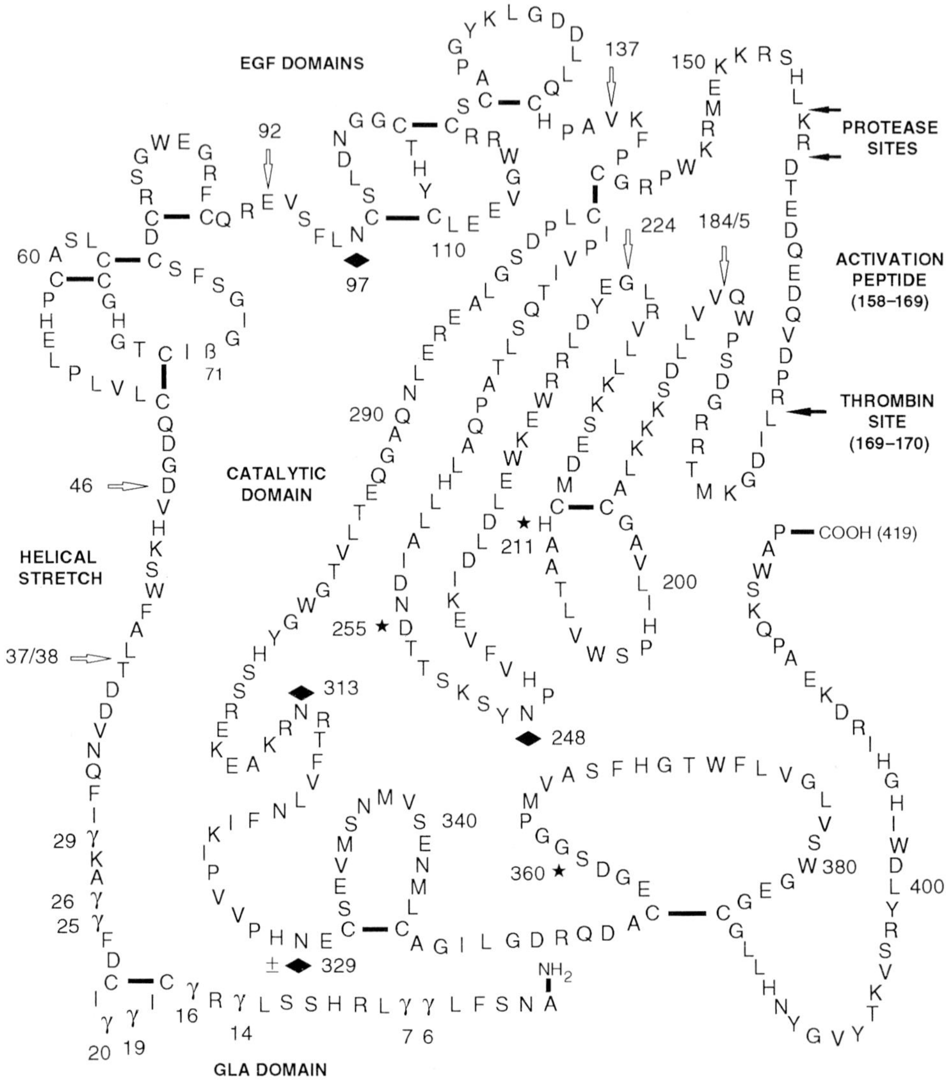

FIG. 1. The amino acid sequence of mature human protein C. Gla residues are labeled as γ and Hya at position 71 is symbolized as β. Positions of splice junctions in the gene are marked by open arrows. Active site residues are indicated by filled stars and the N-linked glycosylation sites by filled diamonds. The domain units of the protein are within the exonic structures.

replaced with other amino acids, and Ca^{2+}-dependent functions of the variant proteins are examined.[7–9] One possible difficulty with this design is that removal of one Gla residue might influence γ-carboxylation at other Gla-precursor E-residues, and such an occurrence could lead to serious misinterpretations of data obtained from study of such mutants. Thus, variant proteins obtained in this manner must be carefully characterized prior to functional analysis.

This chapter summarizes the methodology that we employ for expression, purification, and characterization of the recombinant proteins containing mutations at Gla residues. Properly generated proteins of these types can be of great benefit to studies on the topic of structure–function relationships of individual Gla residues of human PC and APC.

Description of cDNA for Human Protein C and Insertion into the Expression Plasmid[10]

The cDNA encoding wild-type recombinant protein C (wtr-PC) in pUC119 contained, between two *Eco*RI sites, 69 noncoding bases 5′ of the ATG initiation site, 126 bases of leader sequence, 1383 bases coding for the entire PC molecule, and 202 noncoding bases 3′ of the TAG step codon. This latter region also included two poly(A) signals. An *Xba*I restriction endonuclease site in the polylinker region of this plasmid was altered to an *Xho*I site using the synthetic oligonucleotide primer,

5′-GGATCCTCgAGAGTCGA

(the lowercase nucleotide refers to the mismatched base). In addition, an *Nhe*I site was introduced between bases −27 and −26 employing the primer,

5′-AGTATCTCCACGgCtaGCCCCTGTGCCAG

When the cDNA for PC was excised from pUC119 using *Nhe*I as one of the restriction endonucleases, this latter insertion served the purpose of also removing this cDNA downstream of an additional ATG initiation codon situated at bases −62 to −60 in the 5′ noncoding region. These steps resulted in a cDNA for PC containing 26 bases 5′ of the ATG start sequence.

The expression vector employed for PC, pCIS2M, was constructed from the plasmid, pCIS (obtained from Genentech, South San Franciso, CA). This latter plasmid in pML contains the human cytomegalovirus (hCMV)

[7] L. Zhang, A. Jhingan, and F. J. Castellino, *Blood* **80,** 942 (1992).
[8] L. Zhang and F. J. Castellino, *J. Biol. Chem.* **267,** 26078 (1992).
[9] L. Zhang and F. J. Castellino, *J. Biol. Chem.* **268,** 12040 (1993).
[10] L. Zhang and F. J. Castellino, *Biochemistry* **29,** 10828 (1990).

major immediate early promoter–enhancer, which is *cis*-activated by the enhancer[11] and *trans*-activated[12] by adenovirus (Ad) E1 proteins[13] that are present in the Ad-transformed 293 cells. A splice, followed by a polylinker site, which allows insertion of the gene of interest, is present downstream of the promoter–enhancer region. These areas are followed by a poly(A) site, a SV40 (simian virus 40) *ori* and promoter, downstream of which is an amplifiable gene (DHFR) for use in systems wherein amplification can occur. We modified the polylinker site of this vector by excision at the *Cla*I–*Xba*I restriction site and insertion of the following linker:

5′-CGATTGCTAGCT
TAACGATCGAGATC-5′

This procedure provided a unique *Nhe*I restriction site, and restored the *Cla*I and *Xba*I restriction sites. This yielded the new vector, pCIS2M. Commercial plasmids are available, which have similar properties as pCIS2M.

For insertion of wtr-PC and r-PC mutants into the mammalian cell expression vector, the cDNA was excised from pUC119 employing an *Nhe*I–*Xho*I restriction digestion, and inserted into these same restriction sites of plasmid pCIS2M. This vector is diagrammed in Fig. 2.

Procedures for Stable Transfection of wtr-PC and r-PC Mutants into Human Kidney 293 Cells

Numerous procedures are available to transfer the DNA of interest into mammalian cells for the generation of stable transfectants. The most commonly employed method is calcium phosphate coprecipitation,[14] which we have used to obtain consistently high transfection efficiencies.

Reagents

Geneticin (G418, GIBCO, Grand Island, NY)
Vitamin K, sterile solution of 10 mg/ml (AquaMEPHYTON, Merck & Co., West Point, PA)
$CaCl_2$, 2.5 *M*
Normal growth medium (NGM): Dulbecco's modified Eagle's medium (DMEM)/F12 Medium (Sigma Chemical Co., St. Louis, MO), 10% (v/v) fetal bovine serum (FBS) (Gibco-BRL, Gaithersburg, MD)

[11] D. L. Eaton, W. I. Wood, D. Eaton, P. E. Hass, P. Hollingshead, K. Wion, J. Mather, R. E. Lawn, G. A. Vehar, and C. Gorman, *Biochemistry* **25,** 8343 (1986).
[12] M. F. Stinski and T. G. Roehr, *J. Virol.* **55,** 431 (1985).
[13] C. M. Gorman, D. Gies, G. McCray, and M. Huang, *Virology* **171,** 377 (1989).
[14] F. L. Graham and A. J. vam der Eb, *Virology* **52,** 456 (1973).

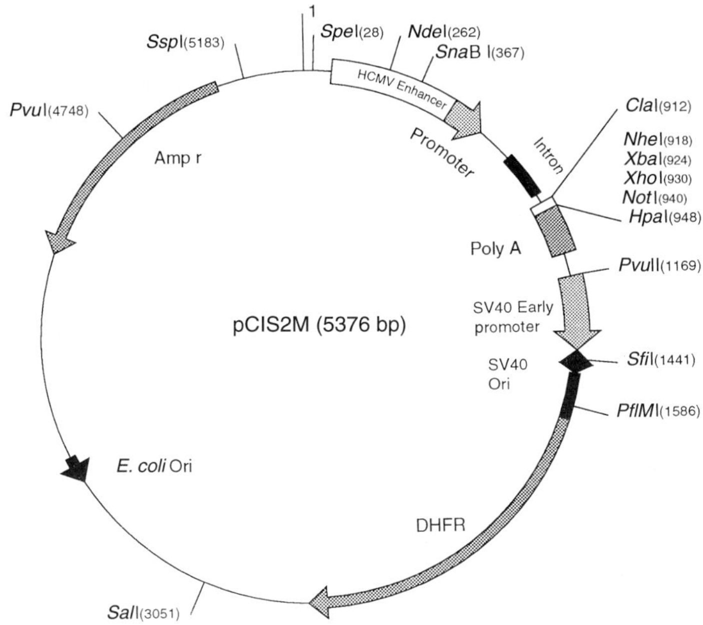

FIG. 2. The pCIS2M expression vector for r-PC. Essential components of the vector are labeled. The cDNA for PC is normally inserted in the multiple cloning region between the *Nhe*I–*Xho*I sites.

Selection medium (SM): DMEM/F12, 10% (v/v) FBS, 0.8 mg/ml G418
Low serum medium plus vitamin K (LSM/VK): DMEM/F12, 1% FBS, 0.8 mg/ml G418, 5 μg/ml vitamin K
Glutamine (Sigma), 200 m*M*
1/10 TE: 1 m*M* Tris-HCl, 0.1 m*M* EDTA, pH 7.1
2 × HBS: 50 m*M* HEPES, 280 m*M* NaCl, 1.5 m*M* NaH_2PO_4, pH 7.1
Phosphate-buffered saline (PBS): 4.3 m*M* Na_2HPO_4, 1.4 m*M* NaH_2PO_4, 137 m*M* NaCl, 2.7 m*M* KCl, pH 7.3
Glycerol, 15% in PBS
Cell culture T-flasks (25, 75, and 150 cm^2, Corning Costar Corporation, Cambridge, MA)
Cell culture dishes (60 and 100 mm, Corning)
Cell culture plates (24 well, Corning)
pRSVneo, neomycin gene in a vector under the control of the Rous sarcoma virus long terminal repeat promoter (obtained from Genentech)

All reagents (except Vitamin K) were passed through a 0.2-μm filter prior to use.

Cells

Human kidney 293 cells (ATCC, Rockville, MD, CRL1573) transformed with sheared adenovirus type 5, are grown in DMEM/F12 medium (pH 7.3, adjusted with $NaHCO_3$ and CO_2), containing phenol red as an indicator, and supplemented with 10% heat-inactivated FBS. The medium is supplemented every 10 days with 1% of the 200 mM glutamine solution. The 293 cells are cultured as adherent monolayers in flasks in an incubator at 3.3% (v/v) CO_2 and a relative humidity of 93%. Healthy cells are spindle-shaped and double every 18–20 hr.

Transfection Protocol

1. A confluent flask of 293 cells is split. Adherent cells are removed by repeat pipetting. Approximately 10^6 cells are seeded into 60-mm culture dishes with 5 ml of NGM the day prior to transfection.
2. The cells are fed with 5 ml of NGM 3 hr prior to the transfection, and should be 70–80% confluent at the time of transfection.
3. An amount of 1 μg of the cotransfectant plasmid, pRSVneo, is added to 10 μg of the cDNA of interest in pCIS2M, followed by addition of 0.05 ml of 2.5 mM $CaCl_2$. The total volume is adjusted to 0.5 ml with 1/10 TE buffer (tube A). Another tube (tube B) is prepared containing 0.5 ml of 2 × HBS. The contents of tube A are added dropwise to tube B, and mixed thoroughly. After observation of a cloudy precipitate, the suspension is added to the 60-mm culture dishes obtained in step 2, and swirled over the cells.
4. The transfected cells are incubated for 3–4 hr at 37°. The cells are then shocked with glycerol. For this, culture medium is aspirated from the dishes and 0.5 ml of 15% glycerol in PBS is added for 30 sec at room temperature.
5. The glycerol is removed from the cells by aspiration and the cells are washed with 5 ml of PBS.
6. A volume of 5 ml of NGM is added and allowed to incubate for 48 hr at 37°.
7. The cells from each 60-mm dish are transferred to three 100-mm dishes with 10 ml of SM. The medium is changed routinely until G418-resistant colonies are sufficiently large to be picked (approximately 18 days). Typically, one 100-mm dish will provide at least 20 suitable colonies.
8. Single colonies are chosen and transferred to individual wells of a 24-well plate. One milliliter of SM is added to each well. The medium is changed routinely until the cells become confluent.

9. After this point, the medium in each well is replaced with LSM/VK and allowed to incubate for an additional 48 hr.

10. The medium is collected and assayed for PC by Western blot analysis.

Procedures for Western Blot Analysis

The presence of r-PC in the wells is screened by Western blot analysis using a monoclonal antibody (MAb) for r-PC. In the absence of such an antibody, a polyclonal antibody (PAb) made against plasma PC also works well. This technique is suitable for rapid screening of supernates from a large number of colonies, and provides reliable results.

Reagents

Transfer buffer: 25 m*M* Tris-HCl, 192 m*M* glycine, 15% methanol, pH 8.3

TBS: 20 m*M* Tris-HCl, 500 m*M* NaCl, pH 7.5

Blocking buffer: 1% gelatin (Bio-Rad, Richmond, CA) in TBS; or 9% (v/v) fat-free milk in TBS

Wash buffer: 0.05% Tween-20 (Sigma) in TBS

Stain buffer: 100 m*M* Tris-HCl, 100 m*M* NaCl, 5 m*M* $MgCl_2$, pH 9.5

Stain solution: 16.5 mg nitro blue tetrazolium (Sigma) is first dissolved in 0.5 ml of 70% DMF; 8.5 mg of 5-bromo-4-chloro-3-indolylphosphate (Sigma) is dissolved in 1 ml of H_2O; both of the preceding solutions are added to stain buffer to a total volume of 50 ml buffer

5 μg/ml of mouse MAb-C3,[15] obtained from J. Griffin, Scripps Research Institute, LaJolla, CA, is dissolved in blocking buffer; goat anti-mouse IgG (Bio-Rad); Immobilon-P membranes (Sigma)

Protocol

1. The chosen conditioned cell media samples are subjected to sodium dodecyl sulfate–polyacrylamide gel electrophoresis (SDS–PAGE) using 12% gels.[16]
2. The protein bands on the resulting gel are transferred to an Immobilon-P membrane by electrophoresis.[17] The transfer is performed at 4° in transfer buffer, at 20 V overnight, or 60 V for 4 hr.

[15] M. J. Heeb, P. Schwartz, T. White, B. Lammle, M. Berrettini, and J. Griffin, *Thromb. Res.* **52,** 33 (1988).

[16] U. K. Laemmli, *Nature* (*London*) **227,** 680 (1970).

[17] W. H. Burnette, *Anal. Biochem.* **112,** 195 (1981).

3. The blotted membrane is removed from the apparatus, and placed in the blocking buffer for 1 hr at room temperature, with gentle swirling on a rotary shaker.
4. The membrane is then rinsed with three changes (15 ml each) of TBS.
5. The MAb-C3 solution is then incubated with the membrane for 2 hr at room temperature using a rotary shaker for gentle agitation.
6. The blotted membrane is then rinsed with wash buffer five times over a 30-min time period and then transferred to a solution containing the blocking buffer/anti-mouse IgG coupled to alkaline phosphatase and incubated for 2 hr at room temperature on the rotary shaker. This step is repeated one additional time.
7. The positive bands are visualized by incubation of the membrane with the stain solution.

The first and second antibody solution can be reused up to five times.

Procedures for Expression of Recombinant Human Protein C

Normally, the highest expressing colony, as indicated by the results of Western blot analysis, is chosen for large-scale expression.

Reagents

DMEM/F12 (Sigma)
Colorless serum-free DMEM/F12-VK: DMEM/F12, 5 μg/ml of VK, without phenol red or FBS
0.5 mg/ml poly(D-lysine) (Sigma)
Cell culture T-flasks (25 and 150 cm^2, Corning)
Cell culture roller bottles (1700 cm^2, Corning)
0.5 *M* benzamidine hydrochloride (Sigma)

Expression Protocol

1. The selected colony is transferred to a 25-cm^2 flask and the cells fed with SM until they reach confluence.
2. The cells are transferred into 150-cm^2 flasks and allowed to reach confluence.
3. The roller bottle is coated with 10 mL of poly(D-lysine) solution for 30 min and then rinsed with H_2O.
4. The cells are transferred from the flask to the roller bottle. Usually, the cells from two 150-cm^2 flasks are placed into the roller bottle and grown for 24 hr in 200 ml of LSM/VK.
5. The FBS level is raised to 8% and the cells allowed to grow for approximately 2 additional days, until 80% confluence is reached.

6. The roller bottle is rinsed with 50 ml of PBS to remove the FBS, and 200 ml of colorless serum-free DMEM/F12-VK is added. The medium is collected every 48 hr and replaced with a fresh solution. Usually three to four collections are made before the cells begin to detach.
7. The cell-conditioned medium is subjected to centrifugation for 10 min at 7000 rpm to eliminate cell debris, after which 2 ml of the benzamidine solution is added.
8. The medium can be frozen for later use, or immediately employed for r-PC purification.

Purification of r-PC

The technique detailed below is an excellent protocol for purification of maximally γ-carboxylated wild-type r-PC,[7] r-PC mutants,[7–9,18] and r-PC chimeric proteins.[19,20] In virtually all cases, the proteins that we purified by this procedure, and that contain no more than two mutated Gla residues, possess the full complement of Gla.

Reagents and Supplies

Fast-flow Q-Sepharose (FFQ, Pharmacia, Piscataway, NJ)
0.5 *M* EDTA-Na_2 (pH 8.0)
Buffer 1: 20 m*M* Tris-HCl, 150 m*M* NaCl, 2 m*M* EDTA-Na_2, pH 7.4
Buffer 2: 20 m*M* Tris-HCl, 150 m*M* NaCl, pH 7.4
Buffer 3: 20 m*M* Tris-HCl, 150 m*M* NaCl, 30 m*M* $CaCl_2$, pH 7.4
Buffer 4: 20 m*M* Tris-HCl, 500 m*M* NaCl, pH 7.4
Dialysis buffer: 20 m*M* Tris-HCl, 150 m*M* NaCl, 5 m*M* benzamidine hydrochloride, pH 7.4
Spectra/por 2 membrane, 12,000–14,000 molecular weight cutoff (Fisher Scientific, Pittsburgh, PA)
Centricon 10 membranes (Amicon Inc., Beverly, MA)

Purification Protocol

1. A FFQ column (1.2 × 4.5 cm) is equilibrated with buffer 1.
2. The cell-conditioned media is supplemented with 0.4 m*M* EDTA-Na_2, and the pH is adjusted to 7.4 and 2 *N* NaOH.

[18] L. Zhang and F. J. Castellino, *J. Biol. Chem.* **269,** 3590 (1994).
[19] S. Yu, L. Zhang, A. Jhingan, W. T. Christiansen, and F. J. Castellino, *Biochemistry* **33,** 823 (1994).
[20] W. T. Christiansen and F. J. Castellino, *Biochemistry* **33,** 5901 (1994).

3. Approximately 1 liter of medium is loaded onto the column at a flow rate of 24 ml/hr at 4°.
4. The column is washed with three column volumes of buffer 1, and then three column volumes of buffer 2.
5. The column is then developed with 120 ml of linear gradient of $CaCl_2$. The start solution is 60 ml of buffer 2 and the limit solution is 60 ml of buffer 3. A typical elution profile is presented in Fig. 3A.
6. The r-PC-containing fractions, which normally constitute the major peak, are identified through absorbance measurements at 280 nm. The fractions are pooled, and equilibrated against the dialysis buffer.
7. A second FFQ column (0.9 × 3 cm) is equilibrated with buffer 2.
8. The dialyzed r-PC-containing pool is loaded onto this column at a flow rate of 15 ml/hr at 4°.
9. The r-PC is eluted with a linear gradient of NaCl. The start solution is 25 ml of buffer 2 and the limit solution is 25 ml of buffer 4. A typical elution profile is provided in Fig. 3B.
10. The r-PC fractions, normally present as the major peak, are identified as above, pooled, and dialyzed against buffer 2.

Chemical Characterization of r-PC

The following procedures are routinely employed to chemically characterize the r-PC mutants that are purified.

Reagents

5.0 *M* KOH/0.1% (v/v) phenol
Saturated aqueous $KHCO_3$
60% (v/v) $HClO_4$
100 mg *o*-phthaldehyde in 5 ml methanol/10 μl of 2-mercaptoethanol/10 ml of 0.15 *M* sodium borate, pH 10.5/0.2% Brij 35 (OPA/ET), flushed with N_2 and stored at −20° in the dark
0.1 *M* sodium acetate, pH 7.2/9.75% methanol/0.25% tetrahydrofuran (NaOAc/THF)
Lithium eluent A (LEA): 0.24 *N* Li^+, pH 2.75 (Pickering, Mountain View, CA)
Lithium eluent B (LEB): 0.64 *N* Li^+, pH 7.50 (Pickering)
Lithium regenerant (LR): 0.3 *N* Li^+ (Pickering)

Procedure for Gla Analyses

Approximately 50 μg of the protein sample in 100 μl of H_2O is mixed with an equal volume of the KOH/phenol solution and hydrolyzed for

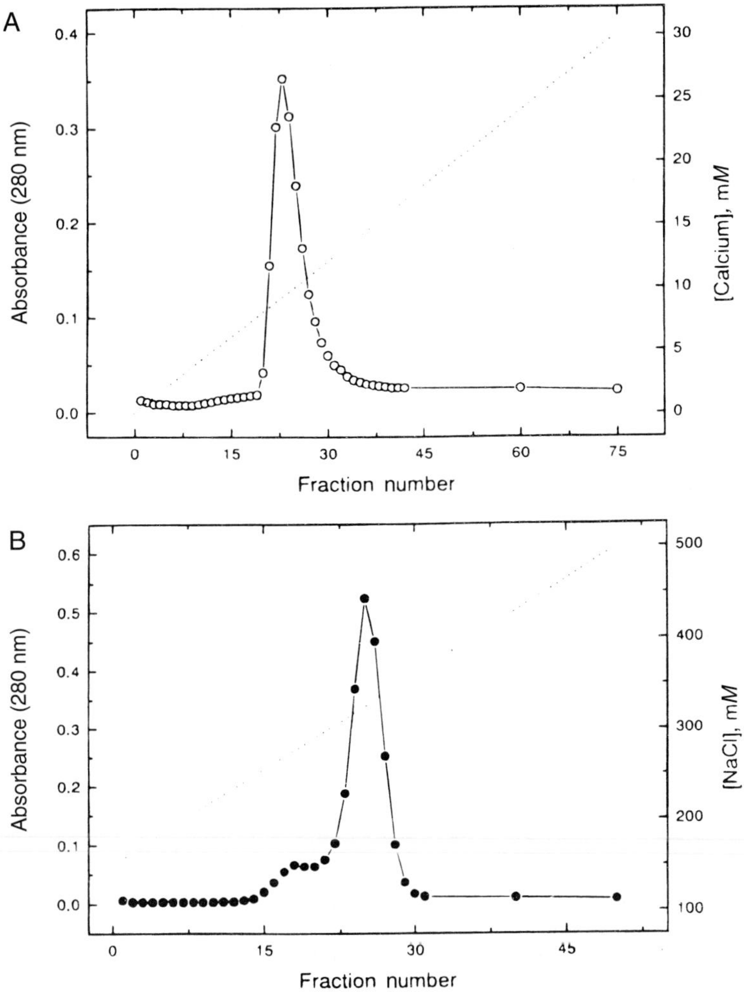

FIG. 3. Chromatography of [Gla^6 → Gln]r-PC on FFQ anion-exchange chromatography at 4°. (A) A volume of 800 ml of conditioned 293 cell culture medium was applied to a 5-ml column of FFQ, equilibrated with 20 m*M* Tris-HCl/150 m*M* NaCl/4 m*M* EDTA, pH 7.4 at 4°. After washing the column with this same buffer, followed by 20 m*M* Tris-HCl/150 m*M* NaCl, pH 7.4, a linear $CaCl_2$ gradient was applied (0 to 30 m*M* $CaCl_2$; 120 ml, total volume). Fractions (1.6 ml) were collected at a flow rate of 0.4 ml/min. Those containing the major peak were pooled. (B) The major peak of part (A) was dialyzed against 20 m*M* Tris-HCl/150 m*M* NaCl, pH 7.4, and reapplied to a 3-ml column of FFQ equilibrated in this same buffer. The indicated linear gradient of NaCl was applied (150–500 m*M* NaCl; 50 ml, total volume) and 1-ml fractions were collected. The flow rate was 0.25 ml/min. The fractions containing the major peak were pooled. (Adapted with permission from Christiansen *et al.*[6] Copyright 1994 American Chemical Society.)

20 hr at 110°. After this time, one part of the $KHCO_3$ solution is mixed with four parts of the hydrolyzate (usually 50 + 200 μl) and the sample centrifuged to remove any precipitate. The resulting supernate is transferred to a fresh tube, which is adjusted to pH 7.0 with the $HClO_4$ solution. The sample is cooled on ice for 30 min, and the supernate is collected after centrifugation.[21]

The protein sample is derivatized by mixing equal volumes of OPA/ET and the hydrolyzate. Amino acids in the sample are then resolved on a reverse-phase column of a Beckman (Palo Alto, CA) Ultrasphere XLS ODS (4.6 mm × 7.0 cm, 3 μm) using NaOAc/THF as the eluant. A typical separation profile displayed Gla (0.67 min), D (1.27 min), E (2.22 min), and Ala (13.77 min).

The ratios of D/Gla, and E/Gla, combined with the standard curves, are used to obtain the number of Gla residues per mole of protein. Reference standards consist of both a commercial *o*-phthaldehyde-derivatized amino acid standard mixture, as well as a peptide, ANSFL$\gamma\gamma$LRHSS, representing the first 12 residues of the light chain of PC. This latter peptide is synthesized by standard Fmoc chemistry (the α-*N*-Fmoc-γ,γ'-di-tert-Bu-L-Gla-OH used in the peptide synthesis for placement of Gla in the peptide is chemically synthesized in this laboratory). Employment of this peptide for this purpose allowed us to obtain recovery factors after alkaline hydrolysis of the peptide and to determine very accurate conversion factors of peak areas to concentrations of Gla residues in the hydrolyzed samples. The Gla/D ratio in this peptide of 2.0 (N is converted to D during the hydrolysis), and the concentration response factor of D from commercial standards are employed to obtain the concentration response factor for a Gla residue. Checks of the method are made by performing Gla residue analyses on human plasma PC and bovine plasma factor IX.

Procedure for β-OH-Asp (Hya) Determinations

Standard preparations of *erythro*(*e*)-Hya and *threo*(*t*)-Hya were gifts of W. T. Jenkins (Bloomington, IN) and M. Miller (Notre Dame, IN). For determination of the amounts of *e*-Hya and *t*-Hya in all samples, measured amounts (approximately 50 μg) of protein are hydrolyzed with 300 μl of 6 *N* HCl at 110° for 20 hr. The hydrolyzate is dried using a speed vacuum system.

Resolution of amino acids is 15 μl of sample is accomplished by HPLC at 42° using a Pickering cation-exchange column (3 × 150 mm, 5 μm) coupled with postcolumn modification of the amino acids with OPA/ET. A Pickering Li^+ guard column (2 × 20 mm) is used to protect the cation-

[21] M. Kuwada and K. Katayama, *Anal. Biochem.* **117,** 259 (1981).

exchange column. The derivatization reaction occurred in a 5-m reaction coil prior to fluorescence detection. Amino acid resolution is accomplished with the following Li^+ gradient,[22] composed of LEA, LEB, and LR, respectively, applied at the following times (t): t = 0–9 min, 100% LEA; t = 9–21 min, 90.5% LEA–9.5% LEB; t = 21–33 min, 80% LEA–20% LEB; t = 33–49 min, 65% LEA–35% LEB; t = 49–88 min, 0.4% LEA–99.6% LEB; t = 88–94.6 min, 0.4% LEA–99.6% LEB; t = 94.6–120 min, 0.4% LEA–93.6% LEB-6% LR; t = 120–120.5 min, 0.4% LEA–93.6% LEB–6% LR; t = 120.5–126.5 min, 0.4% LEA–99.6% LR; t = 126.5–134.5 min, 4% LEA–99.6% LR; t = 134.5–135 min, 100% LEA.

Under the described conditions, Hya and D elute at 7.04 and 11.38 min, respectively. The recovery factor of Hya is 36% of D. The ratio of Hya/D allowed for calculation of the Hya content, with knowledge of the number of D residues in the protein.

Procedure for Amino-Terminal Amino Acid Sequence Analysis

Automated Edman sequence analysis of the proteins is carried out on a Beckman LF3000 automated sequencer, with samples immobilized on sequence filter disks, according to the following protocols:

1. Lyophilized salt-free samples are reconstituted in a minimum amount of 10% acetic acid and applied to the glass fiber filter disk (Beckman) in 10-μl aliquots with drying by N_2 flow between additions to the disk.
2. Samples in solution are treated as above.
3. Electroblotted samples on Immobilon membranes following SDS–PAGE are excised from the wetted (H_2O) membrane and are further cut into small (approximately 1-mm^2) pieces. These are wetted with H_2O, placed in the sample well of the instrument, covered with a Zitex (Beckman) disk, and dried by N_2 prior to sequencing.

Automated sequencing is conducted according to the manufacturer's protocols, and with the recommended reagents, except for substituting 2% phenyl isothiocyanate (PITC) in heptane in place of 0.5% PITC as the coupling reagent.

Cleaved anilinothiazolidine (ATZ) amino acids are automatically converted to phenylthiohydantoin (PTH) derivatives in aqueous 25% trifluoroacetate, dried, redissolved in aqueous 10% acetonitrile, and analyzed by on-line microbore HPLC (System Gold, Beckman Instruments).

Chromatography of generated PTH derivatives separates all the common amino acids (excluding cysteine and cystine), which are detected by

[22] J. A. Grunau and J. M. Swiader, *J. Chromatogr.* **594,** 165 (1992).

monitoring the column eluate for absorbancy at 269 nm. The separated PTH derivatives are identified and quantitated using System Gold software. Solvents for gradient elution of PTH derivatives and the elution conditions are as follows:

Solvent A: 71.67 m*M* sodium acetate, pH 3.85, dissolved in 3.2% THF, 0.15% triethylamine

Solvent B: acetonitrile

The column is equilibrated with a solution of 95% A/5% B. Upon injection of the sample (time 0), the gradient is immediately ramped to 76% A/24% B. At 0.2 min, the solvent is changed to 58% A/42% B. During the next 14 min, the relative percentage of B is altered to 52%, followed by an increase in B to 90% over a 5-min interval, and back to 5% over 3 min, where it is held constant for 1 min. At this point, the cycle is concluded, and the next begins.

Chemical Characterization of Gla Mutants

This laboratory has generated and characterized a large number of Gla domain mutants of r-PC. Yields of fully γ-carboxylated materials have varied from 0.2 mg/liter to 3.0 mg/liter depending on the exact mutant. Usually, the bulk of the r-PC antigen is fully γ-carboxylated, however, some notable exceptions have been found.[23] An example of a series of mutants that we have studied involves mutations at each of the individual Gla residues of r-PC. This entire set of mutants has been purified and chemically characterized by the methods described in this chapter. Nonreduced SDS–PAGE profiles for each mutant of this set, plus two others that were of interest, namely [R15L]r-PC and [F31L, Q32Gla] are presented in Fig. 4. The two bands observed for each of the proteins is due to the presence of two glycoforms of wtr-PC and the r-PC mutants, and the gels demonstrate that highly purified preparations are obtained. Amino acid sequence analysis generally proceeds through 35 residues, and each of the proteins displayed the appropriate amino acids at these sequence positions. Blank cycles are observed for Gla residues, but importantly, $<5\%$ of E was seen at residues containing Gla. The Gla and Hya contents of each of these proteins have been determined separately by the methods described herein, and the values obtained are listed in Table I. In all cases appropriate proteins are obtained, which are useful for other studies that address structure–function relationships of individual Gla residues of r-PC.

[23] L. Zhang and F. J. Castellino, *Biochemistry* **30,** 6696 (1991).

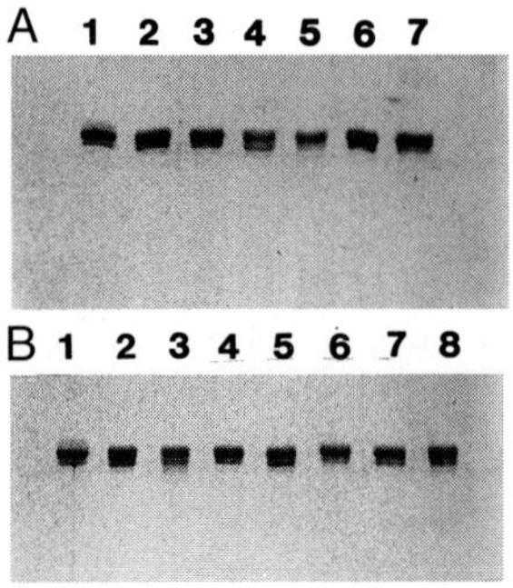

FIG. 4. Nonreduced sodium dodecyl sulfate–polyacrylamide gel (12%) electrophoretograms of all purified r-PC variants employed in this study. The samples displayed are as follows. (A) 1, human plasma PC; 2, 293 cell-expressed wtr-PC; 3, [Gla6D]r-PC; 4, [Gla67D]r-PC; 5, [Gla614D]r-PC; 6, [R15L]r-PC; 7, [Gla616D]r-PC. (B) 1, human plasma PC; 2, 293 cell-expressed wtr-PC; 3, [Gla619D]r-PC; 4, [Gla620D]r-PC; 5, [Gla625D]r-PC; 6, [Gla626V]r-PC; 7, [Gla629D]r-PC; 8, [F31L, Q32Gla6]r-PC. (Adapted with permission from Zhang *et al.*[7])

TABLE I
γ-CARBOXYGLUTAMIC ACID CONTENTS OF VARIOUS R-PC MUTANTS

Protein	Gla[a] (mol/mol)		Hya[b] (mol/mol)	
	Expected	Obtained	Expected	Obtained
Human factor VII			0	0
Bovine factor X	12.0	11.7 ± 0.4		
Human PC	9.0	8.8 ± 0.3	1.0	0.81 ± 0.16
wtr-PC[c]	9.0	8.9 ± 0.3	1.0	0.90 ± 0.13
[Gla6D]r-PC[d]	8.0	8.2 ± 0.2	1.0	0.73 ± 0.17
[Gla7D]r-PC	8.0	8.2 ± 0.3	1.0	0.81 ± 0.09
[Gla14D]r-PC	8.0	7.8 ± 0.3	1.0	1.20 ± 0.10
[R15L]r-PC	9.0	9.0 ± 0.3	1.0	0.95 ± 0.19
[Gla16D]r-PC	8.0	7.9 ± 0.2	1.0	1.10 ± 0.23
[Gla19D]r-PC	8.0	8.0 ± 0.3	1.0	0.90 ± 0.13
[Gla20D]r-PC	8.0	8.2 ± 0.2	1.0	0.78 ± 0.11
[Gla25D]r-PC	8.0	7.7 ± 0.3	1.0	0.85 ± 0.11
[Gla26V]r-PC	8.0	7.9 ± 0.2	1.0	0.87 ± 0.16
[Gla29D]r-PC	8.0	7.9 ± 0.2	1.0	0.91 ± 0.25
[F31L, Q32Gla]r-PC	10.0	10.0 ± 0.3	1.0	0.93 ± 0.27

[a] γ-Carboxyglutamic acid.
[b] β-Hydroxyaspartic acid.
[c] Wild-type recombinant protein C, expressed in human kidney 293 cells.
[d] The mutation convection used is the [normal amino acid, sequence position in PC, single-letter code for the new amino acid placed in that sequence position by mutagenesis], followed by recombinant (r)-activated protein C (APC). (Adapted with permission from Zhang *et al.*[7])

Conclusion

One issue that must be addressed in mutagenesis investigations regarding Gla residues is the extent to which the lack of any single Gla residue might affect γ-carboxylation of other E-precursor Gla residues. Unless proven otherwise in any specific case, it is inappropriate to mutate such E residues and work with protein purified by bulk nonspecific techniques, with the assumption that all other E residues are correctly processed to Gla. Attempts to purify such mutated proteins with Ca^{2+}-dependent antibodies to the Gla domain also present difficulties, since without rigorous chemical characterization of the resulting protein, it is not certain that maximally γ-carboxylated proteins have been obtained. Some of these Ca^{2+}-dependent antibodies do react with undercarboxylated species of these proteins. The chromatography steps described herein have been uniformly successful in allowing purification of appropriate materials for further study, at least in cases of r-PC and r-PC mutants, as well as some fIX and fVII mutants that we have constructed.

Acknowledgments

This work was supported in part by grant HL-19982 from the National Institutes of Health and by the Kleiderer-Pezold family professorship (to F. J. C.).

[32] Determination of Site-Specific γ-Carboxyglutamic Acid Formation by Vitamin K-Dependent Carboxylase Utilizing De-γ-carboxy Bone Gla Protein as Substrate

By MARGARET E. BENTON and J. W. SUTTIE

Introduction

The vitamin K-dependent carboxylase catalyzes the posttranslational conversion of glutamic acid residues to γ-carboxyglutamic acid (Gla) residues in a limited number of proteins. These proteins include the clotting factors II (prothrombin), VII, IX, and X and proteins C, S, and Z, as well as bone Gla protein (BGP) and matrix Gla protein (MGP). Multiple Glu residues, clustered within a given substrate, are targeted by the enzyme for carboxylation. Currently, very little information is available detailing the specificity and directionality of this multisite event.

In this chapter, we address site-specific Gla formation by the vitamin K-dependent carboxylase utilizing de-γ-carboxy bovine BGP (dBGP) as a

0076-6879/97 $25.00

model substrate.[1] BGP is a physiological substrate for the enzyme yet contains very few potential Gla residues so as to simplify interpretation of the results. Bovine BGP, a 49-residue protein, contains three Gla residues located at positions 17, 21, and 24.[2] In our procedure for determining site specificity of the enzyme, heat-decarboxylated BGP (dBGP) is used as an *in vitro* substrate for a semipurified bovine hepatic γ-glutamyl carboxylase.[3] Under conditions of saturating substrate (5 or 20 μM), reduced vitamin KH_2, and $NaH^{14}CO_3$, $^{14}CO_2$ is incorporated into potential Gla residues and the ^{14}C label is used to identify Gla formation in a two-part experimental design (Fig. 1). The first method involves using isoelectric focusing gel electrophoresis (IEF) to separate the carboxylated BGP species on the basis of charge, thereby determining the types of carboxylated species produced. The second method involves analyzing Gla formation by determining the specific activity of $^{14}CO_2$ incorporated at each potential Gla site. Based on this procedure the order of carboxylation of these sites can be determined.

Procedures

Enzyme and Substrate

An immunoaffinity purified preparation of the bovine hepatic carboxylase (500-fold purified over crude microsomes) is obtained from a dicoumarol-treated cow and prepared as previously described.[3] The partially pure preparation contains 100 μM human factor X (hFX) propeptide (residues -18 to $+1$).[4] Purified bovine BGP and dBGP have been provided by Dr. Paul Price. Briefly, the BGP is extracted from bone powders on demineralization and purified to homogeneity by gel filtration and diethylaminoethyl (DEAE) chromatography.[5]

Preparation of Bone Gla Protein Antibody Resin

Antisera to BGP is raised in New Zealand White male rabbits and purified[6] on a resin composed of bovine BGP coupled to CNBr-preactivated Sepharose 4B (2 mg/ml, 4 ml resin) (Pharmacia, Piscataway, NJ). Antisera

[1] M. E. Benton, P. A. Price, and J. W. Suttie, *Biochemistry* **34,** 9541 (1995).
[2] P. A. Price, J. W. Poser, and N. Raman, *Proc. Natl. Acad. Sci. U.S.A.* **73,** 3374 (1976).
[3] M. C. Harbeck, A. Y. Cheung, and J. W. Suttie, *Thromb. Res.* **56,** 317 (1989).
[4] J. E. Knobloch and J. W. Suttie, *J. Biol. Chem.* **262,** 15334 (1987).
[5] P. A. Price, A. S. Otwuka, J. P. Poser, J. Kristaponis, and N. Raman, *Proc. Natl. Acad. Sci. U.S.A.* **73,** 1447 (1976).
[6] J. W. Poser, F. S. Esch, N. C. Ling, and P. A. Price, *J. Biol. Chem.* **255,** 8685 (1980).

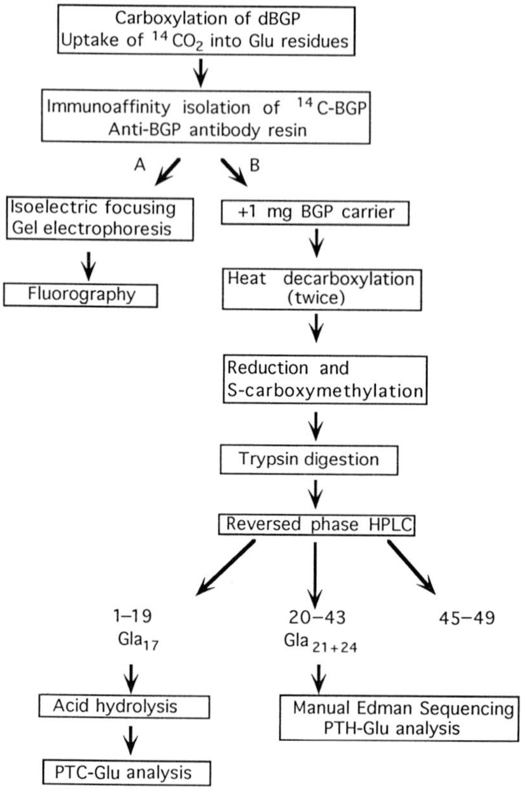

FIG. 1. Experimental design for determining the site(s) of carboxylation in dBGP by the vitamin K-dependent carboxylase. Carboxylase assays are carried out under conditions of saturating dBGP (20 μM unless otherwise noted), reduced vitamin K, and [^{14}C]bicarbonate at 25°. Following a 60-min incubation, approximately 30–50% of the total protein is carboxylated at least once.

is centrifuged at 10,000*g* for 10 min, and the BGP antibodies are isolated by passing 120 ml of antisera through the column of immobilized BGP at a flow rate of 30 ml/hr. The column is washed with five volumes of 0.2 *M* Tris-HCl, 0.5 *M* NaCl, pH 7.5, 0.1% Tween 20 and then with five volumes of the same buffer without detergent. Antibodies are desorbed by the addition of 0.2 *M* glycine hydrochloride, pH 2.7, and the eluate is neutralized by collecting 1.4-ml fractions into 0.6 ml of 1 *M* K_2HPO_4. Protein is monitored at an absorbance of 280 nm and the maximal fractions are pooled and dialyzed into 0.1 *M* $NaHCO_3$, 0.5 *M* NaCl, pH 8.3. The antibodies are coupled to CNBr-preactivated Sepharose 4B (2 mg/ml, 4 ml), and the resin is stored at 4°. The anti-BGP antibody resin is stable for at least 6 months and can be used repetitively without loss in binding capacity.

Carboxylase Assay and Isolation of Carboxylated Products

Each reaction mixture contains 700 μl of enzyme preparation added to an equal volume of other reactants such that the final 1.4-ml volume contains 20 μM dBGP in 0.25 M sucrose, 0.025 M imidazole, 0.5 M KCl (SIK buffer), 50 μCi/ml $NaH^{14}CO_3$, 150 μg/ml vitamin KH_2,[7] 0.6 M $(NH_4)_2SO_4$, and 10 mM dithiothreitol (DTT). The reaction is carried out at 25° in a shaking waterbath for 60 min, and then the mixture is bubbled under a stream of CO_2 for 5 min to remove unincorporated $^{14}CO_2$.

The products and remaining reactants are isolated on an anti-BGP antibody resin (2 mg/ml, 4 ml) by gravity filtration. The resin is washed with two column volumes of 0.2 M Tris-HCl, 0.5 M NaCl, pH 7.5, eluted with 4 M guanidine hydrochloride, and the denaturant is removed on an Econo-Pac 10DG desalting column (Bio-Rad, Richmond, CA) equilibrated in 50 mM NH_4OH. The isolated products are lyophilized to dryness and stored at −20° if not used immediately.

Isoelectric Focusing Gel Electrophoresis and Fluorography

Lyophilized protein is suspended in sample buffer (50 mM Na_2HPO_4, 7.5 M urea, 1% 2-mercaptoethanol, pH 8) and incubated 15 min at 40°. Samples are applied to dry wells of 7.5% polyacrylamide gels containing 7 M urea, 2% pH 2.5–5 Pharmalytes (Sigma, St. Louis, MO), 0.3 mg/ml ammonium persulfate, 0.1% N,N,N',N'-tetramethylethylenediamine (TEMED) and overlayed with 50 mM imidazole, 10% (v/v) glycerol, pH 7. The top and bottom reservoirs are filled with 0.1 M NaOH and 0.1 M H_2SO_4, respectively. Gels are run 18–20 hr at 250 V and then 60 min at 400 V. Protein is fixed 2 hr in 10% (w/v) trichloroacetic acid (TCA), 10% (w/v) sulfosalicylic acid and stained with Coomassie Brilliant blue. After rinsing thoroughly with water, gels are incubated 30 min in Fluoro-Hance (Research Products International, Mt. Prospect, IL), dried on a gel dryer, and subjected to fluorography. Quantitation of individual bands is carried out by scanning gels with a blot analyzer (Betagen Corporation, Waltham, MA). The more commonly available PhosphorImager (Molecular Dynamics, Sunnyvale, CA) can also be used.

Isoelectric Focusing Separation and Elution of Mono- and Dicarboxylated Bone Gla Protein

Isolated BGP reaction products can be separated into mono- and di-Gla fractions or the mixed products can be analyzed directly. In the latter

[7] G. M. Wood and J. W. Suttie, *J. Biol. Chem.* **263,** 3234 (1988).

case, proceed to the next step. To separate the reaction products, samples are loaded in multiple lanes on an IEF gel as described earlier. Partially decarboxylated BGP species are electrophoresed in an adjacent lane and used as markers. The lane containing the markers is detached and fixed 5 min in 10% TCA, 10% sulfosalicylic acid. (*Note:* Gel does not expand significantly during this short incubation.) The markers, which became opaque white bands during fixing, are aligned with the unfixed portion of the gel, and the radiolabeled mono- and dicarboxylated BGP bands are excised from the gel using the fixed markers as a guide. Proteins are eluted from the gel into 600 μl 0.3 *M* pH 8 NH_4HCO_3 per band at 37° for 5 hr. The products from a 700-μl reaction are generally loaded in several lanes, eluted bands of each species are combined, and the volume is adjusted to 3 ml with water. Approximately 0.3 mg BGP is added to each as a carrier, and each sample is applied to an Econo-Pac 10DG desalting column. The eluate, free of urea, is dried under a stream of filtered air.

Heat Decarboxylation of Gla

BGP reaction products are heat decarboxylated by a modified version of the procedure described by Poser and Price.[8] During this procedure there is a 50% chance that the radiolabel will be lost at any given Gla site. One milligram of BGP is added as a carrier to each sample of reaction product, and each mixture is dissolved in 0.5 m*M* EDTA, 0.1 *M* NH_4HCO_3, pH 8, and applied to a 2-ml Sephadex G-25 (20–50 μm) column equilibrated in 0.1 *M* pH 8 NH_4HCO_3 to remove divalent metal ions. The eluate is monitored for radioactivity, and peak fractions are pooled, lyophilized to dryness, and sealed, under vacuum, in 18- × 150-mm glass tubes. Samples are heat decarboxylated for 8 hr at 110°, redissolved in 0.1 *M* pH 8 NH_4HCO_3, lyophilized to dryness, and heat decarboxylated again for 8 hr. Based on IEF of the final product, >90% of the Gla is converted to Glu.

Reduction and S-Carboxymethylation

A single disulfide bond in BGP, Cys^{23}-Cys^{29}, must be reduced to allow for sequencing of the Gla residues. Approximately 1 mg of dBGP/[^{14}C]BGP (i.e., all of the final product from the previous step) is reduced and S-carboxymethylated according to a modification of the procedure of Huq *et al.*[9] The protein is suspended in 100 μl of 25 m*M* Tris-HCl, 5 m*M* EDTA, pH 7.4, and diluted with 900 μl of degassed Tris-urea buffer (500 m*M* Tris-HCl, 8 *M* urea, 1 m*M* EDTA, pH 8.5). The reaction vessel is sealed and

[8] J. W. Poser and P. A. Price, *J. Biol. Chem.* **254,** 431 (1979).
[9] N. L. Huq, L.-C. Teh, D. L. Christie, and G. E. Chapman, *Biochem. Int.* **8,** 521 (1984).

flushed with N_2. Five milligrams of DTT dissolved in 200 μl Tris-urea buffer is added, and reduction of the disulfide bonds proceeds, with stirring, for 2 hr at room temperature. Iodoacetic acid (15 mg), freshly recrystallized from hexane, is dissolved in 200 μl Tris-urea buffer and adjusted to pH 7. The iodoacetic acid and 1 μl 2-mercaptoethanol are added to the reaction vessel, which is then flushed with N_2 and covered with aluminum foil. Carboxymethylation proceeds for 10 min at room temperature. The reaction mixture is immediately transferred to dialysis tubing and dialyzed into 1% NH_4HCO_3, pH 8, at 4° overnight. The final volume is approximately 1.5 ml.

Trypsin Digestion

Reduced and S-carboxymethylated BGP/[^{14}C]BGP, in a 1.5-ml volume of 1% NH_4HCO_3, pH 8, is digested with 1-tosylamide-2-phenylethyl chloromethyl ketone (TPCK)-treated trypsin (Sigma) in a 1:100 weight ratio. Digestion is carried out for 2 hr at room temperature with magnetic stirring. A second aliquot of trypsin, equal to the first, is then added, and digestion proceeds another 14–15 hr at which time the digest is lyophilized to dryness.

Separation of Tryptic Fragments

Reversed-phase high-performance liquid chromatography (HPLC) separation of trypsin-digested BGP/[^{14}C]BGP is accomplished using a Hamilton PRP-1 analytical column (15 × 4.1 mm, 10 μm) equilibrated in buffer A [2% acetonitrile, H_2O, 0.1% trifluoroacetic acid (TFA)]. Reversed-phase HPLC is carried out using a Waters (Milford, MA) system. The system consists of a Model 721 programmable system controller, two Model 6000 chromatography pumps, a Model 450 variable absorbance detector, and a Model U6K injector. A linear gradient is developed on injection from 0 to 100% buffer B [60% acetonitrile, H_2O, 0.1% (v/v) TFA] in 60 min with a flow rate of 1 ml/min. The eluate is monitored at 220 nm, and peptide fragments containing residues 1-19 and 20-43 are collected (Fig. 2) and dried under a stream of filtered air.

Acid Hydrolysis

Tryptic fragments are suspended in 200 μl of 6 *N* constant boiling HCl in Eppendorf tubes and placed in 18- × 150-mm glass tubes, which are evacuated and sealed. After 22 hr at 110° the tubes are opened, and the HCl is evaporated under a stream of air. The samples are resuspended in H_2O, aliquoted to 6- × 50-mm glass tubes, lyophilized to dryness, and stored at −20°. Acid hydrolysis converts Gla residues to Glu residues; therefore, it is not necessary to heat decarboxylate samples prior to this procedure.

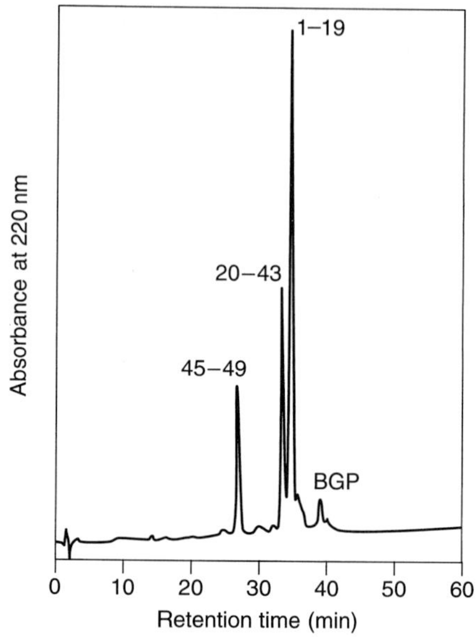

FIG. 2. Reversed-phase HPLC separation of trypsin-digested BGP. Samples are resolved under conditions described in the text. Identification of the peaks is determined by partial manual Edman degradation of fragment 20–43. Peaks containing fragments 1–19 and 20–43 are manually collected for analysis of Gla content. On one occasion the isolated peaks were reapplied to the column individually to check for purity; despite overlap seen in the original peak profile, single peaks were seen for fragments 1–19 and 20–43.[13] The elution profile is representative of typical elution patterns seen with trypsin-digested BGP.

Glutamic Acid Analysis

Derivatization. Acid hydrolyzates are derivatized with 11 μl of coupling solution [200 μl acetonitrile, pyridine (PYR), H_2O, triethylamine (TEA) (10:5:3:2 v/v) plus 20 μl phenylisothiocyanate (PITC)] for 30 min at room temperature. The samples are dried under a stream of N_2 and stored at 4° prior to same-day analysis.

Detection. Phenylthiocarbamyl-Glu (PTC-Glu) is detected using reversed-phase HPLC conditions modified from Bergman *et al.*[10] Each sample is suspended in 150 μl of buffer A [1% (v/v) acetonitrile, 0.03 *M* NaOH, pH 6.6 with 85% (v/v) phosphoric acid] and applied to a Zorbax ODS analytical HPLC column (25-cm × 4.6-mm i.d., Dupont, Wilmington, DE) equilibrated in buffer A. PTC-Glu is eluted under isocratic conditions, at a flow rate of 1 ml/min, within the first 15 min of injection. A linear gradient

[10] T. Bergman, M. Carlquist, and H. Jornvall, *in* "Advanced Methods in Protein Microsequence Analysis" (B. Wittmann-Liebold, ed.), p. 45. Springer-Verlag, Berlin, 1986.

is then developed over 15 min to 100% buffer B (60% acetonitrile, H_2O) and held until the remaining PTC-amino acid derivatives and other reagent-derived by-products are eluted. Elution is monitored at 254 nm and PTC-Glu peaks are collected, dried under a stream of N_2, dissolved in 100 μl of H_2O, and suspended in 700 μl Tissue-Solubilizer 1. (TS-1, Research Products International; TS-1 is no longer available from RPI and TS-2 is the replacement. To avoid chemiluminescence with TS-2 samples must be neutralized with 175 μl of 30% acetic acid after they are dissolved in TS-2.) The mixture is suspended in 3 ml of Bio-Safe II, transferred to a miniscintillation vial and [^{14}C]Glu is quantified in a liquid scintillation spectrometer. The NaOH in these samples does not interfere with counting efficiency.

Determination of Specific Activity. A standard curve of PTC-Glu is developed using Amino Acid Standards H (Pierce, Rockford, IL), derivatized and detected as stated earlier. Linear regression analysis is carried out on the peak areas and nmol PTC-Glu. Specific activity is calculated as disintegrations per minute (dpm) per nanomole of PTC-Glu.

Manual Edman Sequencing

Membrane Attachment. BGP tryptic fragment 20–43 is dissolved in H_2O, transferred to an Eppendorf tube, and lyophilized to dryness. The peptide is coupled to one-half of a membrane supplied in the Sequelon AA Attachment Kit (Milligen/Biosearch, Burlington, MA) as described in the manual. The membranes are rinsed with 30% acetonitrile/H_2O and transferred to 6- × 50-mm glass tubes.

Cycling. The sequencing method employed is adapted from that described by Kuhn and Crabb.[11] BGP fragment 20–43 is cycled five times through residue 24. To each membrane, 20 μl of PYR:TEA:H_2O (5:2:3 v/v) and 20 μl of 20% PITC in PYR is added. The coupling reaction is carried out for 5 min at 50° after which the coupling solution is removed and the membrane is dried under N_2. Each membrane is washed twice with 200 μl ethyl acetate (EA), once with 200 μl heptane (HEP):EA (15:1 v/v), and again with 200 μl EA with subsequent drying under N_2. Cleavage is carried out with 20 μl TFA for 10 min at 50° and the TFA is removed by drying under N_2. Extraction is carried out twice with 100 μl HEP:EA (1:5), and the extracts are combined in a fresh 6- × 50-mm tube and dried under N_2. Extracts are converted to thiohydantoin derivatives at 65° for 60 min in 30 μl of 1 *N* HCl in methanol. The final product is dried under N_2 and stored at 4° prior to HPLC detection.

[11] C. C. Kuhn and J. W. Crabb, *in* "Advanced Methods in Protein Microsequence Analysis" (B. Wittmann-Liebold, ed.), p. 65. Springer-Verlag, Berlin, 1986.

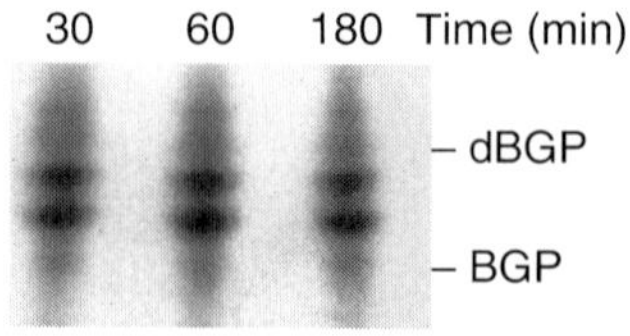

FIG. 3. Isoelectric focusing of carboxylated dBGP. dBGP (20 μM) was carboxylated for 30, 60, and 180 min with 500-fold-purified bovine liver carboxylase. The carboxylated species were isolated as described in the text, resolved on an IEF gel (pH 2.5–5.0), and subjected to fluorography. Positions of BGP and dBGP markers, detected by Coomassie staining of the same gel, are indicated. $^{14}CO_2$ incorporation into the mono- and di-Gla species was quantitated with a Betagen blot analyzer. The ratio of [^{14}C]mono-Gla to [^{14}C]di-Gla was 49:51 at 30 min, 47:53 at 60 min, and 44:56 at 180 min, indicating a 2:1 ratio of mono- to dicarboxylated product.

Detection. Reversed-phase HPLC detection of PTH derivatives is carried out, in part, according to the procedure of Bhown and Bennett.[12] Samples are suspended in buffer A (640 μl acetic acid and 280 μl acetone per 800 ml of 2% methanol, H_2O) and applied to an Alltech Econosil C18 column (25 cm $\times$ 4.6 mm, 5 μm) equilibrated in 77% A + 23% buffer B (450 ml TEA per 500 ml 98% methanol, H_2O) with a flow rate of 1 ml/min. The column is developed for 2 min and then a 5-min linear gradient is applied to final conditions (54% A + 46% B), which is held until the sample elutes. Elution is monitored at 254 nm and PTH-Glu peaks are collected, dried under a stream of filtered air, and suspended in 0.7 ml of TS-1 followed by 3 ml of Bio-Safe II. Radioactivity is quantified in a liquid scintillation spectrometer.

Determination of Specific Activity. A standard curve of PTH-Glu is developed using a PTH amino acid standards kit (Pierce). A measured amount of PTH-Glu is dissolved in methanol and varying amounts are aliquoted to 6- $\times$ 50-mm glass tubes and dried under N_2. Each standard is incubated for 1 hr at 65° in 1 *N* HCl/methanol, which converts PTH-Glu to its methylated ester. Linear regression analysis is carried out on the peak height and amount of each PTH-Glu in nanomoles. Specific activity is calculated as dpm/nmol PTH-Glu.

Results

In part A of the experimental design (Fig. 1), the number of Glu sites carboxylated in a dBGP substrate is determined. Two distinct bands representing mono- and dicarboxylated BGP are seen following 30-, 60-, and

[12] A. S. Bhown and J. C. Bennett, *Anal. Biochem.* **150,** 457 (1985).

TABLE I
DISTRIBUTION OF [^{14}C]GLA IN CARBOXYLATED dBGP[a]

Experiment	Glu site	Specific activity of residue (dpm/nmol)	Percent of total
A	17	98	13
	21	255	33
	24	417	54
B	17	120	16
	21	225	29
	24	422	55

[a] Unfractionated BGP products from two separate reactions were analyzed for [^{14}C]Gla. The specific activity (dpm/nmol) was determined at site 17 by analysis of tryptic fragment 1–19, and specific activities of sites 21 and 24 were determined by manual Edman degradation of tryptic fragment 20–43 as described in the text. The incorporation of $^{14}CO_2$ is also expressed as a percentage of the total combined specific activities per experiment.

180-min incubations (Fig. 3). The use of dBGP and BGP as markers for the fully uncarboxylated and fully carboxylated species, respectively, allows for interpretation of the data. The bands can be quantitated with a Betagen blot analyzer to determine the amount of each band present in the IEF gel. The ratio of mono and di bands is found to be approximately equal, indicating a 2:1 ratio of mono- to dicarboxylated BGP. The same results

TABLE II
DISTRIBUTION OF [^{14}C]GLA IN MONO- AND DICARBOXYLATED BGP[a]

Experiment	BGP species	dpm/nmol in Residue			Percent of total carboxylation		
		17	21	24	17	21	24
A	Mono	62	68	273	15	17	68
	Di	125	342	473	14	37	49
B	Mono	58	54	316	13	13	74
	Di	56	161	205	13	38	49

[a] Carboxylated reaction products were resolved on an IEF gel and the mono- and dicarboxylated BGP species were individually excised and eluted. The specific activities of $^{14}CO_2$ incorporation at each of the three potential Gla sites were determined as described in the text and in Table I.

TABLE III
ANALYSIS OF GLA IN [^{14}C]BGP TRYPTIC FRAGMENTS[a]

Substrate	Concentration (μM)	Incubation time (hr)	% [^{14}C]PTC-Glu	
			Gla site 17	Gla sites 21 + 24
dBGP	5	1	11	89
dBGP	20	1	13	87

[a] Carboxylated dBGP was analyzed as described in the text except the BGP was not S-carboxymethylated or heat decarboxylated. Tryptic fragments 20–43 and 1–19 were acid hydrolyzed and subjected to PTC-Glu analysis. The specific activities of Glu site 17 and Glu sites 21 + 24 were determined and expressed as a percentage of the combined activities. Fragment 20–43 contains four Glu sites of which only two are targeted by the carboxylase. Therefore, the amount of PTC-Glu detected was divided by a factor of 2 to reflect Glu sites 21 and 24 only. In addition, the calculated specific activity of Glu sites 21 + 24 was multiplied by a factor of 2 to calculate total incorporation of $^{14}CO_2$ rather than the average specific activity of both sites.

are seen for the 60-min time point with numerous enzyme preparations of varying carboxylase activity.[13] Fully carboxylated BGP never accumulates to a significant level under the carboxylation conditions used; after a 15-hr incubation the same ratio of mono- and di-Gla species is seen, with a negligible amount of the fully carboxylated species.[13] Using lower, nonsaturating concentrations of dBGP substrate (1 and 0.5 μM), only the di-Gla product is seen following a 60-min incubation.[1] Unfortunately, due to limiting incorporation of label, distribution of products at earlier time points cannot be determined.[13]

Part B of the experimental design (Fig. 1) determines the specific activity of [^{14}C]Gla at carboxylated Glu residues. In two separate experiments, analyzing the mono- and di-Gla species combined, it is found that site 24 contains the highest level of $^{14}CO_2$ incorporation followed by site 21 and then site 17 with the lowest specific activity (Table I). Similar data are obtained when the individual mono- and dicarboxylated BGP species are analyzed for [^{14}C]Gla. The majority of the monocarboxylated species contain Gla at site 24 and the majority of the dicarboxylated species contain Gla at sites 21 and 24 (Table II).

Low recoveries are obtained for PTH-Glu's 21 and 24 with the manual Edman sequencing method. To verify that the data are accurate and can be adequately compared with that obtained for site 17 by amino acid analysis, fragment 20–43 is subjected to amino acid analysis in conjunction with fragment 1–19. Based on data in Table I, the percentage of the total

[13] M. E. Benton, Ph.D. thesis, University of Wisconsin–Madison (1994).

specific activity at all three sites is expected to be, on average, 14% at site 17 and 87% at the combined Glu sites 21 + 24. In two separate experiments using 5 and 20 μM dBGP, 11 and 13% of the total combined specific activities are found at site 17 and 89 and 87% at Glu sites 21 + 24, respectively (Table III).

[33] Purification of Warfarin-Sensitive Vitamin K Epoxide Reductase

By REIDAR WALLIN and THOMAS M. GUENTHNER

Introduction

Vitamin K functions as a cofactor for the vitamin K-dependent carboxylase, an enzyme that resides in the endoplasmic reticulum (ER) and participates in posttranslational modification of secretory proteins.[1] The modification carried out by the carboxylase converts a limited number of glutamic acid residues in targeted proteins to γ-carboxyglutamic acid (Gla) residues. These Gla residues enable the proteins to bind Ca^{2+}.[2] The vitamin K-dependent proteins include the coagulation factors prothrombin,[2] factors VII,[2] IX,[2] X,[2] protein S,[2] and protein C[2] and a number of extrahepatically synthesized proteins some of which have been identified as osteocalcin,[3] matrix Gla protein,[3] protein S,[4] and Gas6.[5]

The naturally occurring form of vitamin K is shown in Fig. 1. It is a naphthoquinone with an isoprenic side chain in the 3 position. On reaching the liver, phylloquinone (vitamin K_1) is reduced to the hydronaphthoquinone (KH_2), which is the active cofactor for the carboxylase. As illustrated in Fig. 2, γ-carboxylation of one Glu residue is coupled stoichiometrically to formation of one molecule of phylloquinone 2,3-epoxide (K>0; see Fig. 2). Phylloquinone 2,3-epoxide is reduced by liver enzyme(s) to the hydronaphthoquinone form, which establishes a reduction oxidation cycle

[1] J. W. Suttie, *Annu. Rev. Biochem.* **54,** 459 (1985).

[2] B. Furie and C. Furie, *Cell* **53,** 505 (1988).

[3] P. V. Haauschka, J. B. Lian, D. E. C. Cole, and C. M. Gundberg, *Physiol. Rev.* **69,** 990 (1989).

[4] C. Maillard, M. Berruyer, C. M. Serre, M. Dechavanne, and P. D. Delmas, *Endocrinology* **103,** 1599 (1992).

[5] T. N. Stitt, G. Conn, M. Gore, C. Lai, J. Bruno, C. Radziefewski, K. Mattsson, J. Fisher, D. R. Gies, and P. F. Jones, *Cell* **80,** 661 (1995).

0076-6879/97 $25.00

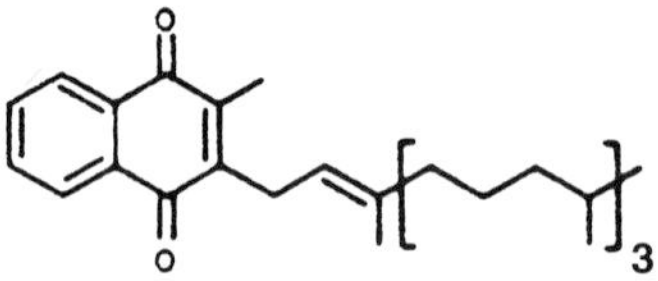

FIG. 1. Chemical structure of phylloquinone.

for vitamin K_1 in liver, known as the vitamin K cycle,[6,7] The cycle is depicted in Fig. 2. Two pathways for vitamin K reduction are associated with the cycle. Pathway I is operative at low tissue concentrations of phylloquinone/phylloquinone 2,3-epoxide and is physiologically the most important pathway of the cycle. The pathway is catalyzed by the enzyme vitamin K epoxide reductase, which reduces phylloquinone 2,3-epoxide and phylloquinone to the active vitamin KH_2 cofactor. Pathway II is catalyzed by NAD(P)H dehydrogenases and requires high tissue concentrations of phylloquinone to provide the carboxylase with sufficient amounts of cofactor for γ-carboxylation.[7] Therefore, pathway II is not operative at normal liver concentrations of the vitamin, but it plays an important role when phylloquinone is used as an antidote to overcome poisoning by coumarin anticoagulants.[7]

It is well documented that the coumarin-type anticoagulants target the vitamin K cycle by preventing reduction of vitamin K and its epoxide by pathway I.[6] The lack of carboxylase cofactor produces nonfunctional vitamin K-dependent coagulation factors, and this is the basis for the clinical use of the drugs in control of thromboembolic disease.[8] The coumarin-type anticoagulants are also used as poisons in rodent pest control.[9] The structure of the drug and the poison warfarin is shown in Fig. 3. One of the biggest setbacks in rodent control was the finding of strains of rats and mice that had become resistant to warfarin. This finding necessitated synthesis of more active compounds to fight the resistant strains of rats. Subsequently the second-generation anticoagulants, difenacoum, bromadiolone, and brodifacoum, were developed.[10] These so-called "superwarfarins" have so

[6] J. W. Suttie, *in* "Handbook of Lipid Research" (H. F. Deluca, ed.), p. 211. Plenum Press, New York, 1978.

[7] R. Wallin and L. F. Martin, *Biochem. J.* **241,** 389 (1987).

[8] J. W. Suttie, *in* "Advances in Experimental Medicine and Biology" (S. Wessler, C. G. Becker, and Y. Nemerson, eds.), Vol. 214, p. 3, Plenum Press, New York, 1987.

[9] W. B. Jackson, A. D. Ashton, and K. Delventhal, *in* "Current Advances in Vitamin K Research" (J. W. Suttie, ed.), p. 38. Elsevier, New York, 1988.

[10] M. R. Hadler and R. S. Shadbolt, *Nature* **253,** 275 (1975).

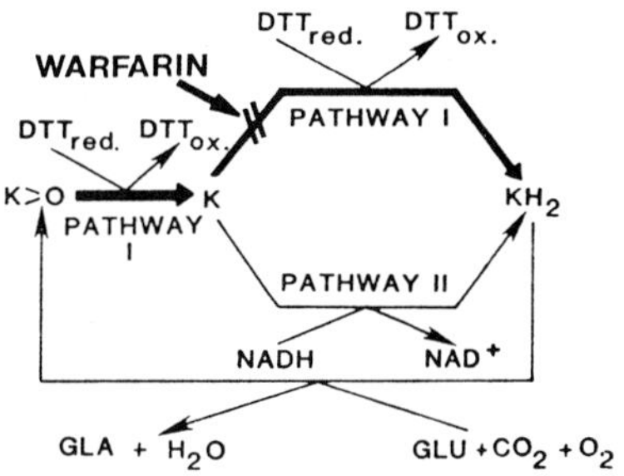

FIG. 2. The Vitamin K cycle. Phylloquinone and its epoxide (K>0) are reduced to the active cofactor hydronaphthoquinone (KH_2) by two pathways. Pathway I is catalyzed by the enzyme vitamin K epoxide reductase, which is inhibited by warfarin and other coumarin-type anticoagulant drugs.

far been quite successful in fighting resistant rodents, but their extreme toxicity is a danger to humans[11] and animals.

Warfarin and the superwarfarins are potent inhibitors of the vitamin K epoxide reductase.[6] To understand how these drugs and poisons interact with the enzyme, and also the genetics of warfarin resistance, numerous attempts to purify this enzyme have been undertaken since its discovery by Bell and Matchiner in 1970.[12] The enzyme has proven to be extremely difficult to purify and most biochemical data have come from studies on the enzyme in crude liver microsomal preparations. From these studies we have learned that thiol groups are involved in the catalytic reduction of phylloquinone 2,3-epoxide.[6] It is also well documented that warfarin prevents reduction of these essential thiol groups.[13,14] *In vitro* studies[7] have suggested that warfarin binds irreversibly to the enzyme, which is consistent with its apparent irreversible effect on the enzyme *in vivo*.[7] The most plausible model for warfarin inhibition of the enzyme has been proposed by Fasco *et al.*[13] Their model predicts that warfarin binds to the oxidized form of the enzyme and prevents reduction of essential cysteines that are located in the warfarin-binding site. This prevents the enzyme from carrying out another round of epoxide reduction.

The consensus view of many researchers has been that expression of vitamin K epoxide reductase requires more than one protein component

[11] C. R. Routh, D. A. Triplett, M. J. Murphy, L. J. Felice, J. A. Sadowsky, and E. G. Bovil, *Am. J. Hematol.* **36,** 50, (1991).

[12] R. G. Bell and J. T. Matchiner, *Arch. Biochem. Biophys.* **141,** 473 (1970).

[13] M. J. Fasco, L. M. Principe, W. A. Walsh, and P. A. Friedman, *Biochemistry* **22,** 5655 (1983).

[14] J. J. Lee and M. J. Fasco, *Biochemistry* **23,** 2246 (1984).

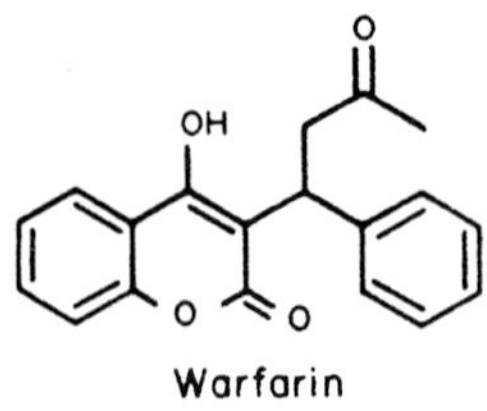

FIG. 3. Chemical structure of warfarin.

of the ER membrane, which certainly would complicate purification of this enzyme. With this background we decided on a strategy to purify the putative enzyme complex from the ER membrane to a state where the individual components of the complex could be separated and visualized by staining on a sequencing membrane after electrophoretic transfer from sodium dodecyl sulfate–polyacrylamide gel electrophoresis (SDS–PAGE) gels. This approach would allow structural analyses of the enzyme components following their affinity labeling or identification by other means. This chapter describes a purification procedure that was developed to achieve this goal. The usefulness of this approach is exemplified by the identification of microsomal epoxide hydrolase (EC 3.3.2.3) as a component of the warfarin-sensitive vitamin K epoxide reductase and the vitamin K cycle in liver.

Enzyme Assay

Assay

Enzyme activity is measured as percent conversion of phylloquinone 2,3-epoxide to phylloquinone. The assay is described in detail in Wallin and Martin[15] and uses reversed phase high-performance liquid chromatography (HPLC) for separation of phylloquinone and the epoxide. The rate of the enzyme reaction is linear for 20 min at 26°. Phylloquinone 2,3-epoxide and dithiothreitol (DTT) are present in the incubations at 10 μM and 5 mM concentrations, respectively. Quantification of phylloquinone and the epoxide is based on external standards.

Preparation of Enzyme Samples for the Assay

Most tissues and isolated cells exhibit warfarin-sensitive vitamin K epoxide reductase activity. However, there is a great variation in enzyme-specific

[15] R. Wallin and L. F. Martin, *J. Clin. Invest.* **76,** 1879 (1985).

activity, with the liver exhibiting the highest activity. The following procedures for preparation of the enzyme from tissues and cells for activity measurements results in high expression of enzyme activity.

From Whole Tissues. Microsomes from the tissue are prepared in 250 m*M* sucrose, 25 m*M* imidazole, 5 m*M* benzamidine, pH 7.6, as described.[7] The microsomal pellets are kept frozen at −85° with no loss of enzyme activity even after several months of storage. For activity measurements the microsomal pellets are resuspended in 25 m*M* imidazole, 0.5% CHAPS, 2 m*M* DFP (diisopropyl fluorophosphate), pH 7.6, in a Dounce homogenizer to give a protein concentration of 7 mg/ml. Two hundred microliters is used in the assay.

From Preparations of Partially Purified Enzyme. Prior to assay, the preparation of partially purified enzyme should be in a 25 m*M* sodium phosphate buffer, pH 7.85, containing 25 m*M* KCl, 20% glycerol, 0.75% CHAPS, 0.2 mg/ml phospholipids and 2 m*M* DFP. The lipids are added to the buffer from a stock solution of liposomes (80 mg/ml) prepared by sonication of soybean phospholipids (asolectin, Fluka, Ronkonkoma, NY) in 10 m*M* sodium phosphate, pH 7.5, containing 10 m*M* glycerol, 250 m*M* sucrose, and 0.5 m*M* EDTA. Buffer exchange can be achieved by gel filtration on Sephadex G-25. The enzyme, when kept frozen at −80° in the phosphate buffer, is preserved with little loss of enzyme activity even after several months of storage.

From Cells. Cells from primary cultures, proliferating cultures and cells transfected with DNA have been exposed to the conditions described later with good results regarding expression of warfarin-sensitive vitamin K epoxide reductase activity.

Cells are washed two times by resuspending them in ice-cold phosphate-buffered saline (PBS) followed by sedimentation at low speed. The final cell pellet is lysed in a small Dounce homogenizer by resuspending it in 25 m*M* sodium phosphate buffer, pH 7.85, containing 25 m*M* KCl, 20% glycerol, 0.75% CHAPS, and 2 m*M* DFP. Protein concentrations ranging from 10 mg/ml to 0.2 mg/ml of lysed cells have been used for enzyme activity measurements with good results. Aliquots of 200 μl are used in the assay. For COS-1 cells, which frequently are used for transient transfections, the assay measures an enzyme activity of 120 pmol phylloquinone formed in 20 min per milligram of protein, which we have found is similar to the activity measured in several different hepatoma cell lines.

Partial Purification of Vitamin K Epoxide Reductase

All steps in the procedure are carried out on ice or at 4°. Frozen microsomes derived from 32 g of liver are thawed on ice and suspended in a

TABLE I
PURIFICATION OF VITAMIN K EPOXIDE REDUCTASE

Fraction	Volume (ml)	Protein (mg)	Activity (nmol/mg)	Total activity (nmol)
Microsomes	18[a]	176.4	0.14	24.7
$(NH_4)_2SO_4$ cut	6	46.8	0.52	24.3
Sepharose 6B	10	6.0	3.9	23.4
Hydroxylapatite	4.5	1.1	16.7	18.4

[a] Data obtained from a starting volume of 18 ml instead of 48 ml as given in the Methods section.

Dounce homogenizer in 48 ml buffer A (50 m*M* Tris, 1 *M* NaCl, 20% ethylene glycol, 5 m*M* benzamidine, 5 m*M* DFP, 1% CHAPS, pH 7.85). The suspension is cleared by centrifugation at 100,000*g* for 30 min and the supernatant made 43% saturated with $(NH_4)_2SO_4$. After 15 min on ice, the precipitate is collected by centrifugation at 10,000*g* for 10 min and dissolved in 16 ml of buffer B [0.1 *M* phosphate, 20% glycerol, 10% $(NH_4)_2SO_4$, 0.4 mg/ml phospholipid, 0.5% CHAPS, pH 7.85]. The phospholipids are added to the buffer from an 80 mg/ml sonicated stock solution as described earlier. The enzyme solution is cleared by centrifugation at 100,000*g* and loaded onto a 70 × 3.5 cm column of Sepharose 6B equilibrated in buffer B. The void volume fraction, which contains the reductase, is made 0.75% in CHAPS and desalted on a column of Sephadex G-25 in buffer C [25 m*M* phosphate, 25 m*M* KCl, 20% (v/v) glycerol, 0.2 mg/ml phospholipid, 0.75% CHAPS, pH 7.85]. The void volume fraction is sonicated on ice for 5 min using a Branson 250 Ultrasonicator at setting 3. The sonicate is passed through a 0.45-μm filter before it is loaded onto a 6 × 1.5 cm column of hydroxylapatite equilibrated in buffer C.

The column is eluted with a 150-ml linear gradient of phosphate and KCl prepared by mixing equal volumes of buffer C and buffer D [250 m*M* phosphate, 250 m*M* KCl, 20% (v/v) glycerol, 0.2 mg/ml phospholipid, 0.75% CHAPS, pH 7.85]. The fractions with reductase activity are pooled and the enzyme transferred into buffer C by buffer exchange on Sephadex G-25 as described earlier. To obtain a concentrated fraction of partially purified enzyme, the hydroxylapatite column can be stepwise eluted with buffer D. The enzyme is kept frozen at −85°.

This procedure results in a 118-fold purification of the warfarin-sensitive vitamin K epoxide reductase from rat liver microsomes with a recovery of 75% of total enzyme activity (see Table I). The most critical step in the procedure is the Sepharose 6B exclusion chromatography step where the

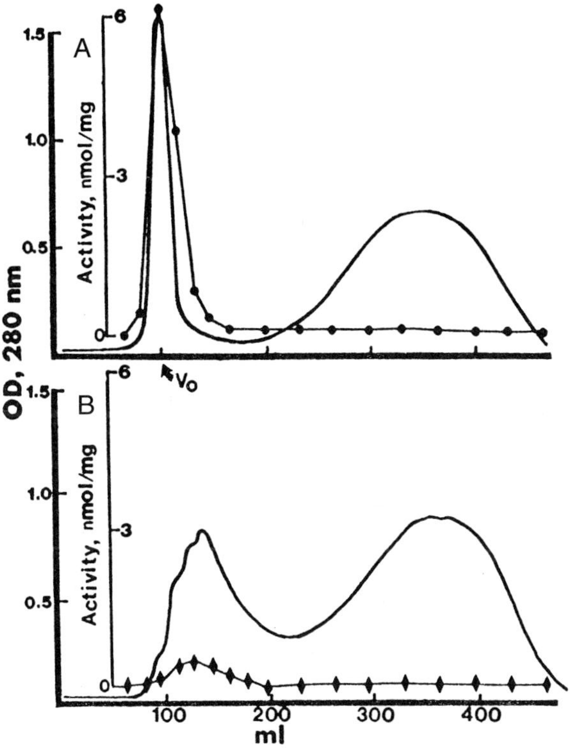

FIG. 4. (A) Gel filtration of vitamin K epoxide reductase on Sepharose 6B. The dotted line shows reductase activity. (B) The elution profile when $(NH_4)_2SO_4$ was excluded from the elution buffer.

active enzyme appears in the void volume fraction only when $(NH_4)_2SO_4$ and phospholipids are included in the elution buffer. Figure 4 shows the elution profiles from the Sepharose 6B column when the precipitate from the $(NH_4)_2SO_4$ cut was gel-filtrated in the presence (Fig. 4A) and absence (Fig. 4B) of 10% $(NH_4)_2SO_4$. Less than 5% of the activity was recovered when $(NH_4)_2SO_4$ was omitted from the elution buffer (Fig. 4B). The last step in the purification protocol is adsorption chromatography on hydroxylapatite (see Fig. 5). Figure 5B shows a silver-stained SDS–PAGE gel containing delipidated proteins from selected column fractions. Vitamin K epoxide reductase activity is "spread out" over many fractions (see Fig. 5A). However no proteins with apparent molecular masses higher than 55 kDa could be detected in fractions with the most enzyme activity (fractions 20–30), which suggests that vitamin K epoxide reductase activity is expressed by proteins of apparent molecular mass around 55 kDa or smaller.

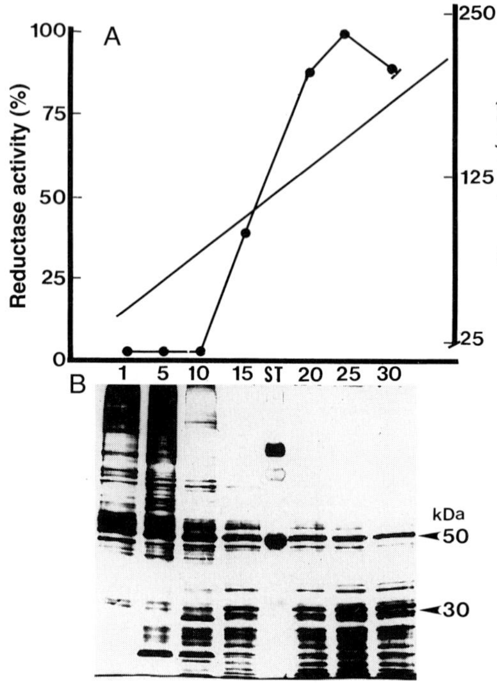

FIG. 5. Hydroxylapatite chromatography of vitamin K epoxide reductase. (A) Linear gradient of phosphate was used to elute the enzyme. The filled circle line shows the activity profile of the eluted enzyme. (B) Silver-stained, SDS–PAGE separated proteins appear in the various column fractions.

Attempts to separate the remaining proteins on ion-exchange columns, hydrophobic and dye resins, and a variety of general affinity columns have resulted in loss of enzyme activity.

Identification by Microsequencing of Proteins Present in Partially Purified Preparation of Vitamin K Epoxide Reductase

Several of the proteins in the enzyme preparation shown in Fig. 5 have been identified by N-terminal sequence analysis following their transfer to PSQ-sequencing membranes from 10% SDS–PAGE gels. Their identities are revealed in Fig. 6 (lane A). Proteins L18 (21 kDa),[16] L30 (15 kDa),[17]

[16] K. R. Gayathri Devi, Y.-L., Chan, and I. G. Wool, *DNA* **7,** 157 (1988).
[17] K. Johnson, *Gene* **123,** 283 (1993).

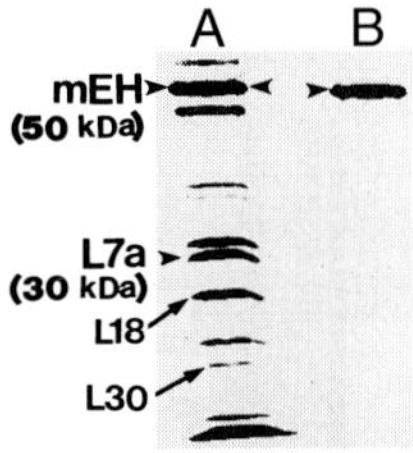

FIG. 6. Identification of proteins appearing in the vitamin K epoxide reductase preparation. Lane A, SDS–PAGE separated and stained proteins; lane B, Western blot of proteins in lane A with anti-mEH antibodies. See text for identification of proteins.

and L7a (30 kDa)[18] are ribosomal proteins derived from the large ribosomal subunit and mEH is microsomal epoxide hydrolase.[19] A polyclonal antiserum raised against rat mEH was used to investigate a possible involvement of this protein in expression of vitamin K epoxide reductase activity. Figure 6 (lane B) shows a Western blot of the enzyme preparation with the anti-mEH antiserum.

Identification of Microsomal Epoxide Hydrolase as a Component of Vitamin K Epoxide Reductase

When incubated with the partially purified vitamin K epoxide reductase enzyme preparation on ice for 3 hr in the absence of phylloquinone 2,3-epoxide (−KO) the anti-rat mEH antiserum inhibited vitamin K epoxide reductase activity (see Fig. 7A). However, when the enzyme preparation was preincubated with phylloquinone 2,3-epoxide before addition of the antiserum (+KO), the antiserum had no effect on enzyme activity.

To ensure that inhibition was an antibody effect and not a serum effect, purified IgG was also used. IgG purified from the mEH antiserum significantly inhibited vitamin K epoxide reductase activity. Control rabbit IgG had no effect on enzyme activity (see Fig. 7B). The ability of purified mEH to block the effect of the mEH antibodies on reductase activity is shown in Fig. 8A. Addition of increasing concentrations of purified mEH competed effectively with the mEH antibodies. Figure 8B demonstrates that the antibodies were able to immunodeplete an antigen responsible for expression of both vitamin K epoxide reductase and mEH activities. Microsomal epoxide hydrolase activity was measured as the rate of conversion of styrene 1′,2′-oxide to the corresponding diol.[20]

[18] A. Giallongo, J. Yon, and M. Fried, *Mol. Cell. Biol.* **9,** 224 (1989).

[19] T. D. Porter, T. W. Beck, and C. B. Kasper, *Arch. Biochem. Biophys.* **284,** 121 (1986).

[20] T. M. Guenthner, P. Bentley, and F. Oesch, *Methods Enzymol.* **77,** 344 (1981).

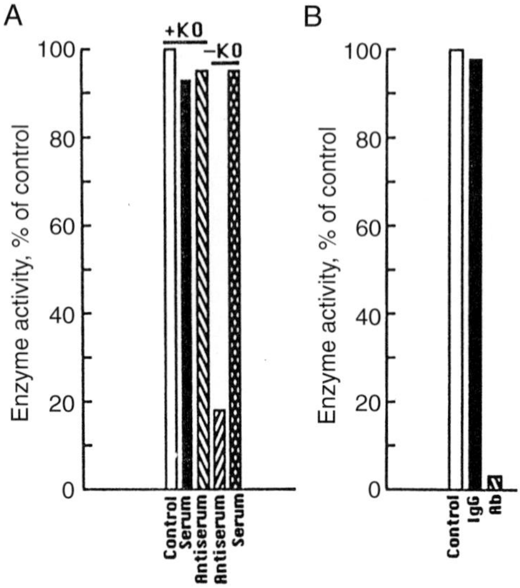

FIG. 7. (A) Effect of mEH antibodies on vitamin K epoxide reductase activity. +KO, phylloquinone 2,3-epoxide was added before addition of antiserum. −KO, No addition of phylloquinone 2,3-epoxide prior to antibody addition. (B) Demonstrates that purified IgG from the mEH antiserum (Ab) recognized and depleted the enzyme from a soluble preparation of the enzyme.

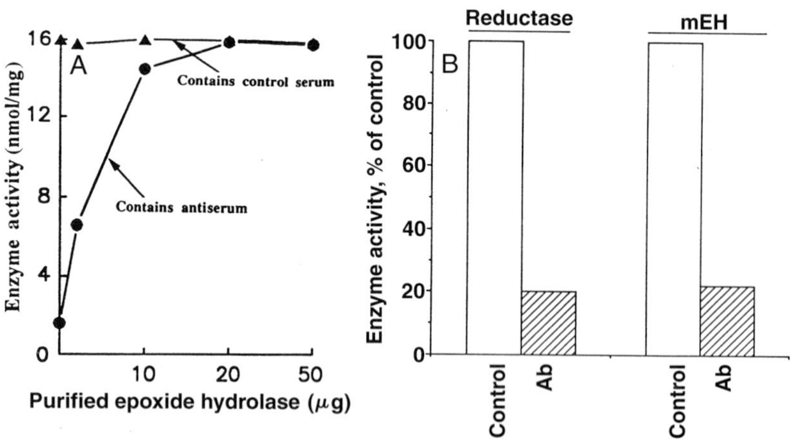

FIG. 8. Purified mEH competes with vitamin K epoxide reductase for mEH antibodies. (A) Purified mEH was added to vitamin K epoxide reductase incubations containing mEH antiserum. (B) mEH antibodies were able to immunodeplete the enzymes responsible for vitamin K epoxide reductase and mEH activities.

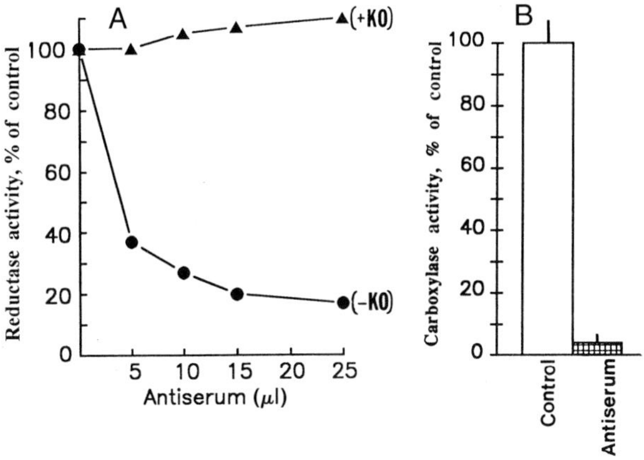

FIG. 9. Effect of mEH antibodies on vitamin K epoxide reductase and γ-carboxylase activities in right-side-out microsomal vesicles. (A) (−KO) demonstrates the inhibitory effect of the antibodies on the activity in intact vesicles. (B) Antibodies inhibited γ-carboxylation of endogenous proteins in the vesicles when the reductase was used to provide the reduced vitamin K cofactor.

Microsomal EH is a transmembrane protein of the ER membrane with >96% of the enzyme structure facing the cytosolic side of the membrane.[21] This model would predict a cytosolic location for the phylloquinone 2,3-epoxide binding site. To test this hypothesis, the effect of the mEH antibodies on vitamin K epoxide reductase activity and vitamin K epoxide reductase supported carboxylase activity in right-side-out vesicles of the ER membrane was determined. As shown in Fig. 9A, the anti-rat mEH antibodies inhibited vitamin K epoxide reductase activity in the intact vesicles (−KO) and inhibition was again prevented by preincubation of the vesicles with phylloquinone 2,3-epoxide (+KO). Because immunoglobulin G (IgG) will be excluded from the luminal side of the ER vesicles, the data suggest a cytosolic location for the phylloquinone 2,3-epoxide binding site. As shown in Fig. 9B, the mEH antibodies also inhibited vitamin K epoxide reductase supported γ-carboxylation of endogenous proteins present in the vesicles. Protein carboxylation was reduced 91% in the vesicles treated with the mEH antiserum (see Fig. 9B). Vitamin K-dependent protein carboxylation was measured as described.[22] These data strongly suggest that mEH participates as a component of the vitamin K cycle in liver and has an essential function in γ-carboxylation of proteins.

[21] J. A. Craft, S. Baird, M. Lamont, and B. Burchell, *Biochim. Biophys. Acta* **1046,** 32 (1990).
[22] R. Wallin, O. Gebhardt, and H. Prydz, *Biochem. J.* **169,** 95 (1978).

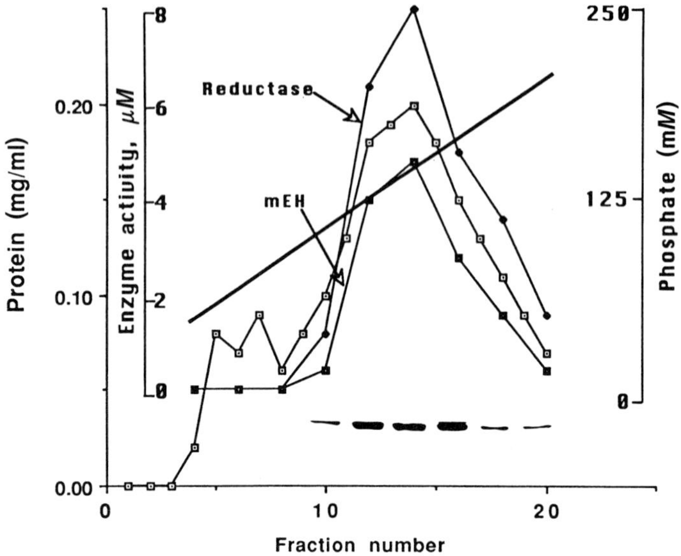

FIG. 10. Co-chromatography of mEH and vitamin K epoxide reductase by gradient elution from hydroxylapatite. Western blot of mEH enzyme in the active fractions is shown at the bottom of the graph.

Further support for the idea that mEH is a component of the warfarin-sensitive vitamin K epoxide reductase is provided in Fig. 10, which shows that the two activities co-chromatograph when the enzyme is eluted by gradient elution from the hydroxylapatite column. The ratio between the two activities is constant in the active fractions. Also Western blots of mEH in the column fractions indicate a correlation between the quantity of mEH enzyme and enzyme activity (see Fig. 10).

Phospholipids are Essential Components of the Enzyme Reaction

Microsomal EH purified from a Lubrol PX extract of microsomes exhibits no vitamin K epoxide reductase activity, which suggests that additional components are needed for expression of the warfarin-sensitive activity. As documented on Fig. 11, phospholipids are important for phylloquinone 2,3-epoxide reduction by the enzyme. When the partially purified enzyme was digested with phospholipase A_2 from *Naja naja* snake venom, this resulted (see Fig. 11A) in a time-dependent inactivation of the reductase. Addition of sonicated phospholipids to the digested samples restored enzyme activity (see Fig. 11B). Phylloquinone and its epoxide are lipophilic

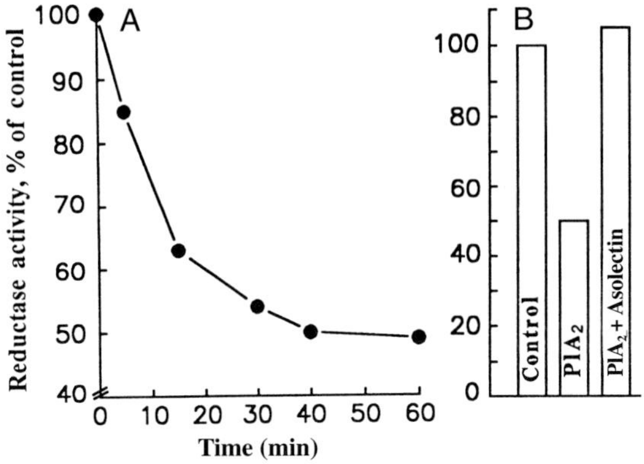

FIG. 11. Phospholipase A_2 (PLA_2) inactivation of vitamin K epoxide reductase. (A) Time-dependent inactivation of reductase activity by the phospholipase. (B) When added to the partially inactivated enzyme, phospholipids restore activity.

components and may therefore require phospholipids for their correct orientation in the enzyme active site.

Future Directions

In the partially purified preparation of the vitamin K epoxide reductase only a minor portion (18%) of the total mEH activity in liver microsomes is recovered. The lipid-associated pool of mEH, which appears in the void volume fraction from the Sepharose 6B column, participates in warfarin-sensitive phylloquinone 2,3-epoxide reduction. Microsomal EH is known to associate with lipids and this association changes its specific activity toward various substrates.[23] This finding led to the assumption that there were several forms of the enzyme,[24] which was later shown to be dictated by its association with components of the ER membrane.[23] The function of mEH in phylloquinone 2,3-epoxide reduction is a new activity associated with this enzyme that would require a second component providing the electrons for the reaction. DTT is an excellent electron donor for the enzyme-catalyzed reduction of phylloquinone 2,3-epoxide. However, because DTT and purified mEH, when incorporated into liposomes, show no vitamin K epoxide reductase activity, this suggests that an intermediate electron carrier exists that participates in the reaction. Several thiol reduction–oxidation components have been suggested as the intermediate carrier

[23] N. J. Bulleid, A. B. Graham, and J. A. Craft, *Biochem. J.* **233,** 607 (1986).

[24] F. P. Guengreich, P. Wang, M. B. Mitchell, and P. S. Mason, *J. Biol. Chem.* **254,** 12248 (1979).

including protein disulfide-isomerase[25] and thioredoxin,[26] but convincing data on their involvement in the reaction have not been provided.[27]

The electron carrier should harbor the warfarin-sensitive thiol reduction–oxidation center. Based on ultrafiltration studies, we believe the carrier is a protein that is part of a lipid–protein enzyme complex. Microsomal epoxide hydrolase is the component that harbors the phylloquinone 2,3-epoxide binding site. Indeed the warfarin-binding protein should be present among the proteins seen on the silver-stained SDS–PAGE gel in Fig. 5. The next goal is to identify this electron carrier, which should be the key to unveiling the genetics of warfarin resistance. A second goal must be reconstitution of warfarin-sensitive enzyme activity following assembly of the components of the enzyme to provide evidence for the architecture of the vitamin K epoxide reductase enzyme system.

Acknowledgments

This work was supported by grant NCR-9403041 94-37313-0740 from the U.S. Department of Agriculture.

[25] B. A. M. Soute, M. M. C. L. Groenen-van Dooren, A. Holmgren, J. Lundstrom, and C. Vermeer, *Biochem. J.* **281,** 255 (1992).

[26] R. B. Silverman and D. L. Nandi, *Biochem. Biophys. Res. Commun.* **155,** 1248 (1988).

[27] P. C. Preusch, *FEBS Lett.* **305,** 257 (1992).

[34] Determination of Vitamin K Compounds in Plasma or Serum by High-Performance Liquid Chromatography Using Postcolumn Chemical Reduction and Fluorimetric Detection

By KENNETH W. DAVIDSON and JAMES A. SADOWSKI

Introduction

Before the advent of high-performance liquid chromatography (HPLC) the determination of vitamin K in biological samples was tedious and relied on, at best, semiquantitative procedures. The extremely low concentration of vitamin K relative to the lipids and lipid-soluble compounds present in plasma and tissues made extraction difficult and the use of large sample volumes was essential. Numerous reports describing HPLC assays for the determination of phylloquinone (vitamin K_1) emerged in the 1980s based

0076-6879/97 $25.00

on various detection systems. Ultraviolet,[1,2] electrochemical,[3,4] electrofluorimetric,[5] and postcolumn, chemical (wet and dry) reduction with fluorescence detection[6–8] were described and for the first time allowed for direct and quantitative measurement of phylloquinone. However, it became increasingly recognized that the applications based on reduction of the vitamin (to its fluorescent hydroquinone) coupled with fluorimetric detection were inherently more selective and sensitive.

Postcolumn, Dry Chemical Reduction with Fluorimetric Detection

The HPLC assays currently being used in our laboratory for the determination of endogenous phylloquinone, phylloquinone 2,3-epoxide, menaquinones (MKs 1–10), and most recently 2′,3′-dihydrophylloquinone have evolved from the postcolumn reduction and fluorimetric detection methodology originally described by Haroon *et al.*[7,8] The analytical system utilized to reduce vitamin K compounds to their fluorescent hydroquinones consists of a postcolumn, dry chemical reactor containing zinc metal. Hydroquinones are produced by chemical reduction over zinc in the presence of zinc ions, which are provided by the mobile phase. This on-line reduction process forms the core of the chromatographic systems we use for the determination of vitamin K and although the column configuration, injector type, mobile-phase composition and flow rates of our analytical systems vary, the on-line, postcolumn reduction is essentially the same. We have applied these assays to the determinations of vitamin K in various biological and nonbiological matrices.[9–12]

In this chapter, we describe two assays based on postcolumn reduction, but with significant enhancements made in the sensitivity of our analytical systems: (1) a simplified method for the determination of fasting plasma or serum concentrations of phylloquinone and (2) an assay for the simultaneous determination of endogenous phylloquinone and phylloquinone 2,3-

[1] M. J. Shearer, *Adv. Chromatogr.* **21,** 243 (1983).
[2] M. F. Lefevere, A. P. De Leenheer, A. E. Claeys, I. V. Claeys, and H. Steyaert, *J. Lipid Res.* **23,** 1068 (1982).
[3] J. P. Hart, M. J. Shearer, P. T. McCarthy, and S. Rahim, *Analyst* **109,** 477 (1984).
[4] T. Ueno and J. W. Suttie, *Anal. Biochem.* **133,** 62 (1983).
[5] J. P. Langenberg and U. R. Tjaden, *J. Chromatogr.* **305,** 61 (1984).
[6] W. E. Lambert, A. P. De Leenheer, and E. J. Baert, *Anal. Biochem.* **158,** 257 (1986).
[7] Y. Haroon, D. S. Bacon, and J. A. Sadowski, *Clin. Chem.* **32,** 1925 (1986).
[8] Y. Haroon, D. S. Bacon, and J. A. Sadowski, *J. Chromatogr.* **384,** 383 (1987).
[9] G. Ferland and J. A. Sadowski, *J. Agric. Food. Chem.* **40,** 1874 (1992).
[10] S. L. Booth, K. W. Davidson, and J. A. Sadowski, *J. Agric. Food Chem.* **42,** 295 (1994).
[11] K. W. Davidson, S. L. Booth, and J. A. Sadowski, *J. Agric. Food Chem.* **44,** 980 (1996).
[12] S. L. Booth, K. W. Davidson, A. H. Lichtenstein, and J. A. Sadowski, *Lipids* **31,** 709 (1996).

epoxide in plasma or serum. These assays have resulted in greater sample throughput making the analyses of sample-intensive populations and metabolic studies more practical.

Simplified Assay for Determination of Phylloquinone

Analytical Improvements

One milliliter of plasma was required for a single determination of phylloquinone as originally described.[7] Sample preparation involved liquid-phase extraction, solid-phase extraction on silica gel, and finally a liquid-phase reductive extraction prior to injection on the HPLC. Detection limits were approximately 50 pg/ml (111 pmol/liter) in plasma. We have since improved our detection limits through the application of various techniques; primarily through the use of a small-bore analytical column (3-mm versus 4.6-mm i.d.) packed with 3-μm particle size material, and by the insertion of a catalytic, platinum-on-alumina, oxygen-scrubber column to reduce residual oxygen in the HPLC system (described in greater detail later in this chapter). Furthermore, a change in excitation from the original 248 to 244 nm has provided an approximate 26% increase in fluorescence. To minimize dispersion in the LC system, we use small-bore stainless steel tubing on the high-pressure side and keep tubing lengths as short as possible. These cumulatively have yielded more than a threefold increase in sensitivity, allowing us to use 0.5 ml of sample for routine analysis. Because of this we have been able to omit the preparative, reductive extraction step, and apply the hexane extract from liquid-phase extraction directly to a reversed-phase, C_{18} column before injection. The extraction on C_{18} is the only preparative step now performed.

Blood Draw and Sample Storage

Samples should be immediately protected from light after blood draw. On separation, plasma or serum samples are either prepared for analysis or stored at $-70°$ prior to use. Vitamin K compounds in control plasma stored in cryogenic vials at $-70°$ and protected from light have shown excellent stability for up to 2 years.

Chemicals and Standard Solutions

The extraction and chromatography solvents used are all HPLC grade (Fisher Scientific, Springfield, NJ). Vitamin $K_{1(20)}$ (2-methyl-3-phytyl-1,4,-naphthoquinone) and the ACS reagent-grade zinc chloride used in the preparation of the mobile phase are purchased from Sigma Chemical Co. (St. Louis, MO). The internal standard $K_{1(25)}$ (synthesized by substitution of a 25-carbon side chain to menadione) was a gift from Hoffman-LaRoche and Co. (Basel, Switzerland). High-purity zinc metal (−200 mesh) used in packing the postcolumn zinc reactor is purchased from Johnson Matthey Co. (Ward Hill, MA), as is the 10% platinum-on-alumina used for packing the catalytic oxygen scrubber. The primary stock solutions are prepared gravimetrically in hexane at concentrations ranging between 0.2 and 0.4 mg/ml. Dilutions of the primary stock are prepared in 100% HPLC-grade methanol at a concentration of 2.0–3.0 μg/ml. These secondary solutions are characterized spectrophotometrically to validate their purity. The concentrations of phylloquinone and $K_{1(25)}$ secondary solutions are calculated from their UV absorption spectra using the following absorptivity values ($E^{1\%}_{1\,cm}$): phylloquinone at 248 nm = 420 and $K_{1(25)}$ at 248 nm = 420. The secondary solutions are then combined and diluted to produce the working calibration standard. Working standards are then characterized chromatographically for validation of their purity and concentration. The concentrations of phylloquinone and $K_{1(25)}$ in the working calibration standard are 5.0 ng/ml (11.1 nmol/liter) and 10.0 ng/ml (19.2 nmol/liter), respectively. The working internal standard [a dilution of the $K_{1(25)}$ secondary stock] is prepared at a concentration of 50.0 ng/ml (96.0 nmol/liter). The working standards are stored at 4° and are shielded from light. Because vitamin K is degradable by photooxidation, samples are protected from UV light. All sample processing and preparation in our lab is performed under yellow lighting. In addition, all glassware used in this assay are rinsed in acetone before use.

Liquid-Phase Extraction

Plasma or serum (0.5 ml) is pipetted into a screw top, 16 × 100 borosilicate glass culture tube, followed by 20 μl of the internal standard $K_{1(25)}$ (2.0 pmol). One milliliter of 100% ethanol is added and the tube is vortexed for 10 sec to precipitate the plasma proteins. This is followed by the addition of deionized H_2O (0.5 ml) and 3.0 ml of 100% hexane. The tubes are then capped with Teflon-lined screw caps. *We have found the ratio of plasma : ethanol : hexane (1 : 2 : 6) to be optimal for extraction efficiency and use this for all plasma and serum extractions.* The samples are mixed for 2 min and centrifuged at 1000*g* (4°) for 5 min. The hexane layer (top) containing the extracted lipids and lipophilic compounds is aspirated and transferred to a clean, acetone rinsed 16 × 100 culture tube. The hexane extract is then evaporated to dryness under vacuum is a centrifugational evaporator (Savant Instruments, Farmingdale, NY).

Solid-Phase Extraction on C_{18}

Two hundred microliters of 100% 2-propanol are pipetted into the sample tube. The contents of the tube are heated to 50° in a heating block until the residue is completely dissolved (approximately 10 min). A 3-ml, 500-mg solid-phase extraction (SPE) C_{18} column (J. T. Baker, Phillipsburg, NJ) is then preconditioned with successive washes of 3.0 ml of dichloromethane–methanol (20:80, v/v), 3.0 ml of 100% methanol and, finally, 3.0 ml of 100% deionized H_2O. The sorbent bed is kept well saturated with H_2O before the application of the sample. The sample in 2-propanol is aspirated and applied directly onto the preconditioned sorbent bed. The sample tube is then rinsed with an additional 200 μl of 2-propanol and the rinse is applied to the sorbent as well. Vacuum is drawn through the column manifold for at least 2 min to dry the sorbent fully after application of the sample extract. The band is then washed with 3.0 ml of water–methanol (5:95, v/v) and with 3.0 ml of 100% acetonitrile. With each wash the sorbent is fully dried under vacuum (2 min). The vitamin K fraction is then eluted with 6.0 ml of dichloromethane–methanol (20:80, v/v) and the collected eluant evaporated to dryness in a centrifugal evaporator with heat. For injection, the final residue is reconstituted initially in 20 μl of 100% dichloromethane (with swirling to dissolve the remaining lipid completely), immediately followed by 180 μl of methanol containing 10 mM zinc chloride, 5 mM acetic acid, and 5 mM sodium acetate (1 liter methanol:5 ml aqueous solution). One hundred microliters of sample is injected into the HPLC.

Chromatographic Instrumentation

The isocratic HPLC system consists of a model 510 reciprocating pump and a model 860 VAX-based data station using Expert-Ease (version 3.1) software for pump control, integration, and quantitation (Waters Chromatography, Milford, MA). Inserted in line between the pump and injector is a platinum-on-alumina oxygen scrubber. The oxygen scrubber consists of a stainless steel column (100- × 4.6-mm i.d.) dry packed with 10% platinum-on-alumina. Because the packing is very fine, 0.062-inch thickness, 0.5-μm porosity frits are used to reduce packing loss under pressure. *Under normal operating conditions the catalytic efficiency of the column is approximately 1 year.* Sample injection is accomplished using either a model 712 B WISP injector (Waters, Milford, MA) or a model 231-401 sample injector (Gilson Medical Electronics, Middleton, WI) fitted with a Rheodyne 7010 injection valve and 100-μl loop. A C_{18} precolumn (15- × 3.2-mm i.d., Brown Lee Labs, Santa Clara, CA) is connected in series to the reversed-phase analytical column. The analytical column (150- × 3.0-mm i.d.) is packed with 3 μm BDS-Hypersil (Keystone Scientific, Bellefonte, PA). The postcolumn zinc reactor is connected in series between the analytical column and the detector. The postcolumn reactor consists of a stainless steel column (50- × 2.0-mm i.d.) dry packed with zinc metal using low-dispersion (0.5 μm porosity) Kel-F frits (Alltech Assoc., Deerfield, IL). Careful attention is given to packing the zinc reactor to minimize cavitation. Fluorescence is monitored with a Spectroflow model 980 fluorescence detector using a 10 μl flow cell (Applied Biosystems, Ramsey, NJ). All high-pressure connections are made with short lengths of 0.007-inch i.d. stainless steel tubing.

Chromatographic Conditions and Quantitation

The mobile phase is comprised of dichloromethane–methanol (10:90, v/v) to which each liter is added 5.0 ml of an aqueous solution containing 2 *M* zinc chloride, 1 *M* glacial acetic acid, and 1 *M* sodium acetate (final concentration: 10 m*M* $ZnCl_2$, 5 m*M* CH_3COOH, and 5 m*M* sodium acetate). The aqueous solution is prepared and filtered using a 0.45-μm filter membrane (Millipore Corp., Bedford, MA). During analyses the mobile phase is continuously sparged with ultrahigh purity nitrogen. Flow rate is maintained at a constant 0.6 ml/min through the run. Excitation is performed at 244 nm and emission is monitored at 418 nm using a long-pass, cutoff filter.

A calibration standard is injected with every six samples in a run to compensate for changes in chromatographic conditions. Standard curves are prepared from each calibration injection. We have found that the fluorescence responses for phylloquinone and $K_{1(25)}$ are linear beyond normal physiological concentrations with the slope of the lines bisecting zero. We therefore routinely perform single-point calibration, forcing the slope of the line through zero. Quantitation is achieved by direct comparison of peak area ratios generated from the sample internal standard and unknowns and the ratios generated from the calibration standard. A representative chromatogram of the separation of phylloquinone and $K_{1(25)}$ from a plasma extract is shown in Fig. 1. Under the conditions described, average retention times for phylloquinone and the internal standard $K_{1(25)}$ are approximately 7.4 and 12.6 min, respectively.

Detection Limits and Precision

The minimal detectable level for phylloquinone is approximately 15.0 pg/ml of plasma (33 pmol/liter). The within-run coefficient of variation (*CV*) for replicates of pooled plasma ($n = 12$) is 5.6% and the between-run *CV* on pooled plasma analyzed over a 3-month period is 11.8% ($n = 14$). Mean (± SD) recoveries of the internal standard are approximately 75.0 ± 5.0%.

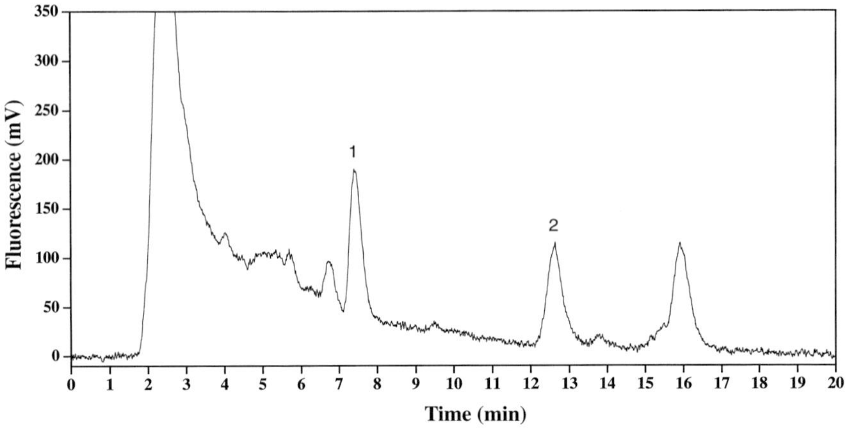

FIG. 1. Isocratic reversed-phase separation of phylloquinone in plasma. Conditions: column, 3 μm BDS-Hypersil; mobile phase, dichloromethane–methanol (10:90, v/v) containing 10 mM zinc chloride, 5 mM acetic acid, and 5 mM sodium acetate; flow rate, 0.6 ml/min. Detection, excitation 244 nm; emission 418. Peak identities: 1, phylloquinone; 2, $K_{1(25)}$.

Application of Assay

This assay has been applied to the analyses of more than 400 samples collected from a metabolic study comparing the relative bioavailability of phylloquinone from a green vegetable and a fortified oil among 10 younger (20–40 years) and 10 older adults (60–80 years). In each of the three 15-day study phases, volunteers were fed a mixed diet containing 100 μg phylloquinone/day. During one phase the volunteers received the mixed diet only. On days 6–11 during the other two phases, volunteers received two servings of broccoli per day in addition to the mixed diet (equivalent to 425 μg/phylloquinone/day) or received the mixed diet with the corn oil fortified with phylloquinone (425 μg/phylloquinone/day). Compared to the mixed diet there was a significant increase in plasma phylloquinone concentrations for both the young and older adults (Fig. 2). There was no difference in absorption of phylloquinone from the broccoli and oil in the young, however, there was a significant increase in plasma phylloquinone during the oil phase when compared to the broccoli phase in the older adults.

Simultaneous Determination of Phylloquinone and Phylloquinone 2,3-Epoxide

Phylloquinone 2,3-epoxide is an intermediate formed during the cyclic interconversion of phylloquinone. With inhibition of the cycle (either spe-

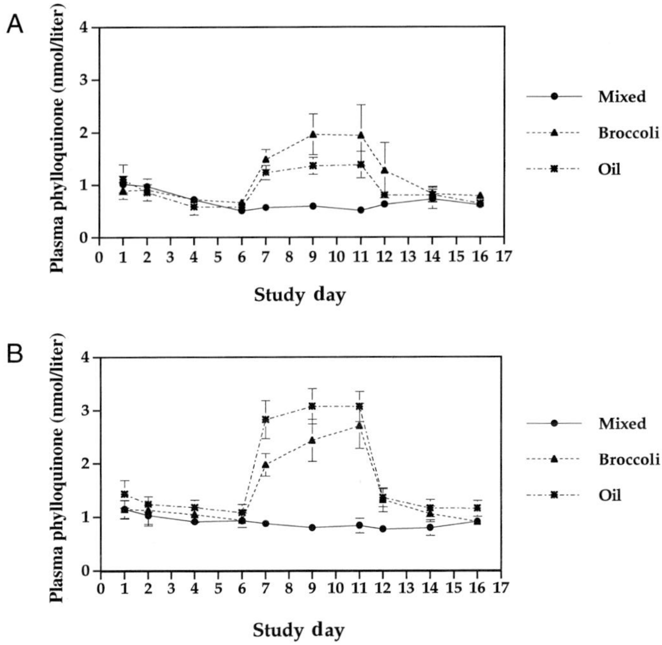

FIG. 2. Plasma concentrations (mean ± SEM) of phylloquinone consuming a mixed diet (100 μg); a mixed diet + broccoli (450 μg/day; days 6–11); and a mixed diet + phylloquinone supplemented oil (450 μg/day; days 6–11). (A) Younger adults (20–40 years). (B) Older adults (60–80 years).

cific or nonspecific) phylloquinone epoxide accumulates in the plasma.[13] The coumarin anticoagulant, warfarin, is one such inhibitor. However, in the nonanticoagulated, fasting state the measurement of phylloquinone epoxide in plasma is difficult because it is normally found at concentrations of less than 1/10 that of phylloquinone. Haroon *et al.*[1] reported preliminary measurements of endogenous phylloquinone epoxide in fasted pooled plasma. The epoxide could only be detected with preparative fractionation of extracts on adsorption HPLC prior to quantitative reversed-phase HPLC. As originally described, the fluorescent yield for phylloquinone epoxide was 35–40% that of phylloquinone, thus making the extraction of larger sample volumes essential for reliable quantitation. The low physiological concentrations compounded by the detection limits of the analytical method made a large population study to develop a reference range unfeasible.

[13] J. N. Weitzel, J. A. Sadowski, B. C. Furie, R. Moroose, H. Kim, M. E. Mount, M. J. Murphy, and B. Furie, *Blood* **76,** 2555 (1990).

Analytical Improvements

Through a series of experiments we found that residual oxygen in the HPLC system and linear velocity of phylloquinone epoxide through the zinc reactor to be the critical factors limiting reduction of the epoxide. We could achieve a fluorescent yield of approximately 70–75% that of phylloquinone by inserting a catalytic oxygen-scrubber column between the HPLC pump and injector and by decreasing transit (0.25 versus 1.0 ml/min) through the zinc reactor. The use of a catalytic oxygen scrubber for HPLC fluorescence detection is described in detail by MacCrehan *et al.*[14] Briefly, the platinum provides the active surface for the reduction of oxygen by methanol in the mobile phase. The reduction reaction is proposed to produce very small amounts of either water and formaldehyde and/or water and formic acid from oxygen and methanol. This can be achieved with as little as 1% methanol added to the mobile phase. The decrease in residual oxygen not only reduces fluorescent quenching overall, but for reduction of the epoxide to the hydroquinone, absolute minimal levels of oxygen are essential for complete reaction to take place. We found that reduction of the epoxide requires first the reduction to phylloquinone and then the reduction of phylloquinone to phylloquinone hydroquinone. This was confirmed by UV photodiode array detection—without the oxygen scrubber more than one reduction product was present in addition to phylloquinone hydroquinone and by inserting the oxygen scrubber the reduction product was homogenous. Furthermore, by installing a narrow-bore analytical column (2.0-mm i.d.) we achieved a 4.3- and 5.6-fold overall increase in sensitivity for both phylloquinone and phylloquinone epoxide, respectively. Selectively, this represented a 30% increase in reduction efficiency for phylloquinone epoxide by lowering velocity through the zinc reactor.

Chemicals and Standard Solutions

Phylloquinone epoxide is synthesized from phylloquinone using the method described by Tishler *et al.*[15] Standard solutions are prepared as previously described. The concentration of the phylloquinone epoxide secondary solution is calculated from absorbance spectra using the following absorptivity value ($E^{1\%}_{1\,\mathrm{cm}}$): phylloquinone epoxide, at 226 nm = 660 (absorptivities for phylloquinone and $K_{1(25)}$ are presented in the previous assay). The concentrations in the working calibration standard are phylloquinone epoxide, 10.0 ng/ml (21.4 nmoles/liter); phylloquinone, 20.0 ng/ml (44.4 nmoles/liter); and $K_{1(25)}$, 20.0 ng/ml (38.3 nmoles/liter). The working inter-

[14] W. A. MacCrehan, S. D. Yang, and B. A. Benner, *Anal. Chem.* **60,** 284 (1988).
[15] M. Tishler, L. F. Fieser, and N. L. Wendler, *J. Am. Chem. Soc.* **62,** 2866 (1940).

nal standard [a dilution of $K_{1(25)}$ secondary stock] is prepared at a concentration of 20.0 ng/ml (38.3 nmol/liter) of $K_{1(25)}$. Working standards are stored at 4° and are shielded from light. We found that a peak eluting at the same retention time as phylloquinone epoxide was the result of a fluorescent contaminant from the borosilicate glassware we were using. This was resolved by washing all glassware in a chromic-sulfuric acid solution and we have continued this practice for all phylloquinone epoxide analyses.

Liquid-Phase Extraction

Sample preparation consists of a liquid-phase extraction followed by solid-phase extraction (SPE) on silica gel and reversed-phase SPE on C_{18}. Plasma or serum (2.0 ml) is pipetted into a screw top, 20 × 125 borosilicate glass culture tube, followed by 50 μl (2.0 pmol) of the internal standard $K_{1(25)}$. Four milliliters of 100% ethanol is added and the tube is vortexed for 10 sec to precipitate the plasma proteins. This is followed by the addition of deionized H_2O (2.0 ml) and 12.0 ml of 100% hexane. The tubes are then capped with Teflon-lined screw caps. The samples are mixed for 2 min and centrifuged at 1000*g* (4°) for 5 min. The hexane layer (top) containing the extracted lipids and lipophilic compounds is aspirated and transferred to a clean, acid-washed 16 × 100 culture tube. The hexane extract is then evaporated to dryness under vacuum in a centrifugational evaporator (Savant Instruments).

Solid-Phase Extraction on Silica Gel

A 300-ml 500-mg SPE silica column (J. T. Baker) is preconditioned by successive washes of 8 ml of hexane–diethyl ether (97 : 3, v/v) and 8 ml of 100% hexane. The lipid residue is reconstituted in 1.0 ml of hexane and applied directly to the sorbent bed of the column. The adsorbed band is washed with 8 ml hexane and the sample eluted with 8 ml of hexane–diethyl ether (97 : 3, v/v). The eluant is collected into acid-washed tubes and evaporated to dryness.

Solid-Phase Extraction on C_{18}

The procedure for SPE on C_{18} follows that outlined in the preceding assay. For injection, the final residue is reconstituted initially in 10 μl of 100% dichloromethane (with swirling to dissolve the residue completely), immediately followed by 90 μl of methanol containing 10 m*M* zinc chloride, 10 m*M* acetic acid, and 5 m*M* sodium acetate (1 liter methanol : 5 ml aqueous solution). Fifty microliters of sample is injected into the HPLC.

Chromatographic Instrumentation

The configuration of the HPLC system is the same as that described previously in this chapter except when indicated. A model 231-401 sample injector (Gilson Medical Electronics) is fitted with a Rheodyne 7010 injection valve and 50-μl loop. This injector is preferred for the phylloquinone epoxide assay due to its lower dead volume. A C_{18} precolumn (30- × 2.1-mm i.d., Brown Lee Labs) is connected in series to the reversed-phase, narrow-bore analytical column (250- × 2.1-mm i.d.) packed with 5 μm BDS-Hypersil (Keystone Scientific). Fluorescence is monitored with the same detector as previously described (a Spectroflow model 980 fluorescence detector, Applied Biosystems), except that a 5-μl flow cell is installed. All high-pressure connections are made with short lengths of 0.005-inch-i.d. stainless steel tubing. Additionally, a stainless steel line (0.040-inch i.d.) is used instead of PTFE tubing for the connection of the mobile-phase delivery cap to the mobile-phase inlet on the HPLC pump to reduce the diffusion of oxygen into the eluant.

Chromatographic Conditions and Quantitation

The mobile-phase is comprised of dichloromethane–methanol (10:90, v/v), to each liter is added 5.0 ml of an aqueous solution containing 2 *M* zinc chloride, 2 *M* glacial acetic acid, and 1 *M* sodium acetate (final concentration: 10 m*M* $ZnCl_2$, 10 m*M* CH_3COOH, and 5 m*M* sodium acetate). However, the concentration of acetic acid in the aqueous solution for phylloquinone epoxide is higher (2×) than the concentration for the phylloquinone assay because we have found reduction of the epoxide is more stable at the higher concentration of acetic acid in the mobile phase. The aqueous solution is prepared and filtered using a 0.45-μm filter membrane (Millipore Corp., Bedford, MA). Once the mobile phase is prepared, it is degassed and under vacuum with sonication for 2 min. During analyses the mobile phase is continuously sparged with ultrahigh purity helium. To shorten the elution time of the internal standard $K_{1(25)}$ on the 250-mm column, the flow rate is changed during the run. The initial flow rate is 0.25 ml/min and is increased to 0.50 ml/min at 16 min. It is then returned to 0.25 ml/min at 29 min to equilibrate the column before the next injection (total run time 30 min). Excitation and emission are performed as previously stated for the phylloquinone assay.

A calibration standard is injected with every three samples in a run to compensate for changes in chromatographic conditions. The column is washed (no sample injected) after every three samples to reduce sample loading and carryover on the narrow-bore column. A standard curve is prepared (as previously described) with single-point calibration performed.

Quantitation is again achieved by direct comparison of peak area ratios generated from the sample internal standard and unknowns and the ratios generated from the calibration standard. Under the conditions outlined, average retention times for phylloquinone epoxide, phylloquinone, and the internal standard $K_{1(25)}$ are 9.0, 12.0 and 18.5 min, respectively. Phylloquinone epoxide at endogenous concentrations from 2.0-ml plasma extracts is well separated and baseline resolved. Figure 3 shows chromatograms of

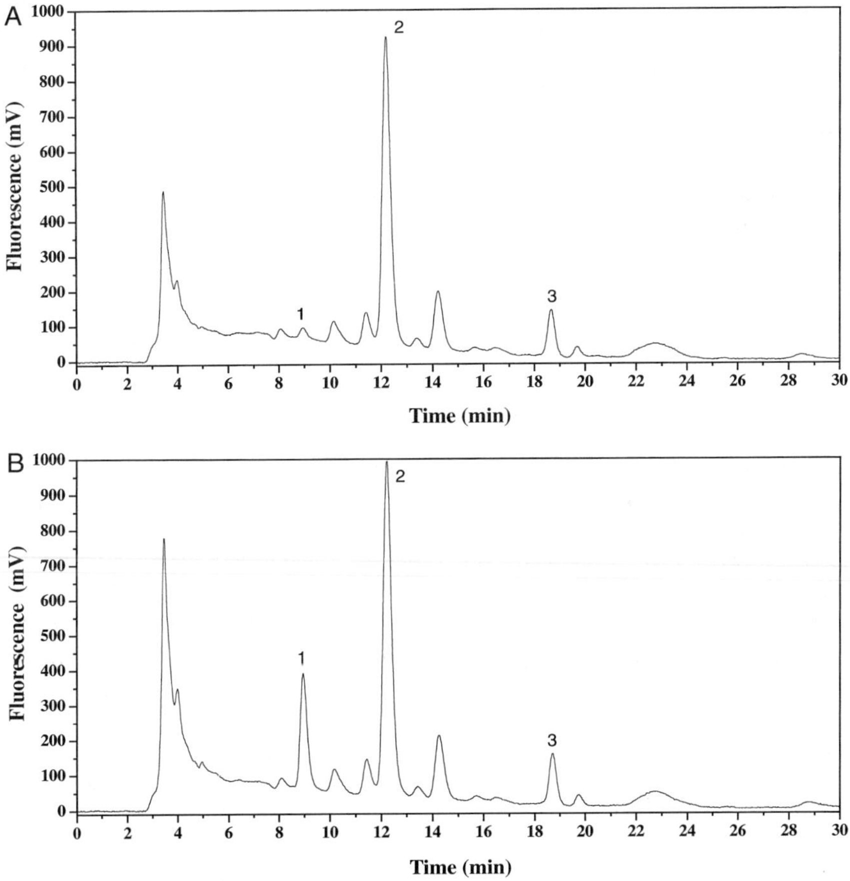

FIG. 3. Isocratic reversed-phase separation of phylloquinone epoxide and phylloquinone in plasma at (A) baseline and (B) 9 hr after administration of warfarin. Conditions: column, 5 μm BDS-Hypersil; mobile phase, methylene chloride-methanol (10:90, v/v) containing 10 m*M* zinc chloride, 10 m*M* acetic acid, and 5 m*M* sodium acetate; flow rate, 0.25 ml/min (0–16.0 min), 0.50 ml/min (16.0–29.0 min) and 0.25 ml/min (29.0–30.0 min). Detection, excitation 244 nm; emission 418. Peak identities; 1, phylloquinone epoxide; 2, phylloquinone; 3, $K_{1(25)}$.

plasma extracts from an individual before and after the administration of warfarin. Under the antagonism of warfarin, phylloquinone epoxide concentrations can increase precipitously. By 9 hr after administration of warfarin the plasma concentration of phylloquinone epoxide had increased to greater than eightfold that of baseline concentrations.

Detection Limits and Precision

The detection limits for the assay under the conditions described are 5.0 pg/injection (10.7 fmol/injection) for phylloquinone epoxide and 4.0 pg/injection (8.9 fmol/injection) for phylloquinone standard solutions. In plasma the detection limits are 10.0 pg/ml (21.4 pmol/liter) and 8.0 pg/ml (17.8 pmol/liter) for phylloquinone epoxide and phylloquinone, respectively. The within-run coefficient of variation (*CV*) for replicates of pooled plasma ($n = 12$) is 8.8% for phylloquinone epoxide and 6.3% for phylloquinone. Mean (± SD) recoveries of the internal standard are approximately 81.0 ± 10.0%.

Application of Assay

This assay was applied to the determination of fasting phylloquinone epoxide and phylloquinone concentrations in a population of 255 healthy volunteers (age range of 19–78 years) to establish a reference range for the epoxide in plasma (Table I). The mean epoxide concentrations were 6.6% that of phylloquinone. We found a strong relationship between phylloquinone epoxide and phylloquinone ($r = 0.66$), with the correlation being

TABLE I
FASTING PLASMA PHYLLOQUINONE EPOXIDE AND PHYLLOQUINONE CONCENTRATIONS IN HEALTHY ADULTS

	Concentration (nmol/liter)		
Vitamin	All ($n = 255$)	Males ($n = 76$)	Females ($n = 179$)
K_1 epoxide	0.078 ± 0.089[a]	0.088 ± 0.102	0.073 ± 0.006
(Range)	(0.004 – 0.682)	(0.004 – 0.679)	(0.004 – 0.682)
Phylloquinone	1.189 ± 0.809	1.297 ± 0.913	1.142 ± 0.057
(Range)	(0.201 – 4.919)	(0.282 – 4.919)	(0.201 – 4.762)
K_1 epoxide/Phylloquinone	0.065 ± 0.049	0.067 ± 0.069	0.063 ± 0.003
(Range)	(0.008 – 0.561)	(0.008 – 0.561)	(0.009 – 0.220)
r	0.66	0.62	0.69

[a] Concentrations represent mean ± SD.

slightly higher in the females ($r = 0.69$) than in the males ($r = 0.62$). There were no statistical differences with respect to age and gender for either phylloquinone epoxide or phylloquinone concentrations, or the phylloquinone epoxide/phylloquinone ratio.

Acknowledgments

The author greatly appreciates the invaluable technical and editorial assistance of Dr. S. L. Booth and M. E. O'Brien-Morse in the preparation of this manuscript.

[35] Assay of Phylloquinone in Plasma by High-Performance Liquid Chromatography with Electrochemical Detection

By P. T. McCarthy, D. J. Harrington, and M. J. Shearer

Introduction

Phylloquinone (vitamin K_1; 2-methyl-3-phytyl-1,4-naphthoquinone) is the most abundant of the K vitamins in normal human plasma. Discernible amounts of menaquinones (MK) 7 and 8, have also been reported by Hodges *et al.*[1] and their detection is described elsewhere in this volume.[1a]

The assay of phylloquinone in biological fluids and tissues by high-performance liquid chromatography (HPLC) remains technically challenging because of the very low concentration of the vitamin usually present. For a more thorough coverage of the principles and techniques used for the isolation, purification, and chromatography of the K vitamins than is possible within the scope of this chapter, the reader is referred to other reviews.[2,3]

Although the fundamental methodology for the extraction and further purification of the K vitamins has remained largely unchanged since the subject was last reviewed by one of us in this series,[4] there have been

[1] S. J. Hodges, M. J. Pilkington, M. J. Shearer, L. Bitensky, and J. Chayen, *Clin. Sci.* **78,** 63 (1990).

[1a] S. J. Hodges, *Methods Enzymol.* **282,** [36] 1997 (this volume).

[2] M. J. Shearer, *Adv. Chromatogr.* **21,** 243 (1983).

[3] W. E. Lambert and A. P. De Leenheer, *in* "Modern Chromatographic Analysis of Vitamins" (A. P. De Leenheer, W. E. Lambert, and H. J. Nelis, eds.), p. 197. Marcel Dekker, New York, 1992.

[4] M. J. Shearer, *Methods Enzymol.* **123,** 235 (1986).

0076-6879/97 $25.00

several significant advances in the detection of vitamin K by HPLC using electrochemical or fluorescence techniques.[3] Despite the fact that these improvements have provided enhanced sensitivity and selectivity for the final chromatographic stage, most methods for the assay of phylloquinone in plasma still involve a fairly laborious multistage procedure. Typical procedures usually encompass a lipid extraction stage, a preliminary HPLC "semipreparative" stage using normal-phase (adsorption) chromatography, and a quantitative HPLC stage using reversed-phase (partition) chromatography.[2,3] The purpose of this chromatographic sequence is to first resolve the K vitamins as a class from other lipid classes using normal-phase HPLC and then to resolve the vitamin K isoprenologs, both from each other and from any remaining interferences, by reversed-phase HPLC.[2,3]

The extraction of vitamin K from plasma is usually achieved by first denaturing the vitamin K–lipoprotein complex with a polar solvent (e.g., ethanol, methanol, or 2-propanol) followed by extraction into a nonpolar solvent (e.g., hexane). Solid-phase extraction (SPE), a more contemporary technique for sample preparation that has been widely adopted in many other fields of clinical analysis, has only occasionally been adopted for the direct extraction of phylloquinone from plasma[5] although it has been used extensively for the purification of vitamin K-rich fractions from solvent extracts.[3,4,6–9]

Three physicochemical techniques have been used to detect vitamin K in HPLC analyses, namely, UV spectrophotometry, electrochemistry, and spectrofluorimetry. The K vitamins exhibit UV absorbance with maxima in the range of 240–250 and 260–270 nm ($\varepsilon = 18.9\ \mathrm{cm}^2/\mathrm{m}M$ at 248 nm). UV spectrophotometry, once the only means of detection available for the HPLC analysis of phylloquinone,[10–12] is now considered to be both too insensitive and too unselective for routine measurements of endogenous levels of phylloquinone in plasma. However, UV spectrophotometric detection is still used to monitor the collection of fractions in methods that include semipreparative HPLC of vitamin K-rich sample

[5] E. M. Kirk and A. F. Fell, *Clin. Chem.* **35,** 1288 (1989).

[6] Y. Haroon and P. V. Hauschka, *J. Lipid Res.* **24,** 481 (1983).

[7] P. M. M. van Haard, R. Engel, and A. L. J. M. Pietersma-de Bruyn, *Clin. Chim. Acta.* **157,** 221 (1986).

[8] M. J. Shearer, *in* "HPLC of Small Molecules—A Practical Approach" (C. K. Lim, ed.), p. 157, IRL Press Oxford, 1986.

[9] K. Hirauchi, T. Sakano, and A. Morimoto, *Chem. Pharm. Bull.* **34,** 845 (1986).

[10] M. F. Lefevere, A. P. De Leenheer, and A. E. Claeys, *J. Chromatogr.* **186,** 749 (1979).

[11] M. J. Shearer, S. Rahim, P. Barkhan, and L. Stimmler, *Lancet* **ii,** 460 (1982).

[12] M. Leclercq, M. Crozet, J. Durand, M. Bourgeay-Causse, and L. Sann, *in* "Chromatography in Biochemistry, Medicine and Environmental Research" (A. Frigerio, ed.), p. 235. Elsevier, Amsterdam, 1983.

extracts prior to quantitative HPLC. In the early 1980s, HPLC with UV detection was superceded briefly by amperometric electrochemical detection (ECD) in the reductive mode, which measured the cathodic current generated by the reduction of phylloquinone to phylloquinone hydroquinone.[13,14] Reductive ECD was found to be about 3-fold more sensitive than UV detection but was in turn abandoned because of the practical problems associated with electrode passivation and the need to remove oxygen completely from the mobile phase to eliminate the high background currents and baseline drift. The introduction of a dual-electrode "coulometric" detector, which enabled the development of ECD based on redox-mode electrochemical detection of phylloquinone[15,16] was responsible for a further 10-fold increase in sensitivity (detection limit 20–50 pg/ml serum) and provided enhanced selectivity and stability compared to the reductive method. In parallel with redox-mode ECD, other methods with similar sensitivity and selectivity were being developed using spectrofluorimetry and have since been widely adopted.[3] HPLC with fluorescence detection of the K vitamins exploits the highly fluorescent nature of vitamin K hydroquinone produced either by postcolumn electrochemical[7,17–19] or chemical reduction. A number of chemical methods for the reduction of vitamin K have been employed including use of sodium borohydrate ($NaBH_4$),[20,21] tetramethylammonium octahydridotriborate [$(CH_3)_4NB_3H_8$][3] contained either in a postcolumn reaction tube[22] or incorporated into the mobile phase,[23] zinc metal in the presence of zinc ions,[24,25] and platinum catalysts.[26,27] The measurement of phylloquinone by HPLC with fluorescence detection is covered by Davidson and Sadowski elsewhere in this volume.[27a]

[13] T. Ueno and J. W. Suttie, *Anal. Biochem.* **133,** 62 (1983).
[14] J. P. Hart, M. J. Shearer, P. T. McCarthy, and S. Rahim, *Analyst* **109,** 477 (1984).
[15] Y. Haroon, C. A. W. Schubert, and P. V. Hauschka, *J. Chromatogr. Sci* **22,** 89 (1984).
[16] J. P. Hart, M. J. Shearer, and P. T. McCarthy, *Analyst* **110,** 1181 (1985).
[17] J. P. Langenberg and U. R. Tjaden, *J. Chromatogr.* **305,** 61 (1984).
[18] L. L. Mummah-Schendel and J. W. Suttie, *Am. J. Clin. Nutr.* **44,** 686 (1986).
[19] M. Guillaumont, M. Leclercq, H. Gosselet, K. Makala, and B. Vignal, *J. Micronutrient Anal.* **4,** 285 (1988).
[20] M. F. Lefevere, R. W. Frei, A. H. M. T. Scholten, and U. A. T. Brinkman, *Chromatographia* **15,** 459 (1982).
[21] B. E. Cham, H. P. Roeser, and T. W. Kamst, *Clin. Chem.* **35,** 2285 (1989).
[22] W. E. Lambert, A. P. De Leenheer, and E. J. Baert, *Anal. Biochem.* **158,** 257 (1986).
[23] W. E. Lambert and A. P. De Leenheer, *Analytica Chimica Acta,* **196,** 247 (1987).
[24] Y. Haroon, D. S. Bacon, and J. A. Sadowski, *Clin. Chem.* **32,** 1925 (1986).
[25] Y. Haroon, D. S. Bacon, and J. A. Sadowski, *J. Chromatogr.* **384,** 383 (1987).
[26] M. Shino, *Analyst* **113,** 393 (1988).
[27] H. Hiraike, M. Kimura, and Y. Itokawa, *J. Chromatogr.* **430,** 143 (1988).
[27a] K. W. Davidson and J. A. Sadowski, *Methods Enzymol.* **282,** [34], 1997 (this volume).

In our laboratory we use a modification of the dual-electrode, redox-mode, coulometric, electrochemical method[28] that was first developed more than a decade ago.[15,16] Redox-mode ECD takes advantage of the reversible nature of the interconversion of phylloquinone and phylloquinone hydroquinone. Phylloquinone is first reduced at a negative potential using a porous graphite flow-through electrode and then reoxidized at a positive potential by a second electrode situated immediately downstream. Interference from oxygen at the detector electrode is reduced because the reduction reaction for this molecule is not reversible at the potentials applied. The principal disadvantages of the method are that adsorption of electrochemical species and metal ions leached from the metal parts of the HPLC column and pump by the mobile phase can significantly reduce the conversion efficiency and hence reduce the assay sensitivity below that required for the measurement of phylloquinone at endogenous concentrations. To improve the degree of electrochemical reduction in day-to-day use, we have adapted our method to use both electrodes of the coulometric cell for the reduction of phylloquinone. An amperometric, "wall jet" electrode positioned downstream and in series with the coulometric cell is now used for the reoxidation of the hydroquinone and its use has been found to increase greatly the assay sensitivity and stability. With this procedure, we have been able to apply this method to the routine measurement of phylloquinone in small volumes (0.2–0.5 ml) of plasma. Details of the extraction of phylloquinone from plasma into hexane and the subsequent preliminary purification of lipid extracts have been described previously.[8,28] In the method described below for the measurement of plasma phylloquinone at endogenous concentrations, the main difference from our previous method of sample preparation[8,28] before subsequent assay by HPLC and ECD is that the initial SPE stage of lipid purification using Sep-Pak silica cartridges is omitted and the lipid extracts are purified directly by semipreparative HPLC. We also describe a simplified method for analysis of phylloquinone at high concentrations (e.g., from high-dose pharmacokinetic and toxicokinetic studies). In this method, solvent extracts are injected directly onto the analytical HPLC column without use of preparative HPLC. In addition details are given of an external quality assurance scheme that has been initiated by our laboratory for the evaluation of interlaboratory performance.

General Precautions

The following precautions have been universally adopted in the assay of phylloquinone and are essential to avoid spurious or inconsistent results.

[28] M. J. Shearer, *in* "Nutritional Status Assessment: A Manual for Population Studies" (F. Fidanza, ed.), p. 214. Chapman and Hall, London, 1991.

Stability

K vitamins undergo photodegradation and are sensitive to alkaline conditions but are quite stable toward heat, oxygen, and acidic conditions. Care should be taken to remove completely all traces of detergent from glassware and to protect samples and ethanolic solutions from strong light. Solvents can be removed under nitrogen at temperatures up to 60°.

Storage

Reports of the stability of phylloquinone in ethanolic solution, dry lipid extracts in isolated fractions at −20° and in serum/plasma at −70° are largely anecdotal but it is generally held that the K vitamers are stable for many months if stored in the dark at these temperatures.

Solvents

HPLC-grade solvents (hexane, methanol, and dichloromethane) should be used in all chromatographic procedures (e.g., Rathburns, Scotland, UK). High-purity deionized water should be used in the preparation of mobile-phase constituents. This should be "polished" by passage through a C_{18} modified silica Sep-Pak-type disposable SPE cartridge preconditioned with methanol as described by the manufacturer. HPLC mobile phases should be filtered under reduced pressure to remove particulates and dissolved gases.

Contamination

Avoid contamination of samples from extraneous sources such as rubber fittings and lubricants. Glassware should be rinsed with acetone or ethanol before use and the sample injection loop of syringe loading injection valves flushed with mobile phase between injections. Syringes for sample and standards injection should not be interchanged.

Extraction of Phylloquinone from Plasma

Phylloquinone is extracted into hexane after flocculation of proteins with ethanol. An internal standard, menaquinone-6 (MK-6, 2-methyl-3-farnesylfarnesyl-1,4-naphthoquinone; gift of Hoffmann La-Roche, Basel, Switzerland), is used to compensate for loss of phylloquinone during the extraction process and subsequent chromatographic procedures.

Phylloquinone at Pharmacological Concentrations

For the assay of phylloquinone at high concentrations (e.g., in samples collected after pharmacological dosing), transfer 0.05–0.10 ml plasma/

serum or quality control sample to a 12-ml stoppered polypropylene test tube. Add 0.4 ml water, 0.8 ml ethanol, 0.2 ml internal standard solution (accurately known concentration of MK-6, ~37.5 μg/liter in ethanol), and 3.0 ml hexane. Vortex-mix (2 min) and centrifuge (1500 *g*, 10 min). Transfer the upper hexane layer containing phylloquinone and the internal standard to a 10-ml borosilicate glass, stoppered, tapered test tube and evaporate the solvent to dryness under nitrogen at 60°. Because this method requires only a 0.05- to 0.10-ml sample, it can be used to "screen" samples where the expected concentration of the vitamin is not known.

Phylloquinone at Normal Endogenous Concentrations

For the measurement of phylloquinone at normal endogenous concentrations, transfer 0.5 ml plasma/serum or quality control sample to a 30-ml borosilicate glass, stoppered, centrifuge tube. Add 0.2 ml internal standard solution (accurately known concentration of MK-6, ~3.75 μg/liter in ethanol), 3.8 ml ethanol, and 12.0 ml hexane. Vortex-mix, centrifuge, and transfer the vitamin K extract as described above and evaporate the solvent to dryness.

High-Performance Liquid Chromatography of Phylloquinone

Equipment for High-Performance Liquid Chromatography

We use a dedicated HPLC system for each stage of the assay. The system used for semipreparative HPLC of solvent extracts composes a standard HPLC pump delivering isocratic flow, variable-wavelength UV photometer, Rheodyne Model 7125 syringe loading injection valve, and a single-channel *xy* analog chart recorder. The analytical system comprises a pulse-free pump, a Coulochem model 5100A controller fitted with a model 5011 dual-electrode coulometric analytical cell (Environmental Science Associates, Chelmsford, MA) for reduction of vitamin K and an Antec VT-03 wall-jet amperometric electrochemical cell and CU-04-AZ controller (Antec, Leyden, The Netherlands) for reoxidation of the hydroquinone. Alternatively, Antec markets a more sophisticated version of this amperometric detector called the Decade (digital electrochemical amperometric detector), which is able to control electronically both Coulochem and Antec electrochemical cells. In the Decade both the coulometric and amperometric electrochemical cells are closely juxtaposed and together with the column are enclosed within an integrated unit that functions as both Faraday shield and oven for precise temperature control. The Decade can also be operated in the pulse mode, which enables the wall-jet electrode to be cleaned *in situ* without the cell being dismantled. A periodic procedure by which the potential is set to oscillate in the pulse mode between −1.0 and +1.0 V has been shown to remove adsorbed species and restore the original sensitivity of the wall-jet electrode. For both systems, an on-line vacuum degassing unit for the removal of oxygen from the mobile phase is essential to maximize detector sensitivity (e.g., Jour X-ACT, Jour Research, Onsala, Sweden). Although a syringe loading injection valve and *xy* analog recorder can be used, an automatic sample injector is useful for unattended injection of samples and a computerized chromatography manager is optimal for collection and processing of chromatographic data, especially from bioanalytical studies conducted to the standards demanded by Good Laboratory Practice (GLP), for example, Waters Millennium 2010 (Milford, MA) or Gynkosoft version 5.32E (Gynkotek, Munich, Germany) chromatography managers. The lower limit of the working range of the analog-to-digital signal converter for these chromatography managers is −0.25 and −1.50 V, respectively, which is ideal for the detection of the reaction signal during baseline drift at the limits of detection sensitivity frequently required.

Semipreparative High-Performance Liquid Chromatography

In the assay of phylloquinone at physiological circulating concentrations, a preliminary semipreparative chromatographic step is used to prepare a vitamin K-rich fraction before quantitative HPLC. Phylloquinone and MK-6 are separated on a Spherisorb (5-μm particle size), cyanopropyl-modified silica HPLC column using a mobile phase containing dichloromethane (50% water saturated) and hexane at a flow rate of 1.0 ml/min. To minimize variability in the retention time of the analytes due to the adventitious hydration of silanol sites on the silica column by water, dichloromethane presaturated with water to a standard composition is used in the preparation of the mobile phase. For the assay of the vitamin at higher than endogenous concentrations, this stage is omitted and samples are injected directly onto the reversed-phase analytical HPLC column.

Column:	Stainless steel containing Spherisorb CN nitrile, 5-μm particle size (Phase Separations, Deeside, UK), dimensions 250- × 5-mm i.d. This is protected by an on-line precolumn filter (2-μm frit size) and a disposable guard column cartridge (25- × 5-mm i.d.) containing the same packing material. The guard column cartridge is replaced every 50–100 injections.
Mobile phase:	3–6% dichloromethane (50% water-saturated) in hexane

Flow rate:	1.0 ml/min
Detection:	UV 270 nm
Sample injection:	70 μl in mobile phase (100-μl loop)
Retention time:	6–10 min (phylloquinone + MK-6, internal standard)

Adjust the composition of dichloromethane in the mobile phase so that the retention of a solution of phylloquinone and MK-6 is constant with the retention of phylloquinone in the range of 6–10 min.

Preparation of 50% Water-Saturated Dichloromethane

To prepare this solution, dichloromethane stored under water is transferred to a separating funnel and the dichloromethane poured sequentially into three clean, dry glass flasks (100% water-saturated solution). Equal volumes of this solution and dry bottled dichloromethane are then mixed to produce a "50% water-saturated" solution, which is used in the preparation of the mobile phase.

Collection of Phylloquinone Fraction

Identify the eluate fraction containing phylloquinone and MK-6 by reference to the retention time of the analytes in a standard solution injected directly on-column [5 μl hexane solution, (2 mg/liter)]. Thoroughly flush the injection valve loop with hexane, dissolve the dried solvent extract in 80–100 μl mobile phase, vortex-mix, and inject about 70 μl onto the column. Collect the smallest fraction possible containing phylloquinone and the internal standard into a clean, stoppered, tapered 12-ml borosilicate glass test tube and evaporate to dryness under nitrogen at 60°. If the fraction collected is too large it often contains extraneous lipids that are difficult to dissolve in alcoholic solvents used in analytical HPLC. Conversely if the fraction is too small, the analytes may be missed. After collecting the fraction, increase the mobile phase flow rate to 2.0–2.5 ml/min to elute more polar interfering lipids from the column (~30 min) before making the next injection.

Analytical High-Performance Liquid Chromatography

For the analysis of phylloquinone at normal circulating concentrations, redissolve the vitamin K fraction isolated by semipreparative HPLC in 50 μl ethanol or mobile phase, warm the test tube, vortex-mix, and inject 10–20 μl onto the analytical HPLC column. For the analysis of the vitamin in plasma samples at pharmacological concentrations, dissolve the dry solvent extract in 100 μl solvent, and inject 10 μl onto the HPLC column.

Column:	PEEK tubing (Jour No-Met) containing Exsil octyl (C_8), 5-μm particle size (Hichrom, Reading, UK), dimensions (250- × 5-mm i.d.)
Mobile phase:	3–5% (v/v) acetate buffer (pH 3.0, final concentration 0.05 *M*) in methanol
Flow rate:	1.0 ml/min
Detection:	ESA series dual-electrode coulometric analytical cell with both electrodes set at −1.2 to −1.6 V (−1.2-V Decade controller) and Antec VT-03 amperometric electrode at +0.15 to +0.20 V (Decade + 0.40 V)
Sample injection:	10–20 μl (ethanol)
Retention time:	~9 min (phylloquinone) and 11 min (MK-6, internal standard)

Although the method we describe is based on the packing material Exsil octyl (manufactured by Exmere, Cheshire, UK), other octyl-bonded phases such as Spherisorb octyl (Phase Separations) give equally satisfactory performance.[8,28]

To prepare the HPLC mobile phase, first prepare aqueous stock solutions of sodium acetate (5 *M*), acetic acid (5 *M*), and EDTA (ethylenediaminetetraacetic acid), disodium salt, dihydrate (0.1 *M*) (all AristaR grade). Prepare an acetate buffer (5 *M*, pH 3.0) by dropwise addition of sodium acetate solution to acetic acid. To prepare 1 liter of mobile phase (3%, 0.05 *M*), mix 10 ml acetate buffer, 19 ml "polished" deionized water, 1 ml EDTA, and add methanol to volume.

Standard Solutions

Prepare stock solutions of phylloquinone and MK-6 in ethanol (25 mg/liter). Determine the concentration accurately by UV spectrophotometry (ε cm^2/m*M* = 18.9 at 248 nm). Dilute these solutions to prepare ethanolic solutions (250 μg/liter) for the subsequent derivation of internal standard solutions and calibration standards. From this solution, prepare two internal standard solutions with accurately known concentrations of around 4.0 and 40 μg/liter for the assay of phylloquinone at endogenous and higher concentrations, respectively.

Calibration

From standard solutions of phylloquinone and MK-6 (250 μg/liter) prepare a series of ethanolic standards containing both analytes with mass ratios of phylloquinone/MK-6 ranging from 0.1 to 4.0 and containing final concentrations of phylloquinone ranging from 4 to 200 μg/liter. Inject 10 μl of each solution directly onto the HPLC and construct a calibration line of the ratio of peak heights of phylloquinone and MK-6 relative to their mass ratio. Use of ethanolic calibration solutions allows determination of phylloquinone over a wide concentration range with the upper limit of detection governed by the ratio of the two analytes in the upper calibration solution. For the two assays at physiological or pharmacological plasma concentrations, the upper limits of detection are usually about 5.0 and 500 μg/liter, respectively. We have also used the assay for the measurement of the vitamin in dog plasma from toxicological studies containing up to 5 mg/liter phylloquinone by adding a more concentrated internal standard solution (350 μg/liter MK-6) to the sample before extraction.

Resolution and Detection of Phylloquinone

The identity of phylloquinone and the internal standard are checked by reference to standards injected concomitantly. Representative chromatograms of an extract of 0.05 ml human plasma containing 150 μg/liter phylloquinone, an ethanolic standard containing phylloquinone and MK-6 (mass ratio 0.36), and plasma from a normal subject who had not received the vitamin are shown in Fig. 1. The sample was assayed by the modified method without need for preparative HPLC. The lower limit of detection of phylloquinone is usually in the range of 2.0–5.0 μg/liter. We have obtained intraassay precision values (*RSD* %) of 4.1, 1.6, and 2.1% for the measurement of phylloquinone in serum containing 14.2 ± 0.6, 61.8 ± 1.0, and 149.7 ± 4.4 μg/liter, respectively ($n = 10$). In comparison, the intraassay precision values ($n = 90$) at these concentrations were 16.7, 8.6, and 9.5%, respectively. A representative chromatogram of 0.5 ml serum extract prepared by our standard procedure for samples containing the vitamin at endogenous concentration is shown in Fig. 2. The lower limit of detection following analysis of 0.5 ml sample by this method is in the range of 0.05–0.10 μg/liter. In our laboratory, we have obtained intraassay precision values for the measurement of the vitamin in quality control plasma samples containing 0.30 ± 0.02, 0.76 ± 0.05, and 3.66 ± 0.12 μg/liter of 5.9, 7.2, and 3.3%, respectively ($n = 6$). The interassay precision values based on repetitive analysis were 11.3 ($n = 26$), 10.2 ($n = 26$), and 9.8% ($n = 38$), respectively.

Interferences from substances coextracted and cochromatographed with either phylloquinone or MK-6 are unusual and when apparent can normally be eliminated by reducing the voltage applied to the oxidation electrode although usually at the expense of a corresponding reduction in detector sensitivity.

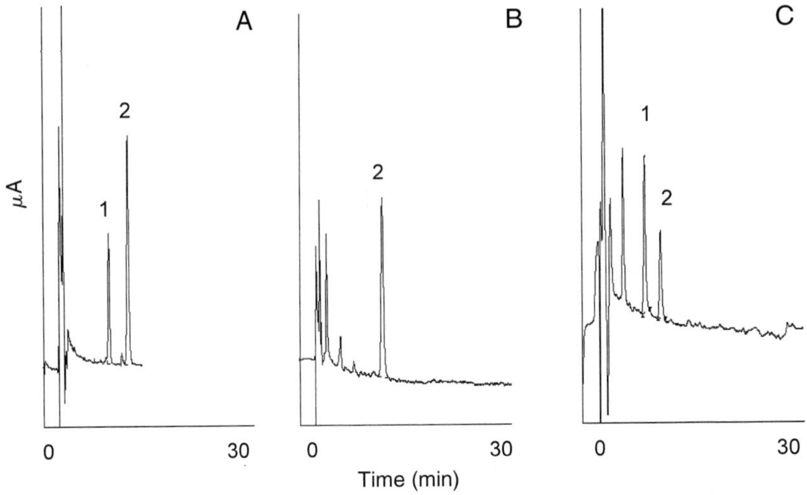

FIG. 1. Representative chromatograms illustrating the assay of phylloquinone in plasma at pharmacological concentrations. (A) Ethanolic standard containing phylloquinone (peak 1) and the internal standard (MK-6, peak 2) at a mass ratio (K_1/MK-6) of 0.36. (B) Extract of 0.05 ml plasma from subject not given phylloquinone. (C) Extract of 0.05 ml plasma from subject given 20 mg phylloquinone (Konakion) orally at a measured phylloquinone plasma concentration of 150 μg/liter. Chromatograms (B) and (C) were obtained following the simplified procedure for the assay of phylloquinone at pharmacological concentrations. Detection: reduction potential −1.6 V, oxidation potential +0.20 V; range 0.5 nA/V. Mobile phase: 3% acetate buffer (pH 3.0) in methanol at a flow rate of 1.0 ml/min.

Calculation

From the chromatograms recorded, calculate the ratio of the peak heights of phylloquinone and internal standard in the sample and use the calibration line to determine the equivalent mass ratio. Calculate the total amount of phylloquinone present by multiplying the calculated mass ratio by the amount of internal standard added and divide by the sample volume to calculate the concentration.

Quality Control and Quality Assurance

There is an increasing demand for quality assurance especially for compliance with laboratory accreditation schemes and GLP. Intra- and interassay performance can be assessed by repetitive assay of serum/plasma samples collected from volunteer donors. Quality control samples for the assay of phylloquinone at high concentrations can be derived from blood samples collected from volunteers who have ingested the vitamin (e.g.,

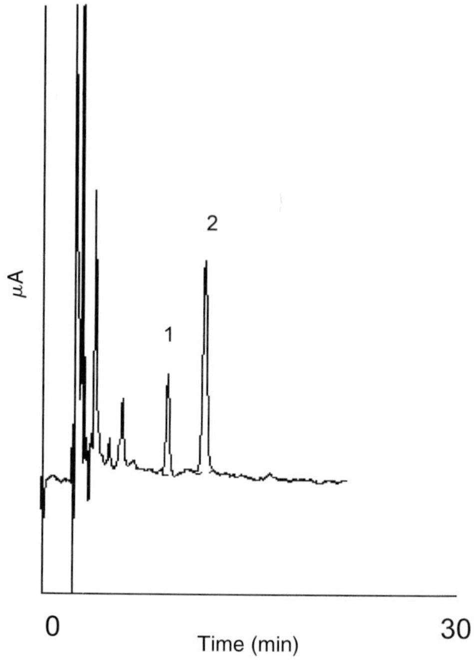

FIG. 2. Representative chromatogram illustrating the assay of phylloquinone at endogenous concentrations. The chromatogram was obtained after extraction and purification of 0.5 ml of plasma by the standard procedure for the assay at normal endogenous concentrations and shows peaks for phylloquinone (peak 1) and the internal standard (MK-6, peak 2). The measured phylloquinone concentration in this sample was 0.5 μg/liter phylloquinone. Detection: reduction potential −1.6 V, oxidation potential +0.2 V; range 0.2 nA/V. Mobile phase: 3% acetate buffer (pH 3.0) in methanol at a flow rate of 1.0 ml/min.

Konakion, Roche). Since the assay of phylloquinone in serum/plasma is largely performed in specialized laboratories as a research tool and has not yet been widely adopted elsewhere, there are no commercially available reference standards or quality assurance schemes by which laboratories can assess their performance. The St. Thomas' Hospital Haemophilia Centre Quality Assurance Scheme is an international pilot project intended to promote interlaboratory uniformity for phylloquinone assessment. The scheme operates by providing at quarterly intervals ethanolic solutions of phylloquinone and samples of human plasma/serum, screened for the absence of human immunodeficiency virus (HIV) 1 and 2 antibodies and hepatitis B and C surface antigen before conveyance. The concentration of the vitamin in each sample is determined by interlaboratory consensus and laboratories out-of-step with their contemporaries may thus be identified. Details of membership in this scheme can be obtained from the authors.

Choice of Internal Standard

It has been difficult to find a suitable internal standard for the measurement of phylloquinone in plasma using the protocol described earlier. An ideal compound should be structurally analogous, absent from the sample matrix, chemically stable, and recovered to the same degree during sample treatment. Furthermore, it should coelute in any preliminary chromatographic step but be separated from the analytes of interest at the analytical stage. In some research studies, we have chosen not to use an internal standard[14,16] but this requires great care by a skilful analyst and is clearly not satisfactory for a multistage assay carried out routinely where the risk of sample losses is compounded at each stage. Phylloquinone epoxide has been used in assays using UV detection[4,11] but it is a natural metabolite and can be readily detected in plasma during oral anticoagulant treatment by some fluorometric assays.[24] However, with the mobile phase and potentials used in our assay, vitamin K epoxide is not reduced and is not detectable.

Some older methods have advocated *trans*-[^{3}H]phylloquinone.[10,13,18] In common with some other laboratories,[19,24] we have used 2′,3′-dihydrophylloquinone but usually for the assay of phylloquinone in liver and other tissues. It is our experience that solutions of this compound tend to be unstable, even at −20°. It has been shown that dihydrophylloquinone is present in foods containing hydrogenated oils[29] and may therefore be present in plasma samples. The internal standard MK-6 used in our current procedures has also been adopted by others[3,7,17] but as a naturally occurring isoprenoloque is also not ideal although our own studies suggest that MK-6 is normally unmeasurable in plasma.[1] A structural analog of phylloquinone with an extrasaturated prenyl unit [$K_{1(25)}$] has been used as an internal standard by Lambert *et al.*[22,23] This compound had the advantage of coeluting with phylloquinone in the semipreparative normal-phase HPLC system, while being well resolved in the analytical reversed-phase system[22] and is not known to occur in nature. None of the above-mentioned internal standards are commercially available and laboratories will usually be dependent on gifts for their supplies.

Acknowledgments

We would like to thank our colleagues, too numerous to mention individually, who have made valuable contributions to the development of the assay, in particular Dr. Alison Cox who also assisted in the preparation of the artwork.

[29] S. L. Booth, K. W. Davidson, and J. A. Sadowski, *J. Agric. Food Chem.* **42,** 295 (1994).

[36] Assay of Menaquinones in Plasma Utilizing Dual-Electrode Electrochemical Detection

By STEPHEN J. HODGES

Introduction

Several chromatographic systems based on high-performance liquid chromatography (HPLC) can be used for the separation of the different forms of vitamin K. These methods are discussed in greater detail elsewhere. Analytical methods for the detection and quantification of menaquinones that can be readily established in the general laboratory are based on either fluorescence or electrochemical detection. Both methods are capable of similar degrees of sensitivity and both have inherent strengths and limitations. Fluorescence detection, which relies on the reduction of the naphthoquinone ring of the menaquinones, can be achieved by electrochemical methods. Therefore, the fluorescence detection systems can have some degree of dependence on, and compatibility with, electrochemical detection methods. This chapter is concerned solely with the discussion of dual-electrode electrochemical detection of the menaquinones. These methods have been successfully used for the detection and quantification of both phylloquinone and the spectrum menaquinones in plasma.[1–4]

Properties of Detection Cell

Having established the chromatographic procedures for the initial purification and final separation of the various forms of the menaquinones, electrochemical detection of the small amounts of the menaquinone congeners found in plasma can be readily achieved using dual-electrode electrochemical detection (Fig. 1). A number of different commercial systems are available for the electrochemical detection of quinone-containing compounds. However, this brief review discusses only the ESA Coulochem detector (Chelmsford, MA) and the Antec Intro electrochemical detector

[1] J. P. Hart, M. J. Shearer, L. Klenerman, A. Catterall, J. Reeve, P. N. Sambrook, R. A. Dodds, L. Bitensky, and J. Chayen, *J. Clin. Endocr. Metab.* **60,** 1268 (1985).

[2] S. J. Hodges, M. J. Pilkington, M. J. Shearer, L. Bitensky, and J. Chayen, *Clin. Sci.* **78,** 63 (1990).

[3] S. J. Hodges, M. J. Pilkington, T. C. B. Stamp, A. Catterall, M. J. Shearer, L. Bitensky, and J. Chayen, *Bone* **12,** 387 (1991).

[4] S. J. Hodges, P. Vergnaud, K. Akesson, K. Obrant, and P. D. Delmas, *J. Bone Miner. Res.* **8,** 1241 (1993).

0076-6879/97 $25.00

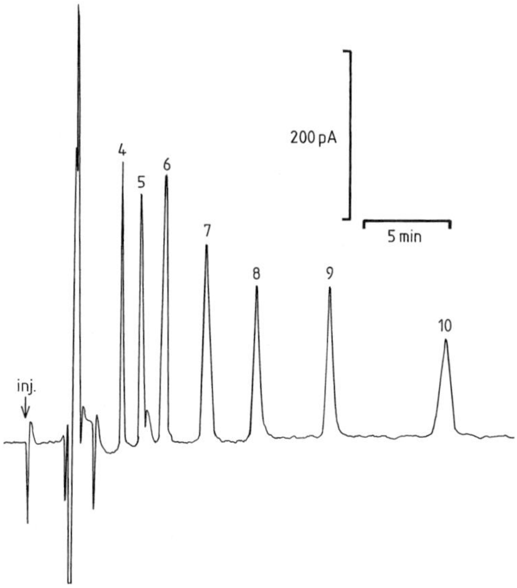

FIG. 1. Reverse-phase HPLC separation and electrochemical detection of authentic menaquinones 4–10.

(Leyden, The Netherlands). This selection is based solely on personal experience.

It is useful to outline briefly some practical features of these systems. The ESA instrument is used to regulate a separate analytical electrochemical cell. This twin, in series, electrode arrangement of the ESA cell allows for independent potential difference settings across the working electrodes and reference electrodes of both upstream and downstream electrode components. Because the reference electrodes in the ESA cell are made of palladium these electrodes act as hydrogen cells and the potential differences required for oxidation reactions are typically lower than those used in conventional amperometric cells. Conversely, for reduction, Coulochem cell potentials are higher than conventional amperometric cells. Additionally, a setting of zero on the Coulochem instrument is equivalent to a +200- to +300-mV setting on a conventional amperometric setting. The only serious disadvantage of the Coulochem system is the cost of electrochemical cell replacement. The electrochemical cell can be regenerated with an appropriate washing program; however, it must still be regarded, ultimately, as a consumable item. Although it is a significant added expense, the Antec Intro system can be coupled to form an ideal partner with the Coulochem instrument for the enhanced detection of menaquinones. The Coulochem is used to effect the reduction of the menaquinones across both electrodes, and the Antec instrument is used to oxidize them. We have found that this arrangement is of sufficient sensitivity to detect phylloquinone levels in the extracellular matrix of milligram fragments of human jaw bone.[5]

Identification of Individual Menaquinones

The single most important issue for the identification of the menaquinones is the need for authentic standard materials. The absence of commercially available sources means that research in this area is potentially restricted to groups with a supply of these compounds. To overcome this limitation, it is necessary to find an appropriate industrial sponsor to donate these materials, undertake to synthesize and purify these compounds, or isolate menaquinones from appropriate bacterial sources. The first of these options is a difficult route to pursue as there is not an excess of potential benefactors. The second requires collaborative links with organic synthetic chemists and access to the correct starting materials. However, the final option is one that is open to a larger number of laboratories, the success of which can be significantly enhanced if these laboratories have collaborative links with microbiology departments.

Isolation of Menaquinone-II from *Prevotella intermedia*

Bacteria that can provide a range of menaquinones have been well characterized for their menaquinone profiles[6,7] and are available from the American Type Culture Collection (Rockville, MD) or the National Collection of Type Cultures (Porton Down, UK).

The bacteria *Prevotella intermedia* NCTC 9336, which was obtained from the National Collection of Type Cultures, contains menaquinone 11 as the principal menaquinone congener. This organism is allowed to grow in liquid culture under anaerobic conditions in standard broth containing added hemin and menadione. After bacterial growth caused the broth to become opaque, the culture media is centrifuged and the supernatant removed before the bacteria are lyophilized. A small aliquot of the culture broth is also examined by secondary culture on blood agar plates as confirmation that the liquid cultures contained a black pigmented bacteria.

The lyophilized bacteria (0.68 g) are resuspended into a slurry with 10 ml of distilled water, 20 ml of absolute ethanol (BDH, Poole, UK) is added, and the resulting suspension is vortexed to ensure complete homogeniza-

[5] S. J. Hodges, D. Harrington, and M. J. Shearer, unpublished results (1994).

[6] M. D. Collins and D. Jones, *Microbiological Rev.* **45,** 316 (1981).

[7] H. N. Shah and M. D. Collins, *J. Appl. Bacteriol.* **55,** 403 (1983).

tion. To this suspension 120 ml of HPLC-grade hexane (BDH) is added before thorough mixing by vortexing for 2 min. The sample is then centrifuged for 10 min at 1000*g* and the hexane layer aspirated into a clean dry vessel. The residue from centrifugation is reextracted with a further 120 ml of hexane, vortexed, and centrifuged. The combined hexane fractions are dried under a stream of oxygen-free nitrogen. The residue is then resuspended in 40 ml hexane and separated into four equal aliquots before further purification on silica Sep-Pak cartridges (Millipore, Bedford, MA). When the hexane extracts are passed through the silica cartridges, the menaquinones bind to the silica matrix and the nonbound and loosely bound hydrophobic materials are eluted with a further two hexane washes (10 ml). A menaquinone-containing fraction is obtained by elution of the silica cartridge with 10 ml of 3% diethyl ether in hexane. This fraction is then dried under oxygen-free nitrogen at 50° and resuspended in 200 μl of hexane before aliquots (50 μl) are injected onto a 5-μm cyanosilane-bonded column (250 × 4.6 mm) (Phenomenex, Ltd., UK) and eluted under isocratic conditions with 7% dichloromethane (BDH) in hexane. The dichloromethane must be 50% water saturated, starting with 100% water-saturated dichloromethane, which has been passed, sequentially, into three separate clean dry glass vessels to remove excess water before an equal volume of pure dry dichloromethane is added to make the final 50% water-saturated

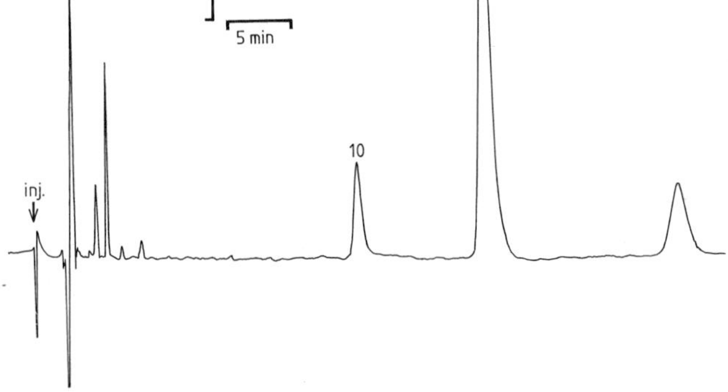

FIG. 2. Reverse-phase HPLC separation and electrochemical detection of putative menaquinones 10 and 11 from *P. intermedia*. See text for isolation and detection methods.

dichloromethane solution. The amount of dichloromethane required to effect an appropriate separation is dependent on the characteristics of the HPLC column. The eluant is continuously monitored at 270 nm. The collection volume for the menaquinone fraction from this stage of the purification is determined by injecting 20 μl of a stock solution of menaquinone 4 (25 μg/ml) (Sigma Chemical Co., Poole, UK) onto the column. The fraction collected after injection of the extract is fixed at 2 ml before and 2 ml after the volume at which the menaquinone 4 standard was found to elute from the column.

The fraction obtained from normal phase HPLC is then purified by reverse-phase HPLC using an octylsilane-bonded column, eluted isocratically with 97% methanol (BDH), 3% sodium acetate buffer (0.05 *M*, pH 3.0) containing 0.01 m*M* EDTA. The eluant is monitored by following the current across the downstream electrode of an analytical Coulochem electrochemical cell operated in the redox mode with an upstream potential of -1.3 V and a downstream electrode potential of $+0.05$ V (Fig. 2). The amount of purified material injected onto the reversed-phase column to obtain the chromatogram shown in Fig. 2 is 3.3×10^{-4} of the total material available. Quantifying the peak of menaquinone 11 against the authentic menaquinone 10 standard allows us to calculate that the total amount of menaquinone 11 recovered from this isolation is 61 μg. Clearly for the isolation of the pure menaquinones the appropriate fractions can be collected by monitoring the eluant with UV detection at 270 nm and characterizing each collected component by UV spectroscopy.

Acknowledgment

I gratefully acknowledge the help of Dr. Michael Wilson, head of the Microbiology Department, Eastman Dental Institute for Oral Healthcare Sciences, University of London, UK, with the growth of *P. intermedia.*

[37] Assay of Phylloquinone and Menaquinones in Human Liver

By YUJI USUI

Introduction

Phylloquinone (vitamin K_1) found in plants and a series of bacterially produced menaquinones (MK-*n*) are natural forms of vitamin K that are

0076-6879/97 $25.00

required for the synthesis of specific plasma clotting factors in the liver.[1–3] The presence of both phylloquinone and menaquinones has been demonstrated in human liver,[4,5] but it has been difficult to measure these vitamins quantitatively in biological samples because of their low concentrations and the many contaminants present in lipid extracts.

Preparation of Liver Samples

Liver specimens are prepared by the method of Hirauchi *et al.*,[6] which incorporates preliminary purification on a Sep-Pak cartridge[7] and thin-layer chromatography (TLC)[8] with slight modification. Briefly, 1 g of liver tissue is homogenized with 5 ml of 66% (v/v) 2-propanol, and then 6 ml of hexane is added. The mixture is shaken and a 5-ml portion of the upper layer is transferred to another brown centrifuge tube and dried under reduced pressure. After the residue has been dissolved in 2 ml of hexane, the solution is applied to a silica gel Sep-Pak cartridge (Waters Assoc., Milford, MA), which had previously been successively washed with 10 ml of 4% of diethyl ether in hexane and 10 ml of hexane. The Sep-Pak cartridge is washed with 10 ml of hexane before the vitamin K compounds are eluted with 5 ml of 4% (v/v) diethyl ether in hexane. The eluate is dried under reduced pressure and dissolved in 150 ml of hexane. The solution is applied to a silica gel 60 F254 plate (20 × 20 cm, E. Merck, Darmstadt, Germany) as a 13-cm-long band. At the same time, 10 ml of hexane containing 1 mg of phylloquinone is spotted as a marker 2 cm from the band on both sides. The plate is developed with a mixture of petroleum ether and diethyl ether (85 : 15, v/v) for 20 min over 12 cm in a developing chamber. After the markers are detected by UV illumination (254 nm), with reference to the RF values of vitamin K compounds, portions of silica gel about 3 cm wide are scraped into a brown centrifuge tube. Vitamin K compounds are extracted with 7 ml of chloroform by shaking. A 5-ml portion of the

[1] J. W. Suttie, *in* "The Fat-Soluble Vitamins" (A. T. Diplock, ed.), p. 225. William Heineman, London, 1985.

[2] R. E. Olson, *in* "Modern Nutrition in Health and Disease" (S. E. Goodhart and M. E. Shils, eds.), p. 170. Lea & Febiger, Philadelphia, 1980.

[3] D. Savage and J. Lindenbaum, *in* "Nutrition in Hematology" (J. Lindenbaum, ed.), p. 271. Churchill Livingstone, New York, 1983.

[4] P. Rietz, U. Gloor, and O. Wiss, *Int. Z. Vitamin Forsch.* **40,** 351 (1970).

[5] T. J. Duello and J. T. Matschiner, *J. Nutr.* **102,** 331 (1972).

[6] K. Hirauchi, T. Sakano, T. Nagaoka, A. Morimoto, and S. Masuda, *in* "Proceedings of the 7th Symposium on Anal. Chem. Biol. Substances," p. 43. Pharmaceutical Society of Japan, Tokyo, 1985.

[7] K. Hirauchi, T. Sakano, and A. Morimoto, *Chem. Pharm. Bull.* **34,** 845 (1986).

[8] T. Sakano, T. Nagaoka, A. Morimoto, and K. Hirauchi, *Chem. Pharm. Bull.* **34,** 4322 (1986).

chloroform layer is transferred into another brown centrifuge tube and dried under reduced pressure. The residue is dissolved in 200 μl of ethanol, and 50 μl of the solution is injected into an HPLC system.

HPLC System

Figure 1 shows a schematic diagram of the system. The high-performance liquid chromatograph (HPLC) consisted of two reciprocating pumps (LC-6A, Shimadzu, Kyoto, Japan) for gradient elution controlled by a system controller (SCL-6A, Shimadzu), a fluorimeter (RF-535, Shimadzu), a column oven (CT0-6A, Shimadzu) and an automatic sample injector (SIL-6A, Shimadzu). A 50-μl sample of the extract is injected onto a reversed-phase column (Nucleosil $5C_{18}$, 250- × 4.6-mm i.d., Macherey-Nagel, Düren, Germany) and eluted at a flow rate of 1.0 ml/min. A stainless-steel column packed with platinum-black powder (RP-10, 10- × 4.0-mm i.d., IRICA, Kyoto, Japan) is used for postcolumn reduction. Detection is at an excitation wavelength of 320 nm and an emission wavelength of 430 nm. Phylloquinone (K_1) was purchased from Wako (Osaka, Japan) and MK-*n* (MK-4 to MK-13), K_1 epoxide and MK-4 epoxide were kindly provided from Eizai (Tokyo, Japan). HPLC-grade solvents were obtained from Kanto (Tokyo, Japan).

Reduction Activity of Platinum-Black Column

Fluorescence intensity decreased when oxygen was present. However, when oxygen was removed by nitrogen bubbling, the reductive activity was stable during continuous analysis. The reductive activity of the platinum-black column has been examined with several solvents after oxygen has

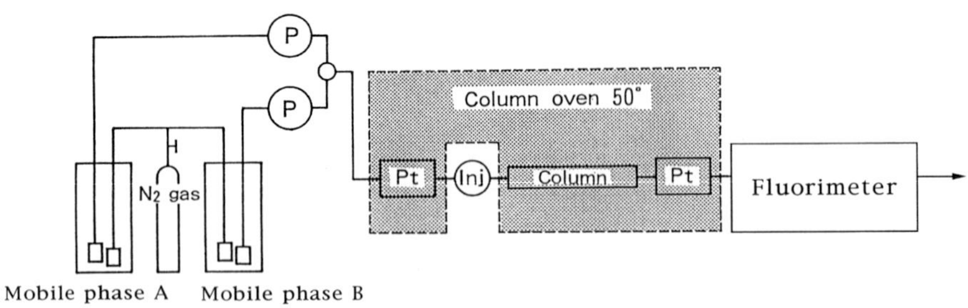

FIG. 1. Schematic diagram of the gradient HPLC system utilized. Details have been described.[9] P, Pump; Inj, injector; Pt, platinum-black reduction catalyst column.

been removed.[9] The reductive activity was stable with methanol, ethanol, 2-propanol, and small amounts of water as the mobile phase. With these solvents, the reductive power of the platinum-black catalyst column was equal to that of the electrochemical method described by Hirauchi *et al.*[7] Stability of the reductive activity was increased when a second platinum-black column was placed between the pumps and the injector, as described by Tejada *et al.*[10] for a platinum–rhodium catalyst column.

Gradient Analysis for Vitamin K Compounds in Liver

Elution times of menaquinones with long isoprenoid chains were long (about 240 min for MK-13) when the vitamin K compounds in human liver were separated by isocratic elution. Gradient elution is therefore needed for the analysis of vitamin K compounds in liver. A large baseline drift was observed during gradient analysis, but when 100% methanol was used as mobile phase A and 2-propanol–ethanol (4 : 1, v/v) as mobile phase B, the baseline drift was minimized. The most efficient gradient conditions for separating vitamin K compounds in the human liver were obtained when the concentration of mobile phase B was increased linearly from 0% (15 min) to 80% (90 min). Under these conditions, the relative standard deviations of the peak areas and retention times were within $\pm 2\%$ and $\pm 1\%$, respectively, during continuous analysis for at least 48 hr ($n = 5$); the day-to-day precision of peak areas and retention times was within $\pm 3\%$ ($n = 5$). In addition, the reduction potential of the platinum column and the resolution of the analytical column were stable after more than 200 analyses, and after 2 years of use. The detection limit of standard compounds was 5 pg for MK-4; 10 pg for K_1, MK-5, and MK-6; 20 pg for MK-7 to MK-9; and 40 pg for MK-10 to MK-13 ($S/N = 3$). They were 10 times higher when measured in 1 g of the liver sample.

Measurement of Vitamin K Compounds in Human Liver

Figure 2 shows chromatograms of a liver sample analyzed by gradient elution with and without reduction. The peaks were identified as vitamin K compounds by the coincident retention times with those of the standards and by the disappearance of the peaks during analysis without reduction removal of platinum-black columns from the system. The unknown peak (*) near that of MK-11 was considered to be MK-9 (H_{10}) based on the

[9] Y. Usui, N. Nishimura, N. Kobayashi, T. Okanoue, M. Kimoto, and K. Ozawa, *J. Chromatogr.* **489,** 291 (1989).

[10] S. B. Tejada, R. B. Zweidinger, and J. E. Sigsby, Jr., *Anal. Chem.* **58,** 1827 (1986).

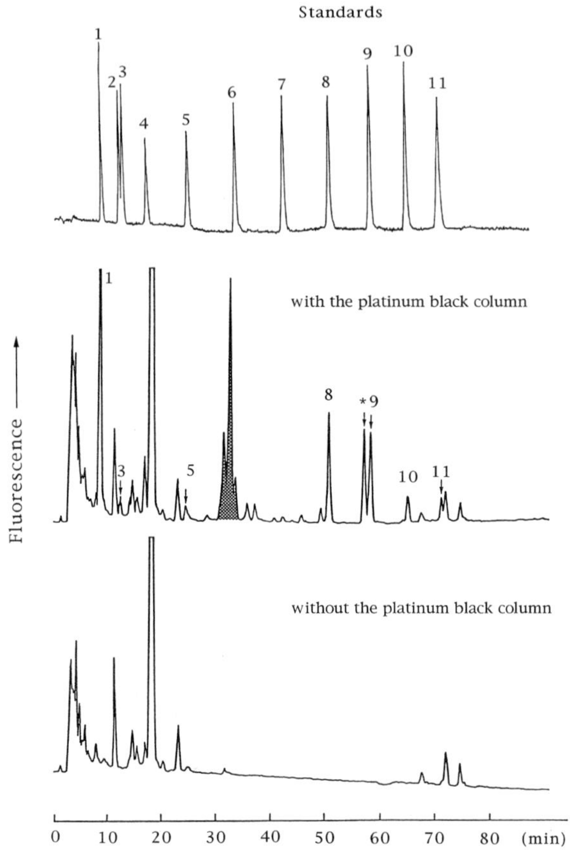

FIG. 2. Gradient elution chromatograms of standards and a human liver sample analyzed with and without reduction. Peaks: 1, MK-4; 2, MK-5; 3, K_1; 4, MK-6; 5, MK-7; 6, MK-8; 7, MK-9; 8, MK-10; 9, MK-11; 10, MK-12; and 11, MK-13. Mobile phase A, 100% methanol; mobile phase B, 2-propanol–ethanol (4 : 1, v/v); temperature, 50°. The concentration of mobile phase B was increased linearly from 0% (15 min) to 80% (90 min) at a flow rate of 1 ml/min. (From Usui *et al.*[9] with modification.)

retention times of MK-*n* reported by Tamaoka *et al.*[11] Neither the peak or MK-11 nor of this unknown substance appeared in the analysis without reduction. At room temperature, it was difficult to separate this peak efficiently from that of MK-11 by gradient elution. However, the two peaks could be separated easily when the analytical column was heated to 50°. MK-5 and MK-6 could not be measured because of interferences from

[11] J. Tamaoka, Y. Katayama, and H. Kuraishi, *J. Appl. Bacteriol.* **54,** 31 (1983).

TABLE I
VITAMIN K CONCENTRATIONS IN HUMAN LIVER[a]

Specimen	n	K1	MK-7	MK-8	MK-9	MK-10	MK-11	MK-12	MK-13
Normal	6	18.7 ± 4.0	123.1 ± 61.6	11.4 ± 2.1	4.2 ± 1.5	96.2 ± 16.6	94.4 ± 26.4	21.6 ± 5.9	7.9 ± 3.2
Chronic hepatitis	10	17.1 ± 4.5	22.6 ± 5.5	13.5 ± 9.6	2.7 ± 2.2	14.4^b ± 2.6	10.7^b ± 2.1	2.4^c ± 0.8	1.5 ± 0.9
Liver cirrhosis	22	24.7 ± 4.9	62.6 ± 26.5	7.7 ± 1.5	1.9 ± 0.5	13.2^b ± 3.0	9.8^b ± 2.8	2.4^c ± 0.7	1.1 ± 0.4

[a] Values are pmol/g wet weight. Quoted from Usui *et al.*[9] with modification.
[b] $p < 0.05$ compared with normal group.
[c] $p < 0.01$ compared with normal group.

other peaks. However, their amounts were small (subnanogram levels) when we employed isocratic analysis and electrochemical reduction using acetonitrile–methanol (4:1, v/v) as the mobile phase. The patient studied received Kaytwo (MK-4, Eizai) at the preoperative stage. The peaks, denoted by a shaded area, have retention times very close to that of MK-8 and disappeared during the analysis without reduction. They are not MK-8, but could be degradation products of MK-4.[9] The concentrations of vitamin K in human liver were calculated by peak–area measurements using calibration curves, each of which showed good linearity. The calibration curves of vitamin K compounds were linear over the range 0.02–100 ng per injection (corresponding to 0.2–1000 ng/g in the liver sample).

Regression equations, injected vitamin versus peak area, for phylloquinone and MK-4 to MK-13 were 0.99995 or greater. Recoveries of phylloquinone and menaquinones added to liver homogenates ranged from 89 to 107% with coefficients of variation from 3.0 to 6.5%.

Table I shows results of the measurement of resected human liver samples from patients with primary or metastatic liver cancer.[9] MK-4 was excluded from Table I because many patients received MK-4 preoperatively. Its concentration was found to be very low in the specimens without MK-4 administration.[12]

Phylloquinone is found in green plant material in large amounts, and MK-4 to MK-8 are found in animal tissues and fermented foods in relatively minute amounts.[13,14] A large amount of MK-7 is also found in certain

[12] Y. Usui, H. Tanimura, N. Nishimura, N. Kobayashi, T. Okanoue, and K. Ozawa, *Am. J. Clin. Nutr.* **51,** 846 (1990).
[13] T. Sakano, S. Notsumoto, T. Nagaoka, A. Morimoto, K. Fujimoto, S. Masuda, Y. Suzuki, and K. Hirauchi, *Vitamins (Japan)* **62,** 393 (1988).
[14] K. Hirauchi, T. Sakano, S. Notsumoto, T. Nagaoka, A. Morimoto, K. Fujimoto, S. Masuda, and Y. Suzuki, *Vitamins (Japan)* **63,** 147 (1989).

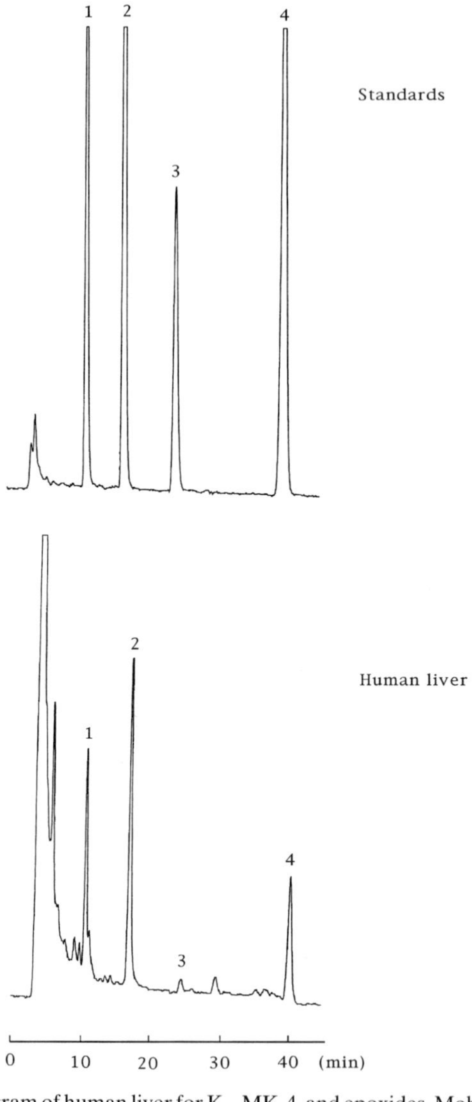

FIG. 3. Chromatogram of human liver for K_1, MK-4, and epoxides. Mobile phase, methanol–acetonitrile–water (60:40:1, v/v) containing 0.25% of sodium percholate; temperature, 50°. A − 1.0-V potential was applied to the analytical cell for reduction of epoxides. Peaks: 1, MK-4 epoxide; 2, MK-4; 3, K_1; 4, K_1 epoxide. The patient received MK-4 preoperatively.

TABLE II
MK-4 AND MK-4 EPOXIDE CONCENTRATIONS IN HUMAN LIVER[a]

Groups	n	MK-4 (nmol/g)	MK-4 epoxide (nmol/g)
Normal	10	2.77 ± 0.24	4.01 ± 0.29
Chronic hepatitis	5	3.79 ± 0.50	3.09 ± 0.33
Liver cirrhosis	9	3.83 ± 0.42[b]	2.62 ± 0.38[c]

[a] Sixty min after intravenous administration of 200 μg/kg MK-4. Values are mean ± SEM. Quoted from Nishimura *et al.*[25] with modification.
[b] $p < 0.05$ compared with normal group.
[c] $p < 0.01$ compared with normal group.

fermented foods, such as natto and butter,[13,15] and these foods are a source of MK-7 for humans. MK-*n* with more than nine isoprenoid units are virtually nonexistent in these foods. However, possible peaks for MK-10 and MK-11 were found in the portal vein samples using this gradient HPLC system. It is likely that they were absorbed passively via the portal vein at very low concentrations.

K_1 and MK-*n* have been found in the human liver using UV detection and mass spectrometry.[4,5] However, it is difficult to measure vitamin K compounds in the liver sample because of their very low concentrations and contaminating substances. Most of the biological contaminants in liver samples can be removed by Sep-Pak and TLC purification. Separations by HPLC with fluorimetric or electrochemical detection (ED) offer high sensitivity and selectivity.[7,15–19] Chromatographic collection, UV, or ED has been used for the measurement of vitamin K compounds in liver,[19–22] but satisfactory results were not obtained for menaquinone measurements. This gradient elution system is capable of measuring a wide range of vitamin K compounds in liver. Other gradient systems have been reported,[17,23] but vitamin K compounds in the liver were not measured. When the electro-

[15] M. Shino, *Analyst* **113,** 393 (1988).
[16] J. P. Langenberg and U. R. Tjaden, *J. Chromatogr.* **305,** 61 (1984).
[17] Y. Haroon, D. S. Bacon, and J. A. Sadowski, *J. Chromatogr.* **384,** 383 (1987).
[18] Y. Haroon, C. A. W. Schubert, and P. V. Hauschka, *J. Chromatogr. Sci.* **22,** 89 (1984).
[19] M. J. Shearer, *Methods Enzymol.* **123,** 235 (1986).
[20] Y. Haroon and P. V. Hauschka, *J. Lipid Res.* **24,** 481 (1983).
[21] M. J. Shearer, *in* "HPLC of Small Molecules: A Practical Approach" (C. K. Lim, ed.), p. 157. IRL Press, Oxford, 1986.
[22] M. J. Shearer, P. T. McCarthy, O. E. Crampton, and M. B. Mattok, *in* "Current Advances in Vitamin K Research" (J. W. Suttie, ed.), p. 437. Elsevier Science Publishing, New York, 1988.
[23] C. Kindberg, J. W. Suttie, K. Uchida, K. Hirauchi, and H. Nakao, *J. Nutr.* **117,** 1032 (1987).

chemical reduction method is used, there is a risk of decreased reductive power, because of the coating of electrodes caused by continuous analysis. The platinum-black column is not consumed and reduces the vitamin K catalytically, like a platinum oxide catalyst column[15] or a platinum–rhodium catalyst column[10] without any modifier.

Vitamin K Epoxides in Human Liver

Liver samples were prepared as described earlier. Vitamin K epoxides were not reduced by the platinum-black catalyst column. Chromatographic determination was done by the method of Hirauchi *et al.*[24] with some modification. An electrochemical reduction cell (analytical cell 5010, ESA, Bedford, MA) was inserted between the analytical column and the platinum-black column in Fig. 1, and an applied potential was -1.0 V for reduction of epoxides. TSKgel ODS-120T (5 μm) 250- $\times$ 4.6-mm i.d. (Tosoh, Japan) was used for analytical column, and methanol–acetonitrile–water (60 : 40 : 1, v/v) containing 0.25% sodium percholate was used as the mobile phase. The detection limit in 1 g of the human liver was 50 pg/g for MK-4 and 100 pg/g for MK-4 epoxide, K_1, and K_1 epoxide. Figure 3 shows a chromatogram of a human liver specimen, which was obtained by wedge biopsy 60 min after intravenous administration of 200 μg/kg body weight of MK-4 during hepatectomy.[25] Liver specimens were histologically classified into normal, chronic hepatitis, or liver cirrhosis group. The results are shown in Table II.

[24] K. Hirauchi, T. Sakano, T. Nagaoka, and A. Morimoto, *J. Chromatogr.* **430,** 21 (1988).
[25] N. Nishimura, Y. Usui, N. Kobayashi, T. Okanoue, and K. Ozawa, *Scand. J. Gastroenterol.* **25,** 1089 (1990).

[38] Determination of Phylloquinone in Foods by High-Performance Liquid Chromatography

By SARAH L. BOOTH and JAMES A. SADOWSKI

Introduction

Vitamin K has been historically identified for its role in coagulation, with frank deficiency rarely observed among adult populations. Circulating plasma phylloquinone (vitamin K_1) is present in very low concentrations

0076-6879/97 $25.00

relative to the other fat-soluble vitamins,[1] and it has been assumed to be widely distributed in the food supply.[2] Furthermore, it has been stated that 50% of the human vitamin K requirement is met by endogenously produced menaquinones (vitamin K_2).[3] Collectively, these observations and assumptions limited the need to develop an accurate food composition database for vitamin K. However, with the expansion of knowledge on the role of vitamin K in the posttranslational synthesis of γ-carboxyglutamic acid (Gla), and the uncertainty about the contribution of menaquinones to human vitamin K nutrition,[3] a need has been expressed for accurate food composition data for phylloquinone, which is the predominant form of vitamin K in the food supply.

Much of the earlier phylloquinone food composition data were obtained using the chick bioassay, which is a qualitative index that lacks the precision for estimating dietary intakes of phylloquinone.[3,4] Subsequent analytical methods used for phylloquinone analysis included thin-layer chromatography (TLC)[5] and gas chromatography (GC).[6] Although gas chromatography is still used for identification purposes, it has been the introduction of high-performance liquid chromatography (HPLC) that has facilitated the routine analysis of phylloquinone in foods.

HPLC determination of phylloquinone in various food matrices is described in this chapter, with an emphasis on an assay developed in our laboratory that uses postcolumn chemical reduction of the quinone, followed by fluorescent detection of the hydroquinone form of the vitamin.[7]

Sampling

Error introduced during the collection and preparation of samples can create large discrepancies in phylloquinone nutrient values. Phylloquinone contents of raw food items vary with geographic locations (which incorporate differences in climatic and soil conditions),[8] stage of maturity, and season.[9] Stability studies demonstrate that phylloquinone is stable to heat

[1] J. A. Sadowski, S. J. Hood, G. E. Dallal, and P. J. Garry, *Am. J. Clin. Nutr.* **50,** 100 (1989).
[2] D. B. Parrish, *CRC Crit. Rev. Food Sci.* **13,** 337 (1980).
[3] J. W. Suttie, *Annu. Rev. Nutr.* **15,** 399 (1995).
[4] M. J. Shearer, V. Allan, Y. Haroon, and P. Barkhan, *in* "Vitamin K Metabolism and Vitamin K-Dependent Proteins" (J. W. Suttie, ed.), p. 317. University Park Press, Baltimore, 1980.
[5] J. D. Manes, H. B. Fluckiger, and D. L. Schneider, *J. Agric. Food Chem.* **20,** 1130 (1972).
[6] R. M. Seifert, *J. Agric. Food Chem.* **27,** 1301 (1979).
[7] S. L. Booth, K. W. Davidson, and J. A. Sadowski, *J. Agric. Food Chem.* **42,** 295 (1994).
[8] G. Ferland and J. A. Sadowski, *J. Agric. Food Chem.* **40,** 1874 (1992).
[9] B. Fournier, L. Sann, M. Guillaumont, and M. Leclerq, *Am. J. Clin. Nutr.* **45,** 441 (1987).

and processing,[10,11] but is rapidly destroyed by both fluorescent light and sunlight.[10] Sampling must be conducted in a manner to reflect foods as they are available to the consumer in the retail market, with adequate description of the storage conditions of the foods prior to sampling. All procedures involving sample processing and preparation should be done under amber lighting because vitamin K compounds are sensitive to photooxidation.

Variation attributable to sampling can be illustrated in a green, leafy vegetable, such as spinach, which would have inherent natural sources of variation such as amount of rainfall, light exposure, and soil conditions. Under the conditions of our assay, the between-run analytical variation defined by the coefficient of variation (*CV*) in the phylloquinone content of spinach as measured on 3 different days in the same sample, was 7.4%.[7] The within-run *CV* for a single spinach sample was 5.0%. In contrast, the total sample variation as defined by the between sample *CV* in the phylloquinone content of spinach ($n = 10$) was 32.6%, which is far greater than that attributable to the assay.

In the absence of detailed information on sampling procedures, sample size, and the portion of the sample analyzed, it is not known if variation in published nutrient data for a given food item is biological and/or analytical in origin. Improvement of the sampling plan description will facilitate comparison of data among different research facilities, which in turn will maximize the efficiency of resources in compiling an extensive phylloquinone database.

Reagents and Standards

All solvents used in sample extraction and chromatography are of HPLC grade (Fisher Scientific, Springfield, NJ). Vitamin $K_{1(20)}$ and the ACS reagent-grade zinc chloride used in the preparation of the mobile phase are purchased from Sigma Chemical Co. (St. Louis, MO). Dihydrophylloquinone is synthesized by hydrogenation of phylloquinone using a modified procedure originally described by Langenberg and Tjaden.[12] The internal standard $K_{1(25)}$ is a gift from Hoffman-La Roche and Co. (Basel, Switzerland). Primary and secondary stock solutions are diluted to known concentrations in 100% methanol and characterized spectrophotometrically and chromatographically. The concentrations of the phylloquinone, dihydro-

[10] G. Ferland and J. A. Sadowski, *J. Agric. Food Chem.* **40,** 1878 (1992).

[11] J. P. Langenberg, U. R. Tjaden, E. M. De Vogel, and D. Langerak, *Acta Aliment.* **15,** 187 (1986).

[12] J. P. Langenberg and U. R. Tjaden, *J. Chromatogr.* **289,** 377 (1984).

phylloquinone, and $K_{1(25)}$ secondary solutions are calculated from their UV absorption spectra using the absorptivity value ($E^{1\%}_{1cm}$) of phylloquinone at 248 nm = 420. The secondary solutions are then combined and diluted to produce the working calibration standard. The concentrations of phylloquinone, dihydrophylloquinone, and vitamin $K_{1(25)}$ in the working calibration standard are 44, 44, and 38 nmol/liter, respectively. The concentration of vitamin $K_{1(25)}$ in the working internal standard is 192 nmol/liter. Working solutions are stored at 4° and shielded from light. Glassware and utensils used in sample processing and extraction are washed in a chromic–sulfuric acid solution (Fisher Scientific) to minimize contamination of samples from carryover fluorescent material.

A number of vitamin K analogs have been used as the internal standard for HPLC analysis of phylloquinone in foods, as summarized in Table I. In the past, we have used dihydrophylloquinone as an internal standard for food analyses. However, we have determined that phylloquinone is converted to dihydrophylloquinone in the commercial hydrogenation process using phylloquinone-rich vegetable oils.[13] Since the dihydro form of the vitamin is not a suitable internal standard for foods containing hydrogenated or partially hydrogenated oils, we now use $K_{1(25)}$, a synthetic analog of phylloquinone produced by the substitution of a 25-carbon side chain to menadione, as the internal standard for all food analyses. The application of this internal standard to simultaneous determination of phylloquinone and dihydrophylloquinone is illustrated in the chromatograms of soybean oil that has been subjected to varying degrees of hydrogenation (Fig. 1). Use of $K_{1(25)}$ precludes use of liquid-phase reductive extraction since the chemistry of the reductive extraction, due to the polarity of the solvents, is such that recovery diminishes as side-chain length increases beyond 20 carbons. To remove the remaining lipids that would otherwise be removed with the reductive extraction step as described in an assay for plasma phylloquinone,[14] successive reversed-phase solid-phase extractions using 3- or 6-ml silica gel and 6-ml C_{18} columns are used.

Whereas under the conditions of our own assay we chose to use $K_{1(25)}$ as an internal standard, other vitamin K analogs can be used such as menaquinone 4 (MK_4), which is one of the few internal standards that is commercially available (Sigma Chemical Co.). However, MK_4 is naturally occurring in a variety of foods.[15] We have not found any evidence to support

[13] K. W. Davidson, S. L. Booth, G. G. Dolnikowski, and J. A. Sadowski. *J. Agric. Food Chem.* **44,** 980 (1996).

[14] Y. Haroon, D. S. Bacon, and J. A. Sadowski, *Clin. Chem.* **32,** 1925 (1986).

[15] T. Sakano, S. Notsumoto, T. Nagaoka, A. Morimoto, K. Fujimoto, S. Masuda, Y. Suzuki, and K. Hirauchi, *Vitamins (Japan)* **62,** 393 (1988).

TABLE I

SAMPLE EXTRACTION METHODS USED FOR HPLC DETERMINATION OF PHYLLOQUINONE IN FOODS

Foods	Sample preparation	Internal standard	Coefficient of variation (CV)	Ref.[a]
Infant formulas	Lipase treatment; extract with hexane; evaporate to dryness; eluate dissolved in methanol–2-propanol–ethyl acetate–water	$K_{1(25)}$, $K_{1(15)}$, cholesteryl phenylacetate	Within run: 1.5% Between run: 3.5%	1
Human milk	Lipid extraction in 2-propanol–hexane; evaporate to dryness; open column chromatography with silica in a nitrogen atmosphere; eluate dissolved in hexane–chloroform	MK-7	Within run: <10% Between run: 32.3%	2
Vegetable oils	Lipid extraction with hexane; evaporate to dryness; silica column cleanup with hexane–2-isopropyl ether; evaporate to dryness; eluate dissolved in acetronitrile–ethanol	$K_{1(25)}$	Within run: 0.9–3.5% Between run: 3.2–7.1%	3
Vegetable oils	Lipase treatment; lipid extraction with hexane; evaporate to dryness; silica column cleanup; evaporate to dryness; eluate dissolved in hexane	MK-4	Between run: 2.6–3.7%	4
Vegetables	Homogenization; lipid extraction in hexane; evaporate to dryness; eluate dissolved in methanol	MK-6	Within run: 3–8%	5
Miscellaneous foods	Lipid extraction with acetone or chloroform–methanol; silica chromatography cleanup; mobile phase of hexane–dichloromethane, in which the dichloromethane component was 50% water soluble	K_1 epoxide	Within run: 2.5–6.9% Between run: 5.4–10.6%	6

[a] *Key to references:* (*1*) H. E. Indyk, V. C. Littlejohn, R. J. Lawrence, and D. C. Woolward. *J. AOAC Int.* **78,** 719 (1995); (*2*) L. M. Canfield, J. M. Hopkinson, A. F. Lima, G. S. Martin, K. Sugimoto, J. Burr, L. Clark, and D. L. McGee, *Lipids* **25,** 406 (1990); (*3*) F. Moussa, F. Depasse, V. Lompret, J.-Y. Hautem, J.-P. Giradet, J.-L. Fontaine, and P. Aymard, *J. Chromatogr.* **664,** 189 (1994); (*4*) Z. H. Gao and R. G. Ackman, *Food Res. Int.* **28,** 61 (1995); (*5*) J. P. Langenberg, U. R. Tjaden, E. M. De Vogel, and D. Is. Langerak, *Acta Aliment* **15,** 187 (1986); and (*6*) M. J. Shearer, V. Allan, Y. Haroon, and P. Barkhan, *in* "Vitamin K Metabolism and Vitamin K-Dependent Proteins" (J. W. Suttie, ed.), p. 317. University Park Press, Baltimore, 1980.

the existence of $K_{1(25)}$ in foods or a peak with the same elution characteristics of $K_{1(25)}$.

Sample Extraction

Phylloquinone is generally present in low concentrations relative to other lipophilic compounds, so crude lipid extracts cannot be used for direct HPLC analysis. Owing to their instability under alkaline conditions, vitamin K compounds cannot be isolated from triglycerides using saponification.

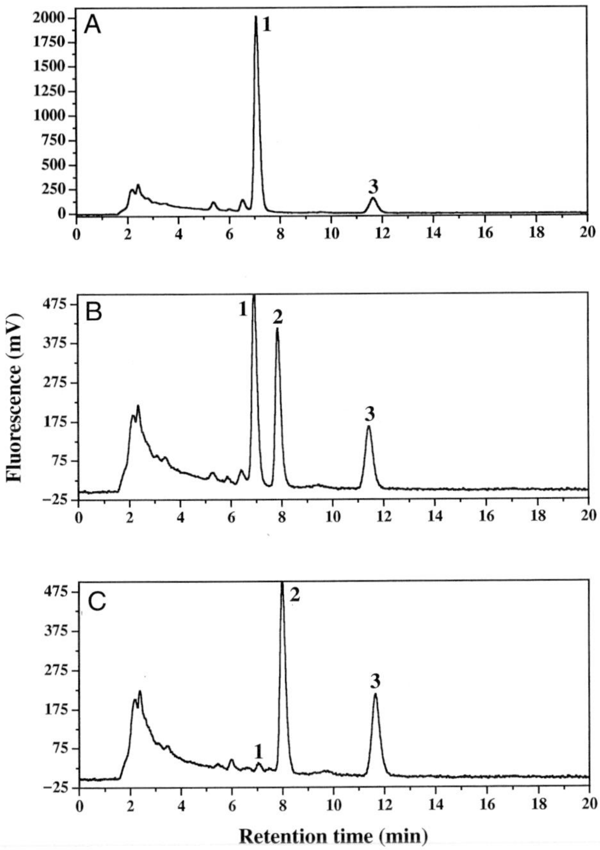

FIG. 1. A representative chromatogram of (A) a refined, bleached, deodorized soybean oil sample, (B) a partially hydrogenated soybean oil sample, and (C) a heavily hydrogenated soybean oil sample. Peaks correspond to (1) phylloquinone, (2) 2′,3′-dihydrophylloquinone, and (3) $K_{1(25)}$. (Reprinted with permission from Davidson *et al.*[13] Copyright 1996 American Chemical Society.)

Instead, a number of sample extraction methods have been successfully used by different laboratories as summarized in Table I. Whereas initial lipase treatment is often used for triglyceride-rich foods, such as infant formula, lipid extraction followed by solid-phase extraction on silica columns has been used by a number of laboratories as an effective and simple, but more expensive, purification step prior to HPLC injection.

A flow chart summarizing the sample preparation procedures developed in our laboratory is presented in Fig. 2. Solid food samples require thorough mixing in a commercial food processor (Waring Products Division, New

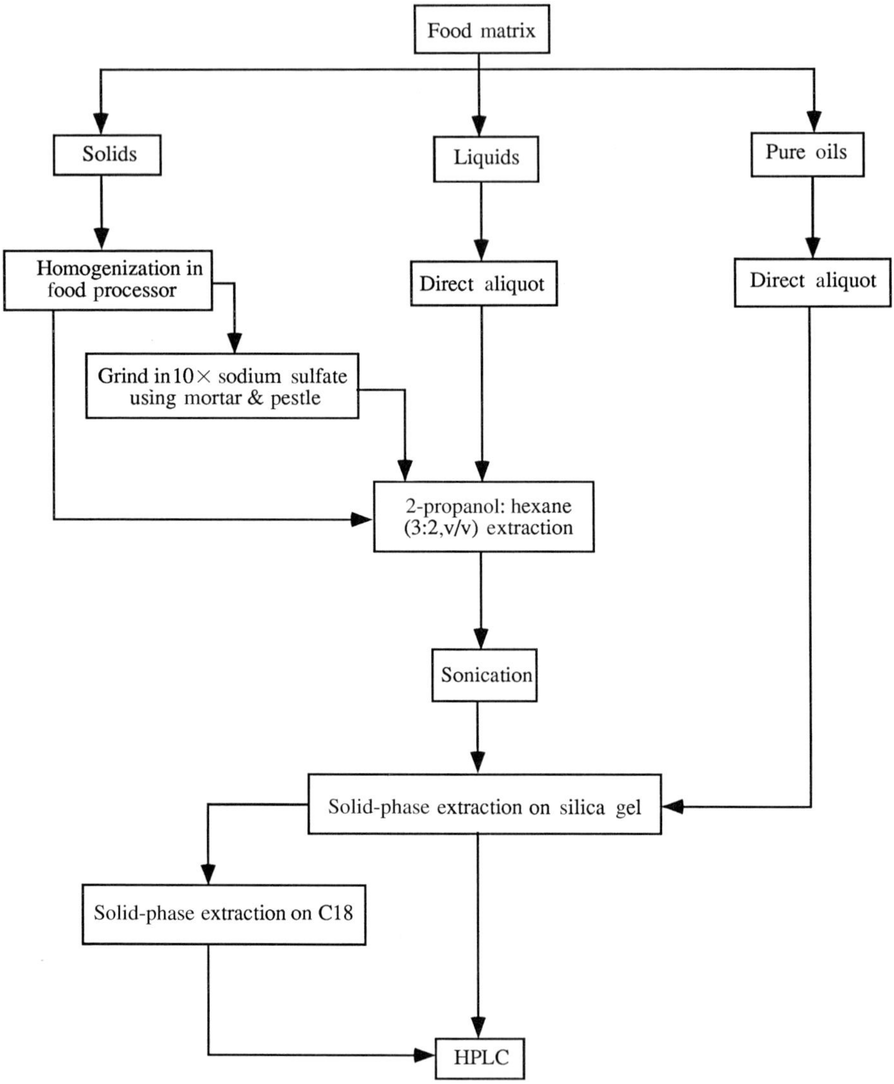

FIG. 2. Flow chart summarizing the analytical scheme of phylloquinone determination in different food matrices.

Hartford, CT) to ensure homogeneity prior to aliquoting. A 0.25 to 2.0-g aliquot of each sample is weighed directly into a 50-ml polypropylene centrifuge tube (Corning, Corning, NY). Aliquots of solid foods that have high moisture contents, such as vegetables, fruits, and meats, are further ground to a fine powder with a mortar and pestle in 10 times their weight of anhydrous sodium sulfate, followed by quantitative transfer to the centrifuge tube. An appropriate amount of the internal standard, $K_{1(25)}$, is added directly to the aliquots. Fifteen milliliters of 2-propanol/hexane (3:2, v/v) are then added, followed by 4 ml of H_2O (32 ml of H_2O are added to those homogenates that were further ground with anhydrous sodium sulfate). The mixtures are then dispersed by sonication (continuous output at 40% duty cycle for 30 seconds) using a Branson Model 350 sonifer-cell disruptor with a 1/8-inch tapered microtip (Branson Ultrasonics Corp., Danbury, CT), vortexed for 10 min, and centrifuged at 1000 *g* (10 min). After phase separation, the upper hexane layer is aspirated into a 16 × 100 culture tube and evaporated to dryness under reduced pressure in a centrifugal evaporator (Savant Instrument, Farmingdale, NY). Residues are reconstituted in 2 ml (for 3-ml SPE silica column) or 4 ml (for 6-ml SPE silica column) of hexane for solid-phase extraction (SPE) on silica gel. Residues of phylloquinone-rich foods, such as green vegetables, are first reconstituted in 10 ml of hexane, vortexed, and a 50-μl aliquot placed in a clean 16 × 100 culture tube and further processed by SPE. For analysis of pure vegetable oil samples, a 0.1-g aliquot is measured into a clean 16 × 100 culture tube, an appropriate amount of the internal standard $K_{1(25)}$ added, and 4 ml of hexane added. Following vortexing, this aliquot can be applied directly to the 6-ml SPE silica column.

The reconstituted lipid extracts are applied to a 3- or 6-ml SPE silica column (J. T. Baker, Chicago, IL) for initial purification. The size of the column is determined by the lipid content of the food, with the 6-ml column most suited for meats, fats, and oils. Each 3- or 6-ml column is preconditioned by successive washes of 8 ml of hexane:diethyl ether (97:3, v/v) and 8 ml of 100% hexane. The 2- (for 3-ml SPE column) or 4-ml (for 6-ml SPE column) extract is applied directly onto the preconditioned column packing, followed by a wash with 8 ml of hexane to remove any hydrocarbons. The phylloquinone-containing fraction is eluted with an 8-ml wash of hexane:diethyl ether (97:3, v/v). The eluants are collected into clean 16 × 100 culture tubes and evaporated to dryness in a centrifugal evaporator. Lipid-rich samples, such as oils, require an additional purification step on 6-ml C_{18} SPE columns to separate the phylloquinone from the large quantities of lipid. After evaporation, the residue from the eluate of the silica gel SPE columns is dissolved in 200 μl of 2-propanol while heating (45°) for 10 min. The 6-ml reversed-phase C_{18} columns (J. T. Baker) are preconditioned by successive washes with 10 ml of methanol:dichloromethane (80:20, v/v), followed by 6 ml of 100% methanol and 6 ml of 100% H_2O. The extracts are applied directly to the preconditioned packing. The column is then washed with 6 ml of methanol:H_2O (95:5, v/v), followed by 6 ml of 100% acetonitrile and the phylloquinone-containing fraction eluted from the column with 10 ml of methanol : methylene chloride (80 : 20, v/v). The C_{18} SPE column eluant is then evaporated to dryness. The final residue from either the silica or the C_{18} columns is reconstituted initially in 30 μl of 100% dichloromethane, immediately followed by 270 μl of methanol containing 10 m*M* zinc chloride, 5 m*M* acetic acid, and 5 m*M* sodium acetate. A sample volume of 150 μl is injected into the HPLC.

Analytical Conditions

Different analytical conditions have been successfully used for HPLC analyses of phylloquinone in foods.[16] Whereas adsorption HPLC is required for separation of the *cis* from the *trans* form of phylloquinone, reversed-phase HPLC separates phylloquinone from the menaquinones.

For the analysis of phylloquinone in foods, we use a reversed-phase HPLC system with postcolumn reduction of the quinone, followed by fluorometric detection as described in detail by Davidson *et al.*[17] Briefly, the chromatographic system consists of a Model 510 pump and Model 712B (WISP) injector (Waters Associates, Milford, MA). Pump control, integration, and quantitation are achieved using a Model 860 VAX based data station with Expert-Ease software (version 3.0) (Waters). Fluorescence was monitored using a Model 980 fluorescence detector (ABI Analytical, Ramsey, NJ) with excitation and emission wavelengths of 244 and 418 nm, respectively (high-voltage setting, 1250 V; photomultiplier tube setting, 1.0 μA).

The analytical column (150 $\times$ 4.6 mm) is packed with 3 μm of Hypersil-ODS (Keystone Scientific, Bellefonte, PA). Fluorescent derivatives of the injected quinone are produced on-line using a postcolumn reactor (2.0 $\times$ 50 mm) packed with zinc metal (-200 mesh, Alpha Products, Ward Hill, MA). The mobile phase consists of methanol/dichloromethane (90 : 10, v/v), to each liter of which we add 5 ml of a solution containing 2 *M* zinc chloride, 1 *M* acetic acid, and 1 *M* sodium acetate, and pumped at a constant flow rate of 1.0 ml/min.

Using a Waters 860 chromatography data system (Waters Chromatography, Milford, MA), quantitation is achieved by direct comparison of peak area ratios of the food samples to those of the calibration standard that contains phylloquinone, dihydrophylloquinone, and the corresponding internal standard, vitamin $K_{1(25)}$. The calibration standard is also injected after every six samples to correct for within-run changes in the retention

[16] A. Rizzolo and S. Polesello, *J. Chromatogr.* **624,** 103 (1992).
[17] K. W. Davidson and J. A. Sadowski, *Methods Enzymol.* **282,** [34], 1997 (this volume).

times for the peaks of interest due to fluctuations in chromatographic conditions (temperature, mobile phase, flow).

Analytical Variation

Repeatability for the different methods of phylloquinone analysis, as defined by the *CV*, is summarized in Table I. Under the conditions of our assay, there is a between-run *CV* of 6.3–12.1% as determined by the type of food matrix (10 runs for the same sample on the same day), and a between-day *CV* of 7.4–13.8% (3 different days of analysis for the same sample). Recovery rates of the internal standard range from 52 to 81%, with the lower recovery rates attributable to high lipid loads and the use of the reversed-phase extraction on C_{18} columns.

Applications of Assay

The application of HPLC in the determination of phylloquinone in various food matrices has facilitated the routine analyses of food items, thereby allowing accurate description of the phylloquinone and, more recently, the dihydrophylloquinone content of common food items. Using this method, we have analyzed 261 foods from the U.S. Food and Drug Administration's (FDA) Total Diet Study (TDS).[18] In the FDA-TDS, more than 250 core foods in the American food supply, representing foods consumed in the United States, are purchased from retail markets four times per year from different geographic regions. Analyses of these foods for phylloquinone and dihydrophylloquinone have identified food classes that are potential contributors to phylloquinone and dihydrophylloquinone intake in the United States and provide food composition data for the estimation of dietary intake of this vitamin across different age and gender groups.

These food composition data can also be applied to the design of metabolic studies and to dietary counselling of patients on long-term warfarin (a vitamin K antagonist) therapy. Based on food composition data for the individual items, we recently developed three individual menus that had a mean (±SD) calculated phylloquinone content of 98 (±6) μg. When the three menus used in a metabolic study were prepared on 10 separate occasions over a period of 12 months, and homogenates analyzed by HPLC, the mean (±SD) phylloquinone concentration was 100 (±12) μg. In our metabolic study, we used frozen vegetables from the same lot to reduce the variation in phylloquinone content. The phylloquinone concentrations

[18] S. L. Booth, J. A. Sadowski, and J. A. T. Pennington, *J. Agric. Food Chem.* **43,** 1574 (1995).

of a free-living diet may have greater daily variation because fresh vegetables have wider margins of biological variation in their nutrient content.

Green, leafy vegetables still appear to be the predominant dietary source of phylloquinone (113–440 μg of phylloquinone/100 g of vegetable), followed by certain vegetable oils that are derived from vegetables or seeds containing large amounts of phylloquinone. Some mixed dishes contain moderate amounts of phylloquinone that are attributable to the vegetable oils used in their preparation. Other foods, such as certain meats, brewed beverages, soft drinks, and alcoholic beverages, contain negligible amounts of phylloquinone.

These values for phylloquinone in foods were then applied to the FDA-TDS consumption model to determine how much phylloquinone is in the American diet.[19] The average intakes among men and women were within the current Recommended Dietary Allowance (RDA) of 65–80 μg of vitamin K_1/day,[20] with the exception of the 25- to 30-year-old adults who had phylloquinone intakes below the RDA. The top food sources of phylloquinone were dark green vegetables, although the fats and oils added to mixed dishes and desserts were also important contributors. However, the proportion of phylloquinone obtained from vegetables increased with age.

An overall good agreement between the FDA-TDS phylloquinone data and those produced by Shearer *et al.*[21] demonstrate that detailed sampling plans coupled with the application of HPLC methodology has facilitated an expansion in quality and quantity of the phylloquinone food database. These data are currently being applied to the development of dietary assessment methods for the evaluation of usual phylloquinone intake and its relationship to optimal vitamin K status.

Acknowledgments

The authors gratefully acknowledge K. W. Davidson and M. E. O'Brien-Morse for their technical assistance and helpful comments.

[19] S. L. Booth, J. A. T. Pennington, and J. A. Sadowski, *J. Am. Diet. Assoc.* **96,** 149 (1996).

[20] Food & Nutrition Board, "Recommended Daily Allowances," 10th ed. National Academy of Sciences, Washington, DC, 1989.

[21] M. J. Shearer, A. Bach, and M. Kohlmeier, *J. Nutr.* **126,** 1181S (1996).

[39] Assay of Menaquinones in Bacterial Cultures, Stool Samples, and Intestinal Contents

By JOHN M. CONLY

Introduction

Vitamin K exists naturally as either phylloquinone (vitamin K_1), which is found in green leafy plants, or as one of the menaquinones (vitamin K_2), which are found in certain bacterial species. Phylloquinone synthesized by plants is associated with chlorophyll[1,2] and is concentrated in the chloroplast lamellae, whereas menaquinones are synthesized by bacteria and are constituents of the plasma membrane. They are associated with the reversible redox components of the electron transport chains in the bacteria, often associated with NADH oxidation[3,4] and function under reduced (anaerobic) conditions.[5,6] The major precursors of menaquinones are shikamate, 2-ketoglutarate, methionine, and mevalonate. Menaquinone synthesis in bacteria is influenced by the degree of anaerobiasis, temperature, and the presence of certain nutritional supplements.[7] A number of reports have illustrated the utility of menaquinones as markers for the classification of bacteria.[5]

Menaquinones (MK) are referred to on the basis of the number of isoprene units (repeating 5 carbon units) in their side chain, and this isoprenoid side chain varies in length giving rise to isoprenologs varying in length from 2 to 15 isoprene units. Menaquinones can also be found unsubstituted at the second carbon position, giving rise to demethylmenaquinone or may have a side chain that is partially saturated in one or more isoprene units. Demethylmenaquinone (DMK) appears to be an intermediate in the synthesis of menaquinones in bacteria, although some organisms may produce DMK alone or in combination with menaquinone isoprenologs.

[1] D. B. Parrish, *Crit. Rev. Food Sci. Nutr.* **13,** 337 (1980).
[2] J. Suttie, *in* "The Fat Soluble Vitamins" (A. T. Diplock, ed.), p. 225. William Heinemann, London, 1985.
[3] M. Fujita, S. Ishikawa, and N. Shimazono, *J. Biochem.* **59,** 104 (1966).
[4] S. Ishikawa and A. L. Lehninger, *J. Biol. Chem.* **237,** 2401 (1962).
[5] M. D. Collins and D. Jones, *Microbiol. Rev.* **45,** 316 (1981).
[6] H. Taber, *in* "Vitamin K Metabolism and Vitamin K-Dependent Proteins" (J. W. Suttie, ed.), p. 177. University Park Press, Baltimore, 1980.
[7] R. Bentley and R. Meganathan, *Micro. Rev.* **46,** 241 (1982).

0076-6879/97 $25.00

Similar to phylloquinone, vitamin K_2 was discovered as a result of studies on the nutrition of chickens, using putrefied fish meal.[8] The antihemorrhagic factor contained in fish meal was crystallized as a pure compound in 1939 by McKee and coworkers and was found to be MK-6.[9] Vitamin K_2 was first isolated from a pure culture of bacteria (*Bacillus brevis*) in 1948 by Tishler and Sampson.[10]

Since these early experiments and those of other authors, the alkaline saponification and acid hydrolysis techniques initially used for extraction and the column and thin-layer chromatography techniques used for qualitative analysis have given way to much more sophisticated methodologies that allow accurate qualitative and quantitative analysis of menaquinones from bacteria, stool, and tissues.

The modern principles of physiochemical detection of menaquinones are similar to those employed for any of the other vitamin K compounds. These principles include extraction of menaquinone, preliminary purification of lipid extracts, and quantitation by analytical reversed-phase high-performance liquid chromatography (HPLC).

This chapter describes procedures that have been employed in my laboratory in the analysis of menaquinones from individual bacteria, stool samples, and intestinal contents. Wherever appropriate, reference will be made to other methodologies that may provide an alternative strategy. The details of the extraction and detection of the K vitamins from tissues using HPLC were reviewed in a previous publication from this series in 1986.[11]

Menaquinone Assay of Bacterial Cultures, Stool Samples, and Intestinal Contents

Sampling

Individual Bacteria. As mentioned previously, bacteria may contain MK only, DMK, or both but organisms containing only DMK are distinctly uncommon. Another major class of bacterial isoprenoid quinones is the benzoquinones of which there are two major types, ubiquinones and plastiquinones. Menaquinones are the predominant class of isoprenoid quinones in bacterial species with biquinones restricted primarily to gram-negative eubacteria and plastiquinones to the cyanobacteria.[5] Individual bacterial

[8] H. J. Almquist and E. L. R. Stokstad, *Nature* (*London*) **136,** 31 (1935).

[9] R. W. McKee, S. B. Binkley, S. A. Thayer, D. W. MacCorquodale, and E. A. Doisy, *J. Biol. Chem.* **131,** 327 (1939).

[10] M. Tishler and W. L. Sampson, *Proc. Soc. Exp. Biol. Med.* **68,** 136 (1948).

[11] M. J. Shearer, *Methods Enzymol.* **123,** 235 (1986).

strains of menaquinone analysis may be obtained from many sources, including the American Type Culture Collection (Rockville, MD), environmental or agricultural sites, and basic or clinical microbiology laboratories depending on the purpose for which the analysis is chosen. We focused on examining representative organisms from stool samples from hospital patients or normal volunteers and clinical isolates from specimens obtained in patients with perforations in the intestinal tract causing intraabdominal sepsis. Individual organisms should be obtained in pure culture and identified to species level if possible. Organisms can be grown in either liquid media or on solid media for subsequent extraction and analysis. If larger quantities of MK are required, use of liquid media may be most appropriate. We have used brain–heart infusion broth supplemented with 5 g yeast extract/liter, 5 mg hemin/ml, and 0.5 g L-cysteine hydrochloride for most bacterial species,[12] although a nutrient both medium containing meat extract 10 g/liter and NaCl pH 7.0 5 g/liter has also been used.[13] Alternately any solid media, including 5% blood agar, MacConkey agar, chocolate agar, or Mueller–Hinton agar may be used as solid media. Any blood agar supplemented with phylloquinone should be avoided due to the potential for picking up a piece of the agar with the organism, which may contain phylloquinone. For subsequent qualitative HPLC analysis, a few colonies scooped up with a wire loop will usually suffice. The optimal growth interval for the use of liquid media varies depending on the organism and the rate of growth, but usually ranges between 2 and 7 days. Harvesting of bacteria for extraction is best done during or at the height of the logarithmic phase of growth since the quantitative determination of menaquinone isoprelologs does vary with the age of the bacterial culture. Following harvesting of bacteria by centrifugation at 6000 *g* at 4° for 10 min, the pellet is washed twice with 0.05 *M* phosphate buffer (pH 7.0) and wet weight/dry weight determinations of aliquots are made. A purity culture is always done of the washed cells.

Stool Samples. Stool samples collected for menaquinone analysis are collected into sterile containers using a wooden spatula for assistance. Refrigeration of samples at 4° is sufficient to prevent further bacterial growth prior to extraction. For optimal results the sample should be processed in an expedient manner since isoprenoid quinones are rapidly photooxidized in the presence of oxygen and strong light. The use of tin foil around the specimen container during transport to the laboratory prevents any direct light exposure. If immediate analysis of stool samples is not possible, freezing of specimens at −70° for periods of a few days to a few

[12] K. Ramotar, J. M. Conly, H. Chubb, and T. J. Louie, *J. Infect. Dis.* **150,** 213 (1984).

[13] J. Tamaoka, Y. Katayama-Fujimura, and H. Kuraishi, *J. Appl. Bacteriol.* **54,** 31 (1983).

weeks does not seem to alter significantly either qualitative or quantitative analysis of menaquinones. Stool specimens collected for menaquinone analysis should be promptly weighed, with aliquots removed for wet weight/dry weight determinations which are necessary to account for individual variations in moisture content of stools and a quantitative culture is inoculated onto both selective and nonselective agar.[12] The cultures serve as a quality control for the menaquinone analysis since a crude correlation exists between the presence of certain intestinal bacteria and their respective menaquinone isoprenologs. Usually, a 1-g aliquot of the stool sample is vortexed and serially diluted in prereduced Virginia Polytechnic Institute balanced salt solution to attain final dilutions of 10^{-3}, 10^{-5}, 10^{-7}, and 10^{-9}. Aliquots of these dilutions may be inoculated onto a number of selective and nonselective media. Storage of stool samples at $-70°$ for periods of a few days to a few weeks will result in a 1–2 log loss of several species of bacteria, especially the anaerobes.

Intestinal Contents. Samples of the intestinal contents from various sites within the bowel may be obtained through a variety of methods including intestinal intubation, colonoscopic examination, or during surgical procedures. Total intestinal contents may be obtained using saline catharsis. Ingestion of 4 liters of an electrolyte solution is routinely used as a preparation for individuals undergoing colonoscopic examination and produces a complete catharsis of large bowel and presumably small bowel contents as well over a period of several hours. To facilitate collection of these contents, we used sterile paint cans that could be sealed firmly once filled. Placement of the cans on a large rotatory shaker at 4° allowed thorough mixing so that aliquots used for assays would be more closely representative of each other. Multiple aliquots of approximately 1 g were obtained and wet weight/dry weight determinations were made and quantitative cultures as described previously were done.

Extraction and Purification of Menaquinones from Bacteria, Stool, and Intestinal Contents

The extraction and purification of vitamin K_2 from bacteria, stool, and intestinal contents are very similar. The major difference relates to variable water content, the presence of fiber and other solid constituents in stool, and larger numbers of bacterial species in stool. The extraction of menaquinone may be accomplished by a number of methods using organic solvents. All the methods are based on the principle of extraction of lipids from the bacterial cells using a single-phase mixture or organic solvents followed by partitioning of the extract into aqueous and organic phases.

The aqueous phase contains the more polar lipids while the organic phase contains more neutral lipids, including the quinones. Various solvents have been used to extract bacterial cells from cultures or stools including ethanol, methanol : ether, isooctane : 2-propanol, acetone : methanol and chloroform : methanol. Procedures that have been used extensively for vitamin K extraction are the methods developed by Folch *et al.*[14] and then modified by Bligh and Dyer.[15] The methods of Folch have also been modified by Collins[16] and have been extensively used for the extraction of dry bacterial cells. Lyophilized cells (50–100 mg) are extracted with small volumes (25–50 ml) of chloroform : methanol (2 : 1, v/v) for approximately 2 hr. The cell/solvent mixture is filtered, collected, and evaporated to dryness. The dried residue is then reconstituted with organic solvents appropriate to the chromatographic procedure being employed.

The method employed is our laboratory is a modified Bligh and Dyer extraction procedure.[17] Cell suspensions of individual bacteria (5–14 g wet weight) are mixed with chloroform and methanol (0.8 : 1 : 2, v/v) in a 250- or 500-ml separatory funnel, and after vigorous shaking for 5 min the mixture is allowed to stand for at least 15 min. One volume of chloroform and one volume of distilled water are added to facilitate partitioning into a biphasic mixture. Separation is facilitated by gentle centrifugation. The lower organic phase is collected and dried over anhydrous Na_2SO_4 and evaporated to dryness *in vacuo* using a flash evaporator. The ratio of water : chloroform : methanol in the initial single-phase extract must be 0.8 : 1 : 2 (v/v) for optimal retrieval of lipids.

Extraction of stool samples and intestinal contents employs a similar method. Samples (~1.0 g wet weight) are suspended in 10 ml of distilled water and chloroform : methanol (12.5/25 ml) added to achieve a concentration of 0.8 : 1 : 2 (v/v). If necessary the mixture is homogenized for approximately 1 min using an Omni-mixer (Sorvall Inc., Newton, CT). The mixture is then allowed to stand at room temperature for 3 hr with intermittent shaking. The mixture is centrifuged at 150*g* for 5 min and the supernatant collected. The residue is resuspended in 10 ml distilled water and reextracted with chloroform : methanol as previously described. The supernatant is again collected after centrifugation and partitioned into a biphasic mixture by the addition of 25 ml of 5% Na_2SO_4 solution (w/v) to achieve a final concentration of water : chloroform : methanol of 1.8 : 2 : 2. Following partitioning, the lower organic phase is collected and evaporated

[14] J. Folch, M. Lees, and G. H. S. Stanley, *J. Biol. Chem.* **226,** 497 (1957).

[15] E. G. Bligh and W. J. Dyer, *Can. J. Biochem. Physiol.* **37,** 911 (1959).

[16] M. D. Collins, *Methods Microbiol.* **18,** 329 (1985).

[17] R. K. Hammond and D. C. White, *J. Chromatogr.* **45,** 446 (1969).

to dryness *in vacuo* using a flash evaporator. Conditions of sample extraction using these methods allow recovery of 80–92% of spiked fecal samples containing MK-4 to MK-10.[18]

An alternate method for the extraction of K vitamins from stool described by Hara and Radin[19] uses a mixture of hexane : 2-propanol and with less polarity than other organic mixtures this method has the advantage of an extract with less contamination with proteins and other nonlipid material.

A preliminary purification step is required to remove excessive amounts of unwanted polar lipids prior to analysis of menaquinones[20] and can be achieved using a number of methods. Although conventional column chromatography or adsorption thin-layer chromatography on silica gel[12,21] can be used, the use of commercially available single-use silica gel minicolumns has simplified the procedure. We used the Sep-Pak (Waters Chromatography Division, Multipore Corporation, Millford, MA) cartridges for the preliminary purification and found them both convenient and effective. Lipid extracts are suspended in 1 ml of hexane and loaded onto the Sep-Pak using a syringe filled with an appropriate Luer-lok tip. The Sep-Pak is eluted with 8–10 ml of diethyl ether : hexane (3 : 97, v/v) and evaporated to dryness under a stream of nitrogen. The residue is suspended in 100–200 μl of ethanol for HPLC analysis.

High-Performance Liquid Chromatographic Analysis of Menaquinones

The advent of HPLC using a reversed-phase partition mode has provided a rapid, sensitive, and reproducible method for generating qualitative and quantitative data of quinone mixtures and can be readily applied to the analysis of menaquinones. The theory and application of HPLC for the analysis of K vitamins has been well described previously[20] and is not repeated here. Identification of menaquinones is done by comparing retention times with menaquinones of known structure. Although menaquinones are not commercially available, standards for detection of isoprenologs of unsaturated MK-4 to MK-10 were provided for our experiments through the generosity of Hoffman-LaRoche (Basel, Switzerland). Menaquinone preparations from bacterial strains of known menaquinone compo-

[18] J. M. Conly and K. Stein, *Am. J. Gastroenterol.* **87,** 311 (1992).
[19] A. Hara and N. S. Radin, *Analyt. Biochem.* **90,** 420 (1978).
[20] M. J. Shearer, *Adv. Chromatogr.* **21,** 243 (1983).
[21] Y. Haroon, M. J. Shearer, and P. Barkhan, *J. Chromatogr.* **206,** 333 (1981).

sition can be used as standards as well.[5] Methods for the preparation of standard mixtures from bacteria have been described previously.[22,23]

We extract MK-11 and MK-12 as described earlier from *Bacteroides* species grown in batch culture in a cell harvester adjusted for anaerobiasis. The retention time for these isoprenologs is determined by plotting the linear relationship between the common logarithms of the K' of a standard mixture (MK-4 to MK-10) versus the number of isoprene units.[22]

After initial sample cleanup using adsorption chromatography reversed-phase HPLC is performed and appropriate fractions are collected where the detector indicated the peak to yield individual isoprenologs. A fraction collector facilitates the collection of eluate containing the desired isoprenolog. Mass spectrometry was used to confirm the identification of individual isoprenologs.

Reversed-phase HPLC for the preparation of menaquinone isoprenologues has been described by several authors using a number of different columns, mobile solvents, elution methods (isocratic and gradient), and with ultraviolet, fluorometric, and electrochemical detection.[5,18,21–27] Menaquinones similar to other vitamin K compounds do not naturally exhibit fluorescence and require a postcolumn reduction to enable detection fluorometrically. Postcolumn reduction may be achieved through chemical[23,26] or electrochemical methods[25] and each has its advantages and disadvantages.

The components of the analytical reversed-phase HPLC system used in our laboratory are all commercially available and are described in the following sections.

HPLC Conditions

U6K universal injection
Double plunger reciprocating pump
C_{18} reversed-phase analytic column; 3.9 mm × 15 cm (Novapak; Waters Scientific Ltd.)
Mobile solvent ethanol : double distilled H_2O [95 : 5 (v/v)]
Flow rate, 0.7 ml/min
Temperature, ambient
Sample injection, 25–100 μl

[22] J. Tamaoka, *Meth. Enzymol.* **123,** 251 (1986).
[23] J. M. Conly and K. Stein, *Clin. Invest. Med.* **16,** 45 (1993).
[24] J. Tamaoka, Y. Katayama-Fujimura, and H. Kuraishi, *J. Appl. Bacteriol.* **54,** 31 (1983).
[25] K. Hirauchi, T. Sakano, and A. Morimoto, *Chem. Pharm. Bull.* **34,** 845 (1986).
[26] T. Sakano, N. Tadayoshi, A. Morimoto, and K. Hirauchi, *Chem. Pharm. Bull.* **34,** 4322 (1986).
[27] Y. Akiyama, K. Hara, A. Matusomoto, S. Takahashi, and T. Tajima, *Biochem. Pharmacol.* **49,** 1801 (1995).

Postcolumn Reduction

0.1% $NaBH_4$ in absolute ethanol (w/v)
Double plunger reciprocating pump
Flow rate, 0.3 ml/min
Mixing coil, ~200 cm

Detection

Model 420-AC fluorescence detector (Waters Scientific) fitted with a 338-nm bandpass excitation filter and a 425-nm-long pass emission filter

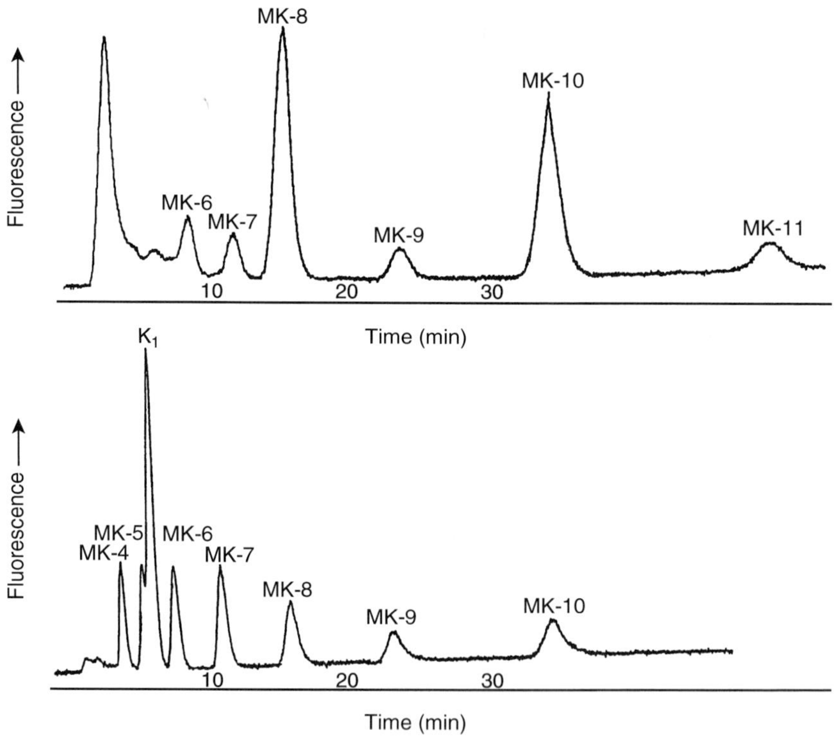

FIG. 1. Isocratic chromatogram of vitamin K isoprenologs found in healthy adult male stool specimen (top) and of standard phylloquinone and K_2 mixture (bottom). Analytical column Novapak C_{18} (150- × 3.9-mm i.d.); mobile-phase ethanol–water (95 : 5 v/v); flow rate 0.7 ml/min; K_1, phylloquinone; MK 4–11, menaquinone isoprenologs 4 through 11. The retention time of MK-11 and other long-chain isoprenologs was determined by plotting the linear relationship between the common logarithms of the K' of a standard mixture (MK-4 to MK-10) versus the number of isoprene units.

TABLE I
REPRESENTATIVE MENAQUINONE PROFILES OF HUMAN BOWEL CONTENTS

	μg/g Dry weight (mean ± SEM)		
Site (*n*)	MK-4 to MK-7	MK-9, MK-10	Total MK
Distal colon (10)	5.55 (±0.27)	14.54[a] (±0.29)	19.85 (±0.36)
Colonstomy (5)	1.64 (±0.68)	4.50[a] (±1.56)	6.10 (±1.95)
Ileostomy (4)	0.69 (±0.33)	1.16[a] (±1.1)	1.85 (±0.93)
Terminal ileum (2)	0.92	7.93[a]	8.85
Jejunum (2)	0.01	0.02[a]	0.03

[a] MK-11 present in 9/10 distal colon, 4/5 colostomy, 2/4 ileostomy, 2/2 terminal ileum, and 0/2 jejunum.

Data Analyzer

Data module model 745 (Waters Scientific) with memory module for storage of chromatograms

Initial experiments are conducted to determine optimal conditions for quinone reduction using the standard mixture of menaquinone standards (MK-4 to MK-10). With a 200-cm-long reaction coil, and a variance of flow rates of the ethanolic $NaBH_4$ solution between 0.1 and 0.8 ml/min and variance of the concentration of $NaBH_4$ between 0.015 and 0.2%, the best response is obtained with a 0.1% $NaBH_4$ solution and a flow rate of 0.3 ml/min. Additional coil length experiments did not alter the response or elution times significantly.

An estimate of the precision of the HPLC assay under these operating conditions is determined and the within-run and between-run coefficients of variation are approximately 10%. Addition of known quantities of standard

TABLE II
MENAQUINONE ANALYSIS OF TOTAL INTESTINAL CONTENTS

		Menaquinone isoprenolog (mg total contents)					
Subject	Dry weight (g)	MK-4	MK-5	MK-6	MK-7	MK-9	MK-10
1	408.1	0.27	0.34	0.24	0.40	0.23	0.74
2	327.6	0.22			0.38	1.82	2.13
3	217.0	0.03		0.08	0.05	0.06	0.03
4	167.6					1.32	3.32
5	97.5	0.07	0.01	0.09	0.04	0.03	1.20

menaquinones to feces provides consistent recovery of approximately 80% of each of the added isoprenologs (MK-4 to MK-10).

Identification by HPLC is based on elution time of the standard menaquinone mixture. Figure 1 shows a representative chromatogram from a fecal sample. To reduce the retention times for long-chain MK isoprenologs (MK-11, -12, -13) a mobile solvent containing 97% (v/v) ethanol is more suitable.

Under the operating conditions described the threshold for detection of menaquinones in fecal samples ranges from 20 ng/g for MK-4 to 50 ng/g for MK-10. Representative results of the menaquinone concentrations of bowel contents from various sites and corresponding quantitative stool culture examinations are shown in Tables I and II.

Author Index

Numbers in parentheses are footnote reference numbers and indicate that an author's work is referred to although the name is not cited in the text.

A

B

C

E

F

G

H

K

L

M

N

O

P

Q

R

S

T

U

V

W

Subject Index

S

W

Y

ISBN 0-12-182183-8

90038

9 780121 821838